Inorganic Two-dimensional Nanomaterials

Fundamental Understanding, Characterizations and Energy Applications

Smart Materials

Series editors:
Hans-Jörg Schneider, *Saarland University, Germany*
Mohsen Shahinpoor, *University of Maine, USA*

Titles in this series:
1: Janus Particle Synthesis, Self-Assembly and Applications
2: Smart Materials for Drug Delivery: Volume 1
3: Smart Materials for Drug Delivery: Volume 2
4: Materials Design Inspired by Nature
5: Responsive Photonic Nanostructures: Smart Nanoscale Optical Materials
6: Magnetorheology: Advances and Applications
7: Functional Nanometer-Sized Clusters of Transition Metals: Synthesis, Properties and Applications
8: Mechanochromic Fluorescent Materials: Phenomena, Materials and Applications
9: Cell Surface Engineering: Fabrication of Functional Nanoshells
10: Biointerfaces: Where Material Meets Biology
11: Semiconductor Nanowires: From Next-Generation Electronics to Sustainable Energy
12: Supramolecular Materials for Opto-Electronics
13: Photocured Materials
14: Chemoresponsive Materials: Stimulation by Chemical and Biological Signals
15: Functional Metallosupramolecular Materials
16: Bio-Synthetic Hybrid Materials and Bionanoparticles: A Biological Chemical Approach Towards Material Science
17: Ionic Polymer Metal Composites (IPMCs): Smart Multi-Functional Materials and Artificial Muscles Volume 1
18: Ionic Polymer Metal Composites (IPMCs): Smart Multi-Functional Materials and Artificial Muscles Volume 2
19: Conducting Polymers: Bioinspired Intelligent Materials and Devices
20: Smart Materials for Advanced Environmental Applications
21: Self-cleaning Coatings: Structure, Fabrication and Application
22: Functional Polymer Composites with Nanoclays
23: Bioactive Glasses: Fundamentals, Technology and Applications
24: Smart Materials for Tissue Engineering: Fundamental Principles

25: Smart Materials for Tissue Engineering: Applications
26: Biobased Smart Polyurethane Nanocomposites: From Synthesis to Applications
27: Inorganic Two-dimensional Nanomaterials: Fundamental Understanding, Characterizations and Energy Applications

How to obtain future titles on publication:
A standing order plan is available for this series. A standing order will bring delivery of each new volume immediately on publication.

For further information please contact:
Book Sales Department, Royal Society of Chemistry, Thomas Graham House, Science Park, Milton Road, Cambridge, CB4 0WF, UK
Telephone: +44 (0)1223 420066, Fax: +44 (0)1223 420247
Email: booksales@rsc.org
Visit our website at www.rsc.org/books

Inorganic Two-dimensional Nanomaterials

Fundamental Understanding, Characterizations and Energy Applications

Edited by

Changzheng Wu
University of Science and Technology of China
Email: czwu@ustc.edu.cn

ROYAL SOCIETY OF CHEMISTRY

THE QUEEN'S AWARDS FOR ENTERPRISE: INTERNATIONAL TRADE 2013

Smart Materials No. 27

Print ISBN: 978-1-78262-465-3
PDF eISBN: 978-1-78801-030-6
EPUB eISBN: 978-1-78801-206-5
ISSN: 2046-0066

A catalogue record for this book is available from the British Library

The Royal Society of Chemistry is a charity, registered in England and Wales, Number 207890, and a company incorporated in England by Royal Charter (Registered No. RC000524), registered office: Burlington House, Piccadilly, London W1J 0BA, UK, Telephone: +44 (0) 207 4378 6556.

For further information see our web site at www.rsc.org

Printed in the United Kingdom by CPI Group (UK) Ltd, Croydon, CR0 4YY, UK

Preface

Inorganic two-dimensional (2D) nanomaterials that possess ultrathin nanosheet morphology, atomic scale thickness and one-dimensional (1D) confinement have attracted tremendous research in the fields of fundamental science and nanotechnology characterization as well as energy applications, since the discovery of mechanically exfoliated graphene. Unlike pure carbon graphene, inorganic 2D nanomaterials have a more abundant elemental composition and various crystallographic structures, giving unique chemical reactivity, tunable physical properties and fascinating applications in the field of energy conversion and storage. The rapid progress in this field calls for a comprehensive review. This book gives an overview of recent developments in inorganic 2D nanomaterials from computational simulation, theoretical understanding, experimental synthesis, modern physical characterization, and their energy applications including electrocatalysts, lithium ion batteries and supercapacitors, water-splitting catalysts and thermoelectrical devices. The book is organized into three main sections: 'Fundamental Understanding', 'Characterizations' and 'Energy Applications'.

The first section gives a fundamental understanding of inorganic 2D nanomaterials. Here, we review recent experimental and theoretical progress on the growth of inorganic 2D nanomaterials, the intrinsic properties of inorganic 2D nanomaterials and a theoretical investigation of nanoscale buckling. We briefly introduce solid-state and liquid-phase synthetic methods for preparing atomically-thick inorganic 2D crystals, covering both top-down and bottom-up strategies. We give a theoretical prediction of characteristic inorganic 2D nanomaterials ranging from single elements to their compounds, as well as possible physical properties. Also, a theoretical review surveyed recent progress on observed nanoscale buckling in a free-standing

Smart Materials No. 27
Inorganic Two-dimensional Nanomaterials: Fundamental Understanding, Characterizations and Energy Applications
Edited by Changzheng Wu

Published by the Royal Society of Chemistry, www.rsc.org

monolayer sheet, in a monolayer sheet on substrates and in a twist bilayer sheet, respectively.

The second section describes the characterization of inorganic 2D nanomaterials. The properties of inorganic 2D nanomaterials show remarkable changes when compared with their 3D bulk. Based on innovative technologies, new and unexpected properties are expected to be observed at the nanoscale and even atomic-scale resolution. Here, we highlight recent experimental progress on the characterization of inorganic 2D nanomaterials. We will give a brief review on the investigations of the inorganic 2D nanomaterials by scanning tunneling microscopy (STM) and spectroscopy (STS), which is a powerful technique for investigating atomic structures and electronic properties of inorganic 2D nanomaterials. In addition, X-ray absorption fine structure (XAFS) can provide precise structural information at the atomic scale around a specific atom, *i.e.* near-neighbor species and distance, coordination number, fluctuation in bond distance for the 2D nanomaterials. Recent advantages of using XAFS spectroscopic techniques for fine structure characterization of two-dimensional nanomaterials will be discussed.

The third, and final, section shows the energy applications of inorganic 2D nanomaterials. When the thickness of the bulk material decreases down to the atomic-scale size, novel atomic and electronic structures emerge, providing new opportunities to manipulate the electronic transport, photon capturing, exciton separation, and surface reaction kinetics. Accordingly, inorganic 2D nanomaterials stand out as an ideal platform applying in various applications in energy conversion and storage. For energy conversion, inorganic 2D nanomaterials have emerged as a new group of catalysts for many catalytic processes. Recent research progress is reviewed here on developing an inorganic 2D electrocatalyst for several important reactions, including the oxygen evolution reaction (OER), oxygen reduction reaction (ORR), hydrogen evolution reaction (HER), and carbon dioxide electrochemical reduction. We also summarize recent progress in inorganic 2D photocatalysts, optimizing the three primary processes of artificial photosynthesis: light-harvesting, charge separation and transport, and surface redox reaction kinetics. Besides, considering interdependence of the factors that thermoelectric conversion efficiency relies on, inorganic 2D nanomaterials can decouple them effectively and thus enhance the thermoelectric merit, which paves the way to improving the conversion efficiency from heat to electrical energy. Inorganic 2D nanomaterials also have great potential in the fields of energy storage. We focus on a different class of inorganic 2D nanomaterials applied as supercapacitors, and further discussion and expectation are included to highlight the intrinsic merits of inorganic 2D nanomaterials for fabricating planar supercapacitors and applying polymer gel electrolyte for flexible supercapacitors. Recent literature on LIBs with inorganic 2D nanomaterials has also been introduced.

We thank all the lead authors (X. J. Wu, Y. Ni, Y. Q. Guo, L. Song, X. D. Zhang, Z. Y. Li, L. Chen, T. Yao, Z. M. Peng, X. Peng, G. Z. Shen, W. T. Bi, J. Peng) as well as their co-authors for their contribution and fruitful collaboration in the preparation of this book. We hope that this book will be useful for a broad readership and stimulate new research in inorganic 2D nanomaterials. This book would be a suitable reference for researchers who are attracted by the exotic physical and chemical properties of inorganic 2D nanomaterials.

Changzheng Wu
USTC, Hefei, China

Contents

Section I: Fundamental Understanding

Smart Materials No. 27
Inorganic Two-dimensional Nanomaterials: Fundamental Understanding, Characterizations and Energy Applications
Edited by Changzheng Wu

Published by the Royal Society of Chemistry, www.rsc.org

Section III: Energy Applications

Section I

Fundamental Understanding

CHAPTER 1

Exploring Two-dimensional Crystals with Atomic Thickness from Molecular Design and Global Structure Search

HAIFENG LV, ZHIWEN ZHUO AND XIAOJUN WU*

CAS Key Laboratory of Materials for Energy Conversion, School of Chemistry and Materials Sciences, CAS Center for Excellence in Nanoscience, Hefei National Laboratory of Physical Sciences at the Microscale, Synergetic Innovation of Quantum Information & Quantum Technology, University of Science and Technology of China, Hefei, Anhui 230026, China
*E-mail: xjwu@ustc.edu.cn

1.1 Introduction

The isolation of graphene has attracted enormous attention due to its novel physical and mechanical properties such as high surface ratio, Dirac behavior with ballistic conductance, ultra-high ideal strength and excellent heat transport.[1–6] These properties have inspired a lot of researchers to devote their efforts to searching for graphene allotropes and other two-dimensional (2D) materials.[7,8] Experimentally, several kinds of 2D materials have been synthesized by means of physical exfoliation form their three-dimensional (3D) structures or chemical methods like epitaxial growth or chemical vapor

Smart Materials No. 27
Inorganic Two-dimensional Nanomaterials: Fundamental Understanding, Characterizations and Energy Applications
Edited by Changzheng Wu

Published by the Royal Society of Chemistry, www.rsc.org

deposition, including wide band gap semiconductor-like boron nitride (BN) monolayers, black phosphorene, ferromagnetic VS_2 *etc.*[9–11] For other materials, their 3D crystals do not exist in layered structures like BeO, III-N, IV-C, Zn-VI compounds, but as the number of layers along the polarization axis decreases to a certain extent, it will transform into a graphene-like structure. Other growth methods like epitaxial growth, chemical vapor deposition or through the addition process of building blocks like small molecular are also used to fabricate 2D materials such as the topological insulator Bi_2Se_3, the MoS_2 monolayer, the superconducting FeSe monolayer, the Rh monolayer and some covalent organic frameworks.[8,12–16] They have similar properties such as large surface ratio and atomic thickness in the non-periodic direction and have great potential in areas such as information, energy and materials.

However, structural prediction based on first-principle calculation has served as a very useful tool in the field of materials research. Exploring 2D materials with atomic thickness from molecular design and global search has been a hotspot. Naturally, a 2D material should have relatively strong intralayer bonding to retain its 2D form, so layered materials with van der Waals interactions are likely to form stable 2D materials through mechanical exfoliation, growth on specific substrates, chemical solvent separation, or functionalization.[17,18] Besides, taking account of the bonding rule and existing geometrical configurations, some novel monolayers with unusual physical and mechanical properties can be designed directly. For instance, we can simply replace some atoms of the already known structures with other atoms or groups, then adjust the bonding length and angle for further calculation. Based on this strategy, 2D structures with tunable band gaps, novel magnetic properties, Dirac cone, superconductivity, or topological insulating have already been developed, which can be used in electronics, spintronics, lithium storage, hydrogen storage, and photocatalytic water splitting *etc.*[19–21]

Based on the bonding characteristics and structural database, we can find a lot of interesting 2D materials, but various 2D sheet structures with close energies have become a challenge for determining their ground state. Furthermore, how to predict the most stable 2D structure just per their elemental composition remains a big problem. The key to solving this dazzling issue is to explore the multi-dimensional potential energy surface (PES) based on the electronic structure calculation. Since the scarcity of calculation resources, more efficient algorithms are necessary to define the optimal structures. In recent years, several global structural search algorithms are widely applied, like the particle swarm optimization algorithm, simulated annealing, genetic algorithms, basin hopping, metadynamic, random sampling, and data mining as well as the machine learning method.[22–24] Among them, two kinds of packages have been widely used to approach the problem of computational crystal structure prediction, referred to as USPEX (Universal Structure Predictor: Evolutionary Xtallography) and CALYPSO (Crystal structure Analysis by Particle Swarm Optimization), respectively.[25,26] USPEX is based on the genetic algorithm and possesses the characteristics of genetic information sharing; CALYPSO is based on the particle swarm optimization algorithm and information sharing is one directional, which is good for convergence. Based on these two packages, a lot of 2D materials have been predicted, such as single element allotropes including B/C/Si/Ge/Sn/Pb; some binary compounds like carbides including Be_2C, B–C, Al–C, Si–C; silicides including B–C, Cu_2Si; nitrides including B–N; C–N; CrN and other compounds Si–P, Bi_4F_4 and so on.[27–29] The research fields have expanded from the strictly planar structure to non-planar, bilayer, multilayer, substrate-limitation, even structures in possession of specific properties. To investigate the properties of 2D structures, it is necessary to ensure the stability dynamically and kinetically in theory, especially for new structures. So, we perform the phonon-mode analysis and molecular dynamic simulation based on first principle calculations. If there are no imaginary frequencies in the Brillouin zone, and the intrinsic bonding type can sustain with a certain temperature range, we can assume it is stable. Next, further calculations will be carried out to study the electronic and mechanical structure. In this chapter, we encapsulate the research development in the prediction of 2D material including single elemental allotropes like B, C, Si, P and the B–C–N compounds. Their unique physical properties and potential applications are also studied.

1.2 Boron

Boron is one of the most fascinating elements in the Periodic Table due to its chemical and structural complexity. It is an electron deficient atom with three valence electrons distributing in four available orbitals, leading to a more complex bonding mechanism compared with typically sp^2 or sp^3 hybridization of carbon.[30,31] This kind of bonding configuration results in as much as 16 bulk allotropes of boron, composed of icosahedral.

B_{12} units, small interstitial clusters, and fused supericosahedra.[32,33] Meanwhile, boron clusters of medium size have been predicted to be planar and quasi-planar, for example B_{12}^-, B_{13}^+, B_{19}^- and B_{36}.[34–36] The above boron molecules have shown carbon-like aromatic and anti-aromatic electronic structures and the recently proposed cage-like B_{40} clusters further extend the parallels between boron and carbon cluster chemistry.[37] Boron has the same short covalent radius and the flexibility to adopt sp^2 hybridization as carbon, which would favor the formation of various low-dimensional allotropes, such as boron nanotubes, fullerenes and 2D boron sheets. Nevertheless, a 2D boron sheet meets a large challenge in experimental preparation. The absence of layered boron materials makes it difficult to mechanically exfoliate the corresponding 2D sheet, as successfully realized in other 2D materials. Instead, numerous theoretical studies have examined 2D boron sheets. Here is a brief review of the 2D boron sheets predicted theoretically.

1.2.1 Theoretical Design

1.2.1.1 α-sheet

The stable honeycomb lattice of graphene is absent in 2D boron sheets because boron lacks an electron. The search for 2D boron sheets started with the first predicted quasi-planar boron clusters by Boustani in 1997 through to first principle calculation methods.[38] Then, the structural stability of nanotubular and quasiplanar boron clusters was investigated and the existence of novel layered, tubular and quasicrystalline boron solids built from elemental subunits were postulated. The 2D honeycomb boron sheet has a much smaller binding energy than that of the planar triangle network, while the buckled triangle lattice has a larger one. Thus, the buckled triangle was thought to be the best building blocks to construct a 2D boron sheet.[39,40] In 2007, a new planar 2D boron sheet mixing the triangular and quadrangular motifs was proposed by Lau *et al.*[41] The binding energy of this kind of motif increased but was still smaller than that of the buckled triangular lattice. In 2009, Tang *et al.* proposed a new route to design atomically thin boron sheets by adding or removing boron atoms in 2D boron nanostructures.[42,43] Each removal produces a hexagonal hole, generating a mixture of hexagons and triangles. This provides us a general design rule to construct more energetically stable 2D boron sheets with hexagonal holes embedded in the planar triangle lattice. The α-sheet with one hexagonal hole in every nine boron triangular atoms and β-sheet with one hexagonal hole in every seven boron triangular atoms were proposed and the binding energy of α-sheet is larger than the buckled triangular lattice. To explain the characteristic of energetically stable α-sheet, they considered the nature of their electronic bonding by dividing the bonding state of the boron sheet into the in-plane and out-of-plane parts. The in-plane bonding states are composed of the sum of s, p_x, p_y states while the out-of-plane state is merely contributed to by the p_z state. Generally, in-plane bonds formed from overlapping sp^2 hybrids are stronger than out-of-plane bonds derived from p_z orbitals. Furthermore, the bonding states gain energy while the antibonding states cost energy. So, the electronic structure that optimally fills the in-plane bonding state should be most preferable. The projected density of state (PDOS) for different boron sheets are shown in Figure 1.1(a). An energy gap emerges between the in-plane bonding and antibonding states, and the Dirac point derived by the crossing of the π band and the π* band is in the gap. It can be inferred that the bonding state of graphene-like honeycomb boron sheet is empty occupied and the anti-bonding state of flat triangular lattice is over occupied, both of which makes the boron sheets unstable. However, the Fermi level of the α-sheet locates at the Dirac point, which balances the cases of the honeycomb and triangular sheet. In addition, a three-center bond is used to describe the stability of α-sheet to replace the traditional two-center bonds.[44] In 2011, Galeev *et al.* also provided a detailed model, where the hexagon holes serve as scavengers of extra electrons from the filled hexagons, to explain the stability of α-sheet and this model was also used to explain the stability of B_{80} fullerenes.[45] Moreover, a self-doping picture, in which the adding or removing of boron atoms is essentially equivalent to simply adding or removing electrons from a fixed electronic structure, was put forward, as illustrated in Figure 1.1(b).[46] Their work demonstrates a novel bonding mechanism in pure boron compounds and opens the door for the rational design of stable 2D boron sheet.

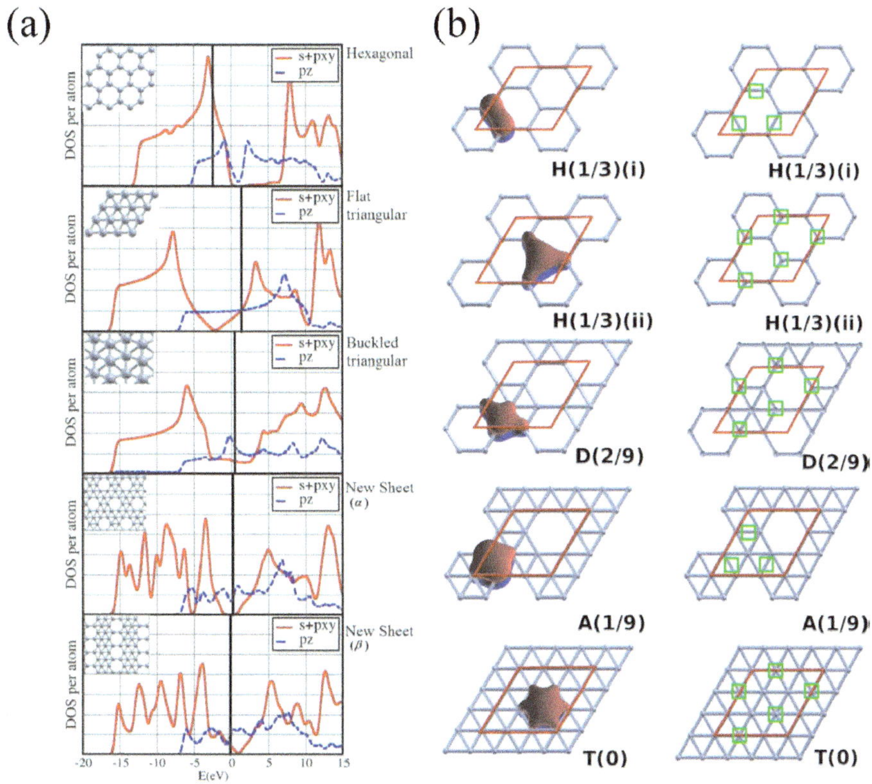

Figure 1.1 (a) Projected density of state (PDOS) for different boron sheets. In-plane states and out-of-plane states are divided. Reprinted with permission from H. Tang, S. Ismail-Beigi, *Physical Review Letters*, **99**, 115501, 2007. Copyright 2007 by the American Physical Society.[42] (b) Isosurface contour plots of π maximally localized Wannier functions for different boron sheets. Reprinted with permission from H. Tang, S. Ismail-Beigi, *Physical Reviews B*, **80**, 134113, 2009. Copyright 2009 by the American Physical Society.[43]

1.2.1.2 Proposal of Polymorphism of the 2D Boron Sheet

The strategies for designing 2D boron sheets proposed by Tang *et al.* can be summarized as the combination of a hexagonal hole and a triangular boron lattice. This will produce a very large number of possible configurations, which is hard to deal with by directly using first-principles calculation methods. In 2012, Penev *et al.* solved this problem by utilizing the cluster expansion method, which leads to a systematic and thorough exploration of the configurational space.[47–49] The boron layers were treated as a pseudoalloy, constituted by closed-pack triangular boron atoms and hexagonal vacancy. It is evident that the different stable structures possessing closely similar formation energy tend to have a similar vacancy concentration in the range 10–15%, suggesting the polymorphism of the 2D boron sheet. In summary, the cluster expansion method in combination with DFT calculations uncover a variety of stable B-layer phases, which are all metallic and the 2D boron sheets tend to form polymorphs.

1.2.1.3 B_{12} Icosahedral Sheet

In 2015, Kah *et al.* designed another strategy to construct 2D boron sheets that contain icosahedra B_{12} as their building units with the semi-empirical Hamiltonian simulation method.[50,51] By arraying B_{12} in a triangular lattice, four new boron allotropic structures have been found and the most stable one called δ_6 sheet has a larger cohesive energy than bulked α sheet by 76 meV per atom in DFT. Besides that, the δ_6 boron sheet has a semiconductor character, which is different from most predicted metallic boron sheets.

1.2.1.4 Global Minimum Searching Boron Sheet

The polymorphism of 2D boron sheet structures with similar energies has become a big challenge for the determination of their ground state. Moreover, some similar planar structures were predicted to have the same or slightly lower energy than the α sheet by using the efficient structure prediction method. Recently, based on the particle swarm optimization techniques implemented in the *ab initio* evolutionary algorithm USPEX and CALYPSO, a variety of 2D boron sheets with novel physical properties were discovered. In 2012, Wu *et al.* firstly conducted the global minimum optimization method to explore stable 2D boron structures, seen in Figure 1.2.[27]

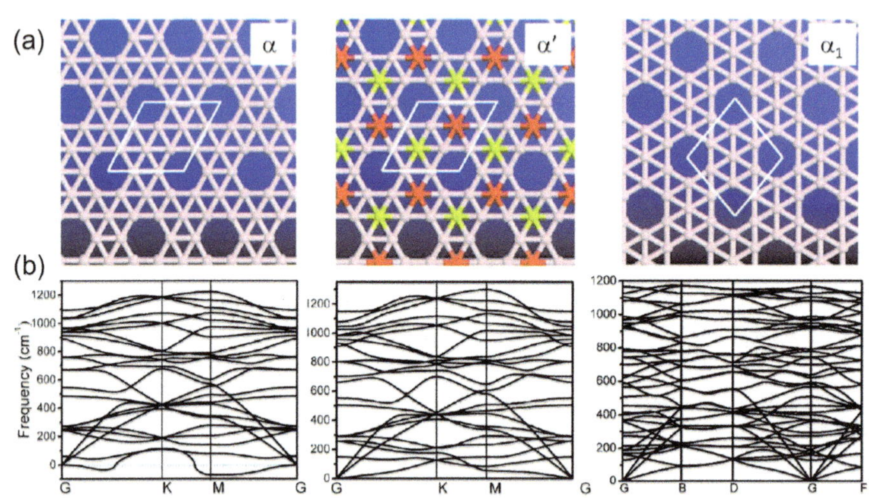

Figure 1.2 The structures of the α-sheet, buckled α-sheet and α_1-sheets and the corresponding phonon dispersions. (a) Top view of α-, α'-, and α_1-sheets. (b) Calculated phonon spectra for α-, α'-, and α_1-sheets. Reprinted with permission from ref. 26. Copyright 2012 American Chemical Society.

The search results showed that the α sheet with a hexagonal hole concentration of 1/9, proposed by Tang *et al.*, was a low-energy structure but it was dynamically unstable since some imaginary frequencies appeared in the Brillouin zone. Further calculation showed that it will transform into a non-planar structure called α′ sheet, analogue to α sheet but dynamically stable as per phonon calculation. The α′ sheet has a larger cohesive energy than the original α sheet and several other low-energy structures were also predicted with very similar cohesive energy of the α sheet (less than 10 meV/atom). The electronic structure calculation with PBE0 functional revealed that the α sheet and α′ sheet should be semiconductors. The most intriguing part is that two of the predicted structures called β1 and β3 were synthesized epitaxially on an Ag (111) substrate by Feng *et al.* in 2016, showing that a global search can act as a powerful tool in structure prediction.[52] After this, several other works of global minima crystal searching were also carried out to obtain more energetically stable 2D boron sheet structures. Yu *et al.* proposed a struc-1/8 boron sheet less stable than an α sheet but still had a close cohesive energy with the α sheet and 1/8 and 2/15 B-layers (less than 10 meV/atom).[53] Lu *et al.* found two types of close-laying stable boron sheets that were composed of two kinds of elementary units with isolated-hexagon and twin-hexagon holes, predicting the possible binary phase of 2D boron sheets.[54] Zhou *et al.* found that a 2D boron structure with non-zero thickness may be more kinetically stable than the planar ones.[55] Through a systematic structure search, two orthorhombic structures with space groups *Pmmm* and *Pmmn* were identified, which is considerably, by 80 meV and 50 meV per atom, lower in energy than the α sheet structure and its analogues from the PBE functional. The non-zero thick 2D boron sheet, which was made of buckled triangular layers, had the similar bonding to the bulk α-boron in geometry, reflecting the frustration of the pure boron system. A more exotic feature is that of the *Pmmn* structure (Figure 1.3(a)) which has a distorted Dirac point located at the Fermi level of the system, like graphene, shown in Figure 1.3(b).

The detailed electronic structure calculation reveals that the Dirac cone in *Pmmn* is direction dependent while that of graphene is isotropical. All the bonding states are fully occupied while the antibonding states are fully empty, which leads to the stability of this structure. Moreover, the physical origin of the Dirac cone comes from the hybrids of in-plane states and out-of-plane states between two sublattices. In 2016, Ma *et al.* reported a low-energy phase called 2D ionic boron.[56] This 2D structure is composed of graphene-like plane and B_2 atom pairs with the *P*6/*mmm* space group and has a lower energy than α sheet by 23 meV/atom from PBE + D3 functional. The charge transfer is identified by Bader charge analysis, which show an average of 0.85 electrons transferred from the graphene-like B-plane to B_2 pairs, showing a slightly but not insignificant ionic character. Another interesting feature of *P*6/*mmm* is that the electronic structure exhibits double distorted Dirac cones and the Fermi velocity ($v_f = {\sim}0.4 \times 10^6$ m s^{-1}) is even higher than that in graphene (10^5 m s^{-1}). A three-layer stacked graphene structure with *C*2/*m*

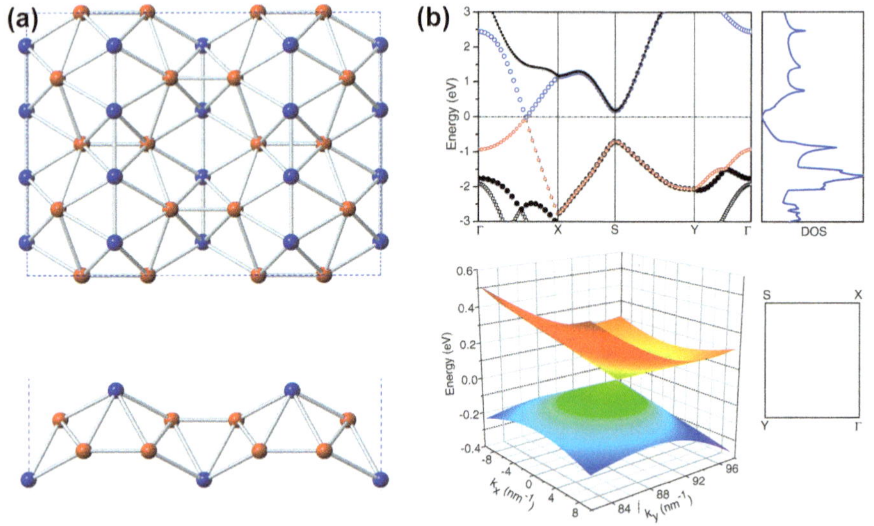

Figure 1.3 (a) projections of 2 × 2 × 1 supercell of *Pmmn*-boron structure along [001] and [100] directions; (b) electronic structure of *Pmmn* boron. Reprinted with permission from X. Zhou, X. Dong, A. R. Oganov, Q. Zhu, Y. Tian, H. Wang, *Physical Review Letters*, **112**, 085502, 2014. Copyright 2014 by the American Physical Society.

space group is also identified and has a larger cohesive energy by 84 meV per atom than α sheet, which shows a metallic character like another boron allotrope. Inspired from CVD growth of graphene on Cu foil, Liu *et al.* firstly studied the substrate restricted formation of 2D boron sheet combined the cluster expansion method with first principle calculation.[57] Pure metal (Cu, Ag and Au) and metal boride (MgB_2, TiB_2) substrates were considered here. Their results showed that boron sheets can be grown on the Ag (111) or Au (111) surfaces by deposition and the high nucleation barrier would help to impede the formation of three-dimensional boron. For metal borides, saturation of boron terminated MgB_2 surface in B rich environment can also lead to the formation of B sheets. The adhesion energy of boron to substrates was weak, suggesting feasible post-synthesis separation into the free-standing forms. In 2015, Zhang *et al.* also conducted similar research about 2D boron sheets mediated by metal by using the cluster expansion method and surface structure-search method.[58] Their results showed that the structure of 2D boron sensitively depends on metal substrate. 2D boron sheet preferred to be non-planar with significant buckling on Au, whereas on more reactive Ag, Cu, a planar structure is found to be the most stable. The icosahedral B_{12} layer was found to be unstable on Cu and Ni but unexpectedly becomes a possible minimum on Au and Ag. The selectivity mechanism of 2D boron on different substrates was also studied and a competition between the strain energy of buckling and chemical energy of electronic hybridization between boron and metal was the major factor. In fact, the 2D boron sheet on Ag (111) predicted in this work was successfully synthesized in 2016, further proving the advantage of substrate restricted search methods.[59]

The successful synthesis of 2D boron has been a great inspiration in the application of 2D boron sheets, further proving the validity of theoretical design in exploring atomic thickness nanostructures. Also, the development of design rules and global searching methods allow more 2D boron sheets with novel physical properties to be anticipated in the future.

1.3 Carbon

Carbon is an essential group-14 element in the Periodic Table that provides the basis for life on Earth. The arrangement of carbon atoms in 3D exists in a variety of allotropic forms and the best known are graphite, diamond and amorphous carbon. Diamond is the hardest natural material and graphite is widely used as a lubricant, and as a heat and electron conductor.

The atomic number of carbon is six with a $2s^2 2p^2$ electron configuration. It tends to form covalent chemical bonds between carbon atoms with sp^1, sp^2 and sp^3 hybridization. Graphite is a typical sp^2 hybridized carbon allotrope and it is a layered structure with intralayer sp^2 hybridization and interlayer van der Waals interactions. Diamond is a characteristic sp^3 hybridization allotrope, with four electrons distributing in four different p orbitals. In 2004, the isolation of graphene opened up the route to explore the crystal structures in 2D form. The carbon monolayer holds a lot of novel properties like Dirac-like band structure with ballistic conductance, excellent heat transportation ability and large surface ratio. Based on these properties, the application of graphene has developed rapidly into a whole industry. Since the polymorphism of element carbon, searching for graphene-like 2D carbon with novel physical properties has attracted enormous attention. Here we look at three representatives of 2D carbon forms: graphene derivatives, the graphyne group and the recently reported penta-graphene.

1.3.1 Graphene Derivatives

A big feature of graphene derivatives is the sp^2 hybridization bonding mechanism like graphene. They may exist in a meta-stable state compared with the cohesive energy of graphene because of the distorted sp^2 bonding, but remain stable both dynamically and kinetically, showing a variety of characteristics like graphene or outperformed graphene. Merz *et al.* did some pioneer research theoretically to predict a series of 'honeycombed' or 'layered' 3, 4-connected nets.[60] Zhu *et al.* presented a series of 2D translationally symmetric structures related to graphite through a kind of local rearrangement on the graphite lattice.[61] Crespi *et al.* proposed a pure-carbon planar structure incorporating pentagons and heptagons into a 2D Bravais lattice, which are metallic with a density of states at the Fermi level of ~0.1 state per eV per atom.[62] The binding energy per carbon atom is comparable to that of C_{60}

with the possibility to roll into tubes in a manner similar to graphite. Tyutyulkov *et al.* examined two classes of infinite polymers consisting of an alternating fusion of cyclobutadiene and benzene rings theoretically with a band gap about 2 eV.[63] Terrones *et al.* proposed a new family of layered sp^2-like carbon crystals, composed of five-, six-, and seven-membered rings with an intrinsic metallic behavior independent of orientation, tube diameter, and chirality.[64] These periodic sheets can be rolled to obtain nanotubes with different diameter and chirality. Further calculations show that these sheets and nanotubes are metastable but still more favorable than C_{60} with similar mechanical properties to those of graphene. Deza *et al.* modified the graphite-like structure with pentaheptites and a unique lattice of *cmm* symmetry was built corresponding to the previously reported pentaheptite carbon metal.[65] Then a series of motifs including the torus, Klein bottle, and cylinder, face-regular tilings of pentagons and b-gons were presented which is analogous to pentagon-heptagon tilings. Andres *et al.* verified the bilayer graphene stacking directly on top of each other (AA stacking) could reconstruct into a quasi-2D carbon allotrope with strong chemical bonds when applied with stress to tune the distance between planes below 0.156 nm.[66] The optimized geometries showed the in-plane C–C bonds are considerably weakened from partial double bond (0.141 nm) to single bond (0.154 nm). The bilayer phase is metastable as per the binding energy calculation with activation energy of 0.16 eV per cell compared with the standard single layer but when it will turn into the global minimum under the 9 GPa in-plane stress. The electronic properties depend on the interlayer distances, first a semimetal, in between metal and a wide gap semiconductor at covalent-like distances. Lusk *et al.* postulated a new strategy to generate new 2D carbon allotropes by the introduction of defects into graphene sheets.[67] The common Stone–Thrower–Wales (STW) defect is formed by the simple rotation of a pair of carbon atoms into opposing five- and seven-membered rings. Through the rearrangement of STW defects, haeckelite can be formed and based exclusively on the inverse Stone–Thrower–Wales (ISTW) defects, they extend the idea to create a new class of membranic carbon allotropes referred to as dimerite, composed exclusively of ad-dimer defects. Appelhans *et al.* found another planar, single-atomic layer material composed entirely of carbon by introducing patterned defects involving DVs into complete graphene.[68] The electronic band structure calculations reveal that it displays an intrinsic semiconductor behavior with a band gap of 1.2 eV. This defect engineering methods can help realize the fabrication of semiconducting ribbons of carbon within graphene, opening the prospect of 2D circuit logic composed entirely of carbon. Brazhe *et al.* estimated the force constants in 2D carbon supracrystals compared to graphene. Luo *et al.* discovered a new carbon allotrope consisting of parallel zigzag and armchair chains alternatively each other, namely pza-C_{10} and it is a semiconductor with band gap of 0.31 eV (0.71) for PBE (HSE06).[69] The binding energy of pza-C_{10} is higher than most of the graphene allotropes like graphdiyne, graphyne and fullerene and the molecular dynamics simulation was conducted to validate the stability of pza-C_{10}. Based on *ab initio* quantum molecular dynamics, Brunetto *et al.* showed a 2D carbon allotrope, called as biphenylene carbon (BPC), which can be obtained from selective dehydrogenation of porous graphene.[70] Further calculations show that BPC possesses a non-zero band gap and well-delocalized frontier orbitals. Wang *et al.* predicted a pure-carbon 2D covalent metal named planar C4 in *P4/mmm* symmetry.[71] Based on phonon-mode analysis, planar C_4 is confirmed to be dynamically stable in the ground state since no imaginary frequency in the Brillion zone. The high density of states (DOS) at the Fermi level shows the strong metallic behavior. Moreover, two distinct types of planar C_4 NRs are identified; one is uniformly metallic regardless of the width and the other exhibits remarkable odd-even metal–semiconductor oscillating behavior depending on the width change. Lu *et al.* predicted four new 2D carbon allotropes including three hexagonal and one tetragonal 2D carbon allotrope: $C65^-$, $C63^-$, $C31^-$ and $C41^-$ sheets.[72] They also calculated the electronic band structure of these carbon sheets and found it π-metallic. Song *et al.* predicted the properties of a new 2D sp^2-carbon network referred as graphenylene, which is meant to be the first example of a non-delocalized sp^2-carbon structure composed of cyclohexatriene units with two quite distinct C–C bonds within a C6 ring and it is a semiconductor with a narrow direct band gap, making it promising for various applications.[73] Mandal *et al.* predicted a new 2D carbon allotrope composed of hexagons, octagons and pentagons and hence named as HOP graphene (HOPG).[74] Its dynamical stability can be confirmed by phonon-mode analysis and molecular dynamic simulations. Sharma *et al.* reported a new sp^2 hybridized 2D allotrope consisting of continuous 5-6-8 rings of carbon atoms, named as 'pentahexoctite' and the sheet is metallic with anisotropic transport properties. By using a structural search method, Xu *et al.* reported three amazing 2D Dirac carbon allotropes, named as S-graphene, D-graphene and E-graphene and through band structure calculation, D-graphene and E-graphene is assumed to be Dirac-type material. Hu *et al.* proposed a 2D carbon allotrope, H-net, which is composed of C_4 distorted squares, C_6 hexagons, and C_8 octagons. Further calculation show that H-net is a semiconductor with an indirect band gap of 2.11 eV and more energetically stable than synthesized graphdiyne. Maruyama *et al.* discovered 2D stable carbon allotropes containing acepentalene structures. Although the total energy is higher than C60 by 0.90 eV per atom, molecular dynamic simulation confirmed that it can retain its planar structure up to 2000 K. Su *et al.* proposed two 2D carbon allotropes comprised of octagons and pentagons, referred as OPG-L and OPG-Z and OPG-L is metallic and OPG-Z is a gapless semi-metal. Particularly, OPG-Z exhibits pronounced electronic anisotropy with highly anisotropic Dirac points at the Fermi level. Maruyama *et al.* investigated geometric and electronic structures of 2D stable carbon allotropes comprised of pentagonal rings and found that it is more stable than C_{60} and can retain its planar structure up to 1000 K. Further calculations indicate that it displays a metallic behavior and the polarized electron spin is ferromagnetically aligned and extends throughout the sheet with a

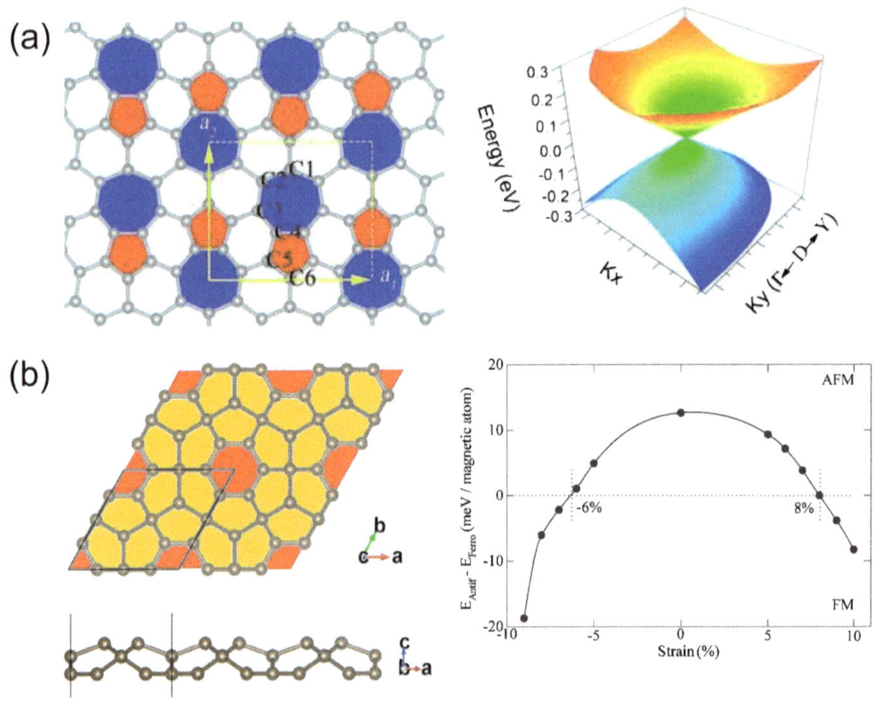

Figure 1.4 (a) Structure of ph-graphene with notable space-inversion symmetry and its 3D band structure. Reprinted with permission from ref. 75. Copyright 2015 American Chemical Society. (b) Atomic structure of a 2 × 2 supercell of monolayer ph-graphene and the transition of monolayer ph-graphene between the AFM and the FM state under strain. Reprinted with permission from Y. Aierken, O. Leenaerts, F. M. Peeters, *Physical Review B*, **94**, 155410, 2016. Copyright 2016 by the American Physical Society.[76]

spin moment of 0.62 μB nm^{-2}. By the means of systematic evolutionary structure searching, Wang *et al.* proposed a novel 2D carbon allotrope, namely phagraphene, which is composed of 5-6-7 carbon rings, as shown in Figure 1.4(a).[75] The calculated total energy per atom from PBE shows that it is more stable than most of the predicted 2D carbon allotropes because of its sp^2 binding features as well as density of atomic packing comparable to graphene. Another feature of this 2D carbon allotrope is the distorted Dirac cones, which is further proved to be orientation-dependent regardless of external strain with tunable Fermi velocities. Recently, Aierken *et al.* proposed a special type of metal-free antiferromagnetic 2D carbon allotrope.[76] It contains one pentagonal carbon ring and one hexagonal carbon ring and the unit cell has eleven atoms. The ground state of this 2D material is antiferromagnetic and can change to ferromagnetic through strain, shown in Figure 1.4(b). Motivated by this work, more research is anticipated in the future.

1.3.2 The Graphyne Group

The distinct feature of graphyne is the hybridized sp^1 bonding in 2D carbon sheets. Narita *et al.* first proposed this series of carbon allotropes, namely graphyne, graphdiyne, graphyne-3, and graphyne-4. This kind of configuration consists of hexagons connected by linear carbon chains and part of the linear carbon chain is composed of acetylenic linkages. The electronic band structure calculation shows a semiconductor behavior with moderate band gaps of 0.5–0.6 eV and the cohesive energies per carbon atom are 7.95 eV for graphyne and 7.78 eV for graphdiyne. Malko *et al.* studied the band structure of graphynes and found it to be a Dirac-like material as per first-principles electronic structure calculations.[77] What is more intriguing is that one of these materials, 6,6,12-graphyne, features two self-doped nonequivalent distorted Dirac cones indicating that it is even more versatile than graphene in some respects, as shown in Figure 1.4. Peng *et al.* investigated the mechanical properties of graphyne monolayers and Yang *et al.* also studied the mechanical properties of graphyne, graphdiyne, graphene-3, graphene-4 and graphene-5, which help to better understand the bonding characteristics of the graphyne group. Zhao *et al.* proposed a 2D carbon topological insulator (δ-graphyne) with superior energetic stability. The electronic band structure is calculated and two Dirac cones at the K and K′ appear in the Brillouin zone. Taking spin-orbital coupling into consideration, the band gap opened at the Dirac points due to SOC is about 0.59 meV and can host the QSHE in an experimentally accessible low temperature regime (7 K). The carrier mobility of in the novel sp–sp^2 hybridization planar 6,6,12-graphyne sheet was calculated to be 4.29×10^5 $cm^2\ V^{-1}\ s^{-1}$ for holes and 5.41×10^5 $cm^2\ V^{-1}\ s^{-1}$ for electrons at room temperature by Chen *et al.* and they found it to be even larger than that of graphene ($\sim 3 \times 10^5$ $cm^2\ V^{-1}\ s^{-1}$) at the first-principles level by using the Boltzmann transport equation coupled with the deformation potential theory. Yin *et al.* proposed a novel 2D carbon allotrope, named rectangular graphyne (R-graphyne), with tetra-rings and acetylenic linkages. R-graphyne is intriguing as the band structure of its nanoribbons could experience semiconductor and metallic transition as the width changes and the nanoribbons with zigzag edges tends to be a Dirac-type material. The graphyne and graphdiyne family is a big part of 2D carbon materials and the successful synthesis of graphdiyne together with its intrinsic semiconductor behavior make it a potential candidate for carbon-based nanoscale electronic devices.

1.3.3 Penta-graphene

Penta-graphene, composed entirely of carbon pentagons and resembling the Cairo pentagonal tiling, was proposed by Zhang *et al.*[78] and it is the first carbon material made exclusively of carbon pentagons. The crystal structure of penta-graphene can be viewed as the 001 face of T12 carbon, one of the 3D carbon allotropic forms that can be acquired by heating an interlocking-hexagon-based metastable carbon phase at high temperature, as shown in

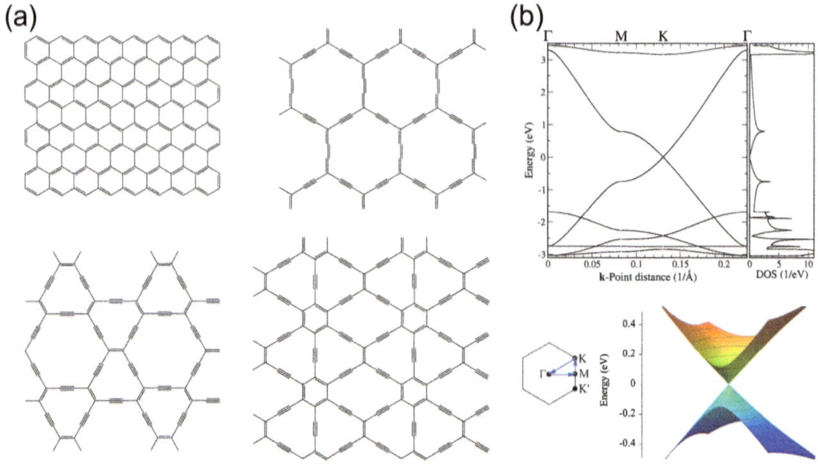

Figure 1.5 (a) Structures of graphene, α-graphyne, β-graphyne and 6,6,12-graphyne; (b) electronic structure of 6,6,12-graphyne. Reprinted with permission from D. Malko, C. Neiss, F. Viñes, A. Görling, *Physical Review Letters*, **108**, 086804, 2012. Copyright 2012 by the American Physical Society.[77]

Figure 1.5(a). It possesses a $P\text{-}42_1m$ space group (No. 113) and is composed of sp^2 hybridization and sp^3 hybridization carbon atoms. Interestingly, it is noted that penta-graphene resembles the structure of the experimentally identified layered silver azide (AgN_3). Next, they talk about the stability of penta-graphene. Firstly, the binding energy per atom is higher than graphene, but still lower than some experimentally identified carbon nanostructures such as the smallest fullerene, C_{20} and the smallest carbon nanotube. Then, they performed the phonon-mode analysis and no imaginary mode in the Brillion zone shows that it is dynamically stable. Finally, they carry out *ab initio* molecular dynamics (AIMD) simulations using canonical ensemble and found that penta-graphene can withstand temperatures as high as 1000 K. As per the standard Voigt notation, a 2D material should obey the Born–Huang criteria. So, they calculated the change of energy due to the in-plane strain to determine the mechanical stability of penta-graphene and the results show that penta-graphene has an unusual negative Poisson's ratio (NPR) and ultra-high ideal strength that can even outperform graphene. The electronic band structure calculation of penta-graphene shows that it possesses a quasi-direct band gap of 3.25 eV, as shown in Figure 1.5(b) from HSE06 functional, close to that of ZnO and GaN.

Finally, the penta-graphene derivatives like rolled-up penta-graphene, stacked penta-graphene layers and functionalized penta-graphene were also investigated. This unusual 2D allotrope of carbon provides us a better understanding of the characteristics of carbon atom and may find it unusual in the application of carbon materials.

1.4 Silicon

Silicon is a member of group-14 in the Periodic Table with a typical tetravalent bonding mechanism, along with carbon above it and germanium, tin, and lead below. Bulk silicon crystallizes in a diamond-like structure and pure silicon is an intrinsic semiconductor, which is widely used in semiconductor industry to produce silicon wafers. Since the isolation of graphene in 2004, the 2D allotropes of silicon have attracted enormous attention for being a candidate 2D material that could bring the exotic electronic structure of graphene to the well-developed silicon-based technology. Recently, single-layer free-standing monolayer of silicon, called silicene, was first to be predicted to be stable and then experimentally synthesized on Ag (111) substrates.[79–86] It is arranged in a honeycomb lattice with alternating silicon atoms located in different planes to attain sp^3-like bonding, forming a low-buckled rather than planar pattern. Many charming properties of graphene can also be observed in silicene, it preserves linearly crossing bands at the Fermi level that leads to a characteristic Dirac-type behavior of its electrons.[87] Further theoretical researches reveal that stronger spin–orbit coupling (SOC) enables silicene with a great potential be a 2D topological insulator and yields quantum spin-Hall effect (QSHE). Due to the resemblance to graphene in electronic structure and potential integration in the well-established silicon technology, searching for more 2D allotropes of silicon remains an urgent challenge. Here, we present the already known 2D silicon and their relevant properties based on the simulation approaches.

1.4.1 Surface Reconstruction

At finite temperature in vacuum, the cleaved Si (111) surface is likely to be constructed into various types of structures including the 7×7 and 2×1 structures due to the existence of dangling bonds on the cleaved surface. In 2011, Morishita *et al.* reported a re-DL-Si structure formed from the reconstruction of bilayer thickness Si (111) surface.[88] It was formed after the 300 K MD run and subsequent relaxation at 0 K with PBE functional implemented in the Vienna *Ab Initio* Simulation Package (VASP). The primitive unit cell consists of 16 atoms and further calculation show that it is a semiconductor with a narrow bandgap of 0.33 eV. The primitive structure (DL-Si) with the ideal corrugated honeycomb layer structure is a metal, so the reconstruction induces a metal–non-metal transition in DL-Si. Compared with the reconstructed Si (111) -2×1 and 7×7 surface, they share the same highly distorted tetrahedral and markedly different electronic band structure. In 2013, Kaltsas *et al.* proposed a Si single-layer with the geometry of the $\sqrt{3} \times \sqrt{3}$ reconstruction, which is another prominent surface reconstruction of silicon.[89] It is derived from the arrangement of atoms in one of the Si (111) surfaces and with this arrangement, the Si adatoms form a characteristic dumbbell structures on an otherwise flat graphene-like layer. As per the so-called dimer-adatom-stacking fault (DAS) model, several other reconstructions like 5×5

and 9 × 9 are also possible. It can be derived that their difference lies in the number of silicon adatoms that can be accommodated between the holes. The results also show the relative stability of the reconstruction single layers and the $\sqrt{3} \times \sqrt{3}$ reconstruction is more stable than honeycomb silicene. The DOS calculation shows that the $\sqrt{3} \times \sqrt{3}$ reconstruction single layer is a metal with strong peaks at the Fermi level. In 2014, Cahangirov *et al.* built the $\sqrt{3} \times \sqrt{3}$ reconstruction silicene model and unveiled the spontaneous and consequential formation of $\sqrt{3} \times \sqrt{3}$ structures from 3 × 3 structures on Ag substrate with the help of first-principles calculations, seen in Figure 1.6(b).[90] When one silicon atom is introduced into silicene and forms a single dumbbell (DB) unit in trigonal lattice, we call it trigonal dumbbell silicene (TDS), with two silicon adatoms for honeycomb dumbbell silicene (HDS), three for full dumbbell silicene (FDS) and two DB units arranged in

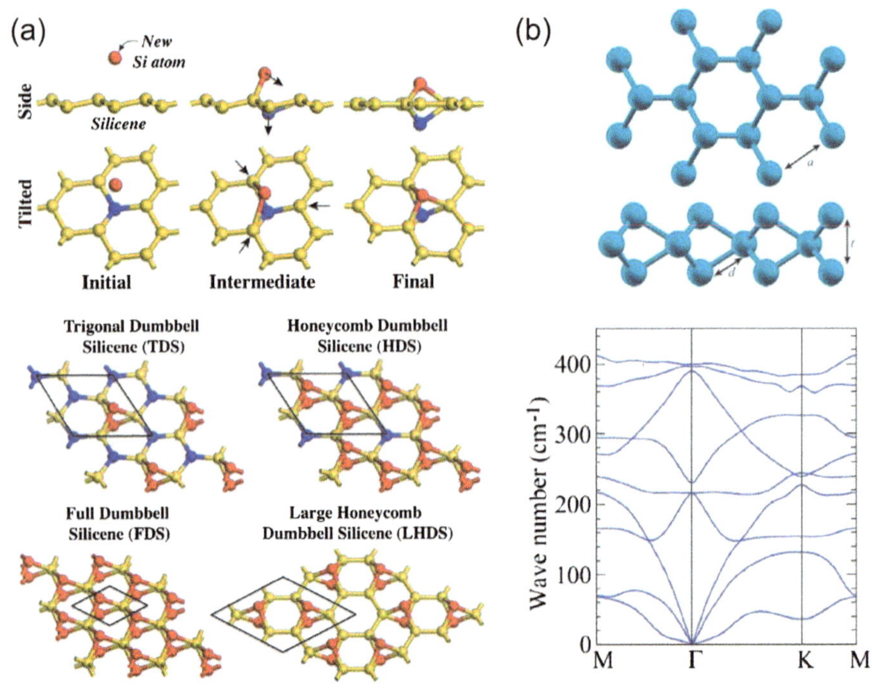

Figure 1.6 (a) Formation of the dumbbell building block units starting from buckled silicene. Atomic structure of $\sqrt{3}$ trigonal dumbbell silicene (TDS), $\sqrt{3} \times \sqrt{3}$ honeycomb dumbbell silicene (HDS), 1 × 1 full dumbbell silicene (FDS), and 2 × 2 large honeycomb dumbbell silicene (LHDS). Reprinted with permission from S. Cahangirov, V. Ongun Özçelik, L. Xian, J. Avila, S. Cho, M. C. Asensio, S. Ciraci, A. Rubio, *Physical Review B*, **90**, 035448, 2014. Copyright 2014 by the American Physical Society.[90] (b) Top- and side-view of the MoS_2-type single layer of Si atoms and phono spectrum. Reprinted with permission from F. Gimbert, C. C. Lee, R. Friedlein, A. Fleurence, Y. Yamada-Takamura, T. Ozaki, *Physical Review B*, **90**, 165423, 2014. Copyright 2014 by the American Physical Society.[91]

a honeycomb lattice in a 2 × 2 unit cell for large honeycomb dumbbell silicene (LHDS). The phonon spectrum is calculated to identify their stability, and the relative cohesive energy of the large honeycomb dumbbell silicene (LHDS) is higher than both TDS and HDS since the maximum coordination of Si atoms in the LHDS is four. The calculated structural parameters of HDS are incommensurate with the Ag (111) surface and stable in its free-standing form. It can be inferred that HDS could dictate the structure of multilayer silicene grown on top of it. Interestingly, Gimbert *et al.* predicted a 2D phase bearing structural resemblance to a single MoS_2 layer, seen in Figure 1.6(b), at the same time[91] and it could be derived from the addition of Si adatoms to pristine silicene to form dumbbell structures. Further calculations show that it has a higher cohesive energy than low-buckled silicene and stability is also checked by calculating the frequencies of their phonon modes. Form the band calculation, it exhibits nematic orbitals with σ bonding with the hex-coordinated middle atoms in the MoS_2 structure and the identification of these nematic orbitals enrich Si bonding configurations, which is different from those of C atoms.

1.4.2 Bilayer and Multilayer Silicene Construction

Since the stacking order and interlayer interactions act as a crucial role to the properties of multilayer graphene, similarly, changing the stacking type may also provide us a simple way to modify the electronic structure of multilayer silicene. In 2010, Bai *et al.* firstly presented a AA stacked bilayer silicene configuration with fully covalent bonded Si atoms by confining the liquid silicon to slit nanopores through MD simulation and the dynamical stability of this quasi-2D bilayer silicon is further confirmed since no imaginary frequencies appeared in the Brillouin zone.[92] In 2013, Liu *et al.* investigated the undoped bilayer silicene and found that the system is a potential chiral topological superconductor and may possess a high superconducting critical temperature mediated *via* strain.[93] Lian *et al.* applied the isotropic tensile strain to silicene bilayers and discovered several new allotropes of 2D silicon derived from different stacking order of bilayer.[94] Unlike graphene bilayers, whose most stable structure is AB (Bernal) stacking with van der Waals interlayer interactions, strong covalent bonds will form between different layers of silicene. Due to the unpaired p_z electron, the similar AB stacking of silicene like graphene is only metastable and a small energy cost may result in a reconstruction. They refer the AB stacking of silicene bilayer as SSBL, which would transform into the constructed silicene bilayer (RSBL), the most stable structure. Then, they studied the phase transition under the isotropic tensile strain for SSBL and RSBL. After the transition, both SSBL and RSBL turn into a graphene-like silicene bilayer (GSBL) phase with no energy barrier. If the applied strain ranging from 15.8% to 16.8%, another new phase occurs with the silicon atoms in the top layer forming a tilted bond with the atoms in the bottom layer, referred as the intermediate silicene bilayer phase (ISBL). In addition to the structural transition, they also studied the

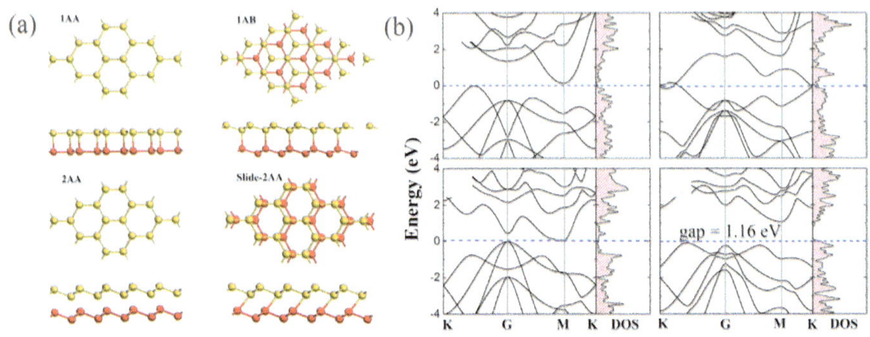

Figure 1.7 (a) Top views and side views of (3 × 3) supercell for the four optimized 1AA, 1AB, 2AA, and slide-2AA structures of BLS. (b) Band structures and DOS calculated for 1AA, 1AB, 2AA, and slide-2AA structures obtained from HSE06. The energy gap of slide-2AA is 1.16 eV in HSE06 (0.69 eV in PBE). Reprinted from ref. 95, with the permission of AIP Publishing.

electronic properties of RSBL and GSBL, most stable states among the four phases. RSBL is a semiconductor with a small bandgap of 0.23 eV and when the strain increases, it will become metallic. As the increase in strain, the energies of the corresponding bands increase but the main feature of dispersion relations of each band remains. The tight-binding results reveal that the hybridization in graphene-like silicene bilayer is characterized by intralayer sp^2 hybridization and the interlayer sp^1 hybridization. Fu *et al.* identified several meta-stable configurations, seen in Figure 1.7(a), including a slightly staggered AA packing structure, named slide-2AA by scanning stacking modes and lattice constants.[95] Moreover, the binding energy of slide-2AA is comparable with that of exact AA and AB-stacking bilayer silicene structures. The small transformation barrier between the slide-2AA and exact AA, AB stacking makes it feasible to control the semiconductor–metal transformation through a small shear force in bilayer silicene nanosheet systems. The van der Waals densities functional, optB86b-vdW, together with hybrid functional, HSE, are applied to attain the correct electronic structure, seen in Figure 1.7(b).

The new morphology of bilayer silicene, slide-2AA is semiconductor with a sizable bandgap 1.16 eV, which is very close to the bulk Si (1.1 eV) and the semiconductor–metal transition may find implications for developing next-generation microelectromechanical system (MEMS) devices. Padilha *et al.* also proposed several possible stacking configurations of bilayer silicene and reaffirmed that planar bilayer silicene with strong covalent bonding, named AA_p, is the lowest energy conformation.[96] Further calculations reveal that extra (σ-like) chemical bonds between the two layers can help stabilize this conformation. This planar bilayer silicene is a metal and the analysis of the 3D band structure shows that the transmission coefficient is highly dependent for some structures. Simulated scanning tunneling microscopy (STM) patterns agree well with experimental STM images of bilayers silicene, which can be used to identify the complex stacking sequences for some structures. The results can provide feasible routes to design new quasi-2D silicon allotropes with novel physical properties. In 2015, Guo *et al.* expanded the bilayer silicene to multi-layer silicene configurations.[97] With the global optimization algorithm, a variety of dynamically stable multilayer silicene phases were discovered, showing the rich polymorphism of silicon. When the silicene thickness is up to four, Si (111) surface structures are obtained, which shows the critical thickness for a structural evolution. An intriguing feature about the multilayer silicene is that π-bonded surfaces presents outstanding optoelectronic properties due to the incorporation of sp^2-type bonds, which makes it more efficient for developing next-generation microelectromechanical system (MEMS) devices.

1.4.3 Hydrogenated Silicene

Silicene, the 2D form of silicon, meets a great challenge since it is far less stable than other 2D materials, such as graphene or MoS_2 due to the unpaired p_z electron. To fulfill their applications in various fields, chemical functionalization especially hydrogenation plays a vital role in modifying their electronic properties. Zhang *et al.* first introduced ferromagnetism into silicene system by surface modification with hydrogenation and opened up the door in functionalization of silicene sheets.[98] Huang *et al.* first conducted a systematical study about the structural and electronic properties of intrinsic and hydrogenated BS by employing first-principles calculations and the cluster-expansion approach.[99] They discovered that hydrogenated bilayer silicene (BS) is one promising optoelectronic material, which has never reported in other hydrogenated structures. Depending on the hydrogenation concentration, we can get different ground states of single and double sided hydrogenated BS characterized by various band structures. At low hydrogen concentrations, there are four ground states of hydrogenated BS with dipole-allowed direct (or quasi-direct) band gaps in the desirable range from 1 to 1.5 eV, suitable for solar applications. At high hydrogen concentrations, there are three well-ordered double-sided hydrogenated BS structures exhibit direct (or quasidirect) band gaps in the color range of red, green, and blue, affording white light-emitting diodes. The corresponding formation with respect to single-layer silicene is calculated, together with the phonon spectrum and molecular-dynamic simulation step at 600 K, which further validate the stability of hydrogenated bilayer silicene structures. The low hydrogenation concentration structures include $Si_{28}H_6$, $Si_{24}H_6$, $Si_{16}H_4$ and $Si_{12}H_4$, which can be used as solar PV devices and high hydrogenation concentration structures include Si_8H_4, $Si_{16}H_{12}$ and $Si_{12}H_{10}$, which can be used as light emitters for white LED. Based on the particle swarm optimization (PSO) technique as implemented in the CALYPSO (crystal structure analysis by particle swarm optimization) code, Luo *et al.* proposed two layered structures, namely, Si_8H_2-*Pm*11 and Si_6H_2-*Pmm*2 with quasi-direct gaps of 0.75 and 1.59 eV, respectively, which can be excellent candidates in light-emitting diodes and photovoltaics.[100]

Also, a new bilayer silicene allotrope with space group *Cmme* is verified through the global search methods and is assumed to be much more stable (by 270 meV per atom) than single-layer silicene. It can be inferred that when more H atoms are added, the band gap tends to increase resulting from the sp^3 hybridization of Si and the hydrogenation can improve the thermal stability of the Si layered structures for the melting point for Si_8H_2-*Pm*11 and Si_6H_2-Pmm_2 are 800–900 K, which is much higher than Si-*Cmme* (500 K).

1.4.4 Group-14 Element Derivatives

Ciraci *et al.* predicted the silicene and germanene derivatives constructed from periodic dumbbell units through adsorption of group-14 elements on 2D honeycomb structures.[101] Kaltsas *et al.* examined honeycomb germanene is less stable than germanium ultrathin films with a $\sqrt{3} \times \sqrt{3}$ or 7×7 arrangement. Luo *et al.* predicted the single-layer and bilayer structures of the silicon related group-14 elements systematically by using a global search method. Matusalem *et al.* study six different 2D allotropes of carbon, silicon, germanium, and tin, namely graphene-like (HEX), large honeycomb dumbbell (LHD), trigonal dumbbell (TD), hexagonal dumbbell (HD), MoS_2-type and kagome lattice (KL) geometries.[102] The results show that for carbon, the hexagonal structure (graphene) is most stable; for silicon and germanium, the dumbbell structures, particularly the large honeycomb dumbbell (LHD) geometries are energetically favored; for tin, the MoS_2-like symmetry to be more stable. These findings shed a new light on chemical bonding and potential applications of group-14 elements. Recently, penta-silicene was proposed, however, it is not stable.[103] Through multilayer stacking, the stable form of penta-silicene can be found with semiconducting or metallic properties, depending on the stacking mode.

In summary, the theoretically predicted silicene allotropes not only help us to verify the poorly understood reconstruction formation mechanism, but also promote the application of silicene in the field of solar energy and photovoltaic industry.

1.5 Phosphorene

Phosphorus can form a variety of allotropes, including α-, β-, γ-phase white, red, violet and black phosphorus, with the color defined by the fundamental band gap.[104–106] It is highly reactive with typical 3-fold coordination through inequivalent sp^3-hybridized orbitals. Phosphorene, 2D form of phosphorus, was firstly exfoliated in layered black phosphorus (BP), with a direct bandgap which can be modulated through thickness and strain. Since phosphorus atoms are arranged in a characteristic pucker structure, it exhibits novel anisotropic physical and mechanical properties, such as anisotropic thermoelectric, topological insulator, negative Possion's ratio, multiferroic, superior gas sensor and photovoltaic, *etc.*[107–113] Therefore, it is assumed to be a formidable competitor in the family of 2D materials and more researchers are committed to find new 2D phosphorus allotropes complementary to phosphorene with tunable properties, both experimentally and theoretically. Here is a short review about the development of 2D phosphorus allotropes.

1.5.1 Theoretical Design

1.5.1.1 Blue Phosphorus

In 2014, Zhu *et al.* proposed an unknown phase of phosphorus with layer stacking and high stability compared with the black phosphorus allotrope in Figure 1.8(a), which is called blue phosphorus, also known as β-P.[114] It possesses the nonplanar hexagonal honeycomb structure, which contains two atoms per layer per unit cell, forming a silicene-like sheet in Figure 1.8(b). The relationship between black phosphorus and blue phosphorus is illustrated in Figure 1.8(c).

It is evident that they share the similar hexagonal structure and if a dislocation is introduced into a monolayer of black phosphorus by flipping specific P atoms from a 'down' to an 'up' position, it will convert a monolayer of

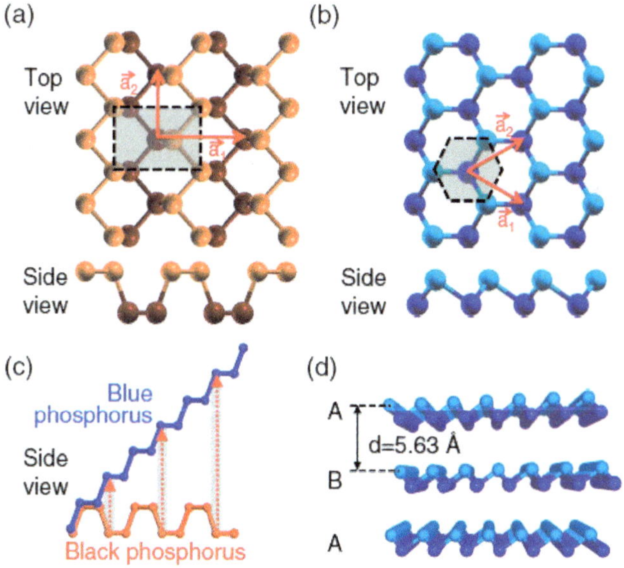

Figure 1.8 The layered structure of (a) black and (b) blue phosphorus in top and side views. Atoms at the top and bottom of the nonplanar layers are distinguished by color and shading, and the Wigner-Seitz cells are shown by the shaded region. (c) Schematic of the conversion of black to blue phosphorus by dislocations, highlighted by the shaded regions and arrows. (d) Vibrational band structure of a monolayer of blue phosphorus. Reprinted with permission from Z. Zhu, D. Tománek, *Physical Review Letters*, **112**, 176802, 2014. Copyright 2014 by the American Physical Society.[114]

black to new-phase blue phosphorus. Further calculation reveals that blue phosphorus have an equivalent cohesive energy with black phosphorus (less than 2 meV) and much larger than other allotropes like white or red phosphorus. The AB and ABC stacking of blue phosphorus showed that the interlayer interaction is very weak with interlayer distance $d = 5.63$ Å, implying the monolayer can be easily exfoliated from the corresponding 3D bulk phase. Then, the stability of blue and black phosphorus is validated by studying the vibration spectrum, seen in Figure 1.8(d). The similar vibration spectra reflect the similar bonding character. The in-plane rigidity of blue phosphorus is smaller than black phosphorus, but the molecular dynamic calculation confirmed the high in-plane rigidity of blue phosphorus. The in-plane stiffness of black phosphorus is anisotropic and the blue phosphorus is isotropic and the finite in-layer compressibility can help when accommodating lattice mismatch during chemical vapor deposition (CVD) growth on a substrate. The electronic structure is also investigated and the bandgap of blue phosphorus is dependent on the number of layers. The layer-dependent band gap value of $\lesssim$3 eV lies just above the photon energy of visible blue light. Strain is also applied to modulate the bandgap and the possibility to change the band gap value by $\lesssim$50% in strained epitaxial geometries on different substrates is one more indication that blue phosphorus may find intriguing applications in nanoelectronics. At last, the phase conversion of black phosphorus into blue was studied by using transition state search method. The activation barrier for the conversion process is 0.47 eV per atom, indicating the relative ease of mechanical conversion from black to blue phosphorus. In 2016, through the epitaxial growth on Au (111), Zhang with co-workers successfully synthesized single layer blue phosphorus, which advanced the application of blue phosphorus and further validated the role of structural prediction.[115]

1.5.1.2 *Phase Coexistence and Metal-insulator Transition in Few-layer Phosphorene*

Besides the α-P and β-P, two additional stable structure phases of layered phosphorus called as γ-P and δ-P were proposed by Guan *et al.* in 2014.[116] The new structures share the three-fold coordination and non-planar layer structures. There are four atoms in the rectangular Wigner-Seitz cell of γ-P and eight atoms in that of δ-P, analogue to that of the anisotropic α-P. The calculated cohesive energy of α-P, β-P, γ-P and δ-P from PBE functional is 3.30, 3.29, 3.22 and 3.23 eV per atom, respectively. The close energy difference (below 0.1 eV) is caused by the similar local environment of the atoms and all the bond lengths are nearly 2.29 Å. The stability of γ-P and δ-P are further verified by calculating their vibration spectra and by performing canonical molecular dynamics (MD) simulations. They also calculated the electronic band structure of γ-P and δ-P monolayer and γ-P has an indirect band gap with 0.5 eV, whereas δ-P has a direct bandgap with 0.45 eV. Since DFT-PBE band gap values are underestimated by ≈1 eV compared with *GW* self-energy calculations, all four phases should display a fundamental band gap in excess of 1 eV in the monolayer. For γ-P, a similar in-plane strain dependent bandgap appears and if stretched beyond 4% can turn a γ-P monolayer metallic. Besides, structure of in-layer connections between α-P and β-P, between β-P and γ-P, and between γ-P and δ-P are also investigated to enrich the possibilities for band structure engineering. The total energy of optimized structure with coexisting phases compared with those of pure, defect-free phases indicates that the energy cost to connect stable but different structural phases is negligible in comparison to the cohesive energy. Thus, it is possible to form multiphase 2D phosphorus structures, realizing the assembly of heterostructures with well-defined metallic and semiconducting regions in one contiguous layer.

1.5.1.3 *Tiling Phosphorus*

In order to study the structural diversity and similarity of different layered phosphorene allotropes, Zhu *et al.* and Zhang *et al.* present a scheme by projecting their non-planar atomic structure onto a two-color 2D triangular tiling pattern.[117] The atoms in the 'top' position is assigned to be dark tiles and *vice versa*. N is the number of surrounding like-color tiles, with $0 \leq N \leq 2$. It is intriguing that both the relative stability and electronic properties depend primarily on the structural index N. The pattern of 2D phosphorene composed of 3–12 membered rings was predicted by Yu *et al.* as Kagome-P.[118] It is an indirect gap semiconductor with a band gap of 1.64 eV and the band gap depends sensitively with the in-plane strain. Li *et al.* proposed a new structure named as the octagonal tiling (OT) structure of 2D allotropes of group-15 elements and further calculation verifies that these allotropes are dynamically stable and are also thermally stable at temperatures up to 600 K with band gap ranging from 0.3 to 2.0 eV, of which the band gap of OT phosphorene is indirect with 2.0 eV.[119] This approach is possible to serve as an effective method to categorize 2D phosphorus sheets with non-hexagonal rings or even 2D quasicrystals with no translational symmetry.

1.5.1.4 *Single-layered Hittorf's Phosphorus*

Hittorf's (or violet) phosphorus is one member in a family of so-called red phosphorus, ranking IV among the six phases. The structure of violet phosphorus consists of layered connected chains of clusters and the corresponding 2D configuration (also called hittorfene) can be expected to exfoliate from bulk phase since binding energy per area per layer is lower than the two-other-known single-layered allotropes, black or A7 phosphorene.[120] Combined with DFT calculation and deformation theory, band structure and the carrier mobility of 2D violet phosphorus are studied. The results indicate that hittorfene can be expected to have a wide band gap of approximately 2.5 eV from HSE functional with dispersion corrections using Tkatchenko and Scheffler (TS) approach. They also find it to have a high and highly anisotropic hole mobility with an upper bound lying between 3000–7000 $cm^2 V^{-1} s^{-1}$. The origin of anisotropic hole mobility comes from the strong anisotropy of the deformation potential, which is also found in black phosphorene, and the smallest deformation potential results in a very large hole mobility up to 7000 $cm^2\ V^{-1}\ s^{-1}$. These combined properties make hittorfene a very good candidate in the field of nanoelectronics. Another new phase in the family of red phosphorus was proposed by Zhao *et al.*, called as tricycle type red phosphorene. The structure can be assumed as the combination of black and blue phosphorenes and it is an indirect band gap semiconductor with a gap of about 1.96 eV by HSE06 calculations.

1.5.1.5 Nine New Phosphorene Polymorphs with Non-honeycomb Structures

The buckled honeycomb lattice was thought to be the best building blocks to construct 2D phosphorus such as in α-P, β-P, γ-P and δ-P, shown in Figure 1.9.

In 2015, an extended family composed of P_4 square or P_5 pentagon units was proposed by Wu *et al.*[121] It contains four new polymorphs, namely, ε-P, ζ-P, η-P, θ-P, together with five additional hybrid polymorphs, greatly enrich the phosphorene structures. Based on the DFT calculation, the phonon spectrum shows no imaginary frequency and these allotropes can remain intact at the temperature of 700 K. The cohesive energy is also calculated and the results show a small energy difference compared with α-P, of which θ-P is shown to be equally stable as the α-P. Further calculation from HSE06 functional was performed and the results showed that except for ε-P with a direct band gap with 0.25 eV at the Γ point, other three monolayers possess indirect

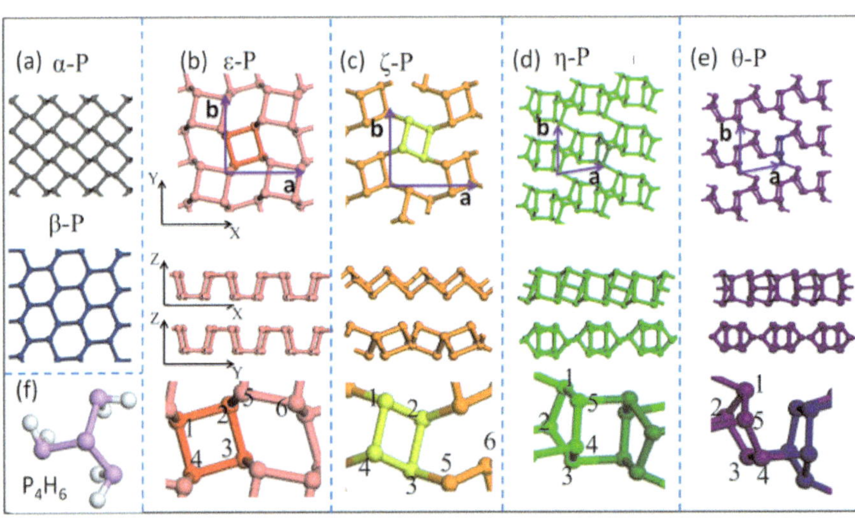

Figure 1.9 Optimized structure of monolayer (a) α-P (black) and β-P (blue), (b) ε-P, (c) ζ-P, (d) η-P, and (e) θ-P, and (f) molecule P4H6. In (b)–(e) we display both top-view and side-views from two horizontal directions, and a zoomed view of the component unit (P4 or P5). Reprinted with permission from ref. 26. Copyright 2012 American Chemical Society.

band gaps around 1 eV. Their hybrid structures called as αε-P, βε-P, γε-P, and δε-P were also considered and their relative energies are still within 0.21 eV per atom from that of α-P. These hybridized phosphorenes possess a variety of distinct bandgaps, in which γε-P is even metallic. These polymorphs can undergo structural transformation between different phases and this kind of property can also be used as ferroelastic switching for shape memory.

1.5.1.6 Porous Polymorphs of 2D Phosphorus

Zhuo *et al.* proposed up to 21 new 2D phosphorus allotropes with porous structure by dividing them into two kinds of building blocks known as 'node' and 'bridge', in which 'nodes' and 'bridges' are represented by phosphorus atoms or clusters, shown in Figure 1.10.[122]

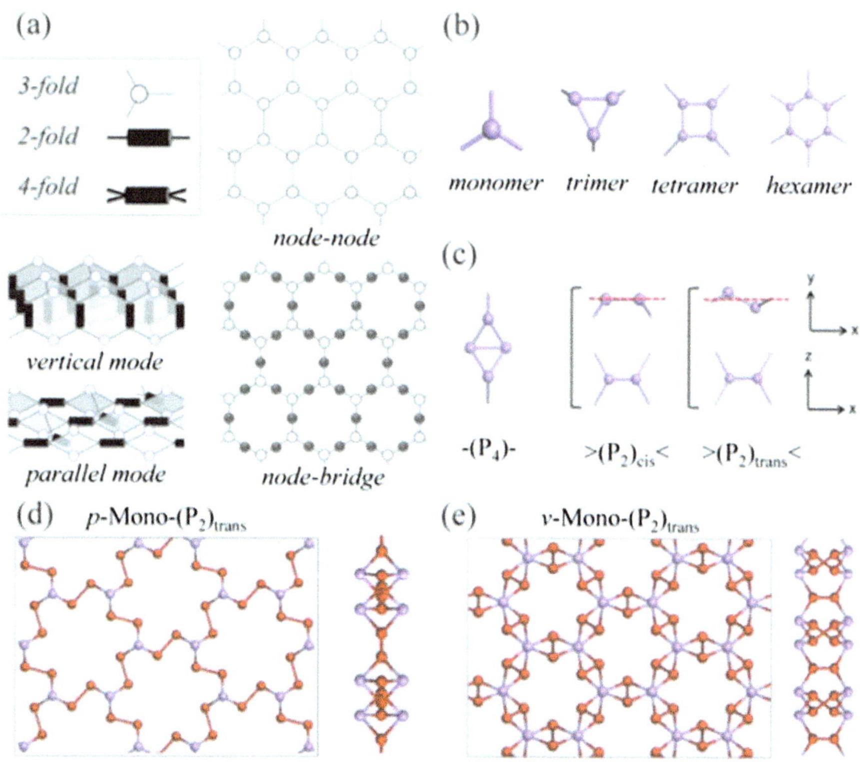

Figure 1.10 (a) Strategy to assemble 'nodes' with three-fold coordination and 'bridges' with either two- or four-fold coordination into 2D topological networks; (b) phosphorus monomer, trimer, tetramer, and hexamer are chosen as four kinds of 'nodes'; (c) phosphorus tetramer and dimer are chosen as two- and four-fold coordinated 'bridges'; (d) and (e) two examples of 2D phosphorus networks obtained by assembling with monomer 'node' and >(P2)trans< 'chain' in parallel or vertical methods. Reprinted with permission from ref. 122. Copyright 2016 American Chemical Society.

There are four kinds of 'node' including monomer, trimer, tetramer, and hexamer and two kinds of 'bridges' including tetramer and dimer. Based on this periodic graphs strategy, they presented two types of 2D phosphorus allotropes, type I only composed of 'node' and type II is formed by introducing 'bridges' into 2D crystalline network. Among them, nine of reported structures are predicted to be more stable than white phosphorus. Vibration spectra calculation and Born–Oppenheimer molecular dynamic simulation at temperature of 1500 K are also performed to validate their stability. Further calculation about the mechanical property and electronic property reveals that phosphorus porous polymorphs have isotropic mechanic properties and they are all characteristic semiconductor with bandgap ranging widely from 0.15 to 3.42 eV from HSE functional. Moreover, the band alignments of eight phosphorus allotropes match well with the chemical reaction potential of both H^+/H_2 and O^2/H_2O, implying the potential for visible-light-driven water splitting. Besides, He *et al.* reported five low-energy new structural stable phosphorene through gene recombination methods and Xu *et al.* proposed a new quasi-2D three-connected network called as Hex-star. The 3-fold coordination is thought to be the best for constructing 2D phosphorus sheets, but Liu *et al.* reported three new phosphorus monolayers (labelled α-P_6, β-P_6, and 558-P_6) having 2-, 3-, and 4-fold coordination and their thermal and dynamic stability were verified by further calculation.[123] These findings can help to enrich the family of 2D phosphorus and bring about novel physical and mechanical properties to the field of atomic thickness nanomaterial.

Here, we reviewed the experimentally synthesized and theoretically predicted 2D phosphorus sheets. Novel structures with exotic properties can be realized by simply arrangement of phosphorus atoms, which may enlighten us in the prediction of 2D materials.

1.6 Compounds

1.6.1 Carbides

1.6.1.1 B–C Compounds

Earlier studies have shown that B–C compounds are a relatively hard material. Yanagisawa *et al.* first predicted that the BC_3 monolayer was composed of six-membered carbon rings and boron atoms, arranging in a close packed honeycomb structure. Graphene doped with boron atoms can realize BC_5 and BC_7 in experiment.[124,125] In 2011, Xiang *et al.* performed a systematic search for the ratio of boron with carbon in 1:5, 1:3, 1:2, 1:1, 2:1, 3:1 and 5:1.[126] Among the 20 structures, structures with boron carbon ratio of 1:3 and 2:1 clearly show a homogeneous phase with boron and carbon in a discrete distribution. Further calculations show that BC_3 is the most stable configuration with a lower energy than BC_5 (51 meV/atom) and BC_2 (49 meV/atom). The results also present that except BC_3-I, all the other structures display metallic behavior. In carbon-rich condition, boron exists in one dimensional zigzag chain (except BC_3-I); When the boron carbon ratio is 1:1, the most stable configuration consists of armchair carbon chains and boron double ring chains, showing a ultrahigh intralayer strength which can withstand temperature up to 2000 K. In boron-rich condition, carbon atom is always tetra-coordinated and boron is in coordination of five or six.

1.6.1.2 Si–C Compounds

The monolayers of SiC, pt-SiC_2 and p-SiC_2 have been investigated before based on their bonding characteristics. Moreover, a series of Si–C monolayers with different ratios have been studied by using the global search method. In 2013, Zhou *et al.* proposed a g-SiC_2 structure and the total energy per atom is lower than previously reported pt-SiC_2 by 0.41 eV.[127,128] The melting point is assumed to 3000–3500 K and the band gap is 1.09 eV. It can form a heterojunction with monolayer GaN and ZnO and the energy conversion efficiency can reach 14.2% and 20%, respectively. The stable configurations with other Si–C compounds like O_1–SiC_3, Si_2C_3–I and SiC_4–I are also explored, showing a similar property with g-SiC_2.[129,130]

1.6.1.3 Be_2C

The structure of the monolayer Be_2C resembles T-VSe_2.[131] It is a semiconductor with direct band gap of 2.34 eV from the HSE06 functional. It includes the features that the carbon atom is quasi-planar hexa-coordinated and the in-plane elastic module is higher than monolayer silicene and MoS_2. Molecular dynamic simulation shows that the melting point is between 1500 K and 2000 K.

1.6.2 Silicates

1.6.2.1 B–Si Compounds

B–Si with ratios of 1:1, 1:2, 1:3, 1:4, 2:1, 3:1, 4:1, 5:1, 6:1 and 7:1 have already been reported and only monolayer BSi_2 and B_4Si are not strictly planar.[132] We summarize the characteristic of monolayer B–Si below: All structures are metallic. In silicon-rich condition, it tends to form a graphene-like structure. In boron-rich condition, it tends to have the trigonal boron fragment like 2D boron sheets; except for BSi_2, B_4Si and B_7Si, silicon is likely to form sp^2 hybridization.

1.6.2.2 Cu_2Si

Cu_2Si is the first predicted hexa-coordinated planar monolayer with metallic behavior.[133] It possesses the special bonding mechanism of 4c-2e, which is helpful for its planar structure. The melting point is assumed to be higher than 900 K and in-plane elastic module is 93 N m^{-1}.

1.6.3 Boron Nitrides and Carbon Nitrides

The most stable structure is the BN monolayer and other B–N compounds like B_2N_3, B_3N_4, B_3N_5 are also investigated.[26] The most stable configurations of B_2N_3 and B_3N_4 possess 5-8-5 BN rings and the most stable configuration of B_3N_5 is composed of rings with odd number of atoms. C–N compounds can be formed by introducing N atoms into graphene, which can help to open the band gap. C_3N is a semiconductor with indirect band gap 0.96 eV and $C_{12}N$ has direct band gap of 0.98 eV.

1.7 Conclusion and Outlook

In summary, the structural prediction based on molecular design as well as global search has achieved great success. It has become the tendency in the field of 2D materials and can provide support in the experimental synthesis progress. However, the prediction still meets a lot of challenges like the limitation of computational resources, calculation accuracy and algorithm optimization. How to design 2D materials with specified functions remains a great challenge, but also brings lots of opportunities. We should gather experience to make structural predictions with specified functions simpler and faster.

Acknowledgements

This work is partially supported by MOST (2016YFA0200602), NSFC (21421063, 51172223, 21573204), by the Strategic Priority Research Program of CAS (XDB01020300), the Fundamental Research Funds for the Central Universities, by the National Program for Support of Top-notch Young Professional, the External Cooperation Program of BIC, and by USTCSCC, SCCAS, Tianjin, and Shanghai Supercomputer Centers.

References

1. K. S. Novoselov, A. K. Geim and S. V. Morozov, *et al.*, *Science*, 2004, **306**, 666–669.
2. F. Schedin, A. K. Geim and S. V. Morozov, *et al.*, *Nat. Mater.*, 2007, **6**, 652.
3. K. S. Novoselov, A. K. Geim and S. V. Morozov, *et al.*, *Nature*, 2005, **438**, 197–200.
4. Y. B. Zhang, Y. W. Tan, H. L. Stormer and P. Kim, *Nature*, 2005, **438**, 201–204.
5. C. Gomez-Navarro, M. Burghard and K. Kern, *Nano Lett.*, 2008, **8**, 2045–2049.
6. I. Calizo, A. A. Balandin, W. Bao, F. Miao and C. N. Lau, *Nano Lett.*, 2007, **7**, 2645–2649.
7. H. Liu, A. T. Neal and Z. Zhu, *et al.*, *ACS Nano*, 2014, **8**, 4033–4041.

8. Y. Min, G. D. Moon and B. S. Kim, *et al.*, *J. Am. Chem. Soc.*, 2012, **134**, 2872–2875.
9. C. Jin, F. Lin and K. Suenaga, *et al.*, *Phys. Rev. Lett.*, 2009, **102**, 195505.
10. L. Li, Y. Yu and G. Ye, *et al.*, *Nat. Nanotechnol.*, 2014, **9**, 372–377.
11. K. Xu, P. Chen and X. Li, *et al.*, *Angew. Chem., Int. Ed.*, 2013, **52**, 10477–10481.
12. C. L. Freeman and F. Claeyssens, *et al.*, *Phys. Rev. Lett.*, 2006, **96**, 066102.
13. X. Xu, W. Yao and D. Xiao, *et al.*, *Nat. Phys.*, 2014, **10**, 343–350.
14. C. L. Song, Y. L. Wang and P. Cheng, *et al.*, *Science*, 2011, **332**, 1410–1413.
15. H. Duan, N. Yan and R. Yu, *et al.*, *Nat. Commun.*, 2014, **5**, 3093.
16. G. Algara-Siller, N. Severin and S. Y. Chong, *et al.*, *Angew. Chem.*, 2014, **126**, 7580–7585.
17. M. Xu, T. Liang and M. Shi, *et al.*, *Chem. Rev.*, 2013, **113**, 3766–3798.
18. S. Z. Butler, S. M. Hollen and L. Cao, *et al.*, *ACS Nano*, 2013, **7**, 2898–2926.
19. X. Wang, Z. Zeng and H. Ahn, *et al.*, *Appl. Phys. Lett.*, 2009, **95**, 183103.
20. X. Li, J. Zhao and J. Yang, *Sci. Rep.*, 2013, **3**, 1858.
21. J. Wang, S. Deng and Z. Liu, *et al.*, *Natl. Sci. Rev.*, 2015, **2**, 22–39.
22. S. Kirkpatrick, C. D. Gelatt and M. P. Vecchi, *Science*, 1983, **220**, 671–680.
23. I. Errea, M. Calandra and C. J. Pickard, *et al.*, *Phys. Rev. Lett.*, 2015, **114**, 157004.
24. D. J. Waled and J. P. K. Dove, *J. Phys. Chem. A*, 1997, **101**, 5111–5116.
25. C. W. Glass, A. R. Oganov and N. Hansen, *Comput. Phys. Commun.*, 2006, **175**, 713–720.
26. Y. Wang, M. Miao and J. Lü, *et al.*, *J. Chem. Phys.*, 2012, **137**, 224108.
27. X. Wu, J. Dai and Y. Zhao, *et al.*, *ACS Nano*, 2012, **6**, 7443–7453.
28. X. Luo, J. Yang and H. Liu, *et al.*, *J. Am. Chem. Soc.*, 2011, **133**, 16285–16290.
29. L. J. Zhou, Y. F. Zhang and L. M. Wu, *Nano Lett.*, 2013, **13**, 5431–5436.
30. J. K. Olson and A. I. Boldyrev, *Chem. Phys. Lett.*, 2012, **523**, 83–86.
31. M. Fujimori, T. Nakata and T. Nakayama, *et al.*, *Phys. Rev. Lett.*, 1999, **82**, 4452–4455.
32. A. R. Oganov, *et al.*, *Nature*, 2009, **457**, 863–867.
33. A. P. Sergeeva, *et al.*, *Acc. Chem. Res.*, 2014, **47**, 1349–1358.
34. J. E. Fowler and J. M. Ugalda, *J. Phys. Chem. A*, 2000, **104**, 397–403.
35. W. Huang, *et al.*, *Nat. Chem.*, 2010, **2**, 202–206.
36. Z. A. Piazza, *et al.*, *Nat. Commun.*, 2013, **5**, 3313.
37. H. J. Zhai, *et al.*, *Nat. Chem.*, 2014, **6**, 727–731.
38. I. Boustani, *Surf. Sci.*, 1997, **370**, 355–363.
39. I. Boustani, A. Quandt, E. Hernandez and A. Rubio, *J. Chem. Phys.*, 1999, **110**, 3176–3185.
40. S. Chacko, D. G. Kanhere and I. Boustani, *Phys. Rev. B: Condens. Matter*, 2003, **68**, 035414.
41. K. C. Lau and R. Pandey, *J. Phys. Chem. C*, 2007, **111**, 2906–2912.
42. H. Tang and S. Ismail-Beigi, *Phys. Rev. Lett.*, 2007, **99**, 115501.
43. H. Tang and S. Ismail-Beigi, *Phys. Rev. B: Condens. Matter*, 2009, **80**, 134113.

44. J. Kunstmann and A. Quandt, *Phys. Rev. B: Condens. Matter*, 2006, **74**, 035413.
45. T. R. Galeev, Q. Chen, J. C. Guo and H. Bai, *et al.*, *Phys. Chem. Chem. Phys.*, 2011, **13**, 11575–11578.
46. N. G. Szwacki, A. Sadrzadeh and B. I. Yakobson, *Phys. Rev. Lett.*, 2007, **98**, 166804.
47. J. M. Sanchez, F. Ducastelle and D. Gratias, *Phys. A*, 1984, **128**, 334–350.
48. A. van de Walle, M. Asta and G. Ceder, *CALPHAD: Comput. Coupling Phase Diagrams Thermochem.*, 2002, **26**, 539–553.
49. E. S. Penev, S. Bhowmick, A. Sadrzadeh and B. I. Yakobson, *Nano Lett.*, 2012, **12**, 2441–2445.
50. C. B. Kah, M. Yu, P. Tandy, C. S. Jayanthi and S. Y. Wu, *Nanotechnology*, 2015, **26**, 405701.
51. C. Leahy, M. Yu, C. S. Jayanthi and S. Y. Wu, *Phys. Rev. B: Condens. Matter*, 2006, **74**, 155408.
52. B. Feng, J. Zhang and Q. Zhong, *et al.*, *Nat. Chem.*, 2016, **8**, 563–568.
53. X. Yu, L. L. Li, X. W. Xu and C. C. Tang, *J. Phys. Chem. C*, 2012, **116**, 20075–20079.
54. H. G. Lu, Y. W. Mu, H. Bai, Q. Chen and S. D. Li, *J. Chem. Phys.*, 2013, **138**, 024701.
55. X. F. Zhou, X. Dong, A. R. Oganov, Q. Zhu, Y. J. Tian and H. T. Wang, *Phys. Rev. Lett.*, 2014, **112**, 085502.
56. F. Ma, Y. Jiao, G. Gao, Y. Gu, A. Bilic, Z. Chen and A. Du, *Nano Lett.*, 2016, **16**, 3022–3028.
57. Y. Liu, E. S. Penev and B. I. Yakobson, *Angew. Chem., Int. Ed.*, 2013, **52**, 3156–3159.
58. Z. Zhang, Y. Yang, G. Gao and B. I. Yakobson, *Angew. Chem., Int. Ed.*, 2015, **127**(44), 13214–13218.
59. A. J. Mannix, X. F. Zhou and B. Kiraly, *et al.*, *Science*, 2015, **350**(6267), 1513–1516.
60. M. Mertz, *et al.*, *J. Am. Chem. Soc.*, 1987, **109**, 6742–6751.
61. H. Zhu, *et al.*, *J. Chem. Phys.*, 1994, **101**, 5281.
62. H. Crespi, *et al.*, *Phys. Rev. B*, 1996, **53**, 20.
63. H. Tyutyulkov, *et al.*, *Chem. Phys. Lett.*, 1997, **272**, 111–114.
64. H. Terrones, *et al.*, *Phys. Rev. Lett.*, 2000, **84**, 1716–1719.
65. M. Deza, *et al.*, *J. Chem. Inf. Comput. Sci.*, 2000, **40**, 6.
66. P. L. de Andres, *et al.*, *Phys. Rev. B*, 2008, **77**, 045403.
67. M. T. Lusk and L. D. Carr, *Carbon*, 2009, **47**, 2226–2232.
68. D. J. Appelhans, *et al.*, *Phys. Rev. B*, 2010, **82**, 073410.
69. X. Luo, *et al.*, *J. Phys. Chem. Lett.*, 2012, **3**, 3373–3378.
70. G. Brunetto, *et al.*, *J. Phys. Chem. C*, 2012, **116**, 12810–12813.
71. J. Wang, *et al.*, *J. Chem. Phys.*, 2013, **139**, 184701.
72. H. Lu, *et al.*, *J. Mater. Chem. C*, 2013, **1**, 3677–3680.
73. Q. Song, *J. Mater. Chem. C*, 2013, **1**, 38–41.
74. B. Mandal, *Phys. Chem. Chem. Phys.*, 2013, **15**, 21001–21006.

75. Z. Wang, *et al.*, *Nano Lett.*, 2015, **15**, 6182–6186.
76. Y. Aierken, *et al.*, *Phys. Rev. B*, 2016, **94**, 155410.
77. D. Malko, *et al.*, *Phys. Rev. Lett.*, 2012, **108**, 086804.
78. S. Zhang, J. Zhou, Q. Wang, X. Chen, Y. Kawazoe and P. Jena, *Proc. Natl. Acad. Sci. U. S. A.*, 2015, **112**, 2372.
79. K. Takeda and K. Shiraishi, *Phys. Rev. B*, 1994, **50**, 14916.
80. G. G. Guzmán-Verri and L. C. Lew Yan Voon, *Phys. Rev. B*, 2007, **76**, 075131.
81. H. Sahaf, L. Masson, C. Léandri, B. Aufray, G. Le Lay and F. Ronci, *Appl. Phys. Lett.*, 2007, **90**, 263110.
82. B. Aufray, A. Kara, S. Vizzini, H. Oughaddou, C. Léandri, B. Ealet and G. Le Lay, *Appl. Phys. Lett.*, 2010, **96**, 183102.
83. B. Lalmi and H. Oughaddou, *et al.*, *Appl. Phys. Lett.*, 2010, **97**, 223109.
84. P. Vogt, *et al.*, *Phys. Rev. Lett.*, 2012, **108**, 155501.
85. P. De Padova, C. Quaresima and C. Ottaviani, *et al.*, *Appl. Phys. Lett.*, 2010, **96**, 261905.
86. P. De Padova, C. Quaresima and P. Perfetti, *et al.*, *Nano Lett.*, 2008, **8**, 271.
87. P. De Padova, C. Quaresima, B. Olivieri, P. Perfetti and G. Le Lay, *Appl. Phys. Lett.*, 2011, **98**, 081909.
88. T. Morishita, M. J. S. Spencer, S. P. Russo, I. K. Snook and M. Mikami, *Chem. Phys. Lett.*, 2011, **506**, 221.
89. D. Kaltsas and L. Tsetseris, *Phys. Chem. Chem. Phys.*, 2013, **15**, 9710.
90. S. Cahangirov, *et al.*, *Phys. Rev. B*, 2014, **90**, 035448.
91. F. Gimbert, *et al.*, *Phys. Rev. B*, 2014, **90**, 165423.
92. J. Bai, H. Tanaka and X. C. Zeng, *Nano Res.*, 2010, **3**(10), 694–700.
93. F. Liu and Y. Yao, *et al.*, *Phys. Rev. Lett.*, 2013, **111**, 066804.
94. C. Lian and J. Ni, *AIP Adv.*, 2013, **3**, 052102.
95. H. Fu, J. Zhang and S. Meng, *et al.*, *Appl. Phys. Lett.*, 2014, **104**, 131904.
96. J. E. Padilha and R. B. Pontes, *J. Phys. Chem. C*, 2015, **119**, 3818–3825.
97. Z. Guo and A. Oshiyama, *et al.*, *Phys. Rev. B*, 2015, **92**, 201413.
98. C. Zhang and S. Yan, *J. Phys. Chem. C*, 2012, **116**, 4163–4166.
99. B. Huang and S. Wei, *et al.*, *Phys. Rev. X*, 2014, **4**, 021029.
100. W. Luo and H. Xiang, *et al.*, *J. Am. Chem. Soc.*, 2014, **136**, 15992–15997.
101. S. Cahangirov and S. Ciraci, *et al.*, *Phys. Rev. Lett.*, 2009, **102**, 236804.
102. F. Matusalem and F. Bechstedt, *et al.*, *Phys. Rev. B*, 2015, **92**, 045436.
103. Y. Aierken and F. M. Peeters, *et al.*, *Phys. Chem. Chem. Phys.*, 2016, **18**, 18486.
104. R. Hultgren, N. S. Gingrich and B. E. Warren, *J. Chem. Phys.*, 1935, **3**, 351.
105. H. Okudera, R. E. Dinnebier and A. Simon, *Z. Kristallogr.-Cryst. Mater.*, 2005, **220**, 259.
106. V. H. Thurn and H. Krebs, *Acta Crystallogr., Sect. B: Struct. Crystallogr. Cryst. Chem.*, 1969, **25**, 125.
107. H. Liu, A. T. Neal, Z. Zhu, Z. Luo, X. Xu, D. Tománek and P. D. Ye, *ACS Nano*, 2014, **8**, 4033.

108. A. S. Rodin, A. Carvalho and A. H. Castro Neto, *Phys. Rev. Lett.*, 2014, **112**, 176801.
109. Q. Wei and X. Peng, *Appl. Phys. Lett.*, 2014, **104**, 251915.
110. Q. Liu, X. Zhang, L. B. Abdalla, A. Fazzio and A. Zunger, *Nano Lett.*, 2015, **15**, 1222.
111. J. W. Jiang and H. S. Park, *Nat. Commun.*, 2014, **5**, 4727.
112. J. Dai and X. C. Zeng, *J. Phys. Chem. Lett.*, 2014, **5**, 1289.
113. V. D. Ganesan, J. Linghu, C. Zhang, Y. P. Feng and L. Shen, *Appl. Phys. Lett.*, 2016, **108**, 122105.
114. Z. Zhu and D. Tománek, *Phys. Rev. Lett.*, 2014, **112**, 176802.
115. J. L. Zhang, S. Zhao and C. Han, *et al.*, *Nano Lett.*, 2016, **16**, 4903–4908.
116. J. Guan, Z. Zhu and D. Tománek, *Phys. Rev. Lett.*, 2014, **113**, 046804.
117. J. Guan, Z. Zhu and D. Tománek, *ACS Nano*, 2014, **8**, 12763.
118. G. Yu, L. Jiang and Y. Zheng, *J. Phys.: Condens. Matter*, 2015, **27**, 255006.
119. P. Li and W. Luo, *Sci. Rep.*, 2016, **6**, 25423.
120. G. Schusteritsch, M. Uhrin and C. Pickard, *Nano Lett.*, 2016, **16**, 2975–2980.
121. M. Wu, H. Fu, L. Zhou, K. Yao and X. C. Zeng, *Nano Lett.*, 2015, **15**, 3557.
122. Z. Zhuo, X. Wu and J. Yang, *J. Am. Chem. Soc.*, 2016, **138**, 7091–7098.
123. J. Liu, Y. Guo, S. Zhang, Q. Wang, Y. Kawazoe and P. Jena, *J. Phys. Chem. C*, 2015, **119**, 24674.
124. H. Yanagisawa, T. Tanaka and Y. Ishida, *et al.*, *Phys. Rev. Lett.*, 2004, **93**, 177003.
125. Y. Ding and J. Ni, *J. Phys. Chem. C*, 2009, **113**, 18468–18472.
126. X. Luo, J. Yang and H. Liu, *et al.*, *J. Am. Chem. Soc.*, 2011, **133**, 16285–16290.
127. Y. Li, F. Li and Z. Zhou, *et al.*, *J. Am. Chem. Soc.*, 2011, **133**, 900–908.
128. L. J. Zhou, Y. F. Zhang and L. M. Wu, *Nano Lett.*, 2013, **13**, 5431–5436.
129. Y. Ding and Y. Wang, *J. Phys. Chem. C*, 2014, **118**, 4509–4515.
130. P. Li, R. Zhou and X. C. Zeng, *Nanoscale*, 2014, **6**, 11685–11691.
131. Y. Li, Y. Liao and Z. Chen, *Angew. Chem., Int. Ed.*, 2014, **53**, 7248–7252.
132. J. Dai, Y. Zhao and X. Wu, *et al.*, *J. Phys. Chem. Lett.*, 2013, **4**, 561–567.
133. L. M. Yang, V. Bačič and I. A. Popov, *et al.*, *J. Am. Chem. Soc.*, 2015, **137**, 2757–2762.

CHAPTER 2

Nanoscale Buckling Mechanics of Ultrathin Sheets

LONG MA AND YONG NI*

CAS Key Laboratory of Mechanical Behavior and Design of Materials, Department of Modern Mechanics, University of Science and Technology of China, Hefei, Anhui 230026, P. R. China
*E-mail: yni@ustc.edu.cn

2.1 Introduction

Graphene as a typical two-dimensional (2D) crystal membrane attracts tremendous interest. Due to its feature of single-atom thickness, the calculated bending stiffness of graphene is found to be of the same order as that of the lipid bilayer,[1] therefore the transition from planar to non-planar configuration may easily occur. The 2D materials can be free-standing or be deposited on substrates, or be reassembled into layered heterostructures or twisted bilayers.[2–4] Differently to the conventional buckling transition, the buckling instabilities in 2D materials show different physical origins, and its buckling pattern could be down to the nanoscale.[4] Various nanoscale buckling morphologies summarized in a recent review,[4] such as ripples,[5,6] wrinkles,[7–9] nanoscale bubble-like blister,[10] soliton-like blister[11,12] are widely observed in these 2D materials.[4]

These out-of-plane geometric distortions are found to significantly influence the mechanical properties of most 2D materials.[13] A molecular

Smart Materials No. 27
Inorganic Two-dimensional Nanomaterials: Fundamental Understanding, Characterizations and Energy Applications
Edited by Changzheng Wu

Published by the Royal Society of Chemistry, www.rsc.org

simulation combined with theoretical analysis demonstrates that the Griffith strength of a graphene containing cracks can be significantly reduced by localized buckling.[14] In contrast, the fracture toughness of the wrinkled graphene by controllable distribution of topological defects can be enhanced notably.[15–17] More effects of the out-of-plane geometric distortions on the mechanics of 2D materials include friction,[18] and negative Poisson's ratio.[19] There is also fast growing interest in the coupling between the mechanical and other physical properties where nanoscale buckles are crucial to modulating the electronic properties.[20]

In this chapter, we limit our review to recent progress on the mechanisms of nanoscale buckles for their formation and controllability. A few examples of nanoscale buckling in graphene-like 2D materials, but not limited to, are introduced to hopefully offer a few commonplace remarks. In a free-standing monolayer sheet the buckled configuration can be induced by mechanical compression and thermal fluctuation. When the monolayer sheet is on substrates, the nanoscale buckling is also influenced by the substrate elasticity, the orientation-dependent substrate surface anchorage, and the interaction between the sheet and the substrate. In a twisted graphene bilayer, the buckling behaviors in the top and the bottom layers are fully coupled and modified by the weak interlayer van der Waals interaction wherein the in-plane misfit relaxation leads to dislocation network.

2.2 Buckling in a Free-standing Monolayer Sheet

The nanoscale buckles usually spontaneously arise in freestanding 2D materials due to their flexibility. By means of developmental experimental methods and theoretical models, it has been found that the appearances of buckling and rippling in free-standing monolayer sheets have different physical origins. Here, several factors that cause buckles in freestanding 2D materials are summarized.

2.2.1 Nanoscale Buckling due to Thermal Fluctuations

It is known that the periodic order of 2D crystal is destroyed by the out-of-plane fluctuation as per the Mermin–Wagner theorem,[21,22] so a perfect 2D crystal does not exist in the free state. The stability of real 2D membranes and layers under finite temperature is maintained by the formation of thermal ripples, which is experimentally observed. Transmission electron microscopy (TEM) images show that in suspended monolayer graphene the long-range crystal order is still retained but in the third dimension there are microscopic ripples and the height of these out-of-plane deformations reaches 1 nm.[23,24] *Ab initio* calculations reveal that the variety of C–C bond lengths produces the intrinsic ripples in suspended graphene.[5] Figure 2.1(a) shows a typical distribution of C–C bond lengths. All the carbon bond lengths are equal at the ground state; however, a finite temperature induces an asymmetric distribution of carbon bonds varies among conjugated bonds

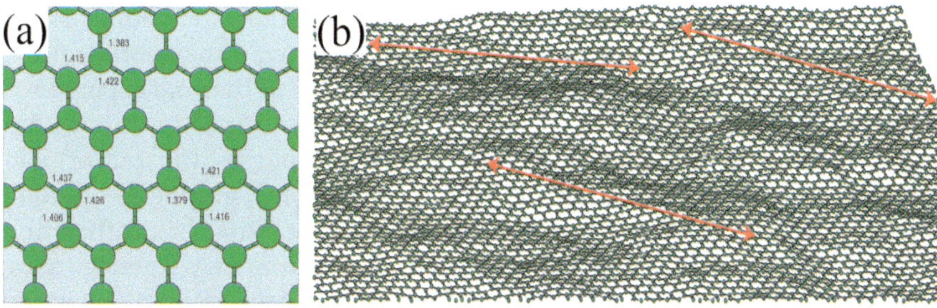

Figure 2.1 (a) A typical distribution of C–C bond lengths, in Å. (b) A typical graphene configuration of thermal ripples, the red arrow represents a spatial scale length, which is about 80 Å. Reprinted with permission from Macmillan Publishers Ltd: *Nature Materials* (ref. 5). Copyright 2007.

(r = 1.42 Å), double bonds (r = 1.31 Å) and single bonds (r = 1.54 Å). The ability of carbon atoms to form different types of bonds raises the fluctuation of graphene from flat state.

In monolayer graphene sheets, the carbon atoms do not have neighboring atoms out of the sheet, so the limit in the perpendicular direction is much weaker than the in-plane directions along the bonds. To release the extra energy created by the thermal effect, out-of-plane deformations are generated at even very low temperature. The spatial scale of the thermal fluctuation can be estimated[5] by the equation $L = 4\pi\kappa\sqrt{\frac{2\pi}{3TB}}$, where κ is the bending rigidity of graphene, B is the two-dimensional bulk modulus and T is the temperature, respectively. From this expression, the distance between the ripples is infinite when $T \sim 0$ K, which means that the flat graphene is stable when temperature approaches 0 K. Moreover, the corrugation of graphene becomes more drastic as the temperature increasing. Figure 2.1(b) displays a typical graphene configuration of thermal ripples where the red arrows represent the spatial scale length L. In addition, further research[25] shows a power-law scaling relationship in thermal ripples of two-dimensional crystalline membranes including graphene as a special case.

Besides the origin and geometric configuration of thermal ripples, influences of thermal fluctuations on the graphene elastic properties are also investigated.[26–28] The temperature dependence of the graphene lattice parameter, as well as the elastic modulus, are calculated by density functional theory[29] and atomistic Monte Carlo simulations.[30] The results reveal that graphene exhibits negative thermal expansion. Furthermore, the coefficient of thermal expansion of suspended graphene experiences a transition from negative to positive between 300 and 400 K, which is observed by scanning electron microscope images[31] and then verified by a statistical mechanics approach and molecular dynamics simulations.[28,32] Thermal ripples are found to be responsible for the transition of coefficient of thermal expansion.[28] At low temperatures, graphene is compressed in the in-plane dimensions because of the out-of-plane thermal ripples, so the coefficient of thermal expansion is negative at a finite temperature. While at high temperatures the anharmonic interactions among in-plane phonon modes[33] suppress the out-of-plane thermal ripples, with the decrease of thermal ripples, graphene gradually shows a positive thermal expansion. Thus, the transition of coefficient of thermal expansion from negative to positive depending on temperature can be understood as a compromise of the competition between in-plane modes and out-of-plane deformations.

Thermal ripples not only affect the coefficient of thermal expansion, but also play an important role in the nonlinear elastic modulus of graphene at a finite temperature.[28] When a positive equibiaxial strain is imposed on the graphene membrane, at first the amplitude of the intrinsic thermal ripples decreases with the increasing strain. Consequently, the contractive membrane begins to expand in-plane and the biaxial modulus will increase with strain. With further increasing strain, the thermal ripples vanish and the elastic modulus of graphene decreases inherently as per its intrinsic non-linear elasticity.[1,34] Finally, the competition between out-of-plane fluctuations of thermal ripples and the intrinsic effect of graphene bring about the non-monotonic biaxial modulus.

2.2.2 Nanoscale Buckling due to Edge Stress

Owing to the specific bonding structures of carbon atoms at the edges of freestanding graphene, intrinsic buckles and scrolls can be observed experimentally near the edges of graphene membranes,[35] which are then verified by theories and simulations[36,37] even without any thermal effect. The configurations of the carbon bonds near the edges of graphene membranes or ribbons can be considered as strings that are different from inner ones in the periodic hexagonal lattice, so edge stresses are generated as tensile stress for longer bonds and compressive stress for short bonds. The excess energies caused by edge stresses can be released by means of hydrogen adsorption, while in the absence of hydrogen the edge buckles emerge to lower the compressive edge stresses.[38,39]

To explore the mechanism of edge stresses induced buckles, Shenoy *et al.* comprehensively studied the edge warping deformation combined with atomistic simulations and finite element calculations.[36] Figure 2.2(a) is the shape of warped freestanding graphene membrane. The out-of-plane deformation modes of the two kinds of edge terminations, zigzag edges and armchair edges, are different because of the different edge stresses and edge modules, which are lately obtained when the intrinsic edge buckling wavelengths are 6.2 nm along the zigzag edge and 8.0 nm along the armchair edge.[37]

To discover the in-depth relationship of the warping geometry and the edge stresses, the competition between in-plane stretching and out-of-plane deformation is considered as an essential route to reduce the edge energy. By

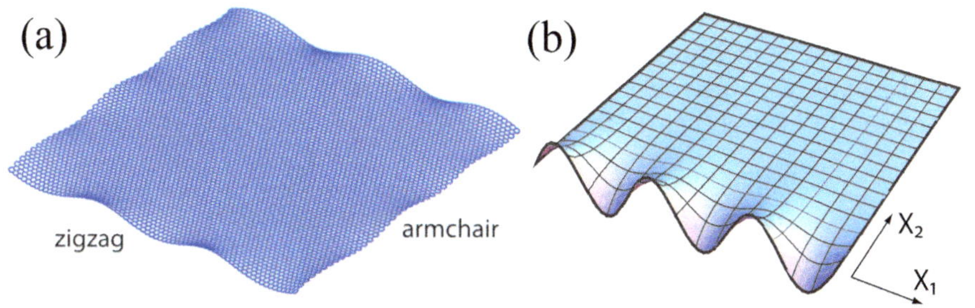

Figure 2.2 (a) A typical configuration of edge buckles with zigzag edges and armchair edges in a freestanding graphene sheet. (b) The schematic diagram of the out-of-plane perturbation analysis near the edge of a semi-infinite sheet (x_1,x_2). Reprinted with permission from V. B. Shenoy, C. D. Reddy, A. Ramasubramaniam, Y. W. Zhang, *Physical Review Letters*, **101**, 245501, 2008. Copyright 2008 by the American Physical Society.[36]

assuming an out-of-plane perturbation in the form $\zeta(x_1,x_2) = A\,sin(kx_1)e^{-x_2/l}$, where ζ is the out-of-plane deformation, A is the amplitude of the wave, k is the wave number and l represents the penetrating length scale of the ripples along the x_2 direction, respectively, to a semi-infinite sheet which $-\infty < x_1 < \infty$ and $0 \le x_2 < \infty$, a perturbation analysis is carried out in terms of the out-of-plane deformation.[36] The total energy per wavelength $\lambda = 2\pi/k$ can be expressed by:

$$U = \frac{\pi\tau_e A^2 k}{2} + \frac{3\pi E_e A^4 k^3}{32} + \frac{\pi E_b A^4\left(3k^4l^4 + 2k^2l^2 + 3\right)}{128kl^3\left(1-\nu^2\right)}, \tag{2.1}$$

where τ_e is the edge stress, ν is the Poisson's ratio, E_e and E_b are the Young's modulus of the graphene edges and sheet, respectively. In eqn (2.1) the amplitude A dominates the total energy in the first term. As a consequence, when edge stresses $\tau_e < 0$, a finite amplitude A will always reduce the total energy. In other words, the formation of edge ripples is energetically favorable when edge stresses are compressive. Furthermore, the scaling laws for the penetration depth l and the amplitude A of the buckles with the wavelength λ can be determined by the perturbation analysis, which l has a linear relationship with λ and A is in proportion to the square root of the wavelength, respectively. It is found that if the penetration depths of the edge ripples are much smaller than the dimensions of sheet as Figure 2.2(a) shows, the ripples are localized near the edges of the membrane and have insignificant influence on the interior of graphene. However, when the penetration depth is about the same size with the width of freestanding graphene ribbon, the stable configurations will have drastic differences with the flat state. Under the domination of edge stress, the narrow graphene nanoribbons show up various geometrical shapes such as scrolling, twisting, arching, and so on, which have significant effects on the electrical and magnetic properties as well as mechanical properties such as Young's modulus of graphene sheet and ribbons.[36,40–42] The edge stresses affect not only the shapes of freestanding graphene membranes or nanoribbons, but the periodic nanoscale buckles in graphene–substrate systems observed by AFM images near the edges of graphene,[43,44] which shares the same mode of releasing extra edge energies.

2.2.3 Nanoscale Buckling in 2D Heterostructures

The previous discussion was about the spontaneous edge warping induced by the relief of edge stresses. Here, another kind of buckling near the interface in 2D materials is presented. Heteroepitaxial growths of 2D lateral heterostructures have attracted research interests for their tunable electronic properties and potentials for tunneling devices. As a representative of 2D heterostructures, the graphene and hexagonal boron nitride (C/BN) superlattice has been synthesized by a controllable 'patterned regrowth' process.[45] In a subsequent study, the topological defects have been observed by STM images at the C/BN interfaces.[46] These discoveries raise the importance of one-dimensional interface between the two materials and several efforts have been made to study the complicated interface with lattice mismatch.[47,48] Both graphene and boron nitride possess hexagonal lattice, however, they exhibit a lattice mismatch about 2% because the unit cell length for graphene is 2.46 Å while for BN is 2.51 Å. Such a lattice mismatch leads to considerable strain energies near the interface and there are two strain-relieving routes: nanoscale misfit dislocations and out-of-plane ripples.

Whether the dislocations are energetically favorable in the coherent interface needs to investigate the competition between the energy raises by introducing dislocations, which are 5,7-membered sets of rings, and the energy reduction that are relaxed by the dislocations. By a means of scale analysis,[47] it is obtained that below a critical C/BN superlattice cell pitch H_{cr}^{dis} (the definition of cell pitch H is shown in Figure 2.3), the coherent interface is energetically stable and dislocations will not appear at the interface.

However, if the superlattice cell pitch is $H > H_{cr}^{dis}$, the introduced energy by misfit dislocation can be offset by the relief of mismatch strain energy so the

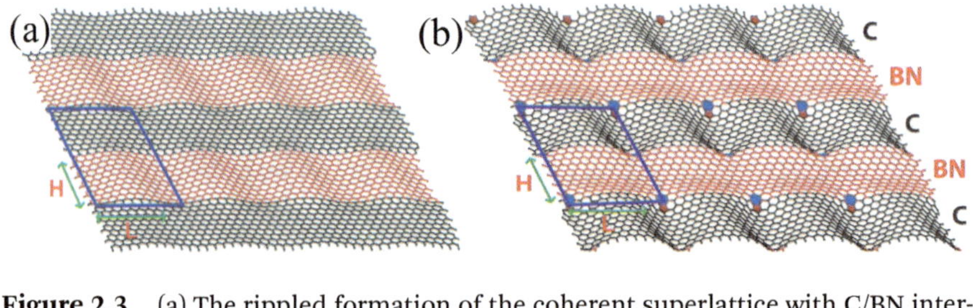

Figure 2.3 (a) The rippled formation of the coherent superlattice with C/BN interface. (b) The rippled formation of the superlattice with C/BN interface decorated by 5,7-membered set of rings. [L,H] denotes the size of superlattice blocks. Adapted with permission from ref. 47. Copyright 2011 American Chemical Society.

dislocations become energetically favorable. Another relief mechanism of out-of-plane ripples is also investigated by a similar analysis. It is found that for coherent interface there also exists a critical pitch H_{cr}^{rip} below which the flat state is stable, otherwise the out-of-plane ripples will emerge to lower the strain energy. The emergence of ripples will lead to a competition between bending energy and transverse stretching energy of the C/BN superlattice. Thus, a set of optimal wavelengths of the ripples are favorable for energy minimization, which are found to depend on the intrinsic lattice misfit and the superlattice cell pitch H in a detailed investigation based on a continuum theory.[48] When ripples emerge at the interface, the total energy densities can be efficiently released by an order of magnitude compared with the flat state. The combined effect of misfit dislocations and ripples formation to lower the mismatch strain energy at the interface is also considered. When the out-of-plane deformations are allowed, ripples formation can maintain the coherency of C/BN interface without the intervention of dislocations, just as Figure 2.3(a) displays. While the misfit dislocations are presupposed at the interface, localized ripples will occur near the cores of dislocations. Figure 2.3(b) shows the rippled formation of the superlattice with C/BN interface decorated by 5,7-membered set of rings. These localized ripples extremely relax the misfit defects induced extra strain energies away from the dislocation core, leave the strain energy densities localized around the core. The cooperation of two energy relief mechanisms, misfit dislocation and out-of-plane ripples formation, generates fascinating configurations for 2D free-standing C/BN superlattices to achieve stable states.

2.2.4 Nanoscale Buckling due to Topological Defects

The CVD based production processes usually introduces topological defects and polycrystallinity in large scale 2D crystals, especially in graphene sheets. These defects have significant influences on the physical and mechanical properties of the 2D materials. In bulk crystals, the defects have been comprehensively studied since the beginning of 1900s, however, due to the special two-dimensional structure, the features of the defects in 2D crystals have distinct differences with those in traditional bulk crystals. As a typical representative of 2D crystals, the effects of defect on freestanding graphene are discussed here.

By aberration-corrected high resolution transmission electron microscopy (AC-HRTEM) experiments,[49,50] it has been observed that out-of-plane deformation can effectively minimize the defect-induced energy. Using density functional theory (DFT) calculations[51–54] it is found that defects such as dislocations, grain boundaries and hole-doped structures will lead to spontaneous nanoscale buckling as well as geometric blisters, bubbles and ridges. These out-of-plane shapes can be artificially constructed by designing defect domains to achieved monolithic nanoengineering.[53,54] To connect the defects in 2D crystals with the classical defect theory together, a continuum elasticity theory is implemented.[55–57] Atomistic simulations and numerical simulations based on continuum theory[56–58] verify that the out-of-plane deformation is energetically favorable near the defects in 2D crystals and can significantly screen the stress fields introduced by defects.

Here, the continuum theory to describe the defect in 2D crystalline membrane is outlined and the governing equations of the out-of-plane deformation of a flexible membrane with topological defects are briefly exhibited. A freestanding flat membrane in the (x_1,x_2) plane is investigated. The in-plane displacement field of the membrane is denoted as $\mathbf{u}(\mathbf{X}) = (u_1,u_2)$ and the out-of-plane displacement is $w(\mathbf{X})$. The in-plane strain tensor contains the contribution of out-of-plane displacement is expressed as:

$$\varepsilon_{ij} = \frac{1}{2}\left(\partial_i u_j + \partial_j u_i + \partial_i w \partial_j w\right), \quad i,j = 1, 2, \tag{2.2}$$

where the third term represents the influence of the out-of-plane displacement. The total energy of a flexible membrane includes two parts: the stretching energy and the bending energy. The former can be expressed by the in-plane strain tensor while the latter involves the mean curvature H and the Gaussian curvature K of the 2D crystalline membrane. In terms of the out-of-plane deformation w, when $|\nabla w|$ is small, these curvatures are expressed as:

$$H = \nabla \cdot \left[\frac{\nabla w}{\sqrt{1 + |\nabla w|^2}}\right] \approx \nabla^2 w, \quad K = \frac{\det\left(\partial_i \partial_j w\right)}{\left(1 + |\nabla w|^2\right)^2} \approx \det\left(\partial_i \partial_j w\right), \tag{2.3}$$

where ∇ denotes the gradient, ∇^2 the Laplace operator and det(**A**) the determinant of a second rank tensor **A**, respectively. So, the total energy contains the stretching energy and bending energy is expressed as:

$$F = \frac{1}{2}\iint\left(2\mu\varepsilon_{ij}^2 + \lambda\varepsilon_{kk}^2\right)\mathrm{d}x_1\mathrm{d}x_2 + \iint\left(\frac{1}{2}\kappa H^2 + \kappa_G K\right)\mathrm{d}x_1\mathrm{d}x_2, \tag{2.4}$$

where μ and λ are the two Lamé coefficients, κ is the bending rigidity and κ_G is the Gaussian rigidity, respectively.

Defects in a crystalline structure change the continuous conditions for the displacement field. The incompatibility condition induced by N disclinations, which can constitute dislocations in the form of disclination dipoles, can be expressed as:[56]

$$e_{ik}e_{jl}\partial_l\partial_k\left(\varepsilon_{ij} - \frac{1}{2}\partial_i w \partial_j w\right) = \sum_{\alpha=1}^{N} s_\alpha \delta\left(\mathbf{r} - \mathbf{r}_\alpha\right), \tag{2.5}$$

where e_{ij} is the permutation tensor, $\mathbf{r}_\alpha$ denotes the location and s_α the strength of α th disclination, respectively. Then the stress tensor σ_{ij} and Airy stress function χ is introduced as:

$$\sigma_{ij} = 2\mu\varepsilon_{ij} + \lambda\varepsilon_{kk}\delta_{ij}, \; \sigma_{ij} = e_{ik}e_{jl}\partial_k\partial_l\chi, \tag{2.6}$$

where δ_{ij} is the Kronecker delta. Skipping the tedious derivation process, finally the von Kármán equations for flexible membrane contains defects are generalized as:[56]

$$\kappa\nabla^4 w + e_{ik}e_{jl}\partial_k\partial_l(\partial_i\chi\partial_j w) = 0, \tag{2.7}$$

$$\frac{1}{K_0}\nabla^4\chi = \sum_{\alpha=1}^{N} s_\alpha \delta(\mathbf{r} - \mathbf{r}_\alpha) - K(\mathbf{r}), \tag{2.8}$$

where $K_0 = \dfrac{4\mu(\mu+\lambda)}{2\mu+\lambda}$ is the two-dimensional Young's modulus and $K(\mathbf{r})$ is the Gaussian curvature, respectively. Numerical simulations are implemented for the above governing equations. Figure 2.4(a) plots the energies for buckled crystalline membrane contains dislocations with unit lattice spacing and Burgers vectors on a semi-logarithmic scale, where R is the distance away from dislocation cores and $\tilde{\kappa} = \dfrac{2}{\sqrt{3}}\kappa$ reflects the bending rigidity of the membrane. The curves show that the out-of-plane buckles significantly lower the dislocation energy compared with the flat state.

Moreover, the value of energy plateau in each curve decreases as the increase of $K_0/\tilde{\kappa}$, which means that the flexible crystalline membrane can release the dislocation energy more efficiently through spontaneous out-of-plane deformation. Figure 2.4(b) displays the simulated configuration of a buckled dislocation with $K_0/\tilde{\kappa} = 100$.

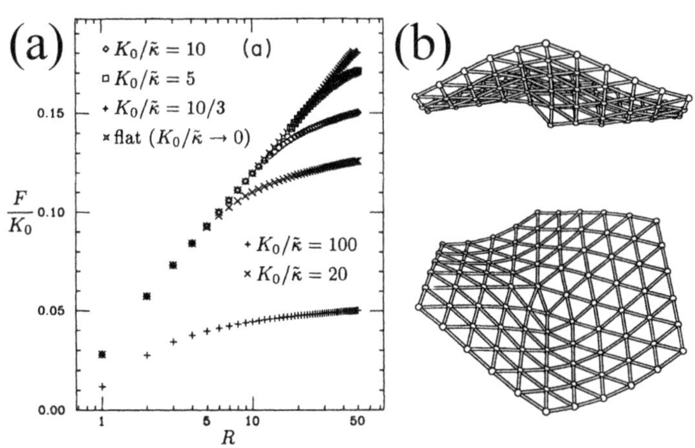

Figure 2.4 (a) The relation between total energy F/K_0 and distance R for buckled dislocation in semi-logarithmic scale. (b) The simulated buckled configuration of dislocation decorated membrane. Reprinted with permission from H. S. Seung, D. R. Nelson, *Physical Review A*, **38**, 1005, 1988. Copyright 1988, The American Physical Society.[56]

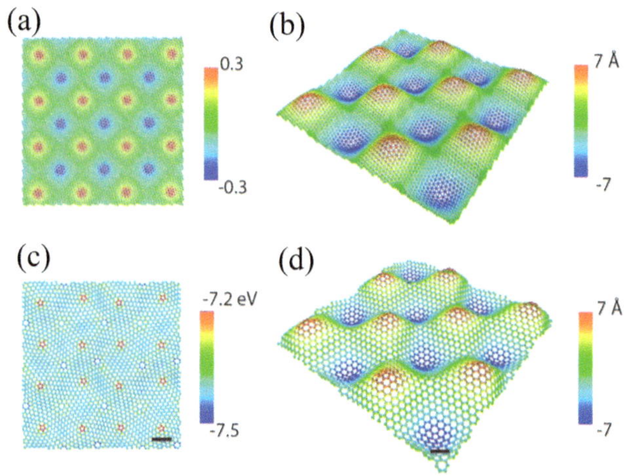

Figure 2.5 Simulated buckled configuration induced by a periodic array of disclinations in freestanding graphene membrane. (a) and (b) are the eigenstrain field of the original flat membrane and the sinusoidal buckled configuration obtained by continuum modeling, respectively; (c) and (d) are the potential energy of disclinations array and the buckled configuration obtained by atomic simulations, respectively. The scale bar is 1 nm. Reprinted from *Journal of the Mechanics and Physics of Solids*, **67**, T. Zhang, X. Li, H. Gao, Defects controlled wrinkling and topological design in graphene, 2–13, Copyright 2014 with permission from Elsevier.[58]

The continuum theory on the buckled dislocation provides guidance to study the complicate coupling relationship between defects and out-of-plane deformations for 2D crystalline materials. The finite element method (FEM) is also employed to predict the atomic scale buckles in disclination and dislocation decorated graphene membranes, whose results are verified accurately compared with the atomic simulations.[58] Figure 2.5 shows the simulated buckled configuration induced by a periodic array of disclinations in freestanding graphene membrane from continuum modeling and atomic simulations. The development of powerful calculation methods which can capture the defect structures as well as out-of-plane deformations will provide suggestions for designing and tailoring the nanoscale structures through introducing defects.

2.3 Buckling of a Monolayer on Substrates

When graphene-like 2D materials are bonded to substrates, the nanoscale thermal rippling may be suppressed due to the geometric confinement of the substrate. However, the nanoscale buckling can still arise due to the compressive stress transmitted by the substrate, *i.e.* the thermal mismatch compressive stress or the applied compression. The nanoscale wrinkling turns out to be localized folding due to the transition of wrinkling-induced delamination when the normal traction along the interface between the monolayer and the substrate is beyond the normal interfacial adhesion strength. In the case of a soft substrate this transition is modified by the substrate elasticity. In the case of a rigid substrate, it is more influenced by the monolayer–substrate–surface interaction energy from van der Waals interactions. In addition, the competition between the elastic energy of the monolayer and the interaction energy can generate other buckling morphologies such as soliton-like buckle-delamination, bubble-like blister and periodically hexagonal ripples in 2D materials. The competition between the orientational bonding energy onto the substrate surface and the elastic energy of the monolayer needed to deform can also lead to intriguing nanoscale localized wrinkling morphology.

2.3.1 Wrinkling Mediated by Substrate Elasticity or van der Waals Interactions

When a thin film deposited on a soft elastic substrate is under compressive stress, the out-of-plane deformation of the thin film is usually associated with the elastic property of the substrate. As is shown in Figure 2.6(a) and (b), the wrinkling and wrinkling-induced delamination of a compressed thin film on soft elastic substrates are illustrated.

If the compressive stress reaches a critical value, wrinkle formation will arise to reduce the compression. The classical continuum theory of von Kármán

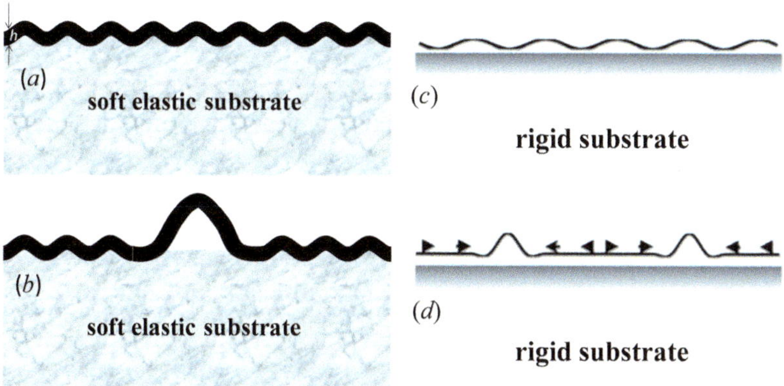

Figure 2.6 Schematic illustration of (a), (c) wrinkling and (b), (d) wrinkle-induced delamination in a compressed monolayer sheet on a soft[62] or rigid substrate,[64] respectively. (a) and (b) reprinted from *International Journal of Solids and Structures*, **51** (21–22), K. Pan, Y. Ni, L. He, R. Huang, Nonlinear analysis of compressed elastic thin films on elastic substrates: From wrinkling to buckle-delamination, 3715–3726, Copyright 2014 with permission from Elsevier.[62] (c) and (d) reprinted from ref. 64 with the permission of AIP Publishing.

equations[59] reveals that the critical wrinkling stress is $\frac{\sigma_C}{\overline{E}_f} = \frac{1}{4}\left(3\frac{\overline{E}_s}{\overline{E}_f}\right)^{2/3}$ with $\overline{E} = E/(1 - \nu^2)$, where E is Young's modulus, ν is the Poisson's ratio and the subscript f and s represent the elastic constants of film and substrate, respectively. In addition, the characteristic wavelength λ and amplitude A of wrinkling are $\frac{\lambda}{2\pi h} = \left(\frac{\overline{E}_f}{3\overline{E}_s}\right)^{1/3}$ and $\frac{A}{h} = \sqrt{4\left(\frac{\sigma}{\sigma_C} - 1\right)}$, where h is the thickness of the film and σ is the compressive stress in the film, respectively. With the increase of the compression, period-doubling bifurcations in film wrinkling occur due to the nonlinear elasticity of the substrate to balance the deformation of film-substrate system and minimize the total energy.[60] In addition, the increase of the compression also increases the normal traction at the film-substrate interface, and the film may undergo a morphological transition from wrinkles to buckling-delamination when the normal traction is beyond the interfacial adhesion strength in the normal direction, σ_{max}. This determines a critical compressive stress in the film for the transition $\frac{\sigma_d^C}{\overline{E}_f} = \frac{1}{4}\left(3\frac{\overline{E}_s}{\overline{E}_f}\right)^{2/3} + \left[\frac{(3 - 4\nu_s)\sigma_{max}}{4(1 - \nu_s)^2 \overline{E}_s}\right]^2$.[61,62] When the substrate is rigid, the wrinkling and wrinkling-induced delamination can still arise in the compressive film as shown in Figure 2.6(c) and (d). Compared with the soft substrate, the rigid substrate cannot deform, however, relaxation of the van der Waals graphene–substrate interaction energy plays a crucial role in the wrinkling and folding formation of graphene.[63–65] The competition between the elastic energy of the monolayer and the interaction energy regulates the transition from wrinkling to folding and the equilibrium separations of localized folds. Minimization of the total free energy of the film-substrate system with a sinusoidal out-of-plane configuration gives the wrinkling wavelength $\lambda_c = 2\pi\left(\frac{Dh^2}{27\Gamma_0}\right)$ and the critical compressive strain $\varepsilon_c = -\frac{6\sqrt{3\Gamma_0 D}}{Ch}$ with D the bending rigidity of the film, Γ_0 the interfacial adhesion energy per unit area and C is the 2D in-plane elastic modulus.[63] Similar analysis is also performed to give the critical compressive strain for the transition from wrinkling to folding.[64]

2.3.2 The Formation of Soliton-like Buckle-delamination and Bubble-like Blister

When the graphene-like monolayer sheet is under a uniaxial compression, the out-of-plane deformation of monolayer sheet usually appears as the form of 1D buckle-delamination that may have directional motion under special

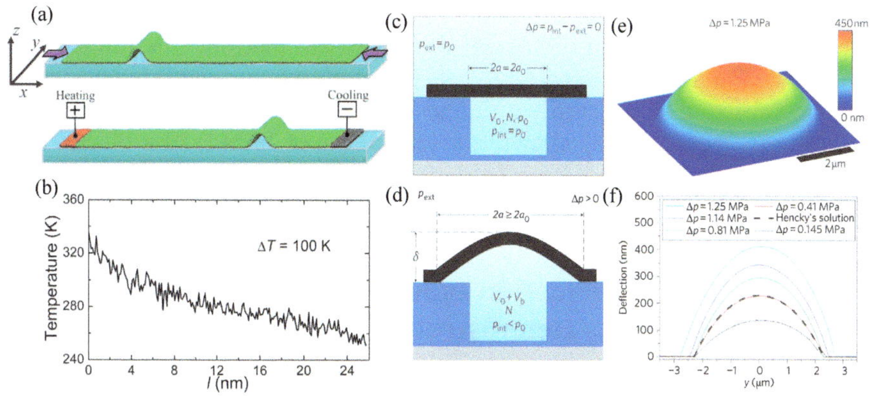

Figure 2.7 (a) Formation of soliton-like buckle-delamination in a graphene by thermophoresis. (b) The distribution of temperature along the graphene sheet. Adapted from ref. 66 with permission from The Royal Society Chemistry; (c) and (d) schematic of the pressurized bubble-like blister in a graphene. (e) The AFM observation of the deformed shape of nanoscale bubble. (f) The deflection profiles of nanoscale bubbles for different values of pressure difference. Reprinted with permission from Macmillan Publishers Ltd: *Nature Nanotechnology* (ref. 70), Copyright 2011.

conditions. Figure 2.7(a) and (b) illustrate the thermophoretic motion of soliton-like buckle-delamination in a graphene sheet on the graphite substrate.

Non-equilibrium molecular dynamics simulations are used to study the motion of such buckle-delamination in supported graphene from high temperature region to low temperature region on a substrate with temperature gradient.[66] It was found that the atoms in the buckle-delamination front root possess lower free energy when they transform into the back foot, which drives the thermophoretic motion from hot to cold region. The Korteweg-de Vries (KdV) equation, first established for the motion of a single-soliton, shows that the motion of buckle-delamination obeys Newton's second law as the classical soliton wave problem describes. The 1D buckle-delamination is also called as ripplocation in multilayer system.[67] Unlike conventional three-dimensional crystal defects which the same-sign dislocations repel due to Frank's rule, however, the same-sign ripplocations attract and tend to merge together. The combined effect of layer bending and van der Waals interaction is responsible for the formation and migration of the ripplocations.

Besides the formation of wrinkles, wrinkling-induced folding, and 1D buckle-delamination, nanoscale bubble-like blisters are experimentally observed as common buckling configurations in supported two-dimension crystals on flat substrates.[10,68] These bubbles have potential applications in manipulating the electronic properties[69] or in measuring the elastic properties of two-dimensional crystals and the adhesion energy of the interface between monolayer sheet and substrate.[70] To have an in-depth understanding of the formation of nanoscale bubbles, both non-linear plate equations with numerical solutions and molecular dynamics simulations are adopted.[71,72] The results show that the van der Waals interaction between monolayer sheet and its substrate significantly influences the center deflection of nanoscale bubble. Figure 2.7(c)–(f) shows the formation of bubble-like blister in a monolayer graphene on silicon oxide substrate induced by the pressure difference across the membrane.

2.3.3 Interlayer Interaction Induced Deformation in 2D van der Waals Heterostructures

From the previous sketches of nanoscale wrinkling and blisters, the van der Waals interaction between monolayer sheet and substrate is the key factor in influencing the out-of-plane deformation in 2D crystals. In fact, recent observations about stacked/layered materials and heterostructures show that lattice mismatch is first accommodated by dislocation structures and then relaxed by buckling the monolayer sheet into the third dimension. The coupling between in-plane and out-of-plane deformations regulates the three-dimensional positions of atoms in the monolayer, which exhibits tunable electronic properties. To consider both the interaction energy and the geometric deformations in a misfitting monolayer on substrates, Kumar *et al.* developed a multi-scale modeling approach which contains continuum deformation theory and *ab initio* calculations.[73] Here are the calculation steps for a case of monolayer A on the top of substrate B (Figure 2.8).

(a) Compute the spatial interaction energy function between the layer A unit cell and the substrate B. The spatial variation of the interaction energy can be expressed as:

$$V_{\mathrm{B}}(x,y,z) = \sum_{m,n\neq 0} u_{mn}(z)\mathrm{e}^{i\vec{G}_{\mathrm{B}mn}\cdot\vec{x}} + u_0(z), \tag{2.9}$$

where $\vec{G}_{\mathrm{B}mn}$ is the 2D reciprocal lattice vector of the substrate surface, and $u_{mn}(z)$ and $u_0(z)$ are the undetermined Fourier coefficients, which can be calculated by the stack energy using density functional theory.

(b) Derive the continuum spatial interaction energy between the layer A and the substrate B within the Moiré unit cell from the above energy in Fourier space using inverse Fourier transformation. It is worth noting that the stacking configurations only have perceptible variations in the scale of Moiré unit cell, so the interlayer energy can be expressed in terms of the Moiré reciprocal lattice vectors:

$$V_{\mathrm{AB}}(x,y,z) = \sum_{m,n} g_{mn}(z)\mathrm{e}^{i\vec{G}^{mn}_{\mathrm{AB}}\cdot\vec{x}} + u_0(z), \tag{2.10}$$

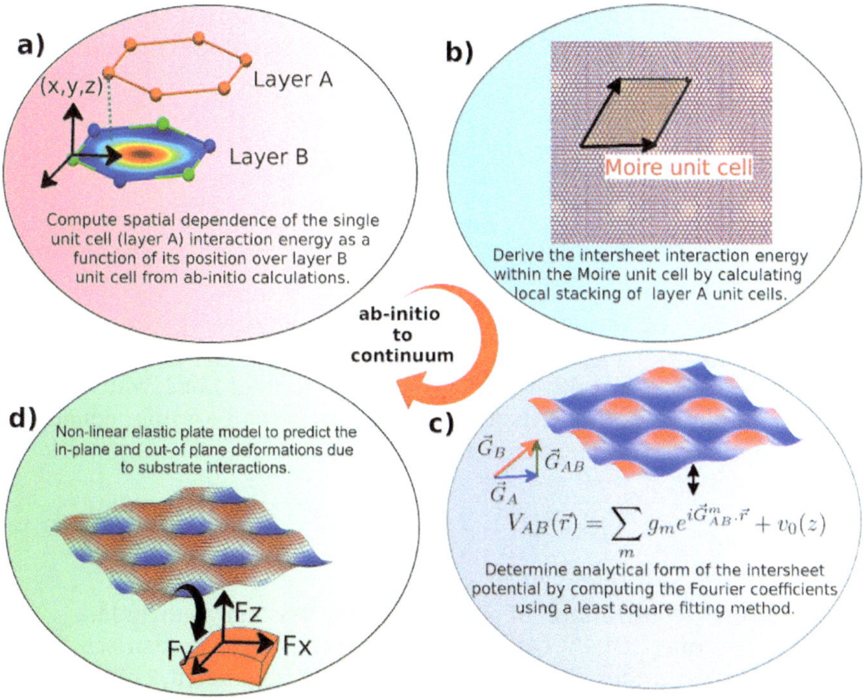

Figure 2.8 Schematic illustration of a multiscale approach to compute nanoscale buckling morphology in 2D van der Waals heterostructure. (a) Computing spatial dependence of the layer A unit cell interaction energy over layer B unit cell. (b) Deriving the intersheet interaction energy within the Moiré unit cell. (c) Determining the intersheet potential by computing Fourier coefficient. (d) Predicting the in-plane and out-of-plane deformation due to substrate interactions. Reprinted with permission from Macmillan Publishers Ltd: *Scientific Reports* (ref. 73), Copyright 2015.

where $\vec{G}_{AB}^{mn}$ are the Moiré reciprocal lattice vectors, g_{mn} are the undetermined Fourier coefficients which can be obtained by density functional theory.

(c) Determine both the in-plane and out-of-plane force fields F_{AB}^{α} (α denotes in-plane direction x and y) and F_{AB}^{z} acting on the layer A due to the substrate B using the above spatial variation of interlayer potential energy. The force fields can be computed by the gradients of the potential energy as:

$$\vec{F}_{AB} = -\nabla V_{AB}(x,y,z). \tag{2.11}$$

(d) Predict the in-plane and out-of-plane deformation in layer A using non-linear plate theory. Given the force fields acting on layer A, it is feasible to calculate the deformation by von Kármán equations. The in-plane displacements **u** and out-of-plane deflection w can be obtained by solving the equations:

$$\frac{Eh^3}{12\left(1-\nu^2\right)}\Delta^2 w - h\frac{\partial}{\partial x_\beta}\left(\sigma_{\alpha\beta}\frac{\partial w}{\partial x_\alpha}\right) = F_{\mathrm{AB}}^z, \tag{2.12}$$

$$\frac{\partial \sigma_{\alpha\beta}}{\partial x_\beta} = F_{\mathrm{AB}}^\alpha, \tag{2.13}$$

where E is the Young's modulus, h is the thickness of layer A, ν is the Poisson's ratio, $\sigma_{\alpha\beta}$ is the stress tensor related to the in-plane strain by Hooke's law, and α, β are indices that take values of 1 or 2, which denote the in-plane components, respectively. In the von Kármán equations, the in-plane strain tensor is expressed by:

$$\varepsilon_{\alpha\beta} = \frac{1}{2}\left(\frac{\partial u_\alpha}{\partial x_\beta} + \frac{\partial u_\beta}{\partial x_\alpha}\right) + \frac{1}{2}\frac{\partial w}{\partial x_\alpha}\frac{\partial w}{\partial x_\beta}, \tag{2.14}$$

where the contribution of out-of-plane deflection w is included.

By solving eqn (2.10)–(2.14), as shown in Figure 2.8, we can see that if the monolayer is under global tension due to the mismatch, emergence of interfacial misfit dislocations driven by the interlayer interactions generates patterned force fields, which contain localized compressive domains. Due to the super flexible property of two-dimensional layers, periodic out-of-plane buckles raise in the compressive domains to release the compressive stresses. These patterned buckles can significantly lower the interaction energy and change the bandgap of heterostructures, offering chances for designing desirable 2D materials with tunable electronic and optical properties. This is the case when Frenkel and Kontorova meet Von Kármán, the in-plane deformation due to interfacial misfit dislocations is strongly coupled to out-of-plane deformation.

2.3.4 Localized Wrinkling by Orientational Binding on the Substrate Surface

Besides the weak van der Waals interaction between the monolayer sheet and the substrate, there may be a strong chemical bond between the edge of the monolayer sheet and the substrate. Sometimes, the bonding shows strong orientational effect. Through *ab initio* calculations, the results in Figure 2.9(a) demonstrate that the interaction energy between the graphene nanoribbon and the Ni(111) surface reaches a minimum when the tilt angle is 90°.

The result in Figure 2.9(b) further shows that when both edges of the graphene nanoribbon are bound to the substrate surface, a size dependent localized wrinkling can form by the competition between orientational bonding energy and the bending elastic energy of the graphene nanoribbon.[74]

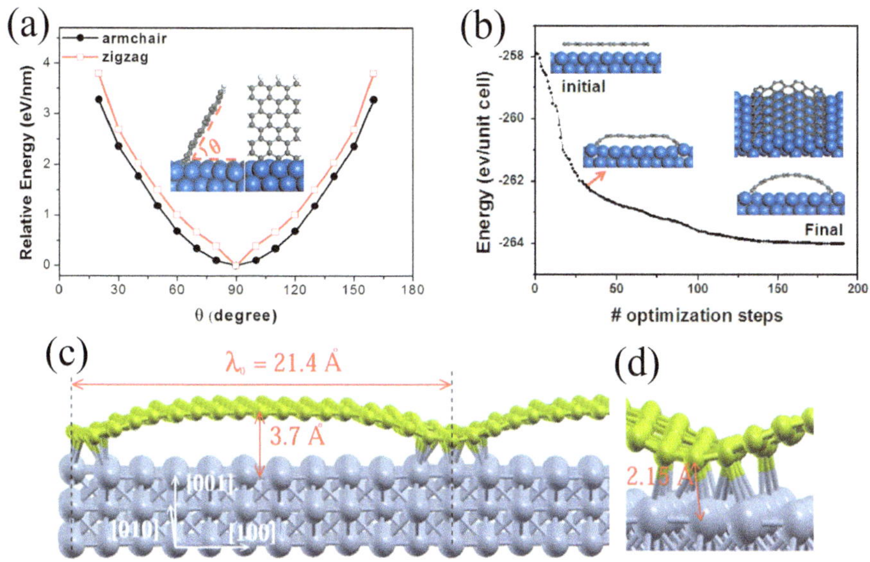

Figure 2.9 Formation of nanoscale rippled Graphene phase on substrates by first principle investigation, (a) bonding energy between the Graphene edge and the Ni substrate, (b) total energy change. Adapted with permission from ref. 74, Copyright 2011 American Chemical Society; (c) and (d) the bonding structures in the rippled Graphene on Ir substrate. Adapted with permission from ref. 75, Copyright 2014 American Chemical Society.

A similar strongly buckled state is also observed in the graphene on Ir substrate, while the formation of the local buckling is determined by the strong chemisorption energy and the bending energy of the graphene,[75] as shown in Figure 2.9(c) and (d). We will expect that the coupling between the strong local chemisorption and the local nanoscale buckle may regulate the chemical interaction by tuning the buckling curvature.

2.4 Buckling of Twisted Bilayer Graphene

In bilayer graphene, a twist between the two layers is found to significantly modify the electronic properties. The twist stacking usually leads to a characteristic interfacial dislocation pattern, known as a so-called Moiré pattern to accommodate the twist mismatch. Recent studies show that emergence of significant buckling in the bilayer graphene governs further relaxation of the Moiré pattern.[11,12,76–79] The results obtained by atomic simulation and atomically informed generalized Peierls-Nabarro modeling in Figure 2.10 show that relaxed Moiré patterns with out-of-plane displacement and twisted dislocation structures can form.[77,78] The interfacial dislocation with allowed out-of-plane deformation is a so-called strain soliton. Further studies to

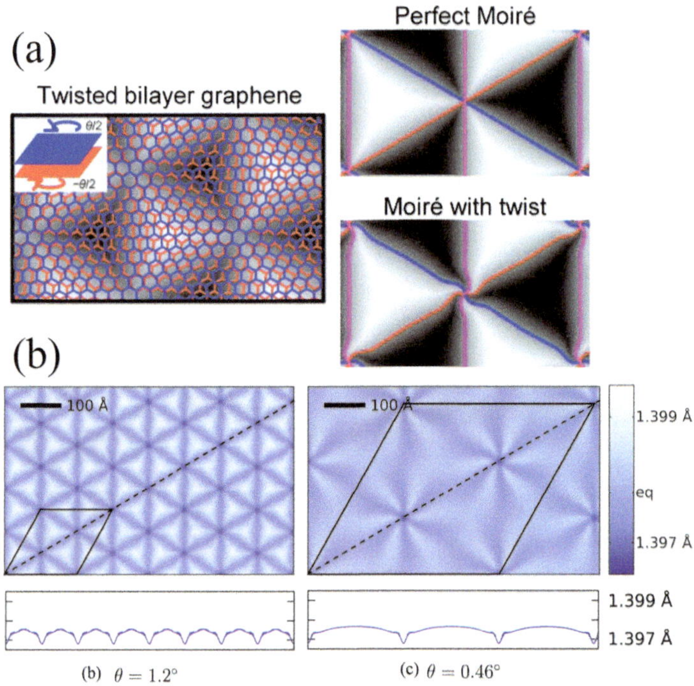

Figure 2.10 Formation of nanoscale buckling in twisted bilayer graphene. (a) Simulated result by a generalized Peierls–Nabarro model. Adapted with permission from ref. 77, Copyright 2016 American Chemical Society. (b) Simulated result by the molecular dynamics code LAMMPS. Reproduced from M. M. van Wijk, A. Schuring, M. I. Katsnelson, A. Fasolino, Relaxation of moiré patterns for slightly misaligned identical lattices: graphene on graphite, 2D Materials, 2(3), 034010, 30/07/2015. https://doi.org/10.1088/2053-1583/2/3/034010. © IOP Publishing. Reproduced with permission. All rights reserved.[78]

manipulate and visualize the motion of such nanoscale buckled structures will attract increasing interest since they have substantial effects on the electromechanical properties of such 2D materials.

2.5 Conclusions

Due to the super flexibility of the 2D materials, non-planar deformation may be easily triggered by various energy dissipations. In this chapter, we have summarized several typical nanoscale bucklings in a free-standing monolayer sheet, in a monolayer sheet bonded on substrates and in a twisted bilayer. Thermal fluctuation, relaxation of edge stress, coherent interface stress and topological defects induced stress may lead to remarkable out-of-plane distortion in a free-stand monolayer sheet. For a monolayer sheet on soft substrate, the substrate elasticity will mediate the morphological formation of nanoscale buckling when the sheet is subjected to compressive stress. The interlayer potential between sheet and substrate as well as the orientational binding on the surface of substrate may mediate localized buckling for sheet/substrate system. In addition, twisted dislocation networks are coupled with out-of-plane deformation during the relaxation of the Moiré patterns in a twisted graphene bilayer. These nanoscale out-of-plane distortions in 2D materials can effectively release the compressive stress compared with the flat counterparts. It usually competes with other stress relieving modes. The different coupling mechanisms may exhibit different physical origins of the observed nanoscale buckles. These buckles and ripples can significantly change the electronical and mechanical properties of 2D materials, which may offer potential applications for functional materials in the future. Further theoretical and experimental investigations into this fascinating phenomenon are to be exploited.

Acknowledgements

Y. N. was supported by the National Natural Science Foundation of China (Grant No. 11672285), the Strategic Priority Research Program of the Chinese Academy of Sciences (Grant No. XDB22040502), the Collaborative Innovation Center of Suzhou Nano Science and Technology.

References

1. Y. Wei, B. Wang, J. Wu, R. Yang and M. L. Dunn, *Nano Lett.*, 2012, **13**, 26–30.
2. A. K. Geim and I. V. Grigorieva, *Nature*, 2013, **499**, 419–425.
3. K. Novoselov, A. Mishchenko, A. Carvalho and A. C. Neto, *Science*, 2016, **353**, aac9439.
4. S. Deng and V. Berry, *Mater. Today*, 2016, **19**, 197–212.
5. A. Fasolino, J. Los and M. I. Katsnelson, *Nat. Mater.*, 2007, **6**, 858–861.
6. A. V. De Parga, F. Calleja, B. Borca, M. Passeggi Jr, J. Hinarejos, F. Guinea and R. Miranda, *Phys. Rev. Lett.*, 2008, **100**, 056807.
7. J. Zang, S. Ryu, N. Pugno, Q. Wang, Q. Tu, M. J. Buehler and X. Zhao, *Nat. Mater.*, 2013, **12**, 321–325.
8. S. J. Kim, D. W. Kim, J. Lim, S.-Y. Cho, S. O. Kim and H.-T. Jung, *ACS Appl. Mater. Interfaces*, 2016, **8**, 13512–13519.
9. B. Lalmi, J. Girard, E. Pallecchi, M. Silly, C. David, S. Latil, F. Sirotti and A. Ouerghi, *Sci. Rep.*, 2014, **4**, 4066.
10. E. Khestanova, F. Guinea, L. Fumagalli, A. Geim and I. Grigorieva, *Nat. Commun.*, 2016, **7**, 12587.
11. J. S. Alden, A. W. Tsen, P. Y. Huang, R. Hovden, L. Brown, J. Park, D. A. Muller and P. L. McEuen, *Proc. Natl. Acad. Sci.*, 2013, **110**, 11256–11260.
12. B. Butz, C. Dolle, F. Niekiel, K. Weber, D. Waldmann, H. B. Weber, B. Meyer and E. Spiecker, *Nature*, 2014, **505**, 533–537.
13. D. Akinwande, C. J. Brennan, J. S. Bunch, P. Egberts, J. R. Felts, H. Gao, R. Huang, J.-S. Kim, T. Li and Y. Li, *Extreme Mech. Lett.*, 2017, **13**, 42–77.
14. Z. Song, Y. Ni and Z. Xu, *Extreme Mech. Lett.*, 2017, **14**, 31–37.

15. T. Zhang and H. Gao, *J. Appl. Mech.*, 2015, **82**, 051001.
16. T. Zhang, X. Li and H. Gao, *Int. J. Fract.*, 2015, **196**, 1–31.
17. H. Qin, Y. Sun, J. Z. Liu and Y. Liu, *Carbon*, 2016, **108**, 204–214.
18. S. Li, Q. Li, R. W. Carpick, P. Gumbsch, X. Z. Liu, X. Ding, J. Sun and J. Li, *Nature*, 2016, **539**, 541–545.
19. H. Qin, Y. Sun, J. Z. Liu, M. Li and Y. Liu, *Nanoscale*, 2017, **9**, 4135–4142.
20. A. C. Neto, F. Guinea, N. M. Peres, K. S. Novoselov and A. K. Geim, *Rev. Mod. Phys.*, 2009, **81**, 109.
21. N. D. Mermin and H. Wagner, *Phys. Rev. Lett.*, 1966, **17**, 1133.
22. N. D. Mermin, *Phys. Rev.*, 1968, **176**, 250.
23. J. C. Meyer, A. K. Geim, M. I. Katsnelson, K. S. Novoselov, T. J. Booth and S. Roth, *Nature*, 2007, **446**, 60–63.
24. R. Zan, C. Muryn, U. Bangert, P. Mattocks, P. Wincott, D. Vaughan, X. Li, L. Colombo, R. S. Ruoff and B. Hamilton, *Nanoscale*, 2012, **4**, 3065–3068.
25. J. Los, M. I. Katsnelson, O. Yazyev, K. Zakharchenko and A. Fasolino, *Phys. Rev. B*, 2009, **80**, 121405.
26. S. Chen and D. Chrzan, *Phys. Rev. B*, 2011, **84**, 195409.
27. M. Pozzo, D. Alfe, P. Lacovig, P. Hofmann, S. Lizzit and A. Baraldi, *Phys. Rev. Lett.*, 2011, **106**, 135501.
28. W. Gao and R. Huang, *J. Mech. Phys. Solids*, 2014, **66**, 42–58.
29. N. Mounet and N. Marzari, *Phys. Rev. B*, 2005, **71**, 205214.
30. K. Zakharchenko, M. Katsnelson and A. Fasolino, *Phys. Rev. Lett.*, 2009, **102**, 046808.
31. W. Bao, F. Miao, Z. Chen, H. Zhang, W. Jang, C. Dames and C. N. Lau, *Nat. Nanotechnol.*, 2009, **4**, 562–566.
32. J.-W. Jiang, J.-S. Wang and B. Li, *Phys. Rev. B*, 2009, **80**, 205429.
33. J. H. Weiner, *Statistical Mechanics of Elasticity*, Courier Corporation, 2012.
34. Q. Lu, M. Arroyo and R. Huang, *J. Phys. D: Appl. Phys.*, 2009, **42**, 102002.
35. M. H. Gass, U. Bangert, A. L. Bleloch, P. Wang, R. R. Nair and A. Geim, *Nat. Nanotechnol.*, 2008, **3**, 676.
36. V. Shenoy, C. Reddy, A. Ramasubramaniam and Y. Zhang, *Phys. Rev. Lett.*, 2008, **101**, 245501.
37. Q. Lu and R. Huang, *Phys. Rev. B*, 2010, **81**, 155410.
38. C. K. Gan and D. J. Srolovitz, *Phys. Rev. B*, 2010, **81**, 125445.
39. P. Wagner, C. P. Ewels, V. V. Ivanovskaya, P. R. Briddon, A. Pateau and B. Humbert, *Phys. Rev. B*, 2011, **84**, 134110.
40. V. B. Shenoy, C. D. Reddy and Y.-W. Zhang, *ACS Nano*, 2010, **4**, 4840–4844.
41. Z. Xu and M. J. Buehler, *ACS Nano*, 2010, **4**, 3869–3876.
42. M. Dewapriya, A. S. Phani and R. Rajapakse, *Modell. Simul. Mater. Sci. Eng.*, 2013, **21**, 065017.
43. Z. Li, Z. Cheng, R. Wang, Q. Li and Y. Fang, *Nano Lett.*, 2009, **9**, 3599–3602.
44. C.-C. Chen, W. Bao, J. Theiss, C. Dames, C. N. Lau and S. B. Cronin, *Nano Lett.*, 2009, **9**, 4172–4176.
45. M. P. Levendorf, C.-J. Kim, L. Brown, P. Y. Huang, R. W. Havener, D. A. Muller and J. Park, *Nature*, 2012, **488**, 627–632.
46. J. Lu, L. C. Gomes, R. W. Nunes, A. Castro Neto and K. P. Loh, *Nano Lett.*, 2014, **14**, 5133–5139.

47. D. Nandwana and E. Ertekin, *Nano Lett.*, 2015, **15**, 1468–1475.
48. D. Nandwana and E. Ertekin, *J. Appl. Phys.*, 2015, **117**, 234304.
49. O. Lehtinen, S. Kurasch, A. Krasheninnikov and U. Kaiser, *Nat. Commun.*, 2013, **4**, 3098.
50. A. W. Robertson, K. He, A. I. Kirkland and J. H. Warner, *Nano Lett.*, 2014, **14**, 908–914.
51. D. Gazit, *Phys. Rev. B*, 2009, **79**, 113411.
52. O. V. Yazyev and S. G. Louie, *Phys. Rev. B*, 2010, **81**, 195420.
53. M. T. Lusk and L. D. Carr, *Phys. Rev. Lett.*, 2008, **100**, 175503.
54. M. T. Lusk, D. T. Wu and L. D. Carr, *Phys. Rev. B*, 2010, **81**, 155444.
55. D. Nelson and L. Peliti, *J. Phys.*, 1987, **48**, 1085–1092.
56. H. Seung and D. R. Nelson, *Phys. Rev. A*, 1988, **38**, 1005.
57. S. Chen and D. Chrzan, *Phys. Rev. B*, 2011, **84**, 214103.
58. T. Zhang, X. Li and H. Gao, *J. Mech. Phys. Solids*, 2014, **67**, 2–13.
59. X. Chen and J. W. Hutchinson, *J. Appl. Mech.*, 2004, **71**, 597–603.
60. J. H. Jung, J. Bae, M.-W. Moon, K.-S. Kim and J. Ihm, *Solid State Commun.*, 2015, **222**, 14–17.
61. H. Mei, C. M. Landis and R. Huang, *Mech. Mater.*, 2011, **43**, 627–642.
62. K. Pan, Y. Ni, L. He and R. Huang, *Int. J. Solids Struct.*, 2014, **51**, 3715–3726.
63. Z. H. Aitken and R. Huang, *J. Appl. Phys.*, 2010, **107**, 123531.
64. K. Zhang and M. Arroyo, *J. Appl. Phys.*, 2013, **113**, 193501.
65. K. Zhang and M. Arroyo, *J. Mech. Phys. Solids*, 2014, **72**, 61–74.
66. Y. Guo and W. Guo, *Nanoscale*, 2013, **5**, 318–323.
67. A. Kushima, X. Qian, P. Zhao, S. Zhang and J. Li, *Nano Lett.*, 2015, **15**, 1302–1308.
68. E. Stolyarova, D. Stolyarov, K. Bolotin, S. Ryu, L. Liu, K. Rim, M. Klima, M. Hybertsen, I. Pogorelsky and I. Pavlishin, *Nano Lett.*, 2008, **9**, 332–337.
69. N. Levy, S. Burke, K. Meaker, M. Panlasigui, A. Zettl, F. Guinea, A. C. Neto and M. Crommie, *Science*, 2010, **329**, 544–547.
70. S. P. Koenig, N. G. Boddeti, M. L. Dunn and J. S. Bunch, *Nat. Nanotechnol.*, 2011, **6**, 543–546.
71. K. Yue, W. Gao, R. Huang and K. M. Liechti, *J. Appl. Phys.*, 2012, **112**, 083512.
72. P. Wang, W. Gao, Z. Cao, K. M. Liechti and R. Huang, *J. Appl. Mech.*, 2013, **80**, 040905.
73. H. Kumar, D. Er, L. Dong, J. Li and V. B. Shenoy, *Sci. Rep.*, 2015, **5**, 10872.
74. Q. Yuan, H. Hu, J. Gao, F. Ding, Z. Liu and B. I. Yakobson, *J. Am. Chem. Soc.*, 2011, **133**, 16072–16079.
75. M. Imam, N. Stojić and N. Binggeli, *J. Phys. Chem. C*, 2014, **118**, 9514–9523.
76. S. Zhou, J. Han, S. Dai, J. Sun and D. J. Srolovitz, *Phys. Rev. B*, 2015, **92**, 155438.
77. S. Dai, Y. Xiang and D. J. Srolovitz, *Nano Lett.*, 2016, **16**, 5923–5927.
78. M. van Wijk, A. Schuring, M. Katsnelson and A. Fasolino, *2D Mater.*, 2015, **2**, 034010.
79. S. K. Jain, V. Juričić and G. T. Barkema, *2D Mater.*, 2016, **4**, 015018.

CHAPTER 3

Surface Modification for Engineering the Intrinsic Magnetic Properties of Inorganic 2D Nanomaterials

YUQIAO GUO, JUNCHI WU AND CHANGZHENG WU*

Department of Chemistry & Hefei National Laboratory for Physical Sciences at the Microscale & Collaborative Innovation Center of Chemistry for Energy Materials, University of Science and Technology of China, No.96, JinZhai Road, Hefei, Anhui, 230026, P. R. China
*E-mail: czwu@ustc.edu.cn

3.1 Introduction

Atomic-scale-thickness 2D nanomaterials have extremely large specific surface areas, making their surface phase as important as bulk counterparts. Therefore, surface chemical modification of inorganic 2D nanomaterials provides a novel way to engineer intrinsic physical properties. Although surface treatment techniques have been traditionally utilized in some 3D bulk materials for better catalysis activity and biocompatibility *etc.*,[1–3] artificial surface phase manipulation is difficult to influence their inner intrinsic properties of entire bulk crystals. Different from that in bulk form, intrinsic physical properties ranging from electronic, magnetic and even structural aspects in

Smart Materials No. 27
Inorganic Two-dimensional Nanomaterials: Fundamental Understanding, Characterizations and Energy Applications
Edited by Changzheng Wu

Published by the Royal Society of Chemistry, www.rsc.org

epitaxial thin film or surface coating layer with ultrathin thickness, which suffers 2D scale dimensionally confined, has been effectively tuned using variant surface engineering techniques, such as gate electric field in transistor, different chemical solution/gas environments on the surface.[4–8] From this point-of-view, inorganic 2D nanomaterials show potential for being an ideal platform for regulating intrinsic physical properties *via* surface modification, as has been done to regulate physicochemical properties in 2D ultrathin nanosheets. In detail, once the dimension reduces, almost all the atoms are exposed on the surface.[9] Surface chemical modifications would therefore influence most of the atoms in the outmost surface layer, and regulate carrier concentrations with a tuned band gap of inorganic 2D nanomaterials by electronic structure engineering.[10] The regulations of electron transport and band structures are essential for better performance of electrode applications and catalysis capability.[11–14] Moreover, for an inorganic 2D nanomaterial, which has transition metal atoms in its lattice structure, unique electrical behaviors such as superconductivity, charge density wave, Mott transition and mosaic phase, which indicate strong electron–electron correlation interaction and electron–phonon correlation interaction, are observed.[15–20] Therefore, surface chemical modification would trigger electron transfer or local atomic lattices disorder that would bring electronic state transformations among semiconductors, semi-metals, metals and even superconductors.[21,22] Spin injection and local magnetic moment can also be realized by means of surface modification and magnetism can be induced and tuned in different 2D materials. More importantly, those physical property changes *via* electron transfer or local lattice distortions do not harm to the original 2D structure. Multiple surface modification strategies have been utilized as facile approaches to tune the intrinsic physical properties of inorganic 2D nanomaterials without changing the layered frameworks. In brief, surface chemical modification strategies, including incorporation, absorption, and defect engineering, paved a new effective way to modulate the intrinsic physical properties whilst keeping the 2D freestanding structure intact, which is vital for mitigating requirements, currently, for 2D nanomaterials with tunable physical properties for a diverse range of practical applications.

For 2D magnetic materials, the quantum confined effect leads to a vast array of unique intrinsic magnetic properties, such as Ising ferromagnet,[23,24] magnetic anisotropy,[25,26] planar spin frustration,[27,28] spin glass state,[29,30] spin density wave[31,32] and 2D superconductor coexisting with magnetic order,[33,34] *etc.* Enormous efforts have also been made on improving the intrinsic magnetic properties of 2D materials for further widening applications in high-anisotropy magnet,[35,36] planar spintronic[37,38] and high-efficient magnetocaloric conversion.[39,40] From a surface modification aspect, when exfoliated into ultrathin nanosheets, the material surface areas dramatically increase, which would not only make their chemical and physical reactivity enhanced, but also result in a 2D electronic structure because of quantum confinement effects.[41–44] Such effects lead to intrinsic magnetic properties modification by the electronic band structure regulation. For instance,[45] mechanically exfoliating the 2H-$NbSe_2$ monolayer exhibiting Ising superconductivity with extreme critical field;[46,47] and zigzag edge state in graphene and phosphorene nanoribbons leading to ferromagnetism.[48–53] Engineering of intrinsic magnetic properties of 2D nanomaterials give them extensive functionalities, such as: surface absorption in non-magnetic 2D materials inducing ferromagnetism, even in supercondutors;[54–56] magnetic atom doping in 2D electrochemical catalysts leading to Jahn-Teller effect to further enhance catalytic performance;[57,58] spin-regulated 2D nanomaterials yielding large magnetoresistance.[59,60] Besides, a wide variety of metal oxides and hydroxides, whether they are layered structure or not, have been successfully synthesized as individual ultrathin nanosheets that show unique magnetic properties when compared to their bulk counterparts and these ultrathin ferromagnetic materials are considered as excellent 2D ferromagnetic candidates.[61–64] By controlling the intrinsic magnetic properties of 2D materials, lots of great potential applications in fields such as electronics, magnetocaloric and spintronics have been realized.[65–69] Methods to regulate the intrinsic magnetic properties in 2D nanomaterials, without destroying the primary freestanding lattices and producing controllable, available intrinsic spin behaviors are expected urgently. In this chapter, we focus on serial surface modification strategies on inorganic 2D nanomaterials, which lead to effectively intrinsic magnetic properties engineering. We also focus on recent research into surface chemical modification strategies for realizing intrinsic magnetic property engineering of inorganic 2D nanomaterials. We highlight chemical modifications of atoms on the surface of 2D nanomaterials, including surface incorporation, defect engineering and structure modulation, which have shown exciting results for tuning intrinsic magnetism properties while maintaining the pristine 2D lattice framework. The chemically-modified surface of 2D structures is significantly expanded with future applications for generating new nano-electronic devices and high-efficiency clean-energy transformation applications.

3.2 Intrinsic Magnetic Properties in Inorganic 2D Nanomaterials

2D magnetic materials have been widely applied as new-generation magnetocaloric convention and spintronic devices. It is well-known that the giant magnetoresistance effect was first found in the Fe–Cr–Fe sandwich thin film, which consisted of alternate ferromagnetic layers and non-magnetic layers.[70] What is more, for almost all ferromagnetic diluted magnetic semiconductors (DMS), only their thin-film form can be prepared, therefore many 2D magnetic materials that have been extensively used are DMSs thin-films.[71–73] Such magnetic thin-films require atomic-precision magnetic doping in the deposited semiconducting film, giving rise to persistent ferromagnetic signatures even at high temperatures. Since freestanding graphene 2D nanosheets are exfoliated from bulk graphite, multiple 2D nanomaterials have been synthesized by a top-down route including both physical type exfoliation and chemical liquid exfoliation or bottom-up chemical gas deposition and confined atom assembly.[74–77] When compared with the DMSs thin-films that have to grow on special substrates, freestanding 2D ferromagnetic materials that can exist stably without a substrate show more potential. Exfoliation or chemical topochemical synthesis of the layered transitional metal oxides/hydroxides with ferromagnetic atoms (Fe, Co, Ni) provide a directly available strategy for 2D ferromagnetic materials.[78,79] Feroxyhyte (δ-FeOOH) is a metastable iron oxide species with two unequally occupied octahedral sites, suggesting valid natural magnetization in the δ-FeOOH structure with a Curie temperature higher than room temperature.[80] Recently, Wu *et al.* developed an atomic-scale topochemical transformation strategy for realizing a large-scale production of 2D δ-FeOOH ultrathin nanosheets, which reveal robust ferromagnetic behavior with high saturation magnetization.[81] What is more, the optical absorption edge of δ-FeOOH ultrathin nanosheets, located near 600 nm, reveals its semiconductor state with a direct band gap of about 2.2 eV. The combination of the ferromagnetism property and semiconductor states makes δ-FeOOH ultrathin nanosheets promising 2D inorganic candidates for future-generation spintronic devices. Iron-based transition metal oxides (Fe_xO_y) are another kind of natural compounds with excellent magnetic properties. Theoretical research has indicated their electronic structure and spin properties are largely dependent on the dimensionality of the materials, which exploits new routes to regulate the ground states by reducing their thickness to atomic scales.[82] However, since the crystal structures of Fe_xO_y are relatively high-symmetry, and the bonds between Fe and O are strong, it is quite difficult to achieve atomically thin TMO nanosheets *via* traditional exfoliation methods. By combining template-assisted oriented growth and topochemical transformation, ultrathin half-unit-cell α-Fe_2O_3 semiconductor nanosheets were synthesized for the first time.[83] Dramatically, the thickness of the half-unit-cell α-Fe_2O_3 nanosheets is in the range of 0.55–0.59 nm, corresponding to two Fe ion layers. The ultrathin α-Fe_2O_3 nanosheets exhibit robust intrinsic ferromagnetism at room temperature. Structural analysis suggests that a unique surface structure distortion exist, producing two sets of Fe ion ($Fe_{5\text{-co}}$ and $Fe_{6\text{-co}}$) located in different chemical environments and inducing distance fluctuation of Fe ion chains. First-principles calculation results reveal ferromagnetic exchange interaction in $Fe_{5\text{-co}}$–O_{bri}–$Fe_{6\text{-co}}$ ion chains were efficiently activated *via* disruption of the degeneracy of Fe 3d energy level. Intrinsic ferromagnetic α-Fe_2O_3 nanosheets will stimulate worldwide interest in modulating room-temperature robust ferromagnetism in 2D transition metal oxides semiconductor materials, providing new candidates for the next-generation small-scale spintronic devices (Figure 3.1).

MPX_3 (M = Fe, Mn, Co, Ni; X = S, Se), a series of layered materials with a strong magnetic phase transition, may be ideal systems as next steps.[84,85] Since bulk MPX_3 materials are layered with weak Van der Waals force between these layers, it is easy to exfoliate MPX_3 layers to few layers *via* physical mechanical exfoliation.[86] Theoretical researchers predicted that robust magnetism may

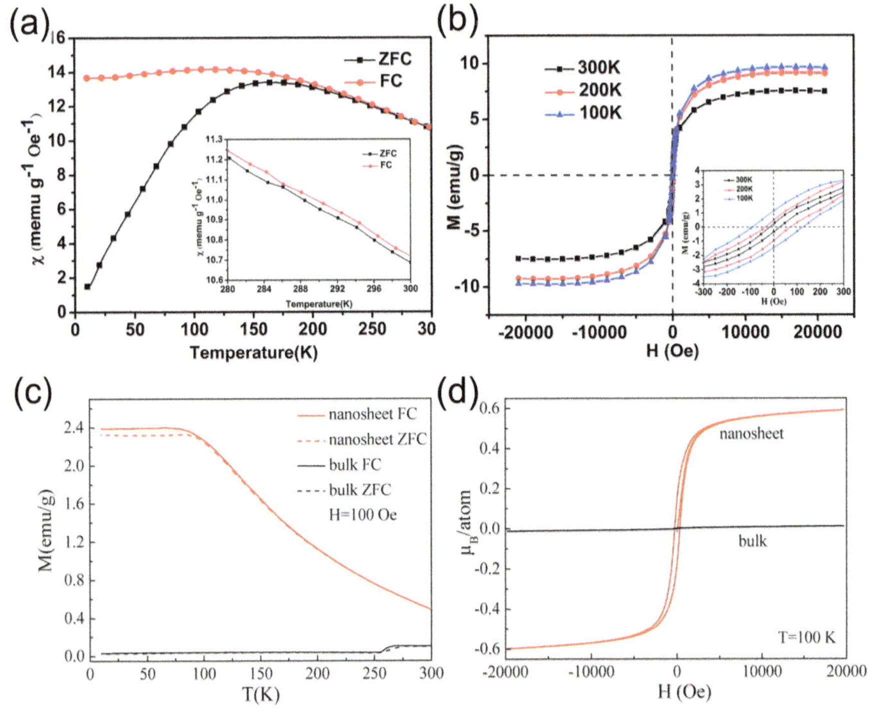

Figure 3.1 (a) Temperature dependence of ZFC and FC magnetization of δ-FeOOH ultrathin nanosheets. Inset: the amplifying part of ZFC and FC curves at a temperature range from 280 K to 300 K. (b) *M–H* curves for δ-FeOOH ultrathin nanosheets measured at different temperatures. Inset: the enlarged *M–H* section in the low applied field, adapted from ref. 82 with permission from The Royal Society of Chemistry. (c) ZFC-FC magnetization curves and (d) hysteresis loops of nanosheet and bulk α-Fe_2O_3, adapted with permission from ref. 83, Copyright 2014 American Chemical Society.

still remain in MPX_3 2D nanosheets even when their thickness reduces down to monolayer.[87,88] But this is in urgent need of being verified experimentally.

Though 2D magnetic films containing a number of ferromagnetic atoms show strong intrinsic magnetic properties, scientists are more interested in classical 2D materials, such as graphene and TMDs, which have shown abundant transport behaviors.[89] Through magnetism introduction, electrical transport can be controlled *via* spin manipulation under exotic magnetic field. Fortunately, unique magnetic structures can emerge in few transition metal dichalcogenides with cations that own odd electrons, *e.g.* VSe_2. It has been confirmed experimentally that VSe_2 is ferromagnetic even at room temperature, making it a 'star material' for future spintronic nanodevices.[90]

However, most other classical 2D materials such as graphene and TMDs are intrinsically non-magnetic, hindering them playing roles in the field of spinelectronic devices. So, it is important for researchers to find ways to induce intrinsic magnetic in those 2D materials. To realize this goal, scientists have made numerous efforts. Theoretical researchers predicted that graphene with point defects, zigzag edges or adatoms shows magnetic order in room temperature and many experiments have proved it.[91–97] This opened the way towards the realization of electronic and spintronic devices based on robust quantum mechanical effects, and it makes flexible spintronic devices become possible when using such 2D magnetic nanomaterials.

After weak ferromagnetism has been found in the zigzag edge of graphene nanoribbons, diamagnetic layered MoS_2, a more powerful competitor in 2D FET due to its ideal band gap, high in-plane mobility and mechanical stability, is thought to be the next chance to develop spintronics *via* inducing tunable magnetic order.[98] As per theoretical calculations, specific edges with different arrangement of termination present diverging magnetic ground states in MoS_2. The armchair edges are more stable in thermodynamics when they locate at a non-magnetic state, while the zigzag edges are stable in a magnetic ground state with a net magnetic moment.[99] From this viewpoint, ferromagnetism should be observed in MoS_2 nanoribbons, nanocrystalline thin films, and even in bulk with enough zigzag edges.

Tongay *et al.* studied the magnetic properties of MoS_2 single crystal experimentally. Their results discovered that the magnetization of MoS_2 are influenced by a temperature-dependent diamagnetic background coming from inner saturated Mo–S bonding and ferromagnetism possibly related to existence of zigzag edges with correlative magnetic moments at domain boundaries.[100] The Kian Ping Loh group synthesized MoS_2 nanosheet-like films with determinate oriented edge through thermal evaporation of the single-source precursor.[101] Such MoS_2 films exhibit weak ferromagnetism with a high Curie temperature of 685 K. When compared with thicker films without ferromagnetism, researchers found a high density of prismatic edges in ferromagnetic samples. Spin-polarized calculations based on triangular-shaped cluster models show that the inherent origin of magnetic ordering is the existence of unsaturated bonding and net spins on the prismatic edges of the nanosheets. Experimental and theoretical results implied this kind of nanostructured films with a high density of magnetic edge can give rise to magnetism in nonmagnetic material systems. Magnetoresistance effect at a low magnetic field was caused by spin polarization at such edge sites, enriching 2D materials for high efficient spintronics devices (Figure 3.2).

Similar results were reported by Gao *et al.* *via* an entirely different ultrasonic exfoliation method.[102] It has been widely proved that layered MoS_2 bulk materials can be exfoliated into nanosheets under ultrasonic conditions *via* breaking the interlayer Van der Waals forces by liquid interaction.[103] During the exfoliation process, in-plane chemical bonds unavoidably get damaged. After ultrasonication for a longer time, smaller nanosheets (even quantum dots) will form. Based on this approach, freestanding MoS_2 nanosheets with a controlled size distribution were obtained. Magnetic measurements revealed that room temperature ferromagnetism with enhanced saturation magnetizations emerged in MoS_2 nanosheets that exfoliated lasting for

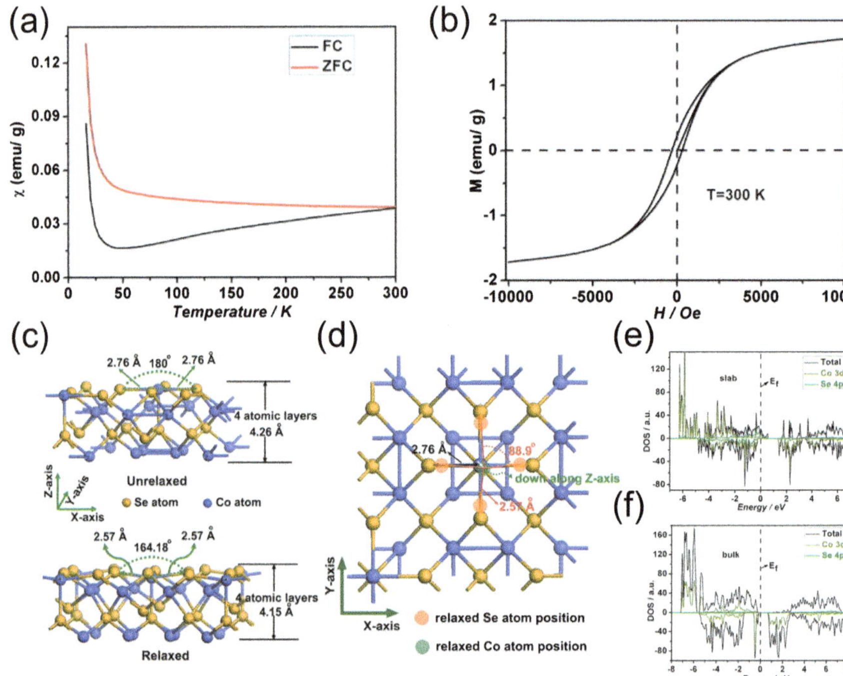

Figure 3.2 (a) Temperature dependence of ZFC and FC magnetization of Co_9Se_8 nanosheets. (b) Magnetic-field dependence of magnetization curve of Co_9Se_8 nanosheets at 300 K. (c) Side view of the (top) unrelaxed and (bottom) relaxed four-atomic-layer Co_9Se_8 slab structure. (d) Structural differences between the unrelaxed and relaxed slab models. (e) and (f) are calculated DOSs of the relaxed Co_9Se_8 slab and bulk Co_9Se_8. The DOSs above zero correspond to the spin-up states and the other to the spin-down states, adapted with permission from ref. 105, Copyright 2012 American Chemical Society.

longer ultrasonic period. Considering the positive correlation relationship between ultrasonication period and size of nanosheets, authors figured out that the witnessed magnetism comes from edge spins existing on the edges of the nanosheets. The edge spin inducing intrinsic ferromagnetism in MoS_2 nanosheets may allow remarkable applications in nanodevices and spintronics by an edge controlling method.

Although the natural compounds are usually stable and full of different abilities, artificial inorganic materials start to exhibit their fascination for their flexible structure and diverse properties. By means of theoretical simulation, scientists can design based on the properties they want and produce many inorganic materials that are not formed spontaneously. Zhang *et al.* reported an artificial 2D half-metallic ferromagnetic Co_9Se_8 single-crystalline nanosheets with atomic thickness and unique lamellar stacking formed by 2D oriented attachment.[104] The flexible ultrathin Co_9Se_8 nanosheets can easily be assembled into large-area films by a vacuum-filtration method and transferred to a transparent flexible substrate. First-principles density functional theory calculations suggested two asymmetric curves: the curves for the spin-down states exhibited a band gap of ~1.5 eV, whereas the curves for the spin-up states showed gapless, which indicated that the Co_9Se_8 ultrathin nanosheet with atomic thickness has a half-metallic nature. Since similar calculations performed on bulk Co_9Se_8 exhibited semiconducting transport property, it is interesting that thickness reduction of the bulk Co_9Se_8 to atomic scale changes the conductive nature from the semiconducting to half-metallic anomalously, while other natural layered semiconductor suffers band gap increasing with decreased layer number.[98] Such DOS polarized half-metallic 2D nanomaterial is an ideal candidate for the new-generation spintronic devices. Obvious room-temperature ferromagnetism was observed by *M–T* curves and hysteresis loop, which agree with the polarized band structure. Intrinsic half-metallicity is very rare in low- dimensional nanomaterials due to its high requirement for crystal structure and atom property. This artificial 2D nanomaterial open a new road to 2D ferromagnetic nanomaterials with unique band structure, providing hopeful prospect for ultrathin, transparent and flexible, paper-like spintronic devices.

Bulk magnetic materials have been widely used in electricity generation, memorizer, medical science, *etc.* To construct faster devices with more miniature volume, freestanding low-dimensional materials with larger intrinsic magnetism are required in haste. Here we introduce some late-model 2D intrinsic ferromagnetic materials and their synthetic method. Though big achievements on 2D intrinsic ferromagnetic materials have been made, we look forward to seeking for better material and their practice in spintronic devices.

3.3 Surface Modification for Engineering the Magnetic Properties of Inorganic 2D Nanomaterials

It is well known that intrinsic electronic properties and magnetic properties are essentially controlled by charge and spin degrees of freedom of relevant electrons, respectively. Unlike those bulk counterparts, 2D ultrathin nanomaterials, in which surface atoms occupied most of the atom content, enabling surface state to be as significant as that in the interior part. Hence, surface chemical modification in 2D nanomaterials has become an effective way to modulate the charge, spin, and coordination environment, further to regulate their intrinsic physical properties. Based on 2D confined systems, surface engineering technologies, such as heteroatom incorporation, small molecular absorption, and multiple defect introductions have been widely used to efficiently adjust intrinsic electrical properties and magnetic behavior. Achievements have been made in the field of highly efficient electrochemical catalysis, sensor devices, magnetocaloric conversion and highly efficient catalyst devices, opening a window for new functional materials design.

In this section, we focus on recent studies of surface modification for engineering the magnetic properties of inorganic 2D nanomaterials.

3.3.1 Introducing Magnetism by Heteroatom Incorporation

2D materials are an ideal platform for electronic devices. Classical 2D materials, such as graphene and MoS_2, reveal fascinating physical phenomena and have been widely used in constructing electronic devices, optical applications, and energy conversion and storage. When used in electronic devices, intrinsic characteristics of electrons need to be precisely controlled by exotic conditions. Spin is one of the most important intrinsic characteristics of electrons and can be controlled by exotic magnetic field. Therefore, endowing functional materials with microcosmic net spins to induce macroscopic magnetism is an alternative method to further modulate its properties. For 2D materials without magnetic atoms, inner net spin is zero and they do not reveal intrinsic magnetism. Much effort has been made to induce magnetism in 3D bulk materials, for example, magnetic ion with net non-paired electron doping into non-magnetic matrix was a valid traditional method to induce and regulate magnetic properties. This method has been extended to 2D dilute magnetic semiconductor films.[71–73] The same process has also been carried out in 2D nanomaterials, such as MoS_2, $TiSe_2$ and even $CoSe_2$ nanosheets, giving rise to controllable magnetism.[106–108] However, simple magnetic cation doping into 2D materials is not the most satisfactory solution, since such spins are much more located near magnetic impurities and result in more electrons scattering in electrical transport. Here, giving several typical 2D materials as examples, we show how to induce magnetism in 2D materials *via* surface modifications without magnetic impurity.

Graphene, the simplest 2D materials with only carbon atoms, is a model material that has unpredicted potential. Perfect graphene has a high-symmetry honeycomb structure and no net spins, making it non-magnetic. Considering that controllable magnetism can enlarge its range of application, researches tried many approaches to combine graphene with magnetism. Among these approaches, constructing graphene ribbons with zigzag edge and graphene with surface heteroatom incorporation are much concerned.[94–97] Theoretical simulations discovered that isolated hydrogen atoms absorbed on graphene surface can break a single out-plane p_z orbital in graphene and are predicted to lead to magnetic moments. With the help of STM, researchers deposited isolated atomic H on graphene and observed the appearance of a spin-polarized state near the Fermi level.[105] Moreover, researchers found the magnetic moments lead to unexpected long-range magnetic interactions. Hydrogenation provides an innovative intriguing strategy to regulate the spin structure of graphene, giving a new opportunity to operate magnetic ordering in 2D metal-free ultrathin nanosheets. However, hydrogenation on graphene, *i.e.* directly hydrogen atoms grafting onto the carbon atoms surface, would suffer tremendous experimental challenges because during the hybridization formation of carbon orbits changes to sp^3 from sp^2, a considerable energy barrier is waiting to be overcome. N atoms with a pair of lone pair electrons can effectively bond with a proton, so N atom doped graphene becomes an ideal object to implement hydrogenation.

Graphite carbon nitride (g-C_3N_4) is a form of highly concentrated N ordering doped graphite, showing superiority for hydrogenation. Inspired by ferromagnetic ordering in hydrogenated graphene, Xu *et al.* detected magnetic property of g-C_3N_4 ultrathin nanosheets with hydrogen dangling bonds and found robust room-temperature ferromagnetism.[110] Besides, the saturation magnetization of g-C3N4 was strongly dependent on the amount of hydrogen dangling bond content. Hydrogen dangling bonding is a convenient way to modulate surface of 2D materials containing base atoms, such as N, O, S, *etc.*[107–110] Except for hydrogen atoms, S atoms and O atoms can also be used to change the magnetism of graphene. Recently, our group developed topochemical strategy to regulate magnetism of reduced graphite oxides. It was found that the content of S and O atoms on rGO can adjust the GO to be ferromagnetic (High-sulfur rGO) or superparamagnetic (Low-sulfur rGO). Such magnetism is tightly connected with electronic transport in exotic magnetic field, so we can also observe a large negative magnetoresistance phenomenon at room temperature in superparamagnetic low-sulfur rGO. This work shows an excellent example of controlling magnetism and constructing magnetoelectric materials in simple 2D systems *via* surface heteroatom absorption (Figure 3.3).

3.3.2 Introducing Magnetism by Molecular Absorption

Atomic absorption on 2D surfaces has successfully induced magnetism in graphene. The strategy has also been extended for use in TMDs. So, beyond simple cation absorption, it is also captures polarized molecules and absorbed polarized molecules can efficiently influence the intrinsic properties of TMD devices.[6] Based on this characteristic, multiple high-sensitivity gas sensors have been developed.[111,112] When considering that absorbed polarized molecules may trigger a slight distortion of covalent bonds between transition metal cations and surface chalcogen anions, magnetism can arise. $NbSe_2$ is a classical 2D superconductor with charge density wave coexistence and recently published research discovered unique Ising superconductivity feature in monolayer $NbSe_2$, guiding a novel hot point in condensed matter physics.[46,47] In $NbSe_2$, Nb ions own odd electrons, therefore high hopes are placed on ferromagnetic $NbSe_2$. The isoelectronic species of $NbSe_2$, VSe_2, has been confirmed to be ferromagnetism *via* both experimental and theoretical researches.[91] However, because of strong hybridization between Nb atoms and Se atoms, a covalent Nb–Se interaction forms and quenches the magnetic moment of the isolated Nb ions.[113,114] Therefore, pristine $NbSe_2$ is non-magnetic. Lately, our group reported that surface molecular adsorption can induce ferromagnetism ordering in ultrathin $NbSe_2$ nanosheets, opening a door to achieve coexistence of superconductivity and ferromagnetics.[56] Absorbed polar hydrazine molecules slight enlarged the covalent Nb–Se bond, weakening the covalent interaction and enhancing the ionicity of the tetravalent Nb^{4+} with unpaired odd electrons, inducing ferromagnetic ordering. Transport results do not only show sharply resistance drop

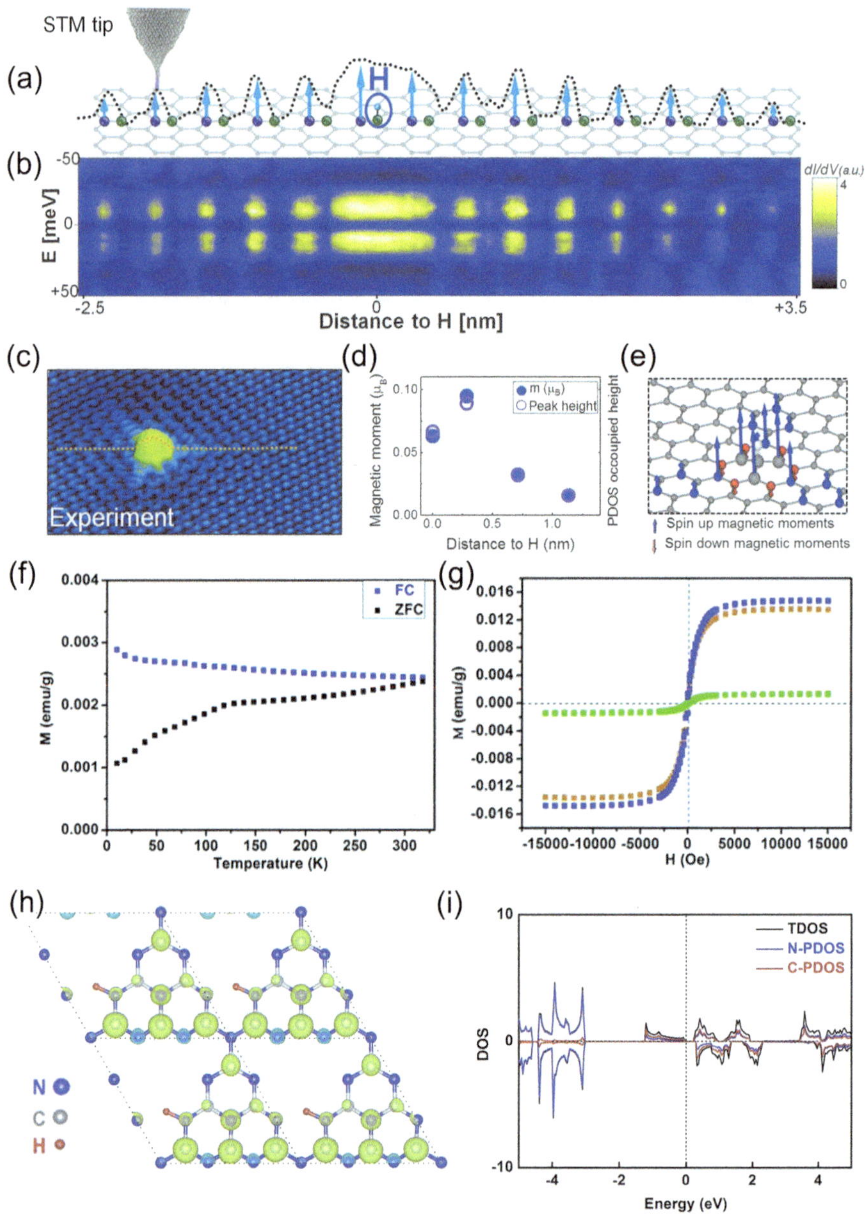

Figure 3.3 (a) Schematic of the graphene structure of single H atoms on graphene. (b) Conductance map [dI/dV(x,E)] along the dashed line in (c), a STM topography of a single H atom on graphene. (d) Comparison between DFT calculations for the local magnetic moment and the height of the occupied projected DOS (PDOS) peak, calculated on different carbon atoms. (e) Calculated magnetic moments induced by H chemisorption (the lengths of the arrows signify the relative magnitudes of the magnetic moments), from ref. 109, reprinted with permission from AAAS.

corresponding to superconductivity, but also intrinsic negative magnetoresistance and the Kondo effect, revealing that superconductivity and ferromagnetism are coupled. Based on the existing experimental fact, the surface molecular adsorption can be a potent implement to modulate spin ordering and inducing magnetic ordering in the 2D system (Figure 3.4).

3.3.3 Introducing Magnetism by Multiple Defects Introduction

In materials with magnetic ions that have odd electrons, surface disorder can induce stretching out or drawing back of covalent bonds, leading to an effective magnetic moment. To achieve magnetism in other materials without odd electrons, new approaches must be found. Theoretical works have suggested that 2H-MoS_2 is an intrinsic diamagnetic material, but when non-metals atoms such as H, B, C, N, and F are absorbed on its surface or it is irradiated by protons, it can become ferromagnetic as revealed by incisive first-principle calculations.[115–117] Those two methods have been confirmed experimentally to be effective and the ferromagnetism is so robust that it presents high Curie temperature far higher than room temperature. Beyond those methods, here we will highlight a distinctive phase incorporation strategy in intrinsic diamagnetic 2H-MoS_2, inducing robust room-temperature ferromagnetism. *Via* a two-step hydrothermal method, Cai *et al.* obtained 2H-MoS_2 ultrathin nanosheets with several sulfur vacancies.[118] Sulfur vacancies can inject charges into the surrounding lattice and accelerate the transition of the involved hexagonal phase (2H-MoS_2) into a trigonal phase (1T-MoS_2) unexpectedly. About a quarter of 2H-MoS_2 undergoes the phase transition. The sequent 1T-MoS_2 phase incorporated in 2H-MoS_2 nanosheets can not only largely increase the electron carrier concentration tenfold, bringing out an internal Mo^{4+} 4d energy state within the bandgap, but also induce robust intrinsic ferromagnetism with a high Curie temperature of 395 K and well-defined hysteresis loops at 300 K. Theoretical simulation asserted that the obtained ferromagnetic ordering comes from the exchange interactions between sulfur vacancies and the Mo^{4+} 4d band gap state. Recently, multiple calculations based on g-C_3N_4 have been done to predict magnetism in g-C_3N_4, suggesting that several methods including external strain and hydrogen dangling bond can effectively induce magnetism.[119,120] Among those methods, vacancies can be simultaneously created during the preparing process, therefore it is thought to be a potential approach. Xue *et al.* synthesized a

---

(f) Temperature dependence of the zero field cooling (ZFC) and field cooling (FC) curves of g-C_3N_4 with much hydrogen dangling bonds under a measuring field of 200 Oe. (g) *M–H* curves of g-C_3N_4 with different amount of at hydrogen dangling bonds 300 K. (h) Spin density distribution of proton modified g-C_3N_4. (i) TDOS and PDOS of the g-C_3N_4 single layer with hydrogen dangling bonds in the N1 site, adapted from ref. 110 with permission from The Royal Society of Chemistry.

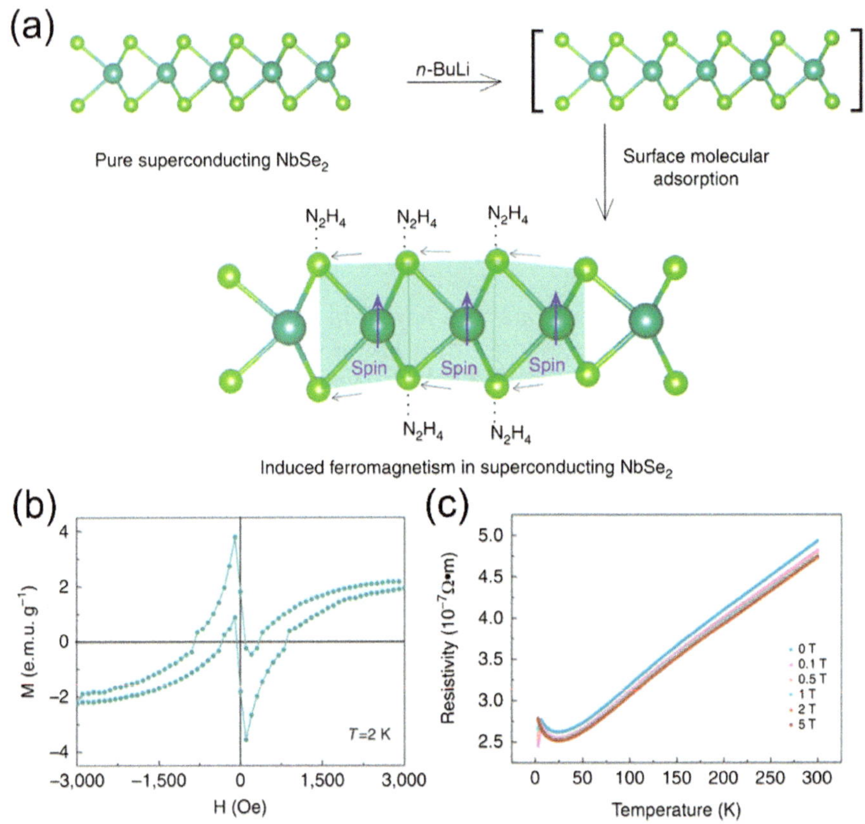

Figure 3.4 (a) Schematic illustration of $NbSe_2$ modified by polar hydrazine molecules. (b) Isothermal magnetizations (*M–H* curves) for hydrazine-treated $NbSe_2$ nanosheets at 2 K. (c) Temperature dependence of the resistivity (*R–T*) under various magnetic fields for hydrazine-treated $NbSe_2$ nanosheets. Reprinted by permission from Macmillan Publishers Ltd: *Nature Communications* (ref. 35), Copyright 2016.

series of ultrathin metal-free g-C_3N_4 with different vacancy concentrations by a one-step thermal pyrolysis method.[121] *Via* prolongation of the pyrolysis time, more vacancies appear. Distinct ferromagnetic signals of hysteresis loops in every defective samples have been observed and the strength rose with increasing vacancy amount. Theoretical calculations revealed that abundant carbon dangling bonds around the vacancy provide amounts of unpaired spin, hence inducing strong ferromagnetism, even at room temperature. By a rough one-step method, tunable magnetism can be induced into defective g-C_3N_4 originating from electron spin-polarization, making metal-free g-C_3N_4 ultrathin nanosheets a promising material for spintronics. This paradigm enlightens new approaches for achieving magnetism creation *via* exchange interactions in 2D materials (Figure 3.5).

In this section, we introduced surface modulation strategies to induce magnetic order in 2D materials in which pristine bulk materials are non-magnetic.

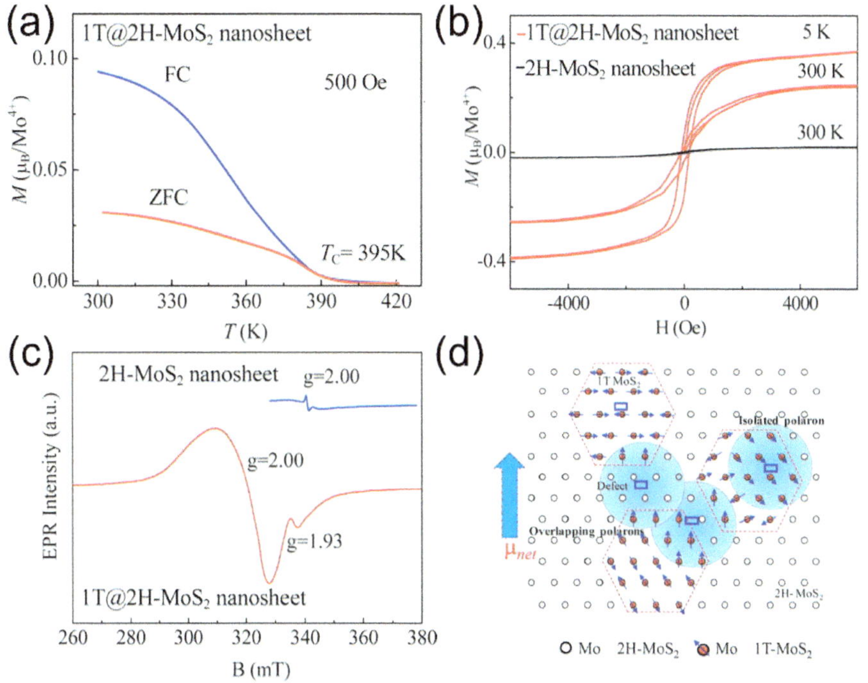

Figure 3.5 (a) Temperature dependence of FC and ZFC magnetization (*M*–*T* curves), (b) magnetization *vs.* magnetic field (*M*–*H*) curves at 5 and 300 K, and (c) EPR spectrum for both 1T@2H-MoS_2 and 2H-MoS_2 nanosheets. (d) Representation of magnetic polarons in 1T@2H-MoS_2, adapted with permission from ref. 122, Copyright 2015 American Chemical Society.

Via strategies of heteroatom incorporation, small molecular absorption, and multiple defect introduction, magnetic order emerges in several 2D materials, including graphene, MoS_2, $NbSe_2$, *etc.* Those methods are universal and can be utilized in other 2D materials. We are optimistic that surface modulation strategies can launch a revolution of creating 2D magnetic materials and understanding the internal mechanism of magnetism generation.

3.4 Application of 2D Magnetic Nanomaterials

3.4.1 2D Magnetic Nanosheets for Spintronics

The 'information revolution' has changed the way we live. Faster computing rates and smaller device sizes are desired. To realize highly integrated devices in a compact device, it is essential to reduce the feature size of transistors. But when the feature size of the transistor is lowered to the nanometer scale, a quantum effect will appear and serious heat generation becomes an issue. As we know, electrons have several features, such as mass, charges, and spins. In the traditional semiconductor industry, only charge is involved. If we take spin, another feature of electrons, into consideration, it seems that this problem can be solved to some extent. Based on this idea, an updated concept, spintronics, was born. With the exception of charge transport, spintronics uses electron spin and magnetic moment together in solid devices.[15] Many revolutionary inventions and products have been widely applied including random-access memory, spin-field effect transistor, spin light-emitting diode, spin resonant tunneling device, *etc.* Though lots of achievements have been made, more efficient MR materials are also demanded for the next generation, especially 2D materials with large MR at low field near room temperature.

Giant Magnetoresistance (GMR) effect, a phenomenon of dramatic change in electrical resistance when external magnetic field changes,[122] has attracted tremendous attention for understanding how charge and spin couple and to design improved spintronics. Original GMR materials are non-magnetic films sandwiched by two ferromagnetic films, which have typical 2D features. Therefore, 2D materials are thought to be promising for realizing GMR.[123] The benefit from these dimensionally-confined electronic structures and homologs to GMR films in dimensionality, 2D nanomaterials, especially inorganic layered materials, have bright prospects of excellent magnetro-transport properties for intriguing spintronic applications.[124] Besides, intrinsic magnetism in 2D nanomaterials, as we elaborated previously, can be easily modulated by surface chemical modification, surface molecule absorption and defect engineering, bringing varied polar transport under an external magnetic field. From this point-of-view, such magnetic-ordering 2D materials with enormous GMR value *via* designing polar structure indicates the possibility to open new opportunities for spintronic devices. Taking advantage of adjustable spin structure in 2D materials, we are hopeful of overcoming remaining challenges for practical applications.

GMR effects are closely linked to polar current in materials. Therefore, to obtain colossal GMR value, materials with regulated magnetic structures are wanted, especially small-sized magnetic domains or weak magnetic interactions in low-dimensional materials, in which spin orientations can arranges along magnetic direction congruously under low magnetic field.

In superparamagnetic (SPM) structure, the magnetic domains of ferromagnetic (FM) regions decrease to a critical size in which they can orient independently.[125] Under low external magnetic-field, all the magnetic domains will be along the magnetic direction, leading to large effective MR. SPM structures have been realized in in 0D nanoparticles, exploiting an avenue to transfer this concept to higher dimensions.[126] Peng and co-workers recently trigger a superparamagnetic structure in 2D graphene by topochemical sulfuration of reduced graphene oxide (rGO).[127] rGO is a derivative of graphene with numerous functional groups covalently bonding both on the basal plane and at the edges of the carbon nanosheets. Many researchers have reported room temperature ferromagnetism in rGO coming from those functional groups. The strength of ferromagnetism is dependent on the categories and a number of functional groups, and hence, we can alter the ferromagnetism by appropriately regulating the functional groups on rGO. If the magnetic domain of ferromagnetism in rGO is decreased to smaller than critical size *via* manipulating the distribution of functional groups, a unique SPM structure can be obtained. Theoretical simulation has demonstrated that thiol groups on rGO can produce localized magnetic moments, therefore thiol-decorated rGO nanosheets will display ferromagnetism at room temperature. Since thiol is highly sensitive to oxygen and steam in air, the surface density of thiol groups is easily adjusted by controlling the corrosion, then the size of ferromagnetic domains could be controlled. Consequently, 2D SPM can be realized in thiol-decorated rGO, paving the way for achieving GMR effect under low magnetic fields at room temperature. Experimental results successfully established a relationship between magnetism and surface modulation, and a large negative MR value (−8.6% under 500 Oe at 300 K) was observed in thiol group decorated rGO. The remarkable MR effect originated from decreasing electron scattering probability across different magnetic domains under exotic magnetic fields. In this case, MR effect is induced in rGO by a constructing superparamagnetic structure through controlling thiol density on the rGO, providing a patent magneto-transport model in a 2D structure with strong coupling of charge and spin. This strategy gives a novel approach for optimizing the intrinsic spin transport properties of 2D materials for next-generation spintronic devices.

Inspired by the achievements that graphene has gained, efforts are also paid to researches of TMDs because they have richer elemental components than graphene and three atomic layer in a single TMD layer, making the structure more variable.[98] Magnetism has been induced in TMDs by way of edge engineering, defect engineering and magnetic ion doping. However, intrinsic magnetism in TMDs induced by edge engineering are always weak and lack of precise regulations; defect engineering causes structural disorder and destroys the complete structure; magnetic ion doping increases electron scattering at the impurity sites and blocks electrons transport. Recently, Guo *et al.* successfully induced weak antiferromagentism in a typical 2D non-magnetic TMD, $TiTe_2$, by anionic doping.[128] Considering the size difference between adulterant iodine (2.20 Å for I^-) and tellurium (2.21 Å for Te^{2-}), I^- could enter the lattice of $TiTe_2$, forming the anionic solid solution system $TiTe_{2-x}I_x$. The higher electronegativity and lower valence state of I^- lead to enhanced electrons localization and appearance of unpaired $3d^1$ electrons in Ti^{3+}, implying the possibility of magnetic ordering. Theoretical simulation revealed the asymmetric density of states for electrons in spin down channel and spin up channel, suggesting that spin-dependent scattering exists. Magneto-transport results verified the superior intrinsic GMR value as −85% under 5 T at 10 K. Compared with Fe/Cr/Fe multilayer superlattice, the initial GMR material whose GMR value is about −20% under 5 T at 10 K,[129] anionic solid solution system $TiTe_{2-x}I_x$ displays better performance. Antiferromagnetic domain scattering effects and external field-tunable spin arrangements

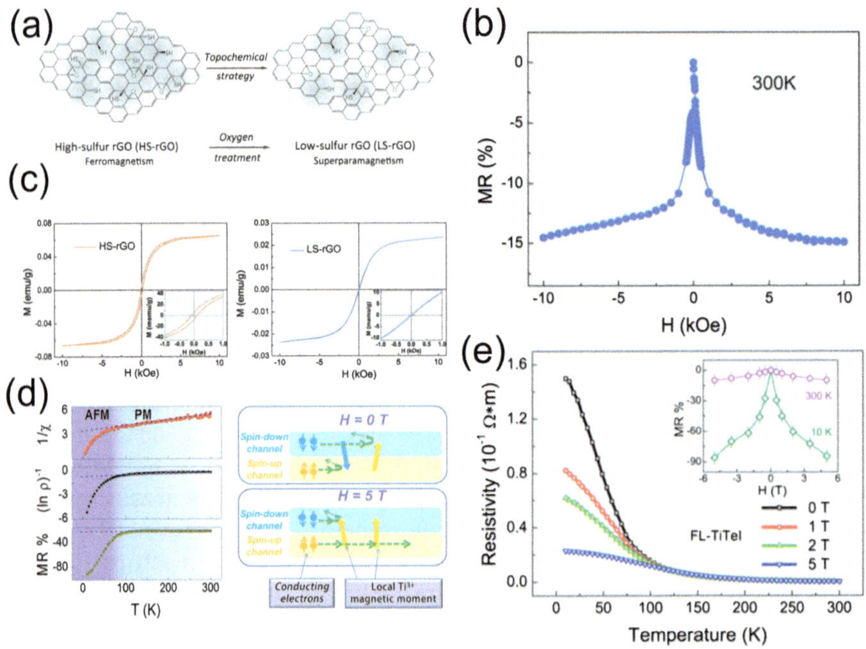

Figure 3.6 (a) Chemical and magnetic structures in high-sulfur-concentration rGO (HS-rGO) and low-sulfur-concentration rGO (LS-rGO). (b) Field-dependent magnetization (*M*–*H* curve) at 300 K for HS-rGO and LS-rGO nanosheets, respectively. (c) Magnetic-field dependence of the MR for LS-rGO at 300 K, reprinted with permission from J. Peng *et al.*, Superparamagnetic reduced graphene oxide with large magnetoresistance: A surface modulation strategy, *Angewandte Chemie International Edition*, John Wiley and Sons, Copyright © 2016 WILEY-VCH Verlag GmbH & Co. KGaA, Weinheim.[127] (d) Left: temperature-dependent inverse susceptibility, fitted resistivity (ρ), and MR for the TiTeI. Right: schematic diagram of the spin-dependent scattering between the conducting electrons and the local Ti^{3+} magnetic moment in the ultrathin TiTeI. (e) The temperature-dependent resistivity under various magnetic fields for TiTeI, reprinted with permission from Y. Guo *et al.*, *Physical Review Letters*, **113**, 157202, 2014. Copyright 2014 by the American Physical Society.[128]

are responsible for the large intrinsic GMR in $TiTe_{2-x}I_x$ as per similar temperature-dependent behavior of magnetism, electric transport and MR. This work highlights the anionic solid solution method for inducing tunable magnetism and efficient GMR for realizing practical application of 2D materials for planar spintronic devices (Figure 3.6).

By active control and manipulation of the charge and spin degrees of the transition metal within the framework of 2D structures, anionic doping is expected to regulate transport and magnetic properties, providing an ideal material platform for achieving unique spin-related effects.

3.4.2 2D Magnetic Superlattice for Enhanced Magnetocaloric Effects

At the start of the twenty-first century, special attention was paid to highly efficient energy conversion and environmental friendly technology to combat the greenhouse effect and energy exhaustion. One of the main culprits that warms the Earth and creates ozonosphere holes is dichlorodifluoromethane, also called Freon, which is used as freezing medium in vapor-cycle refrigerators. Refrigerators designed with harmful chemical-free technology are urgently needed. Magnetic refrigeration based on the magnetocaloric effect (MCE), physical conversion from magnetic energy to heat or *vice versa*, is a powerful candidate as there is high conversion efficiency and no harmful chemicals involved.[130] The MCE is an isothermal energy change between spin entropy and heat achieved tuning magnetization on adiabatic occasions. Cooling may be obtained during the demagnetization process or magnetization process; the former is called positive MCE and the latter is called negative MCE. Huge MCE near room temperature would be benefit for magnetic refrigeration utility and huge MCE at low temperature is significant to low-temperature scientific research.[131] Seeking magnetic materials for outstanding MCE requires them to have a large entropy change at magnetic phase transition from ferromagnetism (FM) or antiferromagnetism (AFM) to paramagnetism (PM) as well as field-induced transition from AFM states to FM states with the lowest hysteresis. While almost available MCE materials reach the highest efficiency in magnetic fields higher than 5 T, novel magnetic materials with large MCEs at lower fields, lower than 2 T, are expected for extending MCE materials to wider applications (Figure 3.7).[132]

Except for enhancement of MCE capacitor at low field, to cater to fashionable planar device design in chips, 2D magnetic materials with high effective MCE attract researchers' attention. As we all know, the superlattice materials, due to the peculiar electronic structure and physical properties brought about by their unique interface, have attracted extensive research interest.[133,134] However, the traditional superlattice materials are usually expensively obtained by pulsed laser deposition or molecular beam epitaxy. This also means that the subsequent practical application will require complex and costly processes. At the same time, the latest research results show that graphene has unique bipolar electronic properties when combined with other materials and can be used as electron donor or electron acceptor to effectively regulate the electrical properties of the composite materials, so-called graphene based superlattices.[135–137] Since the composite superlattice materials possess special interfacial properties and the unique electronic structure of graphene, they may offer a new performance different to that which traditional materials could present. Recently, *via* a new space confinement growth strategy successfully, the perfect stacked graphene based superlattice structure, in the form of graphene monolayer-vanadium oxide monolayer-graphene monolayer, has been prepared to overcome the shortcomings of traditional growth superlattice materials.[138] In such a graphene

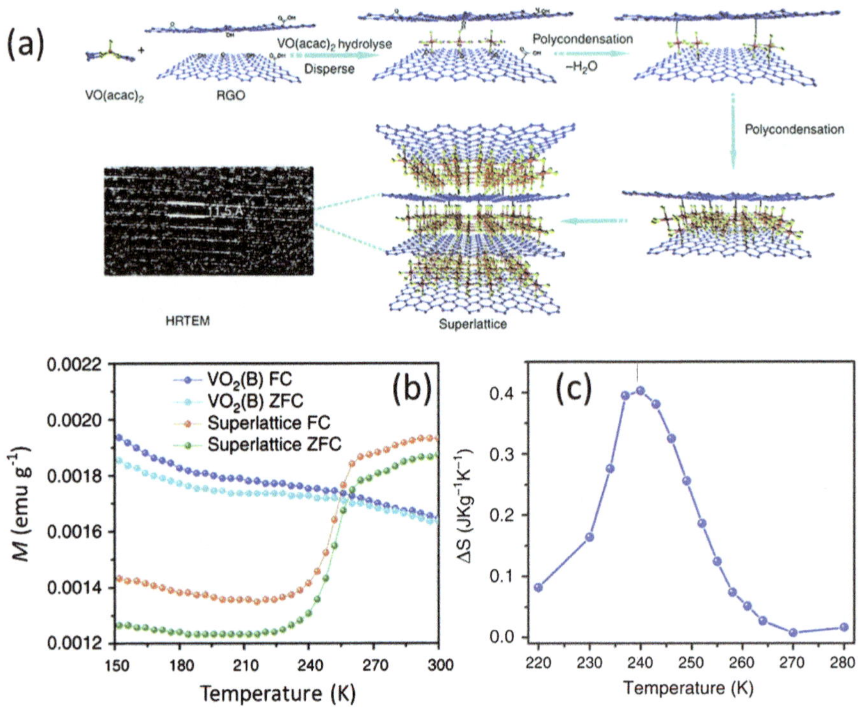

Figure 3.7 (a) Schematic illustration of the constitution process for the vanadium oxide superlattice in a space-confined nanoreactor. (b) ZFC and FC M–T curves for vanadium oxide/graphene superlattice superlattice nanosheets and VO_2(B). (c) Magnetic entropy change of the vanadium oxide/graphene superlattice nanosheets. Reprinted by permission from Macmillan Publishers Ltd: *Nature Communications* (ref. 138), Copyright 2014.

based superlattice material, graphene not only functions as a unique space confinement reactor for the vanadium oxide layer growing within the interval of 2D graphene layers, but also as an electron donor, inducing restructured vanadium oxide layers and reduced formation energy of the high symmetry vanadyl skeleton. Detailed structural characterization of the superlattice shows that the symmetry of vanadium oxide layers in the superlattice is significantly higher than the freestanding VO_2 (B), which led to the emergence of reversible phase transformation accompanied with a large spin entropy change, suggesting huge MCE. This phase transition is similar to VO_2 (M)–VO_2 (R) phase transition, during which the high symmetry changes to the low symmetry transition reported in other works. Considering the stress within the graphene layers and electron injection from hybridized graphene layers, the spin entropy change value for superlattice phase transition improved up to eight times compared with VO_2 (M)–VO_2 (R) phase transition. Importantly, the flexible 2D superlattice material is gained from low-cost solution synthesis, avoiding the traditional complex transfer process, and is adapted to the assembly of the various functions of the device, making it possible for further constructing magnetic-cooling and waste heat utilization units in 2D devices.

3.4.3 2D Magnetic Nanomaterials for Energy-related Application

Electrochemical water splitting, including hydrogen evolution reaction (HER) and oxygen evolution reaction (OER), is now a very hot topic for energy conversion and storage.[139,140] During the splitting process, surface molecule absorption and interface charge transfer are two determinative steps, therefore surface atoms of electrochemical catalysis play critical roles in realizing highly efficient electrochemical catalysis.[141] Surface atoms with appropriate electron configuration favor rapid surface molecule absorption and interaction. Electron configuration at d orbits of metal atoms is strongly related to spin state of corresponding metal ion, constructing a bridge connecting magnetism and electrochemical catalysis performance.

Success has been realized in the surface of perovskite system, a kind of low-cost, environment-friendly and stable oxide for efficient OER.[142–144] Numerous efforts have been made to understand the inner mechanism and factors that should be responsible for the efficiency. Shao-Horn and co-workers have reported a correlation between the OER performance and the e_g electron configuration of transition metal ions in perovskites, which was then proved by several experiments on the OER activity of perovskite.[145] For example, in a typical perovskite, $CaMnO_3$, e_g electrons can be tuned by changing the oxygen vacancy concentration, rare earth (RE) element doping. When the oxygen vacancy concentration increases, and appreciate RE atoms exists, e_g eletrons get close to the best value of 1.2 *via* magnetic measurement, and the OER performance of $CaMnO_3$ is optimized.[146] Recently, Zhou *et al.* found in perovskite $LaCoO_3$ that regulation of particle size can efficiently enhance the OER activity.[147] Magnetic measurement revealed that with different particle sizes, the effective magnetic moments of $LaCoO_3$ were modulated, which suggested the e_g electrons were regulated, reflecting the distribution of high-spin states ($t_{2g}{}^4e_g{}^2$) Co^{3+} ion and low-spin states ($t_{2g}{}^6e_g{}^0$) Co^{3+} ion. The size of the most active sample is 80 nm, corresponding to e_g electrons of 1.2, which agrees with Shao-Horn's principle. Magnetism of perovskite gives the e_g electron amount which is tightly connected with the activity of electrochemical catalysis, pointing to a way to better catalysts.

Magnetic ion doping into a semiconductor matrix creates a dilute magnetic semiconductor which combines charge and spin free degrees of electrons, displaying great potential in the computational realm. Besides, magnetic ion doping can induce Jahn–Teller distortion because of their asymmetric electrons configuration in stereochemistry. Such distortion sites can create more active sites for electrochemical reaction, hence promoting fast material transformation. Recently, Xie *et al.* successfully doped Mn^{2+} ions into

$CoSe_2$ nanosheets, resulting in confined distorted atomic arrangement and more edge active sites, immensely accelerating HER dynamics.[148] In pristine $CoSe_2$ crystal lattice, Co^{2+} ions have an electron configuration of $t_{2g}^6e_g^1$ with unevenly occupied e_g orbit, yielding a strong Jahn–Teller effect in Co–Se coordination octahedra. However, the electron configuration of doping Mn^{2+} ions is $t_{2g}^5e_g^0$, yielding a weak Jahn–Teller effect in Mn–Se coordination octahedra. In synergy with the localized Coulomb interaction around the Mn atoms arising from the introduction of the in-coordinate electron spin, an innovative subtle atomic arrangement distortion has been produced. Taking advantage of the distortion, activation energy barriers during the HER rate-determining step, the Tafel reaction between two adjacent adsorbed H atoms, are lowered. The authors attributed the enhanced HER performance of Mn-doped $CoSe_2$ nanosheets to a distorted atomic arrangement induced by the extraneous spin states. This work presents a paradigm for magnetism inducement in 2D nanomaterials with improving electrochemical catalysis (Figure 3.8).

Magnetism, which is tightly related to spin states, is an important part of material physical properties, but it can also be utilized by chemists for finding ways to regulate material chemical behaviors, especially in 2D materials. In 2D materials, unique spin states can completely reveal the power for electron structure modulation, surface distortion inducing and increased active sites, thus effectively decreasing catalysis energy barriers and promoting quick catalysis dynamics. Therefore, magnetism exploits a new avenue for development of better 2D inorganic catalysts.

3.5 Conclusions and Outlooks

Since the development of graphene, research on 2D nanomaterials has undergone rapid growth during the past decade. A multitude of inorganic 2D nanomaterials are obtained from layered materials ranging from natural graphite, black phosphorous and transition metal dichalcogenides to artificial lamellar hybrid intermediate and they have attracted great interest for their superior intrinsic physical properties that enable promising applications for spintronics and energy-related applications. To take full advantage of those 2D nanomaterials, surface modification strategies which can effectively optimize their intrinsic physical properties are significant to the prospects that the application can reach.

2D nanomaterials, with a thickness down to atomic-scale, have extremely large specific surface area, making their surface phase as important as the bulk counterparts. Therefore, surface treatment techniques provide alternative methods to engineer intrinsic physical properties of inorganic 2D nanomaterials. The surface properties and effects of 2D nanomaterials are completely divergent with those on ordinary bulk nanomaterials. Although surface modifications do include some contributions of bulk solids to raise reaction activity and biocompatibility, *etc.* manual manipulation of surface atom species or electronic configurations cannot affect the intrinsic physical properties of thorough bulk blocks. Differently to the bulk counterparts, 2D

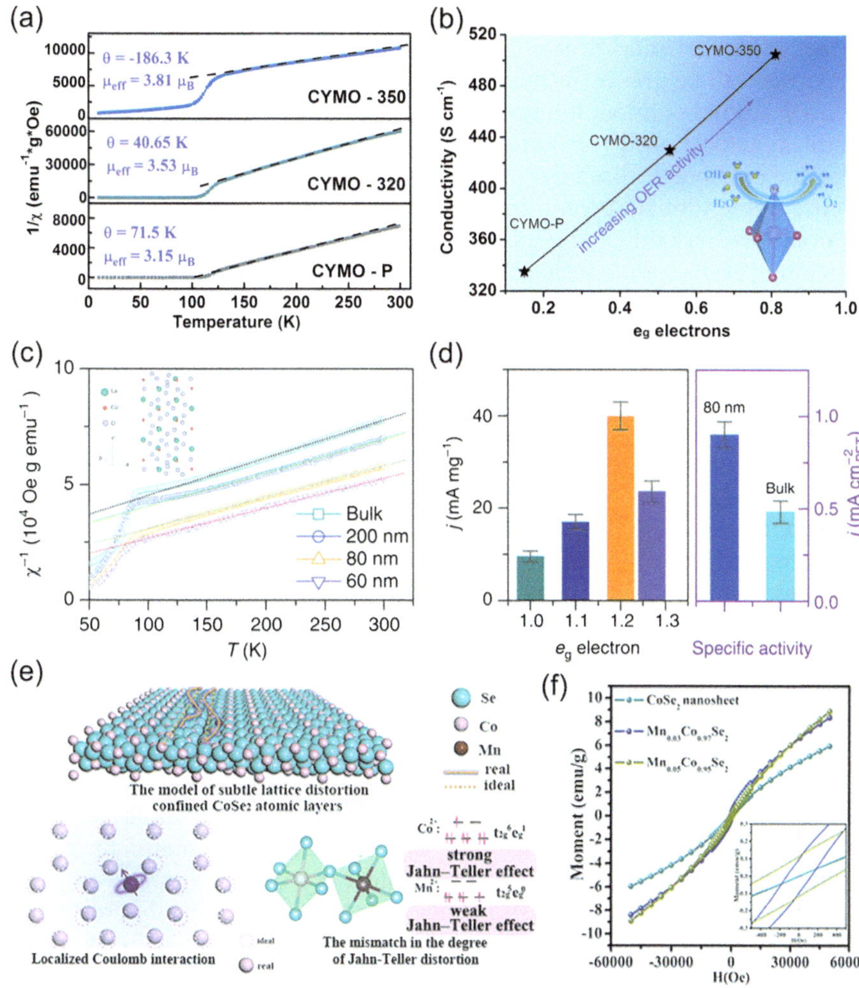

Figure 3.8 (a) Temperature dependent inverse susceptibility for CYMO-P, CYMO-320, CYMO-350. (b) The OER activity increased with conductivity and e_g electrons filling status optimized by hydrogen treatment. Inset: Jahn–Teller distortion promoted the forming of oxygen defects, resulting in optimal Mn e_g filling state and better electrical conductivity, reproduced with permission from Y Guo *et al.*, Engineering the electronic state of a perovskite electrocatalyst for synergistically enhanced oxygen evolution reaction, *Advanced Materials*, John Wiley and Sons, Copyright © 2015 WILEY-VCH Verlag GmbH & Co. KGaA, Weinheim.[146] (c) The temperature dependence inverse susceptibilities for all the LaCoO3 samples. (d) OER activities comparison of the bulk and nanosized LCO, reprinted by permission from Macmillan Publishers Ltd: *Nature Communications* (ref. 147), Copyright 2016. (e) Schematic representations of the formation mechanism for the subtle distortion of atomic arrangement through the incorporated heterogeneous spin states. (f) The magnetic field dependence of magnetization (M–H) curve at 300 K, adapted with permission from ref. 148, Copyright 2016 American Chemical Society.

materials offer enough surface facet atoms and active sites for accomplishing surface chemical modification, and hence their intrinsic physicochemical properties can be greatly influenced. Thanks to the large amount of surface exposed atoms – almost all atoms in 2D inorganic nanomaterials – a high chemical activity, and regulation of electrical transport property and band structure could be realized by surface modifications of low coordinate surface atoms, tremendously promoting applications on electronic and chemical catalysis. Remarkably, as 2D scale confined electrons are always strongly correlated, even slight surface modification has the potential to trigger electron transfer or atomic lattice distortion and leads to reconstitution all over the electronic structure, and it results in effective regulation of conductivity, magnetism, and energy band structures. Thus, chemical modification strategies have the power to engineer multiple physical properties in 2D nanomaterials under the premise of maintaining an unbroken freestanding 2D lattice framework.

In this chapter, we covered intrinsic magnetic 2D materials and magnetic 2D materials whose magnetic properties are successfully introduced *via* lately-reported surface chemical modification strategies, including heteroatom incorporation, small molecular absorption, and multiple defect introduction. We chose one distinguished example from numerous studies to introduce each strategy in detail. From those examples, rational surface modification can effectively regulate intrinsic magnetic properties. We demonstrated two promising applications of 2D magnetic materials and highlighted that regulated magnetism *via* surface modification did indeed improve the ability in spintronic devices, magnetocaloric effect and electrochemical catalysis.

Overall, the essential superior magnetic properties in 2D nanomaterials combined with latest developments in engineering magnetic properties by surface chemical modification stimulate a growing excitement for practical 2D applications. More approaches are still urgently needed to further improve responsible stability and repeatability, which is crucial for thorough comprehension of physical behavior mechanisms as well as promoting them in real industrial applications. We are optimistic enough that surface modification methods will lead to wide application and deeper understanding of the inner concepts of 2D magnetic nanomaterials.

References

1. R. Ferrando, J. Jellinek and R. L. Johnston, *Chem. Rev.*, 2008, **108**, 845–910.
2. V. R. Stamenkovic, B. Fowler, B. S. Mun, G. Wang, P. N. Ross, C. A. Lucas and N. M. Marković, *Science*, 2007, **315**, 493–497.
3. D. M. Brunette, P. Tengvall, M. Textor and P. Thomsen, *Titanium in Medicine: Material Science, Surface Science, Engineering, Biological Responses and Medical Applications*, Springer Science & Business Media, 2012.
4. L. Ju, B. Geng, J. Horng, C. Girit, M. Martin, Z. Hao, H. A. Bechtel, X. Liang, A. Zettl and Y. R. Shen, *Nat. Nanotechnol.*, 2011, **6**, 630–634.

5. Y. Saito, Y. Kasahara, J. Ye, Y. Iwasa and T. Nojima, *Science*, 2015, **350**, 409–413.
6. D. Kiriya, M. Tosun, P. Zhao, J. S. Kang and A. Javey, *J. Am. Chem. Soc.*, 2014, **136**, 7853–7856.
7. Q. He, Z. Zeng, Z. Yin, H. Li, S. Wu, X. Huang and H. Zhang, *Small*, 2012, **8**, 2994–2999.
8. J. Feng, L. Peng, C. Wu, X. Sun, S. Hu, C. Lin, J. Dai, J. Yang and Y. Xie, *Adv. Mater.*, 2012, **24**, 1969–1974.
9. Y. Sun, H. Cheng, S. Gao, Q. Liu, Z. Sun, C. Xiao, C. Wu, S. Wei and Y. Xie, *J. Am. Chem. Soc.*, 2012, **134**, 20294–20297.
10. J. Kim, S. S. Baik, S. H. Ryu, Y. Sohn, S. Park, B.-G. Park, J. Denlinger, Y. Yi, H. J. Choi and K. S. Kim, *Science*, 2015, **349**, 723–726.
11. J. Kibsgaard, Z. Chen, B. N. Reinecke and T. F. Jaramillo, *Nat. Mater.*, 2012, **11**, 963–969.
12. J. Xie, J. Zhang, S. Li, F. Grote, X. Zhang, H. Zhang, R. Wang, Y. Lei, B. Pan and Y. Xie, *J. Am. Chem. Soc.*, 2013, **135**, 17881–17888.
13. X. Zhu, X. Dou, J. Dai, X. An, Y. Guo, L. Zhang, S. Tao, J. Zhao, W. Chu and X. C. Zeng, *Angew. Chem., Int. Ed.*, 2016, **55**, 12465–12469.
14. X. Lu, K. Xu, S. Tao, Z. Shao, X. Peng, W. Bi, P. Chen, H. Ding, W. Chu and C. Wu, *Chem. Sci.*, 2016, **7**, 1462–1467.
15. W. Qing-Yan, L. Zhi, Z. Wen-Hao, Z. Zuo-Cheng, Z. Jin-Song, L. Wei, D. Hao, O. Yun-Bo, D. Peng and C. Kai, *Chin. Phys. Lett.*, 2012, **29**, 037402.
16. E. Navarro-Moratalla, J. O. Island, S. Mañas-Valero, E. Pinilla-Cienfuegos, A. Castellanos-Gomez, J. Quereda, G. Rubio-Bollinger, L. Chirolli, J. A. Silva-Guillén and N. Agraït, *Nat. Commun.*, 2016, **7**, 11043.
17. M. Chhowalla, H. S. Shin, G. Eda, L.-J. Li, K. P. Loh and H. Zhang, *Nat. Chem.*, 2013, **5**, 263–275.
18. K. Sugawara, Y. Nakata, R. Shimizu, P. Han, T. Hitosugi, T. Sato and T. Takahashi, *ACS Nano*, 2015, **10**, 1341–1345.
19. M. Calandra, I. Mazin and F. Mauri, *Phys. Rev. B*, 2009, **80**, 241108.
20. L. Ma, C. Ye, Y. Yu, X. F. Lu, X. Niu, S. Kim, D. Feng, D. Tománek, Y.-W. Son and X. H. Chen, *Nat. Commun.*, 2016, **7**, 10956.
21. V. K. Sangwan, D. Jariwala, I. S. Kim, K.-S. Chen, T. J. Marks, L. J. Lauhon and M. C. Hersam, *Nat. Nanotechnol.*, 2015, **10**, 403–406.
22. D. Xiang, C. Han, J. Wu, S. Zhong, Y. Liu, J. Lin, X.-A. Zhang, W. P. Hu, B. Özyilmaz and A. C. Neto, *Nat. Commun.*, 2015, **6**, 6485.
23. D. Chiba, S. Fukami, K. Shimamura, N. Ishiwata, K. Kobayashi and T. Ono, *Nat. Mater.*, 2011, **10**, 853–856.
24. H. Kabbour, R. David, A. Pautrat, H. J. Koo, M. H. Whangbo, G. André and O. Mentré, *Angew. Chem., Int. Ed.*, 2012, **51**, 11745–11749.
25. S. W. Han, Y. H. Hwang, S.-H. Kim, W. S. Yun, J. Lee, M. G. Park, S. Ryu, J. S. Park, D.-H. Yoo and S.-P. Yoon, *Phys. Rev. Lett.*, 2013, **110**, 247201.
26. M. Osada, Y. Ebina, K. Fukuda, K. Ono, K. Takada, K. Yamaura, E. Takayama-Muromachi and T. Sasaki, *Phys. Rev. B*, 2006, **73**, 153301.
27. B. Berge, H. Diep, A. Ghazali and P. Lallemand, *Phys. Rev. B*, 1986, **34**, 3177.

28. V. T. Ngo and H. Diep, *Phys. Rev. B*, 2007, **75**, 035412.
29. J. M. Kosterlitz and D. J. Thouless, *J. Phys. C: Solid State Phys.*, 1973, **6**, 1181.
30. S. F. Edwards and P. W. Anderson, *J. Phys. F: Met. Phys.*, 1975, **5**, 965.
31. M. Rotter, M. Tegel, D. Johrendt, I. Schellenberg, W. Hermes and R. Pöttgen, *Phys. Rev. B*, 2008, **78**, 020503.
32. M. Rotter, M. Tegel and D. Johrendt, *Phys. Rev. Lett.*, 2008, **101**, 107006.
33. L. Li, C. Richter, J. Mannhart and R. Ashoori, *Nat. Phys.*, 2011, **7**, 762–766.
34. X. Lu, N. Wang, H. Wu, Y. Wu, D. Zhao, X. Zeng, X. Luo, T. Wu, W. Bao and G. Zhang, *Nat. Mater.*, 2015, **14**, 325–329.
35. C. F. Hirjibehedin, C.-Y. Lin, A. F. Otte, M. Ternes, C. P. Lutz, B. A. Jones and A. J. Heinrich, *Science*, 2007, **317**, 1199–1203.
36. A. F. Otte, M. Ternes, K. Von Bergmann, S. Loth, H. Brune, C. P. Lutz, C. F. Hirjibehedin and A. J. Heinrich, *Nat. Phys.*, 2008, **4**, 847–850.
37. S. Wolf, D. Awschalom, R. Buhrman, J. Daughton, S. Von Molnar, M. Roukes, A. Y. Chtchelkanova and D. Treger, *Science*, 2001, **294**, 1488–1495.
38. S. D. Sarma, *Nat. Mater.*, 2003, **2**, 292–294.
39. X. Moya, L. Hueso, F. Maccherozzi, A. Tovstolytkin, D. Podyalovskii, C. Ducati, L. Phillips, M. Ghidini, O. Hovorka and A. Berger, *Nat. Mater.*, 2013, **12**, 52–58.
40. D. Mosca, F. Vidal and V. H. Etgens, *Phys. Rev. Lett.*, 2008, **101**, 125503.
41. Y. Zhang, Y.-W. Tan, H. L. Stormer and P. Kim, *Nature*, 2005, **438**, 201–204.
42. C. Berger, Z. Song, X. Li, X. Wu, N. Brown, C. Naud, D. Mayou, T. Li, J. Hass and A. N. Marchenkov, *Science*, 2006, **312**, 1191–1196.
43. K. S. Novoselov, Z. Jiang, Y. Zhang, S. Morozov, H. Stormer, U. Zeitler, J. Maan, G. Boebinger, P. Kim and A. Geim, *Science*, 2007, **315**, 1379.
44. A. A. Balandin, S. Ghosh, W. Bao, I. Calizo, D. Teweldebrhan, F. Miao and C. N. Lau, *Nano Lett.*, 2008, **8**, 902–907.
45. A. Splendiani, L. Sun, Y. Zhang, T. Li, J. Kim, C.-Y. Chim, G. Galli and F. Wang, *Nano Lett.*, 2010, **10**, 1271–1275.
46. X. Xi, L. Zhao, Z. Wang, H. Berger, L. Forró, J. Shan and K. F. Mak, *Nat. Nanotechnol.*, 2015, **10**, 765–769.
47. X. Xi, Z. Wang, W. Zhao, J.-H. Park, K. T. Law, H. Berger, L. Forró, J. Shan and K. F. Mak, *Nat. Phys.*, 2015, **12**, 139–143.
48. M. Fujita, K. Wakabayashi, K. Nakada and K. Kusakabe, *J. Phys. Soc. Jpn.*, 1996, **65**, 1920–1923.
49. Y.-W. Son, M. L. Cohen and S. G. Louie, *Nature*, 2006, **444**, 347–349.
50. Y. Wang, Y. Huang, Y. Song, X. Zhang, Y. Ma, J. Liang and Y. Chen, *Nano Lett.*, 2008, **9**, 220–224.
51. G. Z. Magda, X. Jin, I. Hagymási, P. Vancsó, Z. Osváth, P. Nemes-Incze, C. Hwang, L. P. Biró and L. Tapasztó, *Nature*, 2014, **514**, 608–611.
52. Y. Du, H. Liu, B. Xu, L. Sheng, J. Yin, C.-G. Duan and X. Wan, *Sci. Rep.*, 2015, **5**, 8921.
53. G. Yang, S. Xu, W. Zhang, T. Ma and C. Wu, *Phys. Rev. B*, 2016, **94**, 075106.

54. A. Krasheninnikov, P. Lehtinen, A. S. Foster, P. Pyykkö and R. M. Nieminen, *Phys. Rev. Lett.*, 2009, **102**, 126807.
55. Y. Liu, N. Tang, X. Wan, Q. Feng, M. Li, Q. Xu, F. Liu and Y. Du, *Sci. Rep.*, 2013, **3**, 2566.
56. X. Zhu, Y. Guo, H. Cheng, J. Dai, X. An, J. Zhao, K. Tian, S. Wei, X. C. Zeng, C. Wu and Y. Xie, *Nat. Commun.*, 2016, **7**, 11210.
57. S. Lee, G. Nam, J. Sun, J. S. Lee, H. W. Lee, W. Chen, J. Cho and Y. Cui, *Angew. Chem., Int. Ed.*, 2016, **55**, 8599–8604.
58. Y. Liu, X. Hua, C. Xiao, T. Zhou, P. Huang, Z. Guo, B. Pan and Y. Xie, *J. Am. Chem. Soc.*, 2016, **138**, 5087–5092.
59. X. Hong, S.-H. Cheng, C. Herding and J. Zhu, *Phys. Rev. B*, 2011, **83**, 085410.
60. J. Barnaś, A. Fuss, R. Camley, P. Grünberg and W. Zinn, *Phys. Rev. B*, 1990, **42**, 8110.
61. R. Ma and T. Sasaki, *Adv. Mater.*, 2010, **22**, 5082–5104.
62. R. Ma, Z. Liu, K. Takada, N. Iyi, Y. Bando and T. Sasaki, *J. Am. Chem. Soc.*, 2007, **129**, 5257–5263.
63. E. Coronado, C. Martí-Gastaldo, E. Navarro-Moratalla, A. Ribera, S. J. Blundell and P. J. Baker, *Nat. Chem.*, 2010, **2**, 1031–1036.
64. X. Dong, M. Osada, H. Ueda, Y. Ebina, Y. Kotani, K. Ono, S. Ueda, K. Kobayashi, K. Takada and T. Sasaki, *Chem. Mater.*, 2009, **21**, 4366–4373.
65. Y. Guo, K. Xu, C. Wu, J. Zhao and Y. Xie, *Chem. Soc. Rev.*, 2015, **44**, 637–646.
66. B. Radisavljevic, A. Radenovic, J. Brivio, V. Giacometti and A. Kis, *Nat. Nanotechnol.*, 2011, **6**, 147–150.
67. Y.-M. Lin, C. Dimitrakopoulos, K. A. Jenkins, D. B. Farmer, H.-Y. Chiu, A. Grill and P. Avouris, *Science*, 2010, **327**, 662.
68. L. Li, Y. Yu, G. J. Ye, Q. Ge, X. Ou, H. Wu, D. Feng, X. H. Chen and Y. Zhang, *Nat. Nanotechnol.*, 2014, **9**, 372–377.
69. D. Pesin and A. H. MacDonald, *Nat. Mater.*, 2012, **11**, 409–416.
70. P. Grünberg, R. Schreiber, Y. Pang, M. Brodsky and H. Sowers, *Phys. Rev. Lett.*, 1986, **57**, 2442.
71. J. K. Furdyna, *J. Appl. Phys.*, 1988, **64**, R29–R64.
72. T. Dietl, *Nat. Mater.*, 2003, **2**, 646–648.
73. T. Dietl, *Nat. Mater.*, 2010, **9**, 965–974.
74. K. S. Novoselov, A. K. Geim, S. V. Morozov, D. Jiang, Y. Zhang, S. V. Dubonos, I. V. Grigorieva and A. A. Firsov, *Science*, 2004, **306**, 666–669.
75. J. N. Coleman, M. Lotya, A. O'Neill, S. D. Bergin, P. J. King, U. Khan, K. Young, A. Gaucher, S. De and R. J. Smith, *Science*, 2011, **331**, 568–571.
76. Y. Shi, H. Li and L.-J. Li, *Chem. Soc. Rev.*, 2015, **44**, 2744–2756.
77. Y. Sun, S. Gao, F. Lei and Y. Xie, *Chem. Soc. Rev.*, 2015, **44**, 623–636.
78. M. A. Woo, M.-S. Song, T. W. Kim, I. Y. Kim, J.-Y. Ju, Y. S. Lee, S. J. Kim, J.-H. Choy and S.-J. Hwang, *J. Mater. Chem.*, 2011, **21**, 4286–4292.
79. Q. Wang and D. O'Hare, *Chem. Rev.*, 2012, **112**, 4124–4155.
80. S. Okamoto, *J. Am. Ceram. Soc.*, 1968, **51**, 594–598.

81. P. Chen, K. Xu, X. Li, Y. Guo, D. Zhou, J. Zhao, X. Wu, C. Wu and Y. Xie, *Chem. Sci.*, 2014, **5**, 2251–2255.
82. L. Vayssieres, C. Sathe, S. M. Butorin, D. K. Shuh, J. Nordgren and J. Guo, *Adv. Mater.*, 2005, **17**, 2320–2323.
83. W. Cheng, J. He, T. Yao, Z. Sun, Y. Jiang, Q. Liu, S. Jiang, F. Hu, Z. Xie and B. He, *J. Am. Chem. Soc.*, 2014, **136**, 10393–10398.
84. K. Okuda, K. Kurosawa, S. Saito, M. Honda, Z. Yu and M. Date, *J. Phys. Soc. Jpn.*, 1986, **55**, 4456–4463.
85. G. Le Flem, R. Brec, G. Ouvard, A. Louisy and P. Segransan, *J. Phys. Chem. Solids*, 1982, **43**, 455–461.
86. Y. Guo, C. Liu, Q. Yin, C. Wei, S. Lin, T. B. Hoffman, Y. Zhao, J. Edgar, Q. Chen and S. P. Lau, *ACS Nano*, 2016, **10**, 8980–8988.
87. B. L. Chittari, Y. Park, D. Lee, M. Han, A. H. MacDonald, E. Hwang and J. Jung, *Phys. Rev. B*, 2016, **94**, 184428.
88. J.-U. Lee, S. Lee, J. H. Ryoo, S. Kang, T. Y. Kim, P. Kim, C.-H. Park, J.-G. Park and H. Cheong, *Nano Lett.*, 2016, **16**, 7433–7438.
89. H. Schmidt, F. Giustiniano and G. Eda, *Chem. Soc. Rev.*, 2015, **44**, 7715–7736.
90. K. Xu, P. Chen, X. Li, C. Wu, Y. Guo, J. Zhao, X. Wu and Y. Xie, *Angew. Chem., Int. Ed.*, 2013, **52**, 10477–10481.
91. C. Rao, H. R. Matte, K. Subrahmanyam and U. Maitra, *Chem. Sci.*, 2012, **3**, 45–52.
92. J. Červenka, M. Katsnelson and C. Flipse, *Nat. Phys.*, 2009, **5**, 840–844.
93. R. Nair, M. Sepioni, I.-L. Tsai, O. Lehtinen, J. Keinonen, A. Krasheninnikov, T. Thomson, A. Geim and I. Grigorieva, *Nat. Phys.*, 2012, **8**, 199–202.
94. R. Singh and P. Kroll, *J. Phys.: Condens. Matter*, 2009, **21**, 196002.
95. M. M. Ugeda, I. Brihuega, F. Guinea and J. M. Gómez-Rodríguez, *Phys. Rev. Lett.*, 2010, **104**, 096804.
96. S. Bhowmick and V. B. Shenoy, *J. Chem. Phys.*, 2008, **128**, 244717.
97. S. Wang, L. Talirz, C. A. Pignedoli, X. Feng, K. Müllen, R. Fasel and P. Ruffieux, *Nat. Commun.*, 2016, **7**, 11507.
98. S. Z. Butler, S. M. Hollen, L. Cao, Y. Cui, J. A. Gupta, H. R. Gutierrez, T. F. Heinz, S. S. Hong, J. Huang and A. F. Ismach, *ACS Nano*, 2013, **7**, 2898–2926.
99. L. Kou, C. Tang, Y. Zhang, T. Heine, C. Chen and T. Frauenheim, *J. Phys. Chem. Lett.*, 2012, **3**, 2934–2941.
100. S. Tongay, S. S. Varnoosfaderani, B. R. Appleton, J. Wu and A. F. Hebard, *Appl. Phys. Lett.*, 2012, **101**, 123105.
101. J. Zhang, J. M. Soon, K. P. Loh, J. Yin, J. Ding, M. B. Sullivian and P. Wu, *Nano Lett.*, 2007, **7**, 2370–2376.
102. D. Gao, M. Si, J. Li, J. Zhang, Z. Zhang, Z. Yang and D. Xue, *Nanoscale Res. Lett.*, 2013, **8**, 1–8.
103. M. A. Lukowski, A. S. Daniel, F. Meng, A. Forticaux, L. Li and S. Jin, *J. Am. Chem. Soc.*, 2013, **135**, 10274–10277.
104. X. Zhang, J. Zhang, J. Zhao, B. Pan, M. Kong, J. Chen and Y. Xie, *J. Am. Chem. Soc.*, 2012, **134**, 11908–11911.

105. H. González-Herrero, J. M. Gómez-Rodríguez, P. Mallet, M. Moaied, J. J. Palacios, C. Salgado, M. M. Ugeda, J.-Y. Veuillen, F. Yndurain and I. Brihuega, *Science*, 2016, **352**, 437–441.
106. K. Xu, X. Li, P. Chen, D. Zhou, C. Wu, Y. Guo, L. Zhang, J. Zhao, X. Wu and Y. Xie, *Chem. Sci.*, 2015, **6**, 283–287.
107. R. Singh and P. Kroll, *J. Phys.: Condens. Matter*, 2009, **21**, 196002.
108. A. K. Geim and K. S. Novoselov, *Nat. Mater.*, 2007, **6**, 183–191.
109. Y. Li, Z. Zhou, P. Shen and Z. Chen, *ACS Nano*, 2009, **3**, 1952–1958.
110. S. Qin, X. Guo, Y. Cao, Z. Ni and Q. Xu, *Carbon*, 2014, **78**, 559–565.
111. K. Lee, R. Gatensby, N. McEvoy, T. Hallam and G. S. Duesberg, *Adv. Mater.*, 2013, **25**, 6699–6702.
112. M. O'Brien, K. Lee, R. Morrish, N. C. Berner, N. McEvoy, C. A. Wolden and G. S. Duesberg, *Chem. Phys. Lett.*, 2014, **615**, 6–10.
113. Y. Zhou, Z. Wang, P. Yang, X. Zu, L. Yang, X. Sun and F. Gao, *ACS Nano*, 2012, **6**, 9727–9736.
114. Y. Ma, Y. Dai, M. Guo, C. Niu, Y. Zhu and B. Huang, *ACS Nano*, 2012, **6**, 1695–1701.
115. J. He, K. Wu, R. Sa, Q. Li and Y. Wei, *Appl. Phys. Lett.*, 2010, **96**, 082504.
116. D. Cao, H. Shu, T. Wu, Z. Jiang, Z. Jiao, M. Cai and W. Hu, *Appl. Surf. Sci.*, 2016, **361**, 199–205.
117. H.-P. Komsa, J. Kotakoski, S. Kurasch, O. Lehtinen, U. Kaiser and A. V. Krasheninnikov, *Phys. Rev. Lett.*, 2012, **109**, 035503.
118. L. Cai, J. He, Q. Liu, T. Yao, L. Chen, W. Yan, F. Hu, Y. Jiang, Y. Zhao and T. Hu, *J. Am. Chem. Soc.*, 2015, **137**, 2622–2627.
119. L. Liu, X. Wu, X. Liu and P. K. Chu, *Appl. Phys. Lett.*, 2015, **106**, 132406.
120. H. Qiu, Z. Wang and X. Sheng, *Phys. B*, 2013, **421**, 46–49.
121. D. Gao, Q. Xu, J. Zhang, Z. Yang, M. Si, Z. Yan and D. Xue, *Nanoscale*, 2014, **6**, 2577–2581.
122. M. N. Baibich, J. M. Broto, A. Fert, F. N. Van Dau, F. Petroff, P. Etienne, G. Creuzet, A. Friederich and J. Chazelas, *Phys. Rev. Lett.*, 1988, **61**, 2472.
123. M. Xu, T. Liang, M. Shi and H. Chen, *Chem. Rev.*, 2013, **113**, 3766–3798.
124. Q. H. Wang, K. Kalantar-Zadeh, A. Kis, J. N. Coleman and M. S. Strano, *Nat. Nanotechnol.*, 2012, **7**, 699–712.
125. A. Berkowitz, J. Mitchell, M. Carey, A. Young, S. Zhang, F. Spada, F. Parker, A. Hutten and G. Thomas, *Phys. Rev. Lett.*, 1992, **68**, 3745.
126. B. Hickey, M. Howson, S. Musa and N. Wiser, *Phys. Rev. B*, 1995, **51**, 667.
127. J. Peng, Y. Guo, H. Lv, X. Dou, Q. Chen, J. Zhao, C. Wu, X. Zhu, Y. Lin and W. Lu, *Angew. Chem., Int. Ed.*, 2016, **128**, 3228–3232.
128. Y. Guo, J. Dai, J. Zhao, C. Wu, D. Li, L. Zhang, W. Ning, M. Tian, X. C. Zeng and Y. Xie, *Phys. Rev. Lett.*, 2014, **113**, 157202.
129. M. Gijs and M. Okada, *Phys. Rev. B*, 1992, **46**, 2908.
130. A. M. Tishin and Y. I. Spichkin, *The Magnetocaloric Effect and Its Applications*, CRC Press, 2016.
131. O. Tegus, E. Brück, K. Buschow and F. De Boer, *Nature*, 2002, **415**, 150–152.
132. T. Krenke, E. Duman, M. Acet, E. F. Wassermann, X. Moya, L. Mañosa and A. Planes, *Nat. Mater.*, 2005, **4**, 450–454.

133. R. Venkatasubramanian, E. Siivola, T. Colpitts and B. O'quinn, *Nature*, 2001, **413**, 597–602.
134. L. Ponomarenko, R. Gorbachev, G. Yu, D. Elias, R. Jalil, A. Patel, A. Mishchenko, A. Mayorov, C. Woods and J. Wallbank, *Nature*, 2013, **497**, 594–597.
135. M. Yankowitz, J. Xue, D. Cormode, J. D. Sanchez-Yamagishi, K. Watanabe, T. Taniguchi, P. Jarillo-Herrero, P. Jacquod and B. J. LeRoy, *Nat. Phys.*, 2012, **8**, 382–386.
136. S. Haigh, A. Gholinia, R. Jalil, S. Romani, L. Britnell, D. Elias, K. Novoselov, L. Ponomarenko, A. Geim and R. Gorbachev, *Nat. Mater.*, 2012, **11**, 764–767.
137. L. Ponomarenko, R. Gorbachev, G. Yu, D. Elias, R. Jalil, A. Patel, A. Mishchenko, A. Mayorov, C. Woods and J. Wallbank, *Nature*, 2013, **497**, 594–597.
138. H. Zhu, C. Xiao, H. Cheng, F. Grote, X. Zhang, T. Yao, Z. Li, C. Wang, S. Wei and Y. Lei, *Nat. Commun.*, 2014, **5**, 3960.
139. M. A. Lukowski, A. S. Daniel, F. Meng, A. Forticaux, L. Li and S. Jin, *J. Am. Chem. Soc.*, 2013, **135**, 10274–10277.
140. C. C. McCrory, S. Jung, J. C. Peters and T. F. Jaramillo, *J. Am. Chem. Soc.*, 2013, **135**, 16977–16987.
141. Y. Li, H. Wang, L. Xie, Y. Liang, G. Hong and H. Dai, *J. Am. Chem. Soc.*, 2011, **133**, 7296–7299.
142. J. Suntivich, K. J. May, H. A. Gasteiger, J. B. Goodenough and Y. Shao-Horn, *Science*, 2011, **334**, 1383–1385.
143. Y. Zhu, W. Zhou, Z. G. Chen, Y. Chen, C. Su, M. O. Tadé and Z. Shao, *Angew. Chem., Int. Ed.*, 2015, **54**, 3897–3901.
144. A. Grimaud, K. J. May, C. E. Carlton, Y.-L. Lee, M. Risch, W. T. Hong, J. Zhou and Y. Shao-Horn, *Nat. Commun.*, 2013, **4**, 2439.
145. J. Suntivich, K. J. May, H. A. Gasteiger, J. B. Goodenough and Y. Shao-Horn, *Science*, 2011, **334**, 1383–1385.
146. Y. Guo, Y. Tong, P. Chen, K. Xu, J. Zhao, Y. Lin, W. Chu, Z. Peng, C. Wu and Y. Xie, *Adv. Mater.*, 2015, **27**, 5989–5994.
147. S. Zhou, X. Miao, X. Zhao, C. Ma, Y. Qiu, Z. Hu, J. Zhao, L. Shi and J. Zeng, *Nat. Commun.*, 2016, **7**, 11510.
148. Y. Liu, X. Hua, C. Xiao, T. Zhou, P. Huang, Z. Guo, B. Pan and Y. Xie, *J. Am. Chem. Soc.*, 2016, **138**, 5087–5092.

CHAPTER 4

Solid-state Synthesis of Two-dimensional Layered Crystals

Z. UR REHMAN, W. ZHU, P. WU AND L. SONG*

National Synchrotron Radiation Laboratory, CAS Center for Excellence in Nanoscience, University of Science and Technology of China, Hezuohua South Road, Hefei, Anhui 230029, China
*E-mail: song2012@ustc.edu.cn

4.1 Introduction

Two-dimensional (2D) layered compounds possess strong lateral chemical bonding in planes but display weak van der Waals interactions between planes. One typical example is graphite that consists of weakly stacked individual sheets forming three-dimensional (3D) bulk crystals. Since the exfoliation and identification of graphene (single carbon atomic layer) in 2004, layered ultrathin 2D materials have been the subject of intensive study over the last decade.[1–5]

Most recently, the remarkable properties of graphene have promoted great interest in two-dimensional materials (2Ds) due to their unique chemical and physical attributes. Among various layered materials, inorganic two-dimensional crystals are considered as promising layered candidates with strong in-plane bonding and weak out-of-plane interactions enabling exfoliation

Smart Materials No. 27
Inorganic Two-dimensional Nanomaterials: Fundamental Understanding, Characterizations and Energy Applications
Edited by Changzheng Wu

Published by the Royal Society of Chemistry, www.rsc.org

into 2D layers of single unit cell thickness, such as transition metal dichalcogenides (TMDs), layered metal oxides, transition metal carbides and nitrides, and layered-double hydroxides.[6–15] For example, bulk MoS_2 is a semiconducor with an indirect band gap of 1.2 eV.[16,17] Each layer in MoS_2 with a honeycomb-like structure typically has a thickness of 6.5 Å, in which the planes of transition metal Mo atoms are sandwiched between two atomic layers of chalcogen S with strong in-plane covalent bonding. The individual MoS_2 layers are vertically stacked together *via* weak van der Waals interactions. Polymorphism, with distinctive characteristics, is the nature of layered MoS_2. Theoretically, it is predicted that MoS_2 layer has two distinct symmetries depending on the arrangement of S atoms, *i.e.*, 2H phase (trigonal prismatic D_{3h}) and 1T phase (octahedral O_h). The octahedral coordination can be obtained from the trigonal prism structure by rotating one of the sulfur basal planes by 60° around the *c*-axis. The two phases exhibit completely different electronic structures. The 2H phase is semiconducting and the 1T phase is metallic. Besides the distinctive electronic properties, it is also noted that the inorganic 2Ds can exhibit ultrahigh specific surface area and strong quantum confinement of electrons in two dimensions, resulting in many unconventional optical, magnetic and chemical properties. Thus, inorganic 2Ds have great potential in various applications such as electronic and optical devices, energy storage and conversion, biomedicine and sensing, *etc.*[18–26]

As the basis of structural investigations and potential explorations, synthesis play a key role in discovering novel inorganic layered materials, especially new layered crystals which do not exist in nature. Tailor-made materials of desired layer-structure and properties is the main goal of the synthesis, but it may not always be possible to do so. Rational synthesis of 2D materials requires knowledge of layered crystal chemistry besides thermodynamics, phase equilibria and reaction kinetics. It is expected that the easiest way to make solid state materials is the direct reaction of their components at high temperatures, for example with the ceramic method. However, the standard ceramic route needs to be improved to resolve the possible problems, *i.e.* diffusion limit, incomplete reaction, inhomogeneous products, non-kinetic control and non-desired microstructure.[27–30] In recent years, many efforts have been focused on the development of solid-state synthesis for realizing various desired inorganic 2Ds. For example, a variety of inorganic layered materials have been prepared by the traditional chemical vapor transport (CVT) reaction method. This solid-state reaction normally involves mixing and evacuating the powders of desired precursors and heating them at high temperatures with intermediate grinding when necessary. Similarly, chemical vapor deposition (CVD), as another popular synthetic way, is a process commonly used to produce high-purity solid layered materials, in which one or more volatile precursors are reacted or decompose on the surface of the catalytic substrate to produce the desired atomistic deposit at fairly low temperatures. Besides, the solid-state decomposition (SSD) method precipitates a monoatomic elemental layer from the solid substrate. Similar to this, the substitution method involves substituting an appropriate amount of pre-deposited elements with additional elements. The produced crystalline atomic layers on solid surfaces are composed of a single building block and unit cell that is copied and stacked together to form the entire 2D. Most recently, the template-driven reaction (TDR) has also been developed to synthesize multifunctional layered materials with tunable dimensions and morphology. In this reaction, the selected template can serve as support within or around which a target material is generated *in situ* and shaped into a nanostructure with a morphology complementary to that of the template. Last but not least, many new techniques with extreme conductions, including molecular beam epitaxy (MBE), atomic layer deposition (ALD) and microwave-assisted growth have also been developed for synthesizing inorganic 2Ds.

In the sections that follow, we briefly discuss the solid-state synthesis of inorganic 2Ds by the various above-mentioned methods with several examples, paying attention to the growth process and the parameter's effects. While graphene is taken as the typical example in the chapter, other classes of inorganic 2Ds such as phosphorene and borophene, heavy group 14 elemental layers (Si, Ge and Sn), layered chalcogenides and oxides, layered carbides and nitrides are also discussed. The advantages and limitations of different synthetic routes are compared and future prospects are given in the last section.

4.2 Solid State Decomposition (SSD)

The decomposition of any material occurs because of its instability under particular conditions. In the decomposition method, a material is heated to its decomposition temperature to obtain the target material. For example, SiC is used as a substrate to produce graphene; it is heated at a temperature above 1150 °C for decomposition to achieve layered graphene.[31] There are several methods to decompose a material (the decomposition process can be initiated by using different sources), including thermal decomposition, microwave assisted decomposition, photochemical decomposition, and γ-radiation decomposition. The choice of a particular source solely depends upon the target material. The decomposition temperature of a compound material is dictated by its redox properties. For chemical compounds that need a high temperature to decompose, generally a photochemical source is used to decompose at room temperature.

4.2.1 Thermal Decomposition Process for Graphene Growth

Thermal decomposition is considered a good option with the decomposable precursor requiring low decomposition temperature *e.g.* metal oxides. This method is not suitable for those decomposable precursors that need a high decomposition temperature due to the fact that if a material is decomposed at high temperature, its sintering temperature will be higher and will lead to large crystallite size and poor surface area. In the decomposition growth of graphene, silicon carbide substrate is heated to high temperatures. At this high temperature, a process called 'surface reconstruction' is started in which the surface atoms start to relocate themselves in order to achieve the lowest surface free energy.[31,32]

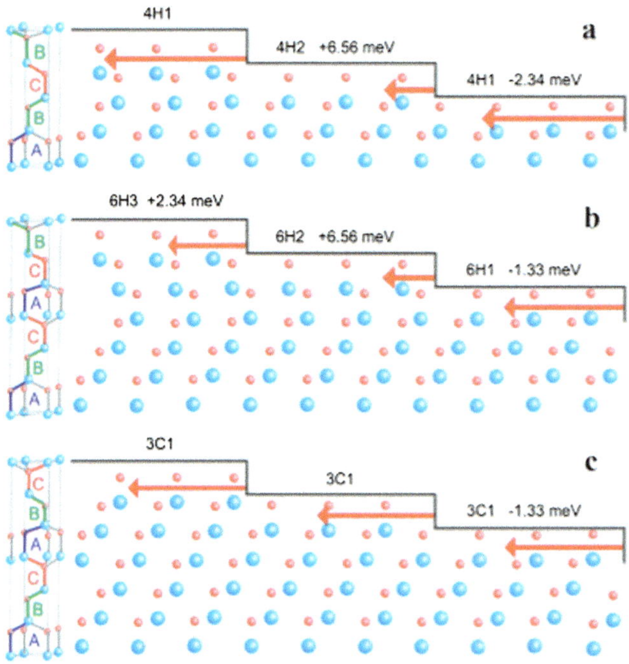

Figure 4.1 During growth, the edges of distinct terrace surfaces of the SiC substrate retract at different velocity due to their specific energy needed for decomposition, resulting in step bunching. (a) 4H-SiC surface, (b) 6H-SiC surface, and (c) 3C-SiC surface. Reprinted from *Carbon*, **57**, G. Reza *et al.*, Growth of large area monolayer graphene on 3C-SiC and a comparison with other SiC polytypes, 477–484, Copyright 2013 with permission from Elsevier.

In particular, SiC starts to decompose at the terrace edges and defects where the binding energy of Si atoms is the lowest. When this happens, small graphene patches start to emerge which eventually grow and merge into a continuous graphene layer. During the process, Si can diffuse through the defects and impurities to cause nucleation site and growth. Such nucleation sites give rise to a bilayer or multilayer graphene. It can be seen in Figure 4.1 that polytype 4H-SiC has two terraces, 6H-SiC has three terraces and 3C-SiC has only one terrace with different decomposition energies.[33]

4.2.2 Parameters Affecting the Growth of Graphene

The quality of graphene produced by SSD method often depends upon many factors; some of them are substrate related (as the substrate is the source of C) and some of them are related to the growth chamber – mainly temperature, pressure, annealing time and presence of inert gases.

Based on the decomposition process, the growth conditions can be improved by controlling the speed of step edge retraction along the surface during their breakage and by using inert gases like Argon (Ar) under ultrahigh vacuum to create an equilibrium condition between sublimation and deposition of Si atoms. The Ar pressure allows a higher process temperature that improves surface kinetics as well as facilitates the formation of long range homogeneously sp^2-bonded graphene network. In addition, pretreatment of the substrate can also enhance the synthesis of graphene. Hydrogen etching of substrate produced high terrace edges during the decomposition growth of graphene, playing a vital role in determining its electronic properties, while pre-annealing of substrate in Ar atmosphere resulted in single layered graphene of sub-nanometer step with nearly isotropic conductivity.[34,35]

Although the decomposition method seems viable for the large-scale production of graphene, the requirements of almost perfect ordered SiC crystal and very high temperature (>1100 °C) and ultrahigh vacuum are the limiting factors for using this method for the large-scale synthesis of graphene. Figure 4.2 shows the growth mechanism of 4H-SiC; we can see that growth of graphene over SiC substrate is not uniform. This is because the bonding energies of Si and C atoms lying near the step edges are very weak, and hence Si atoms can be removed from these areas much more easily than the terraces. The C present in three layers of Si-C bilayers of SiC substrate is enough for the formation of a single layer of graphene. As the decomposition energy in 4H-SiC is less for the 4H1 terrace, its decomposition velocity will be much faster than 4H2. We can see in Figure 4.2 (stage 1) that C atoms are emitted into the terrace when Si atoms are removed from the surface. This is followed by the coalescence and nucleation process in stage 1 and stage 2 to form graphene islands. These graphene islands act as a sink for the successfully released C atoms. As can be seen in Figure 4.2 (stage 2), once the 4H1 terrace is merged to the 4H2 step a new bilayer of SiC is formed, and the two-SiC bilayer height supplies additional C atoms in comparison with the one bi-layer, hence extending the first graphene layer along the step edges. As the process goes on, the source of carbon atoms is increased and hence a second layer of graphene is also formed with the release of some extra carbon atoms (stage 3).[36]

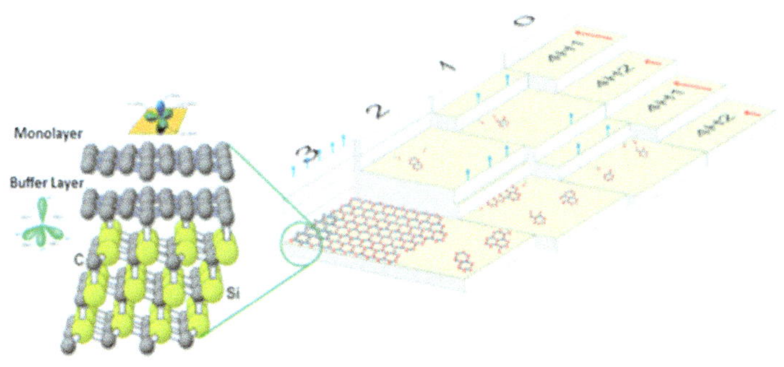

Figure 4.2 Formation of layered graphene *via* decomposition. Reprinted from *Carbon*, **57**, G. Reza *et al.*, Growth of large area monolayer graphene on 3C-SiC and a comparison with other SiC polytypes, 477–484, Copyright 2013 with permission from Elsevier.

4.2.3 The Advantages and Limitations of SSD Method

Since the earliest attempt to fabricate monolayers of graphene in 1975, several methods have been tested to produce high quality graphene to be used for electronic devices, like mechanical exfoliation, which can yield isolated high quality graphene crystals but with a crystal size only of the order of 10 μm, which limits its use in device applications.[37–39] Attempts have been made to grow graphene on the surface of the transition metals Cu, Ni, Pd, Au, or Ru but this method requires the transfer of the graphene sheets/films to an insulating substrate, through methods that are yet to be optimized.[40,41] Another method is to use liquid phase exfoliation that can produce high quality graphene, but the electrical properties of graphene grown by such method has shown poor transport at the contacts between the individual graphene sheets.[42]

So, to date, SSD is extensively used for production of large scale graphene and other related 2D layers. In fact, the SSD growth technique has several advantages:

(a) The samples synthesized by SSD can be directly used for device processing and there is no extra step involved in transfer of film.
(b) The size of film or sheet can virtually be as large as the size of the substrate, which is crucial from a device application point of view.
(c) In the case of graphene growth, SSD is a promising method that can yield high quality graphene with excellent symmetry matching and very interesting physical features, like ballistic transport in nanoribbons,[43] high frequency transistors,[44,45] quantum Hall resistance standard for metrology[46] and half-eV band gap structures.[47]

Although the SSD method is suitable for the large scale production of 2D materials, there are certain limitations associated with this method, such as:

(a) For the precursors that need high temperature to decompose, the SSD method doesn't seem to be a viable option because of the fact that a high decomposition temperature means large crystallite size and a poor surface-to-volume ratio (poor quality of surface).
(b) The SSD method needs ultrahigh vacuum.
(c) The SSD method often requires high quality of substrate, *i.e.* almost perfect ordered SiC crystal for graphene growth.

4.3 Chemical Vapor Transport Reaction (CVT)

4.3.1 The Introduction of CVT

The term chemical vapor transport (CVT) summarizes a variety of reactions that show one common feature: a condensed phase, typically a solid, is volatilized in the presence of a gaseous reactant, the transport agent, and deposited elsewhere, usually in the form of crystals.[48] The study of CVT can be traced back to 1925, when Van Arkel and de Boer used iodine as a transport agent to get more pure metal from the raw metal M (*e.g.*, a metal of the 4th group).[49] Until the 1950s, Schäfer systematically studied and described CVT reactions to demonstrate that the pure and monocrystal of solid state matter could be synthesized by CVT,[50] like metals, alloys, halides, chalcogenides, *etc.*

The transport reactions are usually executed in silica tubes (diameter 10 to 20 mm, length 10 to 20 cm),[50] the tube can be evacuated to the pressure of 5–10 atm or less to reduce the water content.[51,52] The samples and transport agent are transferred to a tube; after this, the tube is evacuated and sealed, then placed at a temperature gradient device in order to set up the gradient in a controlled way, which can be realized by using a two-zone tubular furnace, as shown in Figure 4.3. At the T_1 side, the source materials are transformed into gas state products under the influence of the transport agent, and these gaseous products are deposited on the lower temperature side (T_2) in the form of crystals. The growth of single-crystalline materials by CVT gives a high crystal quality, and can allow the determination of crystal structure by diffraction methods.[48]

4.3.2 Single 2D Materials

For most elementary substances, the CVT method was used to fabricate pure elementary crystals.[53–55] These elements have a common feature: a high melting point and a low vapor pressure. For example, copper has a melting point of 1357.77 K and a vapor pressure of 9×10^{-5} pa, which can be purified by the CVT method; by using iodine as the transport agent, setting the source temperature at 890 °C and the growth zone at 400 °C, and holding this condition for 25h, the pure Cu crystal can be deposited on the low temperature side.

Currently, researchers are using CVT to synthesize Black Phosphorus (BP) crystal.[56–58] BP has been known for about 100 years.[59] In 2014, Li *et al.*[60] found that BP could be exfoliated to 2D material and showed excellent electronic and optical properties. There are many other ways to synthesize BP, like high-pressure synthesis,[59,60] mercury catalysis,[61] recrystallization from Bi,[62] *etc.* In contrast to these methods, CVT is much more convenient and less toxic.

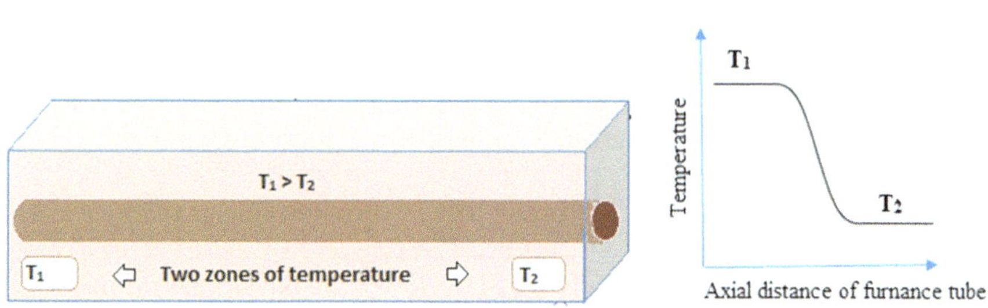

Figure 4.3 The device of a two-zone furnace for the CVT method.

BP was first prepared under low pressure (10^{-5} atm) at 873 K by Nilges *et al.*[56] In 2007, they used red phosphorus as the raw material, SnI_4 as the transport agent, and gold and tin as the additional additive materials; these materials were evacuated and sealed in a silica tube, and heated to 873 K for 5–10 days. The CVT method took a very long time and introduced byproducts like Au_3SnP_7 and AuSn. In 2008, this method was improved to prepare the BP crystals in a short time;[57] red phosphorus was used as the raw material, SnI_4 as the transport agent, and AuSn as the additional additive material. The controlled temperature, and reduced reaction time of about 2–3 days, still yielded BP of a very good crystal quality. The by-products can also be detected, similarly to when using the first method. Until 2014, Nilges's group found that it is not necessary to add Au to the reactants.[58] They removed the Au from the additives meaning that only Sn and iodine from the SnI_4 had the potential to be by-products. The CVT device used was still a one-zone furnace, which was heated to 655 °C at speed of 200 K h^{-1}, then the temperature was maintained for 30 min before the furnace was cooled down to 550 °C in 7.5 h, and held at that temperature for 1 h. Figure 4.4 shows the optical picture of the final product, and the XRD and Raman tests proved the excellent crystal quality of this CVT grown BP crystal.

4.3.2.1 Binary 2D Materials

Since the discovery of graphene, many layered material like hexagonal boron, black phosphorus and transition metal dichalcogenides (TMDs) have been the focus of intense study by scientists.[60,63–66] Among these 2D materials, TMDs have received a lot of attention for their tunable electronic properties and various band gaps.[67,68] Among these, binary TMDs have the form of MX_2, which M is transition metal (Ti, Zr, Hf, V, Nb, Ta, Mo, W, *etc.*) and X is the chalcogenide (S, Se, Te).[69] These materials could be exfoliated down to a two dimensional structure (several layers or monolayer) which shows very different optical and electronic properties from the bulk crystals, and have great applications in optical and electronic devices.[70–74] Here, we will present some typical ways to prepare the bulk crystals of binary TMDs by CVT methods.

For CVD growth of the single crystal MoS_2, Rajiv *et al.*[75] used molybdenum powder and sulfur powder (the molar rate Mo : S = 1 : 2) as the source materials. First, materials were sealed in an evacuated quartz ampoule, then this ampoule was heated to 700 °C for 36 h. Second, this synthetic compound was transferred into another ampoule and added with the iodine. The ampoule was evacuated and sealed again and placed in the two-zone furnace with the high temperature for the reaction zone and low temperature for the crystal growth.

This method also worked for other MX_2 if the suitable transport agent and temperature conditions were found. The typical process of synthesizing MX_2 by CVT is shown in Figure 4.5. Keiji Ueno's group summarized this kind of method and synthesized MoS_2, $MoSe_2$, $MoTe_2$, WS_2, WSe_2, WTe_2. Some results are presented in Figure 4.5.[76]

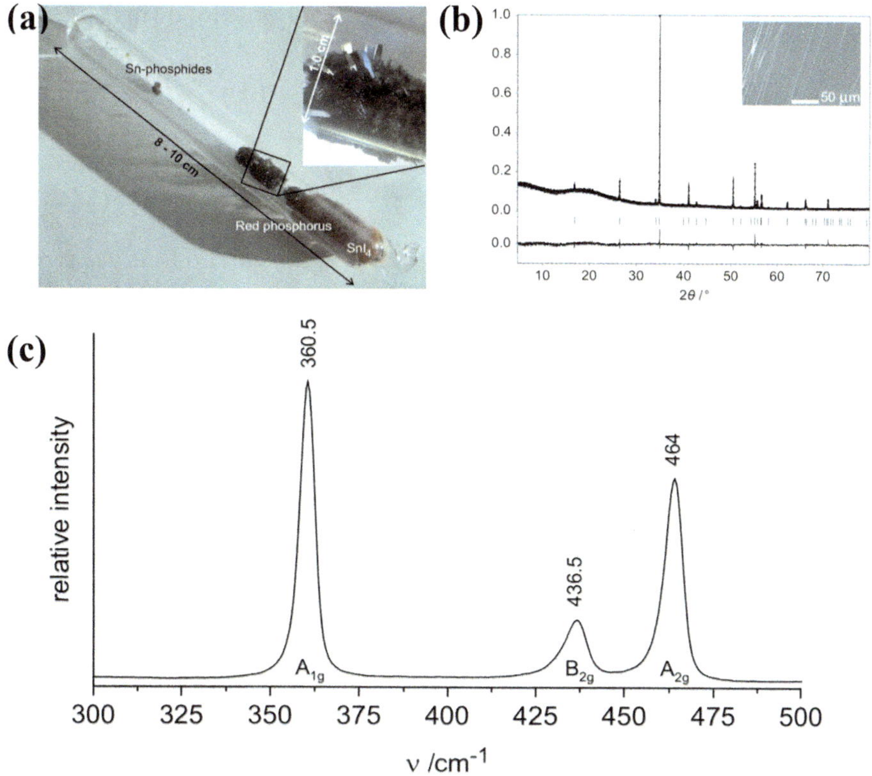

Figure 4.4 (a) The optical picture of reaction products; (b) the XRD of BP; (c) the Raman spectrum of black phosphorus. Reprinted from *Journal of Crystal Growth*, **405**, M. Köpf *et al.*, Access and *in-situ* growth of phosphorene-precursor black phosphorus, 6–10. Copyright 2014 with permission from Elsevier.

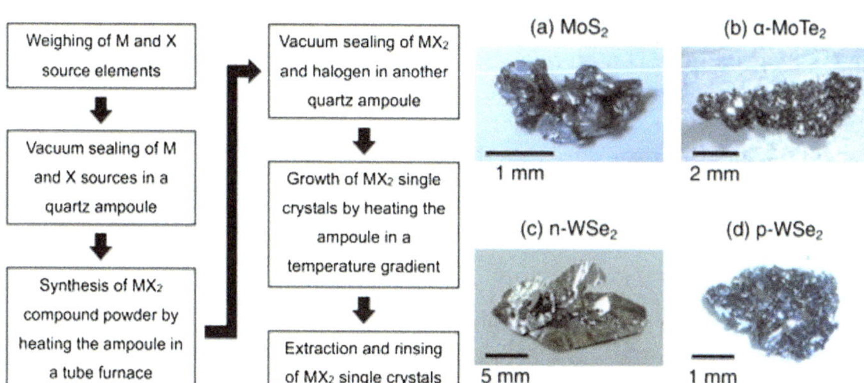

Figure 4.5 Left: The typical process of synthesizing MX2 by CVT. Right: (a)–(d) Some photographs of MX2 crystals. Reproduced with the permission of *J. Phy. Soc. Japn.*, 2015, **84**, 121015. Copyright (2015), The Physical Society of Japan.

This kind of transport reaction can be described by the following equation:

$$MX_2(s) + B_2(g) \underset{T1}{\overset{T2}{\rightleftarrows}} MB_2(g) + \frac{1}{2}X_2(g) \quad (B = I, Br)$$

4.3.2.2 *Ternary 2D Materials*

The family of MPX_3 (M = Fe, Mn, Co, Ni; X = S, Se) compounds are typical ternary layered materials;[77,78] the interactions between the layers are van der Waals forces and the calculated cleavage energies of most MPX_3 is smaller than that of graphite,[79] which means that MPX_3 can be exfoliated to atomic-scale. The magnetic behavior of $FePS_3$, $MnPS_3$, $NiPS_3$ has already been revealed by Joy and Vasudevan,[80] but with the rise of 2D materials, it is very exciting to study this kind of layered materials at atomic level. Lee *et al.*[81] have explored the magnetism of atomically thin $FePS_3$, and found that $FePS_3$ still has Ising-type antiferromagnetic ordering, even down to the monolayer. Li *et al.*[82] calculated the physical properties of 2D $MnPSe_3$ with carrier doping, and showed that the doping could induce the transition from antiferromagnetic semiconductor to ferromagnetic half-metal, which might have applications in spintronic devices.

Du *et al.*[79] proved that CVT is an effective way to synthesize ternary layered materials; they obtained many good quality crystals like $FePS_3$, $MnPS_3$, $NiPS_3$, $CoPS_3$, $FePSe_3$, $MnPSe_3$, *etc.* The CVT method used here is almost same as the method to synthesize binary materials; the source materials with a mole rate M : P : chalcogen = 1 : 1 : 3 (total weight around 1 g) and iodine (10–20 mg) as the transport agent were evacuated and sealed into a quartz ampoule, then this ampoule was placed into a two zone furnace with a high temperature of 650 °C and a low temperature of 600 °C, and the reaction lasted for a week. The synthesized bulk crystals are shown in Figure 4.6, which suggests that the $FePS_3$ can be exfoliated to monolayer or few-layer-thick samples by a mechanical cleavage method on SiO_2/Si.

The compound 2D materials are difficult to synthesize by other methods; the CVT method provides enclosed spaces with low pressure ($<10^{-5}$ atm), which can isolate the source materials from the atmospheric environment. The long reaction time can make sure that these source materials can sufficiently interact with each other, which will result in the formation of a good quality crystal.

4.3.2.3 *Doping/Hybridizing of 2D Materials*

In case of semiconductors, doping can result in n-type or p-type semiconductors that have different electronic properties and are used in many semiconductor devices. By using the CVT method, it is convenient to dope 2D materials with metal ions, which might change the electronic, optical and magnetic properties of the pure 2D materials.

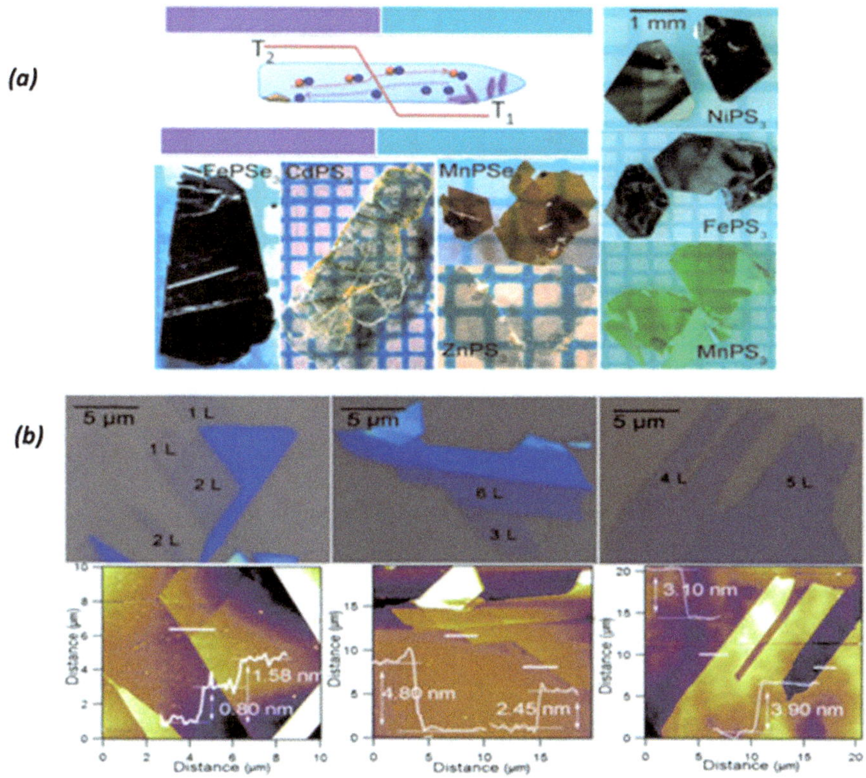

Figure 4.6 (a) Photos of bulk 2D crystals. (b) Optical microscopy photograph and the corresponding AFM topography of FePS3 from monolayer to 6 layers. Reprinted with permission from *ACS Nano*, 2015, **10**, 1738. Copyright (2015) American Chemical Society.

Lin *et al.*[83] used the CVT method to dope MoS_2 with Au and Re, and presented *in-situ* observation of the dopants migration progress; they also modeled the Re-doped MoS_2 by the DFT calculations and concluded that the doping sites can enhance local areas, and chemical reactivity, which will provide catalytically active sites. By using the same kind of method, they conducted another experiment to *in-situ* observe the structure transformation in single layered Re-doped MoS_2.[84] Doping can change the native unipolar propensity of most TMDs materials, and this strategy can produce many n-type or p-type 2D materials, which can be active materials to fabricate various devices.[85] Suh *et al.*[85] synthesized the Nb-doped MoS_2 crystal by CVT method: the source materials were Mo, S, Nb (molar ratio = 1 : 2 : 0.005), the transport agent was iodine, the temperature of reaction zone was 1050 °C and the temperature of the growth zone was 935 °C; the setup is shown in Figure 4.7. These conditions were held over 500 h, which could provide

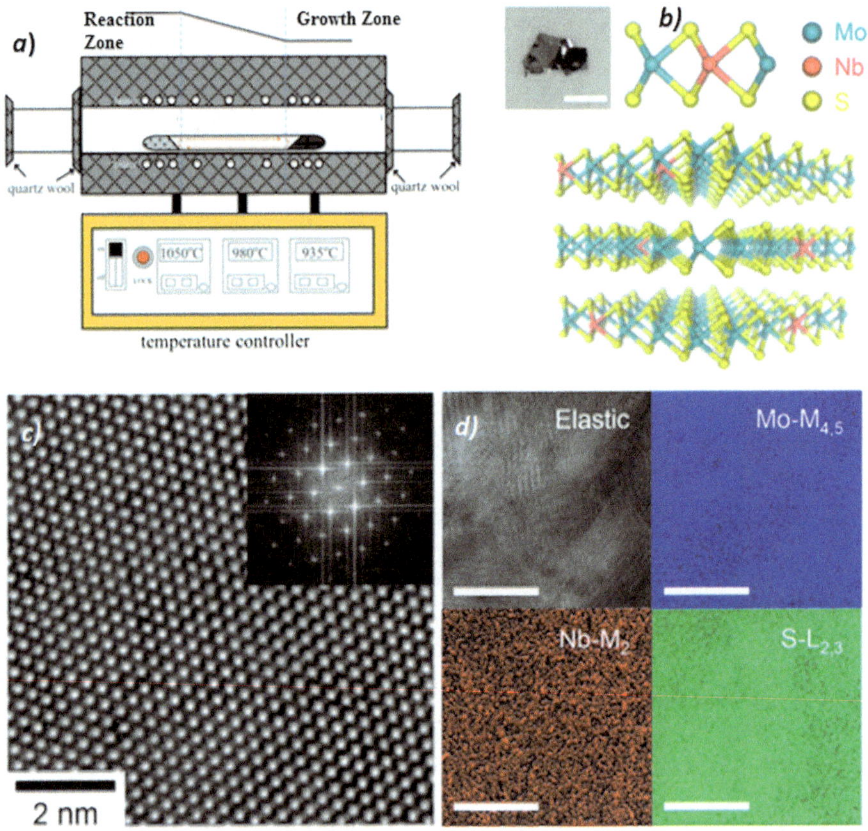

Figure 4.7 CVT synthesis of niobium-doped MoS_2 crystals. Reprinted with permission from *Nano Lett.*, 2014, **14**, 6976. Copyright (2014), American Chemical Society.

sufficient reaction for synthesizing the Nb-doped MoS_2 crystal. The structure and chemical analyses of Nb-doped MoS_2 is presented in Figure 4.7; a vertical stack of Nb-doped MoS_2/MoS_2 formed a p–n junction device, and *I*–*V* characteristics of this device also exhibit excellent rectification characteristics.

In a 2D material system, alloying two materials with different band gaps is an exciting aspect, and in this way some tunable properties (like band gap, conductivity) can be achieved in the 2D alloys. Chen *et al.*[86] have synthesized $Mo_{1-x}W_xS_2$ crystals by CVT, and obtained $Mo_{1-x}W_xS_2$ monolayers by mechanical exfoliation; the photoluminescence characterization showed that this alloy can achieve a tunable band gap emission from 1.82 eV ($x = 0.2$) to 1.99 eV ($x = 1$). Dumcenco *et al.*[87] used annular dark-field (ADF) scanning transmission electron microscopy (STEM) to get a direct view of the atomic species Mo and W in $Mo_{1-x}W_xS_2$ compounds and proved that Mo and W can be mixed uniformly in a single layer. By using the CVT method, alloys like $Mo_{1-x}W_xSe_2$,[88] $Mo_xW_{1-x}Te_2$[89] with tunable chemical composition have been realized and the specific properties of these 2D alloys materials have been studied. Recently, Yu *et al.*[90] have synthesized a single crystals of $WSe_{2(1-x)}Te_{2x}$ (x = 0–1.0) alloys by CVT. The stoichiometric amounts of W, Se, Te were added into a quartz tube as source materials (total weight about 0.3 g), 40 mg iodide was added as a transport agent. After evacuating and sealing, this tube was placed in three-zone furnace, with the reaction zone temperature at 850 °C for 30 h, and the growth zone temperature at 900 °C, which can prevent transport progress. When the pretreatment was over, the reaction zone and growth zone temperatures were readjusted to 1010 °C and 900 °C, respectively, for a reaction time of 3 d. This experiment proved the phase transition of TMD monolayer from semiconductor to metal, and indicated that the CVT is a very important tool to prepare the crystal of 2D alloys.

4.3.3 Effects of Growth Parameters

CVT experiments are always conducted in a closed system, usually a sealed quartz tube, which has a length of 10–20 cm, and a diameter of 10–20 mm (there is another kind of CVT with material transport in open systems, which is not discussed here). When employing CVT for synthesis of 2D materials, the first thing one should consider is whether this material can be prepared by CVT reactions or not. If CVT growth of a certain material is possible, then a suitable source and transport agent should be selected, and the last things one should decide is the temperature (including the source and growth zone temperature, and the temperature gradient) and the time. There are certain parameters that affect the growth and outcome of the CVT synthesis; some of the key factors are discussed in the following section.

4.3.3.1 Transport Agents and Temperature

A suitable transport agent can transfer all solid materials into gas phase, and form solid material in the growth zone.[50] In general, the chlorine, bromine, iodine and halogen compounds (like HX, X = Cl, Br, I) are suitable as transport agents. Transport agents can affect the CVT reaction, which is described by the following equation:[48]

$$iA(\mathrm{s}) + kB(\mathrm{g}) \underset{T1}{\overset{T2}{\rightleftarrows}} jC(\mathrm{g}) + \ldots$$

B(g) is the transport agent. The suitable average temperature $T = (T_1 + T_2)/2$, which can result from the following equation with $\Delta_r G^0 \approx 0$:

$$\Delta_r G_T{}^0 = \Delta_r H_T{}^0 - T\Delta_r S_T{}^0$$

T can be described as:

$$T = \frac{\Delta_r H_T{}^0}{\Delta_r S_T{}^0}$$

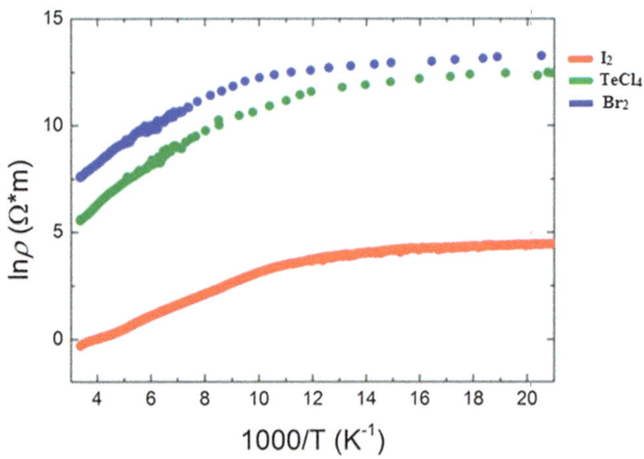

Figure 4.8 Change of resistivity with temperature in MoS_2 single crystals by using different transport agents. Reprinted with permission from *J. Phys. Chem. C*, 2015, **119**, 3918. Copyright (2015), American Chemical Society.

For a reaction described by the above mentioned equations, the equilibrium constant K is expected to be in the range of 10^{-4} to 10^{4}, and the Gibbs energies $\Delta_r G_T^{0}$ in the range −100 to 100 kJ mol^{-1}.[48]

The CVT method might cause doping in the grown crystal by the transport agents. Pisoni *et al.* have measured the conductivity of MoS_2 crystals which were grown by the CVT method using different transport agents: I_2, Br_2 and $TeCl_4$.[91] The resistivity results of different transport agent synthesized MoS_2 are shown in Figure 4.8. Measurements clearly reveal the presence of defects and inclusions inside the material, and the transport agent played an important role in the doping effect.

4.3.3.2 *The Rate of Mass Transport*

The transport rate means the amount of the crystal deposited at growth zone per time. The transport rate can be described by Schäfer's transport equation under a low total pressure (10^{-3} to 3 bar).[50,92]

$$n(\mathrm{A}) = \frac{n(\mathrm{A})}{t'} = \frac{i}{j}\cdot\frac{\Delta p(\mathrm{c})}{\Sigma p}\cdot\frac{T^{0.75}\cdot q}{s}\cdot 0.6\times 10^{-4}\ (\mathrm{mol\ L^{-1}})$$

n(A) Transport rate (mol h^{-1})
t' duration of the transport experiments (h)
i,j stoichiometric coefficients in the transport equation
Δp(c) partial pressure difference of the transport-effective species C (bar)
Σp total pressure (bar)
T average of T_1 and T_2 (K)
q cross section of the diffusion path (cm^2)
s length of the diffusion path (cm)

4.4 Chemical Vapor Deposition (CVD)

Chemical vapor deposition is the term used to describe heterogeneous reactions in which both solid and volatile products are formed from a volatile precursor through chemical reactions, and the solid products are deposited on a substrate. Development of the CVD method dates back to the 1960s for the production of carbon fiber, and since its inception, this route is considered one of the most viable manufacturing way for various types in industrial as well as academic research sectors.[93–99] The CVD method enables the growth of a thin solid film on a substrate *via* chemical reaction of precursor gases (carrier gases). A typical system for the CVD growth is shown in Figure 4.9.

For CVD growth, the precursor gases are supplied to the reaction chamber or heating zone at a predefined temperature. A substrate is placed inside the reaction chamber, and when these precursor gases are supplied to this chamber, these gases react and form a thin layer over the substrate. Many parameters can influence the synthesized material including substrates, phase of source (gas, liquid, solid), pressure, temperature, and so on. Among them, pressure and temperature are the most important parameters that can dictate the quality and characteristics of the grown films. In the last decade, CVD was very popular in the growth of carbon nanomaterials,[100,101] and recently has been further utilized to synthesis 2D materials, including graphene, h-BN, transition metal dichalcogenides (TMDs), heterostructures *etc.*

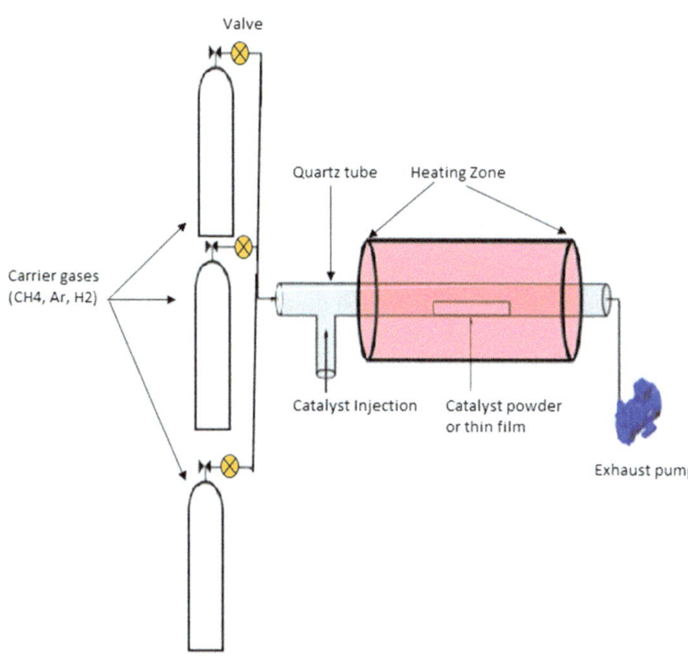

Figure 4.9 A typical CVD system.

4.4.1 CVD Growth of Graphene

The epitaxial growth of graphene *via* CVD has been intensively studied with the goal of high potential applications in the field of electronics and devices. Usually, graphene is grown on metallic substrates in a CVD system. Among metal substrates, copper is the most promising as it provides an opportunity to synthesize monolayer graphene.[102–104]

In a typical growth mechanism of graphene film on copper foil by using CVD, methane is used as a source of carbon and the growth is carried out in a three-zone high temperature furnace with a tubular vacuum CVD reactor. The synthesis is often performed in atmospheric pressure or low vacuum maintaining a characteristic temperature of 800–1000 °C and the mixture of hydrogen, methane and argon gases is used for 10–30 min. The heat anneals the copper and hence increases the domain size. After the growth of graphene film, this film is transferred to the fluorine-doped tin oxide (FTO) or SiO_2 substrate and the primary substrate (copper foil) is etched out. This transfer process is a crucial step as it can considerably affect the characteristics of the transferred film of graphene.[105–108] Figure 4.10 shows the characterization results of the graphene film grown on copper and then transferred to the SiO_2 and glass substrate. Grain boundary and layered graphene flakes can be observed as well as wrinkles of graphene (Figure 4.10B). Folded graphene edges of single and two layers can also be witnessed in the TEM image in Figure Figure 4.10B.[107]

Apart from copper, several other transition metals like nickel and cobalt are also used as substrates in CVD.[109] All these metal substrates differ in their ability to absorb carbon. The absorbency of carbon in Ni and Co is more when compared to Cu, which limits their use as substrate in comparison with the copper substrate. In the case of Ni or Co substrate, because of their large absorbency of carbon, excess of carbon on the substrate foil results in discrete graphene islands instead of a single-layer graphene.[110] However, in the case of Cu, carbon can only be absorbed at the surface of the foil but not into the bulk of material. To absorb carbon, vacant bonding sites are needed to establish weak bonds, to hold carbon that are only available at the surface of Cu foil. This property of copper makes it an attractive and the most common option to be used in CVD as metal substrate.[107] The quality of CVD graphene can be adjusted by changing the composition and alloy of metal substrate. For examples, few-layer graphene with high crystallinity was synthesized on poly-Ni.[111] Single-crystal graphene of one centimeter in size was successfully obtained on Cu–Ni alloy and the Hall mobility at room temperature was between 10 000 and 20 000 $cm^2\ V^{-1}\ s^{-1}$.[112] Using methane as a carbon source, graphene was also grown on h-BN by plasma-enhanced CVD.[113] Besides, both solid and liquid carbon sources have been used to grow graphene. For example, Zeng *et al.* reported graphene growth at very low temperature using PMMA, polystyrene and benzene.[114] Graphene was reported to grow on Cu (111) by decomposing ethylene, as shown in Figure 4.10.[115] Moreover, liquid metals can also be explored to grow strictly one-layer graphene.[116]

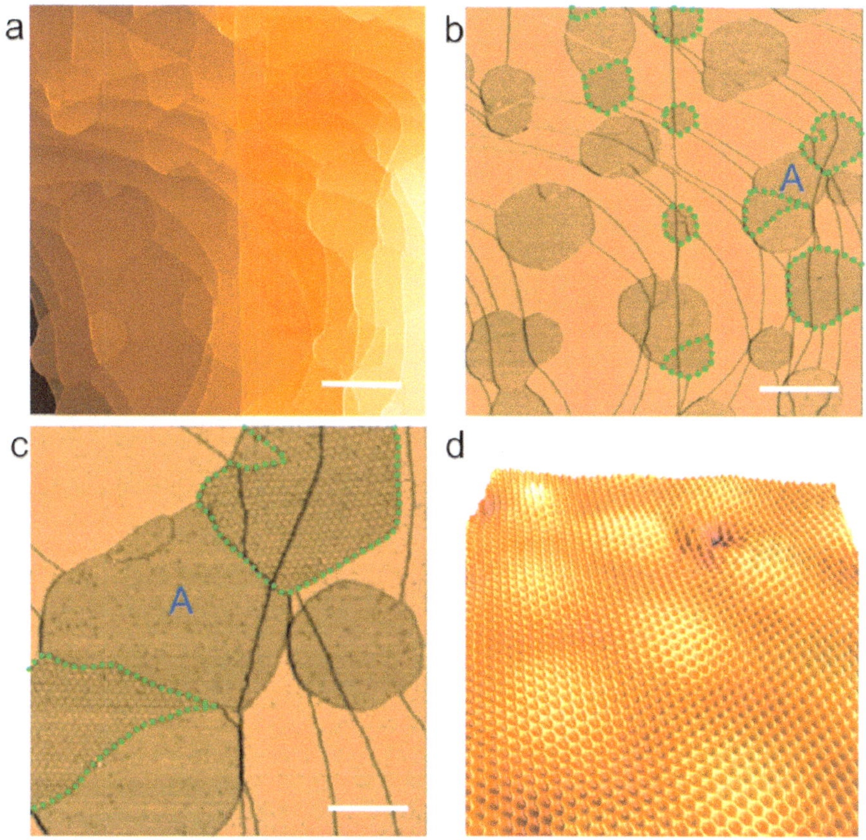

Figure 4.10 STM analysis of CVD-grown graphene islands on Cu(111) *via* thermal decomposition of ethylene. Reprinted with permission from *Nano Lett.*, 2010, **10**, 3512. Copyright (2010), American Chemical Society.

4.4.2 CVD Growth of Transition Metal Dichalcogenides (TMDs)

MoS_2 is one of the most promising TMDs especially for its application in nano-electronics. CVD can be used to grow large area MoS_2 sheets as well. Although epitaxial growth of MoS_2 in ultrahigh vacuum using highly crystalline Au template has been reported,[117] highly crystalline MoS_2 films of several millimeters in size can successfully be grown by CVD on SiO_2 substrate without using high vacuum.[118,119]

For the growth of MoS_2, SiO_2/Si is used as the most common substrate. Lou *et al.* reported that a Mo layer on SiO_2/Si deposited by electron-beam evaporation serves as a Mo source. In the growth procedure, the Mo deposited SiO_2/Si substrate was kept in the center of the reaction chamber and pure sulfur was placed in the low temperature zone of the reaction chamber with the temperature just above 113 °C. N_2 purging was performed for

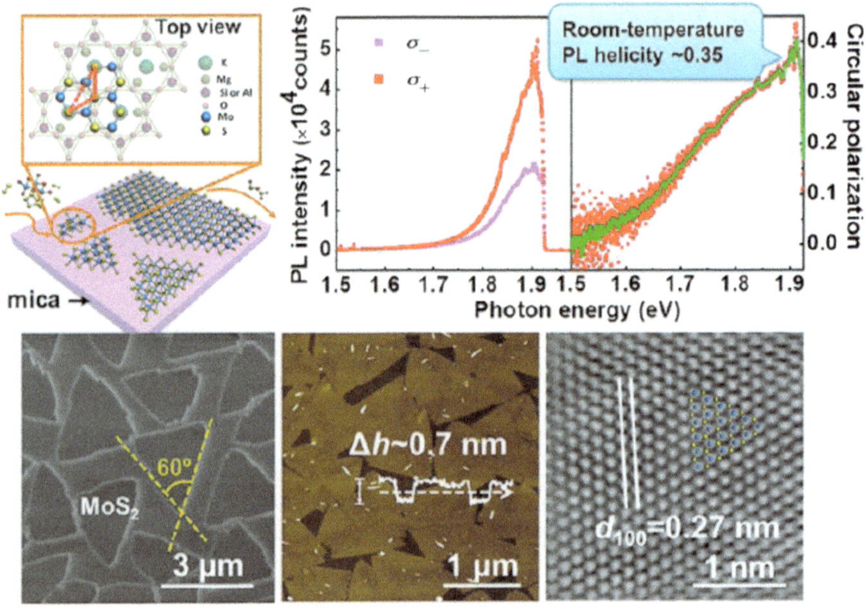

Figure 4.11 Epitaxial monolayer MoS_2 on Mica substrate *via* CVD method. Reprinted with permission from *Nano Lett.*, 2013, **13**, 3870. Copyright (2013), American Chemical Society.

15 min at the flow rate of 150–200 sccm and subsequently the temperature was raised firstly from room temperature to 500 °C for 30 min and then from 500 °C to 750 °C in the next 90 min. This temperature was maintained for the next 10 min before starting the cooling process to bring the furnace back to room temperature within 120 min. The domain size was about 20 nm, which means that the density of grain boundaries is high, thus the mobility stayed at a low level (0.004–0.04 cm^2 V^{-1} s^{-1}).[118] To get a high quality of monolayer MoS_2, many attempts were made and much progress has been made. Other substrates including graphene,[120] mica (as shown in Figure 4.11),[121] and sapphire[122] can also serve as good supporting materials to synthesise MoS_2 and explore its properties.

In another approach for the synthesis of MoS_2, the substrate was firstly prepared by spinning a droplet of aqueous reduced graphene oxide (RGO) on the substrate and then drying the substrate at 50 °C. After preparation of substrate, 0.4 g of MoO_3 powder was placed in a ceramic boat with SiO_2 substrate placed on top of it facing down to the MoO_3 powder. Another ceramic boat loaded with 0.8 g of sulfur powder was placed alongside the first boat. Nitrogen was used as carrier gas and the temperature of the reaction zone was maintained at 650 °C, which caused the reduction of MoO_3 powder by sulfur vapors to form MoO_{3-x} suboxide.[123] The growth of MoS_2 film was the result of diffusion of these suboxides on the substrate and their subsequent reaction with sulfur. As the density of seeds gets higher, the star shaped MoS_2

can be amalgamated to form a continuous film of MoS_2. It is evident that the layered structure of MoS_2 has been created, manifested by the smooth surface morphology, showing the layer thickness of grown MoS_2 layers.

4.4.3 CVD Growth of 2D Hetrostructures

2D materials possess layered structures, hence it is possible to fabricate vertical or lateral heterostructures. CVD can be used to grow one 2D material on top of another to form a vertically stacked hetrostructure. In the presence of weak van der Waals forces, the significant task is to overcome the lattice mismatch. For example, Yang *et al.* reported this kind of lattice mismatch problem in the growth of graphene on h-BN substrate.[124] CVD has also been used to grow heterostructures where epitaxial graphene grown on SiC was used to grow MoS_2 and WS_2 as well as h-BN.[125] In this case, the surface morphology of the underlying graphene layers plays a pivotal role in deciding the properties of the grown heterostructure. Any wrinkle or defect present on the surface of graphene can act as a nucleation site for the growth of material.

Recently, the CVD growth of TMDs heterostructures has also gained attention among researchers. In this regard, WS_2/MoS_2 heterostructures with 2H stacking structures have been grown by using CVD.[127] The synthesis procedure for WS_2/MoS_2 heterostructures is demonstrated in Figure 4.12; SiO_2/Si has been used as substrate and tungsten powder is placed on this substrate

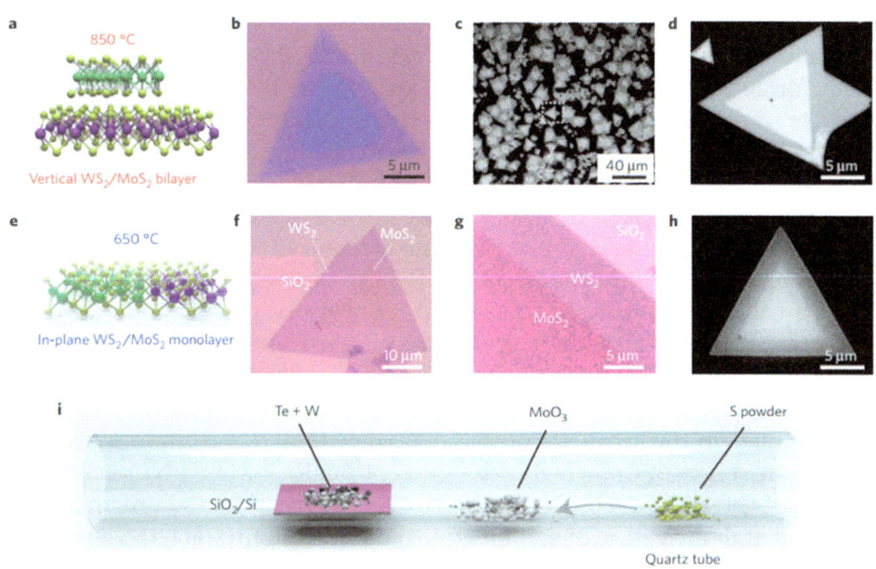

Figure 4.12 Schematic of the synthesis and the overall morphologies of the vertically stacked and in-plane WS_2/MoS_2 heterostructures. Reprinted by permission from Macmillan Publishers Ltd: *Nature Materials* (ref. 127), Copyright 2014.

along with a small quantity of Te powder that helps the melting process of the tungsten. MoO_3 and sulphur powder is also placed inside the reaction chamber as shown in Figure 4.12 keeping the sulphur powder to the low temperature zone of the reaction chamber. Argon is used as carrier gas in this synthesis. As the nucleation and growth rates of WS_2 and MoS_2 are different, it results in their sequential growth rather than formation of a $Mo_xW_{1-x}S_2$ alloy. The structure of the synthesized heterostructure depends upon the reaction temperature. If the temperature is of the order of 850 °C, vertically stacked bilayers are preferred, as is visible schematically in Figure 4.12. Through SEM and optical images, we can see the formation of bilayers of WS_2/MoS_2 triangular heterostructures. If the synthesis temperature is ~650 °C then growth of in-plane lateral heterostructures is dominant. This effect can be observed schematically. One can see the abrupt and sharp change in the color contrast at the interface of the layers. In both the schematics, green, purple and yellow spheres represent W, Mo and S atoms, respectively. Similarly, direct *ex-situ* growth of MoS_2/WS_2/graphene and WS_2/MoS_2/graphene heterostructures by MOCVD has also been reported.[128]

4.4.4 Factors Affecting the Growth in CVD

There are several factors that affect the growth of 2D material in CVD. These factors are important for synthesis of any 2D material in CVD, but we have used the growth of graphene by CVD to emphasize the effect of different parameters and factors on CVD. Some of the key factors that influence the growth of material by using CVD are listed here.

First of all, temperature plays a pivotal role in CVD growth of graphene, especially for defining thickness and film coverage. The temperature also affects the rate of reaction, as does the rate of changes in temperature. It has been reported that the thickness of graphene layer is highly dependent on temperature. At low temperatures (~900–1000 °C) bilayer and few layer graphene is preferred. This is because of the fact that the growth rate of carbon decreases at a lower temperature hence more methane is required to meet the continuous growth of graphene films that results in thicker graphene films.[129–141]

Meanwhile, the substrate used for the growth of the 2D material plays an important role too. There are variety of substrates that are being used as a substrate for the synthesis of graphene, *e.g.* TMDs, alloys like Ni–Au, dielectric substrates like SiO_2 *etc*. Although TMDs like Ni, Co are also used as substrates in the synthesis of graphene they tend to absorb more carbon in comparison with Cu, which shows exceptionally lower carbon solubility. So, use of Cu substrate prohibits the surface segregation and enhances the diffusion and nucleation of carbon atoms. Apart from this fact, the lower solubility of carbon in Cu also means that the film thickness will be much less, which enables the growth of monolayers of graphene as well.[142] Despite all these advantages of using Cu as substrate, these graphene films exhibit poor electron mobility if compared with the film grown using SiC, which may be because of the requirement of chemical etching of Cu before transfer. So, the choice of substrate has primary significance in CVD.[143]

Besides, precursors are basically the source of carbon for the growth of graphene by CVD. A precursor should be volatile but stable enough to be able to be delivered to the reactor. It can be in gas, liquid or solid form. For example, in the case of gas, the amount and type of gases used in CVD process play a major role in deciding the outcome of the synthesis. Methane has been extensively used as the precursor for the growth of graphene. Excessive use of methane gas results in availability of more carbon atoms, which means there will be an increased number of sites for nucleation and an increased number of domains. All this means that the process will yield an increased number of carbon atoms available for deposition. On the other hand, increased use of hydrogen gas stimulates the reaction and ensures increased chemical processes on the copper substrate. The key parameter in deciding the precursor for the growth of graphene is the amount of energy that is needed during the process. This requirement of energy depends upon the type of precursor.[144–146] This energy requirement can be met either by using a C–H bond energy precursor or by using a catalyst. A crucial parameter that should be considered while choosing a precursor is the dehydrogenation energy of the (CH_x to CH_{x-1}) precursor.

4.4.5 The Advantages and Limitations of CVD Method

Salient advantages of using CVD for the synthesis of 2D materials are listed below:

(a) CVD is a simple and low cost method.
(b) CVD synthesis occurs at relatively lower temperature ranges compared to the solid state decomposition method (SSD).
(c) The CVD process offers good control over the morphology of 2D materials.
(d) In the case of graphene, a film can be grown that is virtually as large as the size of the substrate.
(e) Temperature and pressure can be controlled effectively.

Apart from the advantages of CVD, there are certain limitations of this method too:

(a) Material synthesized by CVD contain impurities because of the fact that many materials are used to carry out CVD synthesis.
(b) A typical CVD process involves exfoliation of the deposited film after growth, which is a major limitation of this method because of its complexity and due to the fact that this exfoliation process has major effects on the parameters of transferred film.
(c) In case of graphene, the films deposited by CVD are polycrystalline and they suffer defects of grain boundaries.
(d) In case of graphene, wrinkles may appear at the surface of graphene due to the difference of thermal expansion of graphene and substrate *i.e.* copper.

4.5 Template Driven Growth (TDG)

Template driven growth is a route to realize the synthesis of solid state materials *via* various templates.[147,148] In this process, template can serve as support within or around which a target material is generated *in-situ* and shaped into a nanostructure with morphology complementary to that of template. Post-synthesis treatment is needed in this approach to remove the template to harvest the target material. There are several methods of removing the template that involve chemical etching and calcination. This approach can yield nanostructures with high aspect ratio and large area that is difficult to realize using contemporary lithographic techniques.

Since the last decade, template confined growth has found wide application in the synthesis of nanomaterials because of its suitability to achieve tunable dimensions and morphology. This method has been extensively used in the synthesis of carbon nanotubes (CNTs) by using an anodic aluminum oxide (AAO) template. It is possible to alter the morphology of CNTs merely by changing the anodization of the template. As expected, the template synthesis can provide the luxury of crystal growth confined in a specific dimension. Thus, many efforts have been focused on the template driven growth method to realize 2D materials. For example, this method is found to eradicate the problem encountered in controlling the pore size distribution especially in the case of 2D porous graphene. Although template-free methods are generally less complicated, the control of the pore size distribution within the 3D porous graphene is possible, but accurate control is challenging. Interestingly, this issue can be successfully resolved by the template method with optimized design.[149,150]

4.5.1 Types of Templates

Two types of template are used in this method of growth:

(a) Hard Template
(b) Soft Template

Both hard templates (silica and polystyrene spheres or metal foams) and soft templates (polymeric or amphiphilic compounds) are widely used to ensure uniform pore size. Both hard and soft templates ensure uniform pore creation based on engineered templates that are easily removable after synthesis.[151–161]

In hard template synthesis, the precursor either grows on the surface of the template or it fills the voids present within the template. When the template is removed at the end, a replica structure of template is obtained. For example, template assisted growth has been reported in the case of TMOs as well, for example a half unit cell of α-Fe_2O_3 semiconductor nanosheet containing only two Fe ion layers and having strong intrinsic ferromagnetism of 0.6 μB saturation magnetic moment at 100 K. Particularly, the synthesis of α-Fe_2O_3 was carried out *via* a simple template assisted growth mechanism by using a CuO template followed by hydrothermal process. The preliminary seed for the nanosheets nucleation at the surface of template occurs because of the interfacial reactions of Fe^{2+} hydrolysis. After that, the subsequent absorbance of Fe^{2+} ions prompts the oriented growth at the interface of assisted template by the jointly promoted slow interfacial reaction of hydrolysis of $Fe^{2+}$and etching of CuO template at a low temperature of 25 °C for one day. Once etching of CuO is completed, stable and freestanding α-Fe_2O_3 nanosheets are formed by dehydrogenation of Fe hydroxide nanosheets by heat treatment, as shown in Figure 4.13.[162]

In the case of soft template synthesis, the templates do not have a solid structure like those normally used in hard template synthesis, but they can ensure the self-assembly of the precursor in liquid phase. This method also requires the removal of template to acquire the required material.[163,164] Carbon materials are widely used as soft templates for porous carbon and carbon nanotubes CNTs for producing functional materials, metals and metal oxides.[165–170]

In template driven growth, the main affecting parameters are following:

1. Anodization of template (*e.g.* in case of CNTs, it can dictate the morphology).
2. Temperature of the reactor (plays a pivotal role in determining the grain size as well as the surface morphology of the material).
3. Introduction of a suitable catalyst (*e.g.* Ni in case of CNTs) as it acts as a seed for the intended growth.

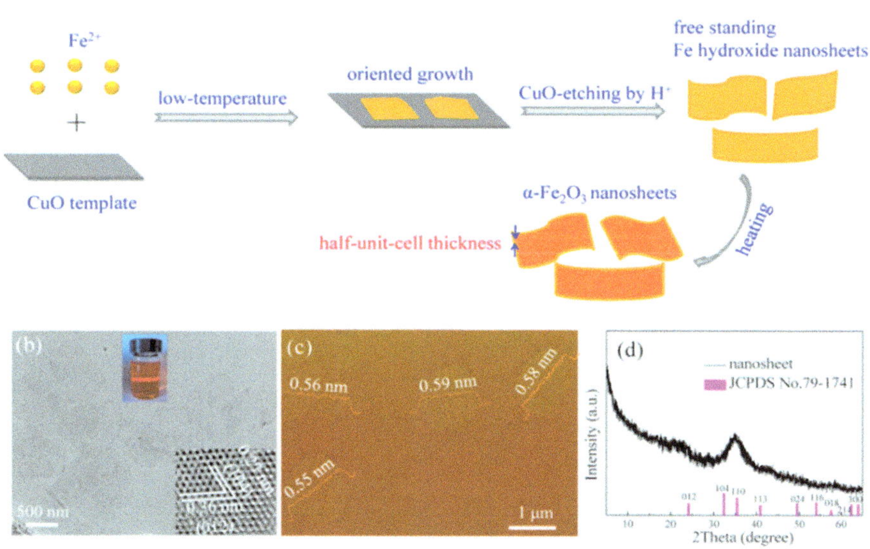

Figure 4.13 Schematic diagram depicting the synthesis, morphology and structure of half unit cell α-Fe_2O_3 nanosheets. (b) TEM image of α-Fe_2O_3 nanosheets. (c) AFM image. (d) XRD of α-Fe_2O_3 nanosheets. Reprinted with permission from *J. Am. Chem. Soc.*, 2014, **136**, 10393. Copyright (2014), American Chemical Society.

4.5.2 Graphene-based Template Growth

Contrary to the traditional soft and hard templates, graphene and its derivatives are considered a novel (2D) template with unique properties. As we know, graphene tends to randomly restack due to van der Waals attraction. In the case of layered stacking, an interlayer nanospace is formed that can be used for the confined reaction and oriented growth of 2D materials. This idea was first utilized by Kyotani in 1988 for template synthesis of graphene (Figure 4.14).[171–177]

In this method, 2D openings between the lamellae of montmorillonite (MONT) were used as template to produce carbon. In this method, first a MONT (Ca form)-acrylonitrile complex was prepared by using an intercalated monomer and then the resulting complex was exposed to γ-radiation at room temperature in a nitrogen environment in order to polymerize acrylonitrile between the lamellae. A similar method was also repeated for poly-acrylonitrile, PAN, as well. The MONT-PAN was heat-treated and it was found through XRD that interlamellar spacing of the complex was 1.3 nm below 300 °C which ensured the formation of monolayers of PAN. Further heat treatment to 400–550 °C resulted in the decomposition of polymer to form a flat carbonaceous molecule, which caused a slight decrease in spacing to 1.1 nm above 600 °C. Subsequent heat treatment at elevated temperatures and successive washing by acidic solutions ensured the liberation of carbon from MONT. In this experiment, two powdered carbon samples were prepared, one by using the space between the lamellae of montmorillonite and the second by using PAN. The first was termed IPC (Intercalated Polymer-derived Carbon) and the second Free Polymer drive Carbon (FPC). Such a method is also utilized to synthesize 2D material by confining the evolution of the precursor within this nano-space between the graphene layers.

Detailed results of the heat treatment are shown in Table 4.1. Heat treatment temperature in this table is indicated alongside IPC or FPC.

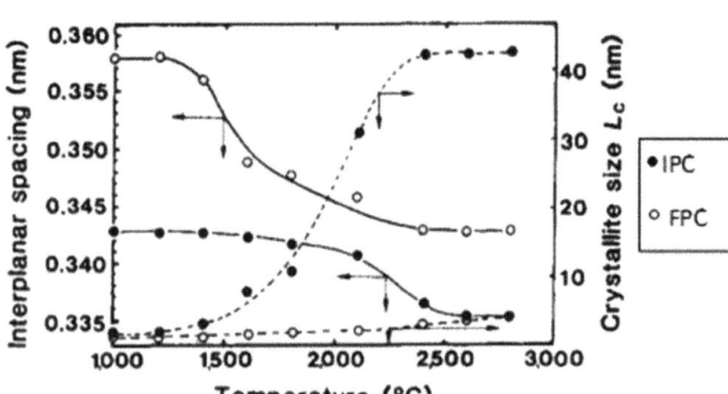

Figure 4.14 The change of inter-planer spacing and crystallite size of IPC and FPC as a function of HTT. Reprinted by permission from Macmillan Publishers: *Nature* (ref. 177), Copyright 1988.

For example, IPC12 means IPC that was heat treated at 1200 °C. This method can be realized by using a pre-stacking technique as described above or a post stacking route in which a mixture of functionalized graphene and precursor is used as the starting material for the layer structure formation followed by the confined reaction of precursors to achieve low dimensional materials.

Similarly, a graphene oxide (GO) template can also be used to synthesize Au nanosheets. For example, hexagonal close-packed (hcp) Au square nanosheets can be synthesized using a GO template. This method can yield Au nanosheets of ultra-small dimensions (~6 nm thickness) with stable and isolable hcp phase at ambient conditions, as shown in Figure 4.15. In this synthesis procedure, a 2 mL growth solution containing GO sheets, 4 mM $HAuCl_4$ and 140 mM 1-amino-9-octadecene, $CH_3(CH_2)_7CH{=}CH(CH_2)_8NH_2$, in a solution of hexane and ethanol was heated at 55 °C for 16 h. This process yields square sheets of Au on GO template. The obtained square-shaped nanostructures display an edge length of several 100 s of nanometers (200–500 nm). The presence of Au in the composition of these as-prepared nanostructures can be confirmed by EDX or XPS.[178]

Table 4.1 Weight percentage of Elemental analysis of heat-treated Carbon. Adapted with the permission of *J. Phys. Chem.*, 1988, **92**, 7029. Copyright (1988), American Chemical Society.

Sample	C	H	N	O(Diff)
IPC	66	3	16	15
IPC10	84	0	10	6
IPC12	90	0	04	6
IPC14	96	0	02	2
IPC16	99	0	01	0
FPC	73	1	19	7
FPC10	91	0	09	0
FPC12	96	0	03	1
FPC14	98	0	01	1
FPC16	99	0	01	0

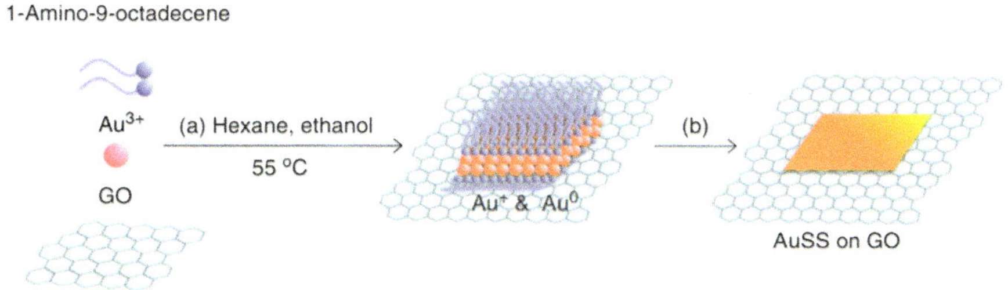

Figure 4.15 Synthesis of 2D Au nanosheets on GO template. Reprinted by permission from Macmillan Publishers Ltd: *Nature Communications* (ref. 178), Copyright 2011.

The synthesis process is completed in two steps:

(a) Heating 1-amino-9-octadecene, Au^{3+} and GO sheets in a mixed solution of ethanol and hexane at 55 °C. This yields the partial reduction of Au^{3+} to Au^{2+} and complexing with 1-amino-9-octadecene.
(b) The complex formed in a first step absorbed on a GO template surface is finally converted to square sheets of Au

4.5.3 The Advantages and Limitations of Template Confined Growth

This method is expected to be a promising route for the synthesis of many 2D layers; some points for using graphene-based template are:

(a) Graphene can be used both as a substrate and template to provide surface or edge nucleation sites for the direct formation of multidimensional nanomaterials.
(b) Graphene derivatives like graphene oxide can be utilized both as soft templates as well as hard templates due to its sheet-like 2D diblock copolymer or surfactant property.
(c) Restacked and 3D graphene nanosheets act as multifunctional templates because of the confinement effect and reactive sites.
(d) Graphene films and hydrogels can be utilized as templates or scaffolds for the incorporation of functional materials for the fabrication of new materials with preferred morphologies and desired properties.
(e) Graphene template can be removed by heating at 600 °C in air which avoids the complex and costly method of template removal, in addition to avoiding the chances of being contaminated.

Major drawbacks of this method include:

(a) Nanomaterials synthesized by using this method are generally polycrystalline.
(b) Yield in each run of synthesis is generally very limited.
(c) In the case of hard template, removal of template is a difficult task because of the fact that most of these templates are often thermally stable and chemically inert to most acids and bases.
(d) The soft template method requires extra processes to etch the template and therefore this method is prone to the addition of impurities.

4.6 Other Methods

4.6.1 Molecular Beam Epitaxy (MBE)

As a transitional epitaxy method for thin-film deposition, MBE synthesis normally takes place in a high vacuum chamber on various crystal wafers with precise deposition rate, which allows thin films to grow epitaxially.[179]

In the MBE process, the ultra-pure solid elemental sources are firstly heated to slowly sublime, then the gaseous elements are condensed on the crystal wafer surface where various elements react with each other. Adjusting the temperature of the evaporated source and the temperature of the substrate can control the rate of material impinging on the substrate surface or the rate of hopping or desorption.[180] In fact, early MBE synthesis in the 1990s may have obtained 2D-layered structures, but no clear structural characterization was identified in atomic scale resolution.[181] Recently, MBE has been further developed for fabricating 2D materials, such as borophene (B),[182–194] silicene (Si),[195,196] germanene (Ge),[196,198] stanene (Sn) and so on.[199,200]

Similar to carbon, 2D boron sheets with different atomic arrangements were predicted by several theoretical studies. In fact, the synthesis of such layers is very challenging because of the extreme activity of boron in air atmosphere. Many recent efforts have been reported for the synthesis of 2D boron layers *via* MBE. For example, 2D boron sheets were successfully grown *via* a MBE process.[189] The epitaxial growth was performed in an ultra-high vacuum chamber combining a MBE system and a low-temperature STM with base pressure of 2×10^{-11} torr. Pure boron (99.9999%) was evaporated from an electron-beam evaporator onto the clean Ag(111) substrate. The substrate temperature during growth was found to be very critical for the preparation of boron sheets. Two types of boron sheet (β_{12} phase and χ_3 phase) with triangular lattices were obtained. Mannix *et al.* reported an atomically thin and crystalline 2D boron sheet (*i.e.* borophene) on silver surfaces grown under ultrahigh-vacuum conditions.[190] Their *in-situ* scanning tunneling microscopy (STM) image confirmed the emergence of a planar structure exhibiting anisotropic corrugation. Interestingly, in contrast to bulk boron allotropes, the obtained borophene exhibited metallic characteristics that are consistent with theoretical predictions of a highly anisotropic 2D metal. Meanwhile, most recent theoretical simulations have indicated possible observations of intrinsic phonon-mediate superconductivity in such 2D layered boron.[192–194]

As another example, silicene is the silicon equivalent of graphene which can act as a buckled honeycomb arrangement of Si atoms and has an electronic dispersion resembling that of relativistic Dirac fermions. Recently, a one-atom-thick Si sheet was synthesized on a Ag (111) surface *via* the MBE method by Vogt *et al.*[195] In the synthesis, clean and well-ordered Ag (111) surfaces were prepared by Ar^+ bombardment and subsequent annealing at around 530 °C for half hour. Si was deposited by evaporation from a Si-wafer piece. The Ag substrate was kept at temperatures in the range of 220–260 °C in ultrahigh vacuum chamber. The MBE-prepared siliene was characterized by STM and angular-resolved photoemission spectroscopy (ARPES). Both structural and electronic properties of the 2D silicene adlayers were demonstrated by STM and ARPES results, which is quite similar to those for graphene and in agreement with the theoretical expectations.

Most recently, 2D Sn-based stanene has been successfully synthesized on the Bi_2Te_3(111) crystal substrate by MBE.[199] The stanene is composed of a biatomic layer of α-Sn(111), in which two triangular sublattices stack together, forming a buckled honeycomb lattice. In the MBE growth, Bi_2Te_3

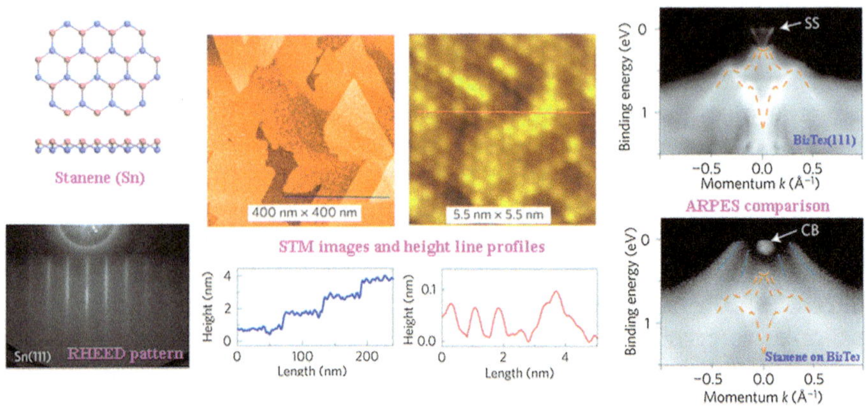

Figure 4.16 Atomic structures and RHEED/STM/ARPES/characterizations of stanene on Bi_2Te_3 synthesized *via* MBE. Reprinted by permission from Macmillan Publishers: *Nature Materials* (ref. 199), Copyright 2015.

bulk single crystals and films were used as the substrate, and high-purity Sn (99.999%) was evaporated from the effusion cells and deposited on the surface of Bi_2Te_3 with rate of 0.4 monolayer/min. The Bi_2Te_3 substrate was kept at room temperature during the MBE process. The thickness and growth mode of stanene were *in-situ* monitored by the reflected high-energy electron diffraction (RHEED), as shown in Figure 4.16. The STM images were acquired at 77 K and 4 K, which confirmed the deposition of stanene layers. The measured height line profile revealed the single-layer thickness of around 0.35, which is consistent with the space distance between biatomic layers along the <111> direction in the α-Sn. The electronic structures of stanene on Bi_2Te_3 were studied by ARPES. In contrast to pure Bi_2Te_3 substrate, the ARPES spectra of stanene/Bi_2Te_3 revealed that the Dirac cone of Bi_2Te_3 disappeared and two hole bands around the Γ point appeared (as shown in Figure 4.16) due to the presence of deposited stanene. In addition, spectra of electron bands very closed to the Γ point centred at about 150 meV below the Fermi level appeared. Notably, the shift of the valence bands of Bi_2Te_3 to higher binding energy after the stanene growth indicated an electron transfer from the stanene to the Bi_2Te_3 substrate.

On the theoretical side, recent simulations by Zeng *et al.* revealed that both black phosphorene (BlackP) and blue phosphorene (BlueP) monolayers can be readily stabilized on Cu(111), Au(111) and GaN(001) substrates.[200] This indicates the possibility of making BlackP and BlueP *via* epitaxial growth in the MBE system. Interestingly, the semiconducting GaN(001) was expected to be superior for fabricating BlueP through an intriguing half-layer-by-half-layer growth mechanism. Such theoretical predications provide useful insight for future epitaxial synthesis of other 2D inorganic materials *via* the MBE method.

Despite the fact that some efforts have been developed to realize single elemental 2D materials by MBE, direct growth of either binary or ternary 2D inorganic layers still remains a daunting challenge. The MBE synthesis is largely affected by various parameters, such as evaporated source, substrate temperature, pressure condition, desirable surface treatments and cleaning processes.

4.6.2 Atomic Layer Deposition (ALD)

Similar to CVD, atomic layer deposition (ALD) is a self-limiting and sequential surface chemistry synthetic method, in which the amount of layers deposition in each reaction cycle is constant.[201,202] During the deposition process, the ALD reaction keeps the precursors separate to achieve the atomic-layer control of film growth as fine as 10 pm per cycle. Such separation of precursors normally is realized by pulsing a purge gas (such N_2 and Ar) after each precursor pulse in order to remove the excess precursor from the reaction chamber. *Via* sequential introduction of gaseous precursors and selective surface chemistry, the ALD reaction can achieve atomic-scale controlled growth of inorganic thin films coated on various substrates at reduced temperatures ranging from 100 to 400 °C. During last decade, the conventional ALD method has been been employed to deposit high-κ thin layers (such as Al_2O_3, HfO_2) as a dielectric platform for realizing high-performance devices. Recently, ALD has been developed as a promising technique to grow inorganic 2Ds for ultra-scaled nano-electronic devices, such as WS_2 and MoS_2.[203–207]

For example, Song *et al.* demonstrated the wafer scale growth of WS_2 *via* the ALD growth of tungsten oxide with subsequent conversion *via* hydrogen sulfide annealing, as shown in Figure 4.17.[204] Besides, Delabie *et al.* directly compared the redox and nucleation mechanisms of 2D WS_2 by ALD and CVD methods.[207] In the ALD growth, the deposition was performed on 300 mm Si wafer covered with 30 nm ALD-grown Al_2O_3 layers. The growth was carried by H_2-plasma-enhanced ALD process at 300 °C with WF_6 and H_2S as precursors. The reaction sequences consisted of WF_6 chemisorption, H_2 plasma reduction of the chemisorbed WF_{6-x} surface species, and WS_2 formation by H_2S reaction. It was found that both plasma power and exposure time have to be controlled carefully to confine the reactions to the surface species. The characterizations revealed that the preferential interaction between chemisorption agents and Al_2O_3 can ensure rapid layer closure at a WS_2 content of two monolayers. After layer closure, inhomogeneous re-nucleation of WS_2 on the WS_2 surface resulted in three-dimensional islands. The high nucleation density in combination with both lateral and vertical growth contributions finally yielded nano-crystalline WS_2 layers with around 10 nm grains, which is comparably smaller than the domain size of ALD-WS_2 deposited on sapphire substrate. In CVD-WS_2 growth, the WF_6 and H_2S were simultaneously flowing for reaction, and most of the obtained samples were triangular WS_2 crystals by a selective lateral growth mode. The crystal domain size of CVD-WS_2 increased as function of the deposition. In contrast to ALD growth, the vertical growth was minimized, providing a possible low-temperature path for large area crystal growth.

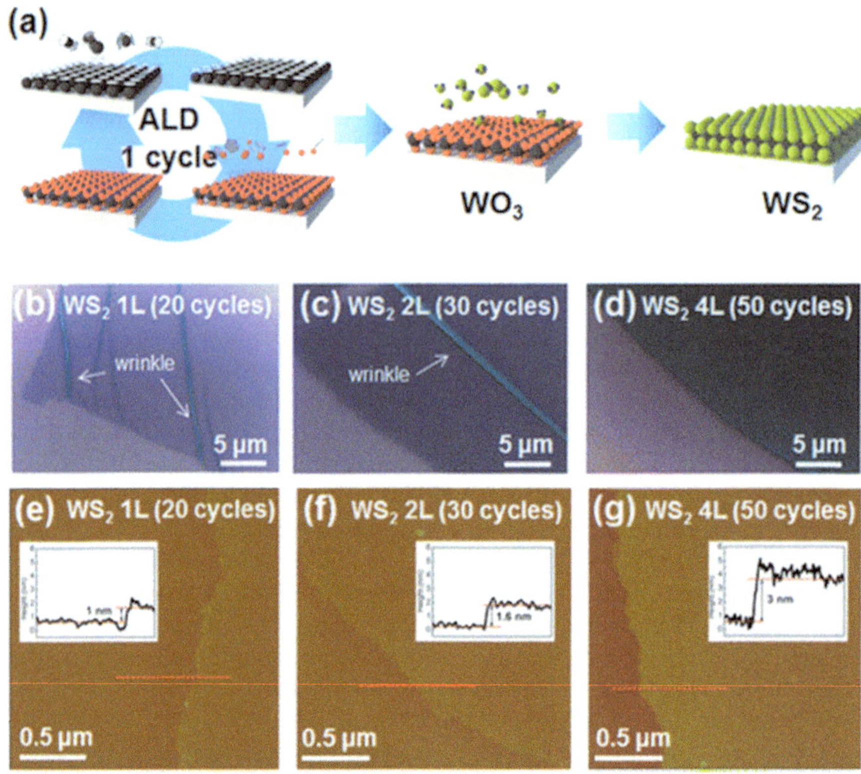

Figure 4.17 Synthesis procedure and optical/AFM characterizations of WS_2 nanosheets *via* ALD method. Reprinted with permission from *ACS Nano*, 2013, 7, 11333. Copyright (2013), American Chemical Society.

Similarly, ALD can be also used to generate ultrathin layers of nitrides and chalcogenides on large area substrates at comparable low temperatures. However, the feasibility of growing high-quality desired inorganic 2Ds and the growth control down to single monolayer are still a big challenge. More importantly, better understanding of the surface chemistry in the ALD process is highly critical for producing high quality layers, in particular the dynamic control of the partial reduction or oxidation of the metal.

4.6.3 Microwave-assisted Synthesis

In the last decade, the microwave-assisted route has been employed as a faster, simple and energy-efficient method for the synthesis of inorganic materials.[208,209] As one typical electromagnetic radiation, microwaves with wavelength range of 1 mm to 1 m (frequency range of 0.3–300 GHz) can efficiently transfer energy to the materials through resonance or relaxation, resulting in very rapid heating. Recently, several 2D inorganic materials have been synthesized by a microwave-assisted route, such as graphene-based layers and chalcogenides.[210–214]

Due to the rich functional groups, graphene oxide (GO) is electrically insulating, and can be easily reduced into high conductive graphene through direct thermal treatment. As a convenient and rapid heating source, microwave irradiation has been used to prepare exfoliated graphite from a wide range of graphite intercalation compounds.[210–214] In the microwave synthesis method, GO powders made from the modified Hummers process were put in a microwave oven in ambient conditions and treated at 700 W for 1 min. A large volume expansion of GO powders was found, accompanied by violent fuming, resulting in a black and fluffy graphite material with high specific surface area and good conductivity.[214] Most recently, Chhowalla's group used microwave methods to reduce GO into pristine graphene *via* 1- to 2-second pulses of microwaves.[215] Notably, the obtained materials show excellent structural properties with electronic mobility values of 1000 centimeter squared per volt per second, along with high catalytic activity for oxygen evolution reaction. This suggests that microwave-assisted reduction of GO may realize the goal of achieving high-quality graphene with a mass-produce method.

Meanwhile, various important chalcogenides have also been synthesized by using microwave irradiation, such as ZnS, ZnSe, Bi_2Se_3, PbSe, PbTe, TaS_2 and SnS_2.[216–218] In principle, stoichiometric quantities of respective metals and chalcogens were firstly sealed in evacuated quartz ampules. The sealed tubes with mixture were irradiated with microwaves for several minutes in a microwave oven at a power around of 900 W. It was noted that emission of light flashes often occurred during the melting process of reactants and the formation reaction. After reaction, the microwave oven was turned off and the sealed tube was naturally cooled down to room temperature. The final products can be obtained by breaking the tube. Many binary chalcogenides with good phase purity and crystallinity were synthesized recently by irradiation of the elemental mixtures for less than 10 min. Besides, ternary chalcogenides such as $CuInS_2$ and $CuInSe_2$ have also been selectively prepared for high potential application in solar cells.[219]

Considering the above, it is concluded that microwave assisted-synthesis may provide the following advantages compared to conventional methods: (1) it can produce higher heating rates resulting in increased reaction rates for mass growth, (2) it can avoid the direct contact between the heating source and the reactants, (3) it allows selective heating of the wanted precursors for the reaction mixture containing compounds *via* adjusting differing microwave-absorbing resonances, (4) it allows easy control of the reaction parameters and better reproducibility.

4.7 Summary and Outlook

In summary, the synthesis of high-quality inorganic layers is interesting for both fundamental and technologically relevant research. With a unique combination of many reactions, the solid-state synthesis has emerged as a promising and feasible route for realizing various desired inorganic 2D layers. Extraordinary progress has been made in developing the suitable synthesis of inorganic 2D layers through solid state decomposition, chemical vapor transport reaction, chemical vapor deposition, template driven growth, as well as other solid-state related reactions (molecular beam epitaxy, atomic layer deposition, microwave-assisted and substitution reactions). Different reactions exhibited distinctive growth processes and parameter effects, along with special advantages and limitations.

In fact, inorganic 2D materials consist of hundreds of types. The current inorganic 2D materials seem to be a small part of the research on 2D materials. As new layered compositions become part of the diverse set of the 2D material selection in future, the importance of controlled growth of their atomic layers will become paramount. It can be easily predicted that the future synthetic routes for synthesizing new inorganic 2D materials will attract more intense efforts because of the critical role of the materials for most fundamental research or any potential applications.

Promoted by the success of graphene, single elemental layer (phosphorus, borophene, silicene, germanene, stane *etc.*), binary and ternary inorganic 2D layers have been synthesized. In future, more complex inorganic layers and their hybrids will be proposed because of specific applications in the fields of electronics, optics, energy, environment, and others. Most current 2Ds are often polycrystalline due to the nucleation, growth and coalescence of grains. To resolve the boundary issue, it is still a great challenge to control the early stages of nucleation and growth of the materials on preferred solid surfaces.[220]

Meanwhile, there are various *in-situ* synthesis approaches for tuning the composition of 2D layers, such as doping with heteroatoms and chemical functionalization. Especially, doping heteroatoms into layered lattices will be promising and feasible way to change the chemical composition of the layer. Meanwhile, the space between interlayers also provides the possibility for producing unique 2D intercalated compounds. To synthesize inorganic 2D intercalations with different types of intercalated agents is expected to engineer the electronic structure at their atomic scale and tune their intrinsic physical and chemical properties (such as carrier concentration, mobility, magnetic, optical and others).

Besides their compositional versatility, the ability to selectively grow specific microstructures and engineer various phases will be an additional variable that can be employed to tune their properties. For example, the controllable synthesis with 1T-2H phase engineering will offer unique opportunities for studying and realizing many novel phenomena in 2D transition-metal dichalcogenides.[221]

Before processing of exotic 2D materials, it is necessary to develop theoretical models to predict stability and potential properties. In spite of many theoretical efforts on isolated 2Ds, future research efforts in the experimental synthesis areas of novel free-standing crystalline structure of materials are still of great interest as they could open up unique opportunities for electronics and photonics that are not yet realized. It is anticipated that future efforts will combine various 2Ds in one set synthesis for decreasing or increasing the inter-coupling between layers, such as van der Waals heterostructures.[222]

Therefore, for better exploration of 2Ds, seeking more suitable synthetic strategies (*e.g.* scaling domain size using limited growth, high crystalline using CVT, mass-production using CVD, and precisely stacking layers using MBE) to improve the current quantity and quality is still a critical task for future studies.

Acknowledgements

We acknowledge the financial support form National Basic Research Program of China (2014CB848900), NSF (U1232131, U1532112, 11375198, 11574280) and CUSF (WK2310000053). L. S. thanks the recruitment program of global experts and the CAS Hundred Talent Program. Z. R. is thankful for the CSC fellowship.

References

1. A. K. Geim and K. S. Novoselov, *Nat. Mater.*, 2007, **6**, 183.
2. J. Warner, F. Schaffel, M. Rummeli and A. Bachmatiuk, *Graphene: Fundamentals and Emergent Applications*, Elsevier, Amsterdam, 2012, ISBN: 978-012-394-827-4.
3. W. Choi and J. Lee, *Graphene: Synthesis and Applications*, CRC Press, Taylor & Francis group, 2012, ISBN: 978-143-986-187-5.
4. S. Mikhailov, *Physics and Applications of Graphene - Experiments*, InTech, 2011, ISBN: 978-953-307-217-3.
5. C. N. R. Rao and A. K. Sood, *Graphene: Synthesis, Properties, and Phenomena*, Wiley-VCH Verlag GmbH, Weinheim, 2012, ISBN: 978-3-527-33258-8.
6. M. Aliofkhazraei and N. Ali, *Two-dimensional Nanostructures*, CRC Press, Taylor & Francis group, 2012, ISBN: 978-143-986-665-8.
7. R. Mas-Balleste, C. Gomez-Navarro, J. Gomez-Herrero and F. Zamora, *Nanoscale*, 2001, **3**, 20.
8. M. Chhowalla, H. S. Shin, G. Eda, L. Li, K. P. Loh and H. Zhang, *Nat. Chem.*, 2013, **5**, 263.
9. M. Xu, T. Liang, M. Shi and H. Chen, *Chem. Rev.*, 2013, **113**, 3766.
10. Q. Tang and Z. Zhou, *Prog. Mater. Sci.*, 2013, **58**, 1244.
11. O. V. Yazyev and Y. P. Chen, *Nat. Nanotechnol.*, 2014, **9**, 755.
12. R. Wen, C. G. Granqvist and G. A. Niklasson, *Nat. Mater.*, 2015, **14**, 996.
13. J. Liu, Y. Liu, N. Liu, Y. Han, X. Zhang, H. Huang, Y. Lifshitz, S. T. Lee, J. Zhong and Z. Kang, *Science*, 2015, **347**, 970.
14. G. Fan, F. Li, D. G. Evans and X. Duan, *Chem. Soc. Rev.*, 2014, **43**, 7040.
15. M. Shao, R. Zhang, Z. Li, M. Wei, D. G. Evans and X. Duan, *Chem. Commun.*, 2015, **51**, 15880.
16. L. F. Matthesis, *Phys. Rev. B*, 1973, **8**, 3719.
17. C. B. Roxlo, R. R. Chianelli, H. W. Deckman, A. F. Ruppert and P. P. Wong, *J. Vac. Sci. Technol., A*, 1987, **5**, 555.
18. Z. Zeng, Z. Yin, X. Huang, H. Li, Q. He, G. Lu, F. Boey and H. Zhang, *Angew. Chem., Int. Ed.*, 2011, **50**, 11093.

19. R. Kappera, D. Voiry, S. E. Yalcin, B. Branch, G. Gupta, A. D. Mohite and M. Chhowalla, *Nat. Mater.*, 2014, **13**, 1128.
20. Q. Liu, X. Li, Z. Xiao, Y. Zhou, H. Chen, A. Khalil, T. Xiang, J. Xu, W. Chu, X. Wu, J. Yang, C. Wang, Y. Xiong, C. Jin, P. Ajayan and L. Song, *Adv. Mater.*, 2015, **27**, 4837.
21. Q. Liu, X. Li, Q. He, A. Khalil, D. Liu, T. Xiang, X. Wu and L. Song, *Small*, 2015, **11**, 5556.
22. M. Pumera, Z. Sofer and A. Ambrosi, *J. Mater. Chem. A*, 2014, **2**, 8981.
23. H. Wang, H. Feng and J. Li, *Small*, 2014, **10**, 2165.
24. D. J. Late, C. S. Rout, D. Chakravarty and S. Ratha, *Can. Chem. Trans.*, 2015, **3**, 118.
25. M. Chhowalla, D. Voiry, J. Yang, H. S. Shin and K. P. Loh, *MRS Bull.*, 2015, **40**, 585.
26. A. A. Tedstone, D. J. Lewis and P. O'Brien, *Chem. Mater.*, 2016, **28**, 1965.
27. C. N. R. Rao and J. Gopalakrishnan, *New Directions in Solid State Chemistry*, Cambridge University Press, Cambridge, 1997, ISBN: 978-051-162-314-1.
28. A. K. Cheetham and P. Day, *Solid State Chemistry–Techniques and Compounds*, Clarendon Press, Oxford, 1987 and 1992.
29. A. R. West, *Solid State Chemistry and Applications*, John Wiley&Sons, Chichester, 1984.
30. C. N. R. Rao and K. Biswas, *Essentials of Inorganic Materials Synthesis*, John Wiley&Sons, New Jersey, 2015.
31. G. R. Yazdi, T. Iakimov and R. Yakimova, *Crystals*, 2016, **6**, 53.
32. F. R. Chien, S. R. Nutt, W. S. Yoo, T. Kimoto and H. Matsunami, *J. Mater. Res.*, 1994, **9**, 940.
33. G. R. Yazdia, R. Vasiliauskasa, T. Iakimova, A. Zakharovb, M. Syväjärvia and R. Yakimovaa, *Carbon*, 2013, **57**, 477.
34. K. V. Emtsev, A. Bostwick, K. Horn, J. Jobst, G. L. Kellogg, L. Ley, J. L. McChesney, T. Ohta, S. a. Reshanov, J. Röhrl, E. Rotenberg, A. K. Schmid, D. Waldmann, H. B. Weber and T. Seyller, *Nat. Mater.*, 2009, **8**, 203.
35. M. H. Oliveira, T. Schumann, M. Ramsteiner, J. M. J. Lopes and H. Riechert, *Appl. Phys. Lett.*, 2011, **99**, 111901.
36. T. Ohta, N. C. Bartelt, S. Nie, K. Thürmer and G. L. Kellogg, *Phys. Rev. B*, 2010, **81**, 121411.
37. B. Lang, *Surf. Sci.*, 1975, **53**, 317.
38. K. S. Novoselov, D. Jiang, F. Schedin, T. J. Booth, V. V. Khotkevich, S. V. Morozov and A. K. Geim, *Proc. Natl. Acad. Sci. U. S. A.*, 2005, **102**, 10451.
39. K. S. Novoselov, A. K. Geim, S. V. Morozov, D. Jiang, Y. Zhang, S. V. Dubonos, I. V. Grigorieva and A. A. Firsov, *Science*, 2004, **306**, 666.
40. P. W. Sutter, J. I. Flege and E. A. Sutter, *Nat. Mater.*, 2008, **7**, 406.
41. X. Li, W. Cai, L. Colombo and R. S. Ruoff, *Nano Lett.*, 2009, **9**, 4268.
42. Y. Hernandez, *et al.*, *Nat. Nanotechnol.*, 2008, **3**, 563.
43. J. Baringhaus, *et al.*, *Nature*, 2014, **506**, 349.
44. Z. Guo, R. Dong, P. S. Chakraborty, N. Lourenco, J. Palmer, Y. Hu, M. Ruan, J. Hankinson, J. Kunc, J. D. Cressler, C. Berger and W. A. de Heer, *Nano Lett.*, 2013, **13**, 942.

45. Y. M. Lin, C. Dimitrakopoulos, K. A. Jenkins, D. B. Farmer, H. Y. Chiu, A. Grill and Ph. Avouris, *Science*, 2010, **327**, 662.
46. A. Tzalenchuk, S. Lara-Avila, A. Kalaboukhov, S. Paolillo, M. Syväjärvi, R. Yakimova, O. Kazakova, T. J. B. M. Janssen, V. Fal'ko and S. Kubatkin, *Nat. Nanotechnol.*, 2010, **5**, 186.
47. J. Hicks, A. Tejeda, A. Taleb-Ibrahimi, M. S. Nevius, F. Wang, K. Shepperd, J. Palmer, F. Bertran, P. Le Fèvre, J. Kunc, W. A. de Heer, C. Berger and E. H. Conrad, *Nat. Phys.*, 2013, **9**, 49.
48. M. Binnewies, R. Glaum, M. Schmidt, and P. Schmidt, *Chemical Vapor Transport Reactions*, Walter de Gruyter, 2012, ISBN: 978-3-11-025465-5.
49. A. E. van Arkel, *Reine Metalle*, Springer, Berlin, 1939, ISBN: 978-3-642-99695-5.
50. H. Schäfer, *Chemische Transportreaktionen*, Verlag Chemie, Weinheim, 1962.
51. G. Schmidt and R. Gruehn, *J. Cryst. Growth*, 1982, **57**, 585.
52. H. Schäfer, *Z. Anorg. Allg. Chem.*, 1986, **543**, 217.
53. R. F. Rolsten, *Z. Anorg. Allg. Chem.*, 1960, **305**, 25.
54. R. F. Rolsten, *Iodide Metals and Metal Iodides*, J. Wiley, New York, 1961.
55. H. Schäfer, *J. Cryst. Growth*, 1975, **31**, 31.
56. S. Lange, P. Schmidt and T. Nilges, *Inorg. Chem.*, 2007, **46**, 4028.
57. T. Nilges, M. Kersting and T. Pfeifer, *J. Solid State Chem.*, 2008, **181**, 1707.
58. M. Köpf, *et al.*, *J. Cryst. Growth*, 2014, **405**, 6.
59. P. W. Bridgman, *J. Am. Chem. Soc.*, 1914, **36**, 1344.
60. L. Li, Y. Yu, G. J. Ye, Q. Ge, X. Ou, H. Wu, D. Feng, X. H. Chen and Y. Zhang, *Nat. Nanotechnol.*, 2014, **9**, 372.
61. H. Krebs, H. Weitz and K. H. Worms, *Z. Anorg. Allg. Chem.*, 1955, **280**, 119.
62. Y. Maruyama, S. Suzuki, K. Kobayashi and S. Tanuma, *Physica B+C*, 1981, **105**, 99.
63. R. Kappera, D. Voiry, S. E. Yalcin, B. Branch, G. Gupta, A. D. Mohite and M. Chhowalla, *Nat. Mater.*, 2014, **13**, 1128.
64. F. Schwierz, *Nat. Nanotechnol.*, 2010, **5**, 487–496.
65. Q. H. Wang, K. Kalantar-Zadeh, A. Kis, J. N. Coleman and M. S. Strano, *Nat. Nanotechnol.*, 2012, **7**, 699.
66. G. Fiori, F. Bonaccorso, G. Iannaccone, T. Palacios, D. Neumaier, A. Seabaugh, S. K. Banerjee and L. Colombo, *Nat. Nanotechnol.*, 2013, **9**, 768.
67. Z. Yin, H. Li, H. Li, L. Jiang, Y. Shi, Y. Sun, G. Lu, Q. Zhang, X. Chen and H. Zhang, *ACS Nano*, 2012, **6**, 74.
68. B. Radisavljevic, A. Radenovic, J. Brivio, V. Giacometti and A. Kis, *Nat. Nanotechnol.*, 2011, **6**, 147.
69. J. A. Wilson and A. D. Yoffe, *Adv. Phys.*, 1969, **18**, 193.
70. B. Radisavljevic, A. Radenovic, J. Brivio, V. Giacometti and A. Kis, *Nat. Nanotechnol.*, 2011, **6**, 147.
71. M. K. L. Man, A. Margiolakis, S. Deckoff-Jones, T. Harada, E. L. Wong, M. B. M. Krishna, J. Madéo, A. Winchester, S. Lei, R. Vajtai, P. M. Ajayan and K. M. Dani, *Nat. Nanotechnol.*, 2017, **12**, 36.
72. B. Radisavljevic, A. Radenovic, J. Brivio, V. Giacometti and A. Kis, *Nat. Nanotechnol.*, 2011, **6**, 147.

73. S. Lei, *et al.*, *Nat. Nanotechnol.*, 2016, **11**, 465.
74. J. S. Ross, *et al.*, *Nat. Nanotechnol.*, 2014, **9**, 268.
75. R. Vaidya, M. Dave, S. S. Patel, S. G. Patel and A. R. Jani, *Pramana-J. Phys.*, 2004, **63**, 611–616.
76. K. Ueno, *J. Phys. Soc. Jpn.*, 2015, **84**, 121015.
77. R. Brec, *Solid State Ionics*, 1986, **22**, 3–30.
78. V. Grasso and L. Silipigni, *Riv. Nuovo Cimento*, 2002, **25**, 1.
79. K. Du, X. Wang, Y. Liu, P. Hu, M. I. B. Utama, C. K. Gan, Q. Xiong and C. Klo, *ACS Nano*, 2015, **10**, 1738.
80. P. A. Joy and S. Vasudevan, *Phys. Rev. B*, 1992, **46**, 5425.
81. J. Lee, S. Lee, J. H. Ryoo, S. Kang, T. Y. Kim, P. Kim, C. Park, J. Park and H. Cheong, *Nano Lett.*, 2016, **16**, 7433.
82. X. Li, X. Wu and J. Yang, *J. Am. Chem. Soc.*, 2014, **136**, 11065.
83. Y. Lin, D. O. Dumcenco, H. Komsa, Y. Niimi, A. V. Krasheninnikov, Y. Huang and K. Suenaga, *Adv. Mater.*, 2014, **26**, 2857.
84. Y. Lin, D. O. Dumcenco, Y. Huang and K. Suenaga, *Nat. Nanotechnol.*, 2014, **9**, 391.
85. J. Suh, *et al.*, *Nano Lett.*, 2014, **14**, 6976.
86. Y. Chen, J. Xi, D. O. Dumcenco, Z. Liu, K. Suenaga, D. Wang, Z. Shuai, Y. Huang and L. Xie, *ACS Nano*, 2013, **7**, 4610.
87. D. O. Dumcenco, H. Kobayashi, Z. Liu, Y. Huang and K. Suenaga, *Nat. Commun.*, 2013, **4**, 1351.
88. M. Zhang, *et al.*, *ACS Nano*, 2014, **8**, 7130.
89. I. Belopolski, *et al.*, *Nat. Commun.*, 2016, **7**, 10639.
90. P. Yu, *et al.*, *Adv. Mater.*, 2016, **29**, 1603991.
91. A. Pisoni, J. Jacimovic, O. S. Barišić, A. Walter, B. Náfrádi, P. Bugnon, A. Magrez, H. Berger, Z. Revay and L. Forró, *J. Phys. Chem. C*, 2015, **119**, 3918.
92. P. Schmidt, M. Binnewies, R. Glaum and M. Schmidt, *Chemical Vapor Transport Reactions–Methods, Materials, Modeling, in Advanced Topics on Crystal Growth*, InTech, Rijeka, 2013, ISBN 980-953-307-791-3.
93. G. G. Tibbetts, *J. Cryst. Growth*, 1984, **66**, 632.
94. G. G. Tibbetts, *Carbon*, 1989, **27**, 745.
95. G. G. Tibbetts, *J. Cryst. Growth*, 1985, **73**, 431.
96. G. G. Tibbetts, M. G. Devour and E. J. Rodda, *Carbon*, 1987, **25**, 367.
97. R. T. K. Baker, *Carbon*, 1989, **27**, 315.
98. P. L. Walker, J. F. Rakszawski and G. R. Imperial, *J. Phys. Chem.*, 1959, **63**, 133.
99. M. S. Dresselhaus, *et al.*, Graphite fibers and filaments, Springer Science & Business Media, 2013, vol. 5.
100. L. J. Ci, Z. G. Zhao and J. B. Dai, *Carbon*, 2005, **43**, 883.
101. R. E. Smalley, *Carbon nanotubes: synthesis, structure, properties, and applications*, ed. M. S. Dresselhaus, G. Dresselhaus and P. Avouris, Springer Science & Business Media, 2003, vol. 80.
102. N. C. Bartelt and K. F. McCarty, *MRS Bull.*, 2012, **37**, 1158.
103. A. V. Eletskii, I. M. Iskandarova, A. A. Knizhnik and D. N. Krasikov, *Phys.-Usp.*, 2011, **54**, 227.

104. D. R. Cooper, B. D'Anjou, N. Ghattamaneni, B. Harack, M. Hilke, A. Horth, N. Majlis, M. Massicotte, L. Vandsburger, E. Whiteway and V. Yu, *ISRN Condens. Matter Phys.*, 2012, 501686.
105. Z. Luo, Y. Lu, D. W. Singer, M. E. Berck, L. A. Somers, B. R. Goldsmith and A. T. C. Johnson, *Chem. Mater.*, 2011, **23**, 1441.
106. W. Liu, H. Li, C. Xu, Y. Khatami and K. Banerjee, *Carbon*, 2011, **49**, 4122.
107. X. Li, W. Cai, J. An, S. Kim, J. Nah, D. Yang, R. Piner, A. Velamakanni, I. Jung, E. Tutuc, S. K. Banerjee, L. Colombo and R. S. Ruoff, *Science*, 2009, **324**, 1312.
108. D. Wei, Y. Liu, Y. Wang, H. Zhang, L. Huang and G. Yu, *Nano Lett.*, 2009, **9**, 1752.
109. S. Bhaviripudi, X. Jia, M. S. Dresselhaus and J. Kong, *Nano Lett.*, 2010, **10**, 4128.
110. K. S. Kim, Y. Zhao, H. Jang, S. Y. Lee, J. M. Kim, K. S. Kim, J. Ahn, P. Kim, J. Choi and B. H. Hong, *Nature*, 2009, **457**, 706.
111. S. J. Chae, F. Günes, K. K. Kim, E. S. Kim, G. H. Han, S. M. Kim, H. Shin, S.-M. Yoon, J. Choi, M. H. Park, C. W. Yang, D. Pribat and Y. H. Lee, *Adv. Mater.*, 2009, **21**, 2328.
112. T. Wu, X. Zhang, Q. Yuan, J. Xue, G. Lu, Z. Liu, H. Wang, H. Wang, F. Ding, Q. Yu, X. Xie and M. Jiang, *Nat. Mater.*, 2016, **15**, 43.
113. W. Yang, *et al.*, *Nat. Mater.*, 2013, **12**, 792.
114. Z. Li, P. Wu, C. Wang, X. Fan, W. Zhang, X. Zhai, C. Zeng, Z. Li, J. Yang and J. Hou, *ACS Nano*, 2011, **5**, 3385.
115. L. Gao, J. R. Guest and N. P. Guisinger, *Nano Lett.*, 2010, **10**, 3512.
116. M. Zeng, L. Tan, J. Wang, L. Chen, M. H. Rümmeli and L. Fu, *Chem. Mater.*, 2014, **26**, 3637.
117. S. Helveg, J. V. Lauritsen, E. Læ gsgaard, I. Stensgaard, J. K. Nørskov, B. S. Clausen, H. Topsøe and F. Besenbacher, *Phys. Rev. Lett.*, 2000, **84**, 951.
118. Y. Zhan, Z. Liu, S. Najmaei, P. M. Ajayan and J. Lou, *Small*, 2012, **8**, 966.
119. S. Najmaei, Z. Liu, W. Zhou, X. Zou, G. Shi, S. Lei, B. I. Yakobson, J. Idrobo, P. M. Ajayan and J. Lou, *Nat. Mater.*, 2013, **12**, 754.
120. Y. Shi, W. Zhou, A. Lu, W. Fang, Y. Lee, A. L. Hsu, S. M. Kim, K. K. Kim, H. Y. Yang, L. Li, J. Idrobo and J. Kong, *Nano Lett.*, 2012, **12**, 2784.
121. Q. Ji, Y. Zhang, T. Gao, Y. Zhang, D. Ma, M. Liu, Y. Chen, X. Qiao, P. Tan, M. Kan, J. Feng, Q. Sun and Z. Liu, *Nano Lett.*, 2013, **13**, 3870.
122. K. Liu, W. Zhang, Y. Lee, Y. Lin, M. Chang, C. Su, C. Chang, H. Li, Y. Shi, H. Zhang, C. Lai and L. Li, *Nano Lett.*, 2012, **12**, 1538.
123. X. L. Li and Y. D. Li, *Chem.–Eur. J.*, 2003, **9**, 2726.
124. Y. Shi, W. Zhou, A. Lu, W. Fang, Y. Lee, A. L. Hsu, S. M. Ki, K. K. Kim, H. Y. Yang, L. Li, J. Idrobo and J. Kong, *Nano Lett.*, 2012, **12**, 2784.
125. W. Yang, G. Chen, Z. Shi, C. Liu, L. Zhang, G. Xie, M. Cheng, D. Wang, R. Yang, D. Shi, K. Watanabe, T. Taniguchi, Y. Yao, Y. Zhang and G. Zhang, *Nat. Mater.*, 2013, **12**, 792.

126. Y. Lin, N. Lu, N. Perea-Lopez, J. Li, Z. Lin, X. Peng, C. H. Lee, C. Sun, L. Calderin, P. N. Browning, M. S. Bresnehan, M. J. Kim, T. S. Mayer, M. Terrones and J. A. Robinson, *ACS Nano*, 2014, **8**, 3715.
127. Y. Gong, J. Lin, X. Wang, G. Shi, S. Lei, Z. Lin, X. Zou, G. Ye, R. Vajtai, B. I. Yakobson, H. Terrones, M. Terrones, B. K. Tay, J. Lou, S. T. Pantelides, Z. Liu, W. Zhou and P. M. Ajayan, *Nat. Mater.*, 2014, **13**, 1135.
128. Y. Lin, R. K. Ghosh, R. Addou, N. Lu, S. M. Eichfeld, H. Zhu, M. Li, X. Peng, M. J. Kim, L. Li, R. M. Wallace, S. Datta and J. A. Robinson, *Nat. Commun.*, 2015, **6**, 7311.
129. X. S. Li, C. W. Magnuson, A. Venugopal, J. H. An, J. W. Suk, B. Y. Han, M. Borysiak, W. W. Cai, A. Velamakanni, Y. W. Zhu, L. F. Fu, E. M. Vogel, E. Voelkl, L. Colombo and R. S. Ruoff, *Nano Lett.*, 2010, **10**, 4328.
130. S. Bhaviripudi, X. T. Jia, M. S. Dresselhaus and J. Kong, *Nano Lett.*, 2010, **10**, 4128.
131. Y. M. A. Wu, Y. Fan, S. Speller, G. L. Creeth, J. T. Sadowski, K. He, A. W. Robertson, C. S. Allen and J. H. Warner, *ACS Nano*, 2012, **6**, 5010.
132. X. S. Li, W. W. Cai, L. Colombo and R. S. Ruoff, *Nano Lett.*, 2009, **9**, 4268.
133. A. W. Robertson and J. H. Warner, *Nano Lett.*, 2011, **11**, 1182.
134. G. T. de Laissardiere, D. Mayou and L. Magaud, *Nano Lett.*, 2010, **10**, 804.
135. A. H. MacDonald and R. Bistritzer, *Nature*, 2011, **474**, 453.
136. T. Ohta, A. Bostwick, T. Seyller, K. Horn and E. Rotenberg, *Science*, 2006, **313**, 951.
137. Y. B. Zhang, T. T. Tang, C. Girit, Z. Hao, M. C. Martin, A. Zettl, M. F. Crommie, Y. R. Shen and F. Wang, *Nature*, 2009, **459**, 820.
138. K. Kim, S. Coh, L. Z. Tan, W. Regan, J. M. Yuk, E. Chatterjee, M. F. Crommie, M. L. Cohen, S. G. Louie and A. Zettl, *Phys. Rev. Lett.*, 2012, **108**, 246103.
139. L. Brown, R. Hovden, P. Huang, M. Wojcik, D. A. Muller and J. Park, *Nano Lett.*, 2012, **12**, 1609.
140. R. W. Havener, H. Zhuang, L. Brown, R. G. Hennig and J. Park, *Nano Lett.*, 2012, **12**, 3162.
141. I. Vlassiouk, M. Regmi, P. F. Fulvio, S. Dai, P. Datskos, G. Eres and S. Smirnov, *ACS Nano*, 2011, **5**, 6069.
142. R. B. Mclellan, *Scr. Metall.*, 1969, **3**, 389.
143. B. Zhang, W. H. Lee, R. Piner, I. Kholmanov, Y. Wu, H. Li, H. Ji and R. S Ruoff, *ACS Nano*, 2012, **6**, 2471.
144. W. An, X. C. Zeng and C. H. Turner, *Chem. Phys.*, 2009, **131**, 174702.
145. W. Zhang, P. Wu, Z. Li and J. Yang, *J. Phys. Chem.*, 2011, **115**, 17782.
146. C. Au, C. Ng and M. Liao, *J. Catal.*, 1999, **185**, 12.
147. S. Fullam, D. Cottell, H. Rensmo and D. Fitzmaurice, *Adv. Mater.*, 2000, **12**, 1430.
148. G. Cao and D. Liu, *Adv. Colloid Interface Sci.*, 2008, **136**, 45.
149. R. C. Furneaux, W. R. Rigby and A. P. Davidson, *Nature*, 1989, **337**, 147.
150. R. L. Fleischer, P. B. Price and R. M. Walker, *Nuclear Tracks in Solids: Principles and Applications*, University of California Press, Berkeley, Calif, 1975, vol. 42, p. 306.

151. Z. Chen, W. Ren, L. Gao, B. Liu, S. Pei and H. M. Cheng, *Nat. Mater.*, 2011, **10**, 424.
152. G. Chen, Y. Liu, F. Liu and X. Zhang, *Appl. Surf. Sci.*, 2014, **311**, 808.
153. H. Ji, L. Zhang, M. T. Pettes, H. Li, S. Chen, L. Shi, R. Piner and R. S. Ruoff, *Nano Lett.*, 2012, **12**, 2446.
154. J. Ji, H. Ji, L. L. Zhang, X. Zhao, X. Bai, X. Fan, F. Zhang and R. S. Ruoff, *Adv. Mater.*, 2013, **25**, 4673.
155. X. Huang, K. Qian, J. Yang, J. Zhang, L. Li, C. Yu and D. Zhao, *Adv. Mater.*, 2012, **24**, 4419.
156. D. Cai, L. Ding, S. Wang, Z. Li, M. Zhu and H. Wang, *Electrochim. Acta.*, 2014, **139**, 96.
157. I. Kondratowicz, K. Żelechowska and W. Sadowski, *Silica-templated Three-dimensional Graphene Xerogels*, EYEC Monograph, 2015, ISBN: 978-83-936575-1-3.
158. J. L. Vickery, A. J. Patil and S. Mann, *Adv. Mater.*, 2009, **21**, 2180.
159. H. Bai, C. Li, X. Wang and G. Shi, *J. Phys. Chem. C*, 2011, **115**, 5545.
160. S. Zu and B. Han, *J. Phys. Chem. C*, 2009, **113**, 13651.
161. Q. Y. Cheng, D. Zhou, Y. Gao, Q. Chen, Z. Zhang and B. H. Han, *Langmuir*, 2012, **28**, 3005.
162. W. Cheng, J. He, T. Yao, Z. Sun, Y. Jinag, Q. Liu, S. Jiang, F. Hu, Z. Xie, B. He, W. Yan and S. Wei, *J. Am. Chem. Soc.*, 2014, **136**, 10393.
163. T. L. Wade and J. E. Wegrowe, *Eur. Phys. J.: Appl. Phys.*, 2005, **29**, 3.
164. N. D. Petkovich and A. Stein, *Chem. Soc. Rev.*, 2013, **42**, 3721.
165. I. Hijazi, T. Bourgeteau, R. Cornut, A. Morozan, A. Filoramo, J. Leroy, V. Derycke, B. Jousselme and S. Campidelli, *J. Am. Chem. Soc.*, 2014, **136**, 6348.
166. S. C. Tsang, Y. K. Chen, P. J. F. Harris and M. L. H. Green, *Nature*, 1994, **372**, 159.
167. K. Matsui, T. Kyotani and A. Tomita, *Adv. Mater.*, 2002, **14**, 1216.
168. E. Philp, J. Sloan, A. I. Kirkland, R. R. Meyer, S. Friedrichs, J. L. Hutchison and M. L. H. Green, *Nat. Mater.*, 2003, **2**, 788.
169. P. M. Ajayan, O. Stephan, P. Redlich and C. Colliex, *Nature*, 1995, **375**, 564.
170. Y. H. Shi, T. Hasan, N. H. Babu, F. Torrisi, S. Milana, A. C. Ferrari and D. A. Cardwell, *ACS Nano*, 2012, **6**, 5395.
171. X. Huang, S. Z. Li, S. X. Wu, Y. Z. Huang, F. Boey, C. L. Gan and H. Zhang, *Adv. Mater.*, 2012, **24**, 979.
172. S. Wang, B. M. Goh, K. K. Manga, Q. L. Bao, P. Yang and K. P. Loh, *ACS Nano*, 2010, **4**, 6180.
173. J. Kim, L. J. Cote and J. X. Huang, *Acc. Chem. Res.*, 2012, **45**, 1356.
174. Z. J. Li, W. Lv, C. Zhang, J. W. Qin, W. Wei, J. J. Shao, D. W. Wang, B. H. Li, F. Y. Kang and Q. H. Yang, *Nanoscale*, 2014, **6**, 9554.
175. X. Y. Xie, C. Zhang, M. B. Wu, Y. Tao, W. Lv and Q. H. Yang, *Chem. Commun.*, 2013, **49**, 11092.
176. Z. Y. Lu, J. X. Zhu, D. H. Sim, W. W. Zhou, W. H. Ship, H. H. Hng and Q. Y. Yan, *Chem. Mater.*, 2011, **23**, 5293.

177. T. Kyotani, N. Sonobe and A. Tomita, *Nature*, 1988, **331**, 331.
178. X. Huang, S. Li, Y. Huang, S. Wu, X. Zhou, S. Li, C. L. Gan, F. Boey, C. A. Mirkin and H. Zhang, *Nat. Commun.*, 2011, **2**, 292.
179. A. Y. Cho and J. R. Arthur, *Solid State Chem.*, 1975, **10**, 157.
180. P. Frigeri, L. Seravalli, G. Trevisi and S. Franchi in *Molecular Beam Epitaxy: An Overview*, ed. P. Bhattacharya, R. Fornari and H. Mamimura, Elsevier, Amsterdam, 2011, p. 480.
181. T. Osaka, H. Omi, K. Yamamoto and A. Ohtake, *Phys. Rev. B*, 1994, **50**, 7567.
182. I. Boustani, *Phys. Rev. B*, 1997, **55**, 16426.
183. K. Lau and R. Pandey, *J. Phys. Chem. C*, 2007, **111**, 2906.
184. E. Penev, S. Bhowmick, A. Sadrzadeh and B. Yakobson, *Nano Lett.*, 2012, **12**, 2441.
185. Y. Liu, E. Penev and B. Yakobson, *Angew. Chem., Int. Ed.*, 2013, **52**, 3156.
186. H. Liu, J. Gao and J. Zhao, *Sci. Rep.*, 2013, **3**, 3238.
187. Z. Piazza, H. Hu, W. Li, Y. Zhao, J. Li and L. Wang, *Nat. Commun.*, 2014, **5**, 3113.
188. X. Zhou, X. Dong, A. Oganov, Q. Zhu, Y. Tian and H. Wang, *Phys. Rev. Lett.*, 2014, **112**, 085502.
189. B. Feng, J. Zhang, Q. Zhong, W. Li, S. Li, H. Li, P. Cheng, S. Meng, L. Chen and K. Wu, *Nat. Chem.*, 2016, **8**, 563.
190. A. Mannix, X. Zhou, B. Kiraly, J. Wood, D. Alducin, B. Mayers, X. Liu, B. Fisher, U. Santiago, J. Guest, M. Yacaman, A. Ponce, A. Oganov, M. Hersam and N. Guisinger, *Science*, 2015, **350**, 1513.
191. B. Feng, J. Zhang, Q. Zhong, W. Li, S. Li, H. Li, P. Cheng, S. Meng, L. Chen and K. Wu, *Nat. Chem.*, 2016, **8**, 563.
192. Z. Zhang, Y. Yang, G. Gao and B. Yakobson, *Angew. Chem., Int. Ed.*, 2015, **54**, 13022.
193. E. Penev, A. Kutana and B. Yakobson, *Nano Lett.*, 2016, **16**, 2522.
194. M. Gao, Q. Li, X. Yan and J. Wang, *Phys. Rev. B*, 2017, **95**, 024505.
195. P. Vogt, P. De Padova, C. Quaresima, J. Avila, E. Frantzeskakis, M. C. Asensio, A. Resta, B. Ealet and G. Le Lay, *Phys. Rev. Lett.*, 2012, **108**, 155501.
196. Z. Liu, M. Wang, J. Xu, J. Ge, G. Lay, P. Vogt, D. Qian, C. Gao, C. Liu and J. Jia, *New J. Phys.*, 2014, **16**, 095002.
197. E. Bianco, S. Butler, S. Jiang, O. Restrepo, W. Windl and J. Goldberger, *ACS Nano*, 2013, **7**, 4414.
198. S. Jiang, S. Butler, E. Bianco, O. Restrepo, W. Windl and J. Goldberger, *Nat. Commum.*, 2014, **5**, 3389.
199. F. Zhu, W. Chen, Y. Xu, C. Gao, D. Guan, C. Liu, D. Qian, S. Zhang and J. Jia, *Nat. Mater.*, 2015, **14**, 1020.
200. J. Zeng, P. Cui and Z. Zhang, *Phys. Rev. Lett.*, 2017, **118**, 046101.
201. M. Ritala and M. Leskela in *Handbook of Thin Films*, ed. H. S. Nalwa, Academic Press, San Diego (CA), 2002, p. 103.
202. S. M. George, *Chem. Rev.*, 2010, **110**, 111.

203. T. W. Scharf, S. V. Prasad, M. T. Dugger, P. G. Kotula, R. S. Goeke and R. K. Grubbs, *Acta Mater.*, 2006, **54**, 4731.
204. J. Song, J. Park, W. Lee, T. Choi, H. Jung, C. Lee, S. Hwang, J. Myoung, J. Jung, S. Kim, C. Lansalot-Matras and H. Kim, *ACS Nano*, 2013, **12**, 11333.
205. D. Nandi, U. Sen, D. Choudhury, S. Mitra and S. Sarkar, *Electrochim. Acta*, 2014, **146**, 706.
206. L. Tan, B. Liu, J. Teng, S. Guo, H. Low and K. Loh, *Nanoscale*, 2014, **6**, 10584.
207. A. Delabie, M. Caymax, B. Groven, M. Heyne, K. Haesevoets, J. Meersschaut, T. Nuytten, H. Bender, T. Conard, P. Verdonck, S. van Elshocht, S. De Gendt, M. Heyns, K. Barla, I. Radu and A. Thean, *Chem. Commun.*, 2015, **51**, 15692.
208. K. Rao, B. Vaidhyanathan, M. Ganguli and P. Ramakrishnan, *Chem. Mater.*, 1999, **11**, 882.
209. I. Bilecka and M. Niederberger, *Nanoscale*, 2010, **2**, 1358.
210. D. Chuang, *J. Mater. Sci.*, 1987, **22**, 4190.
211. B. Tryba, A. Morawski and M. Inagaki, *Carbon*, 2005, **43**, 2417.
212. E. Falcao, R. Blair, J. Mack, A. Morawski and M. Inagaki, *Carbon*, 2005, **43**, 2417.
213. T. Wei, Z. Fan, G. Luo, C. Zheng and D. Xie, *Carbon*, 2009, **47**, 337.
214. Y. Zhu, S. Murali, M. Stoller, A. Velamakanni, R. Piner and R. Rouff, *Carbon*, 2010, **48**, 2118.
215. D. Voiry, J. Yang, J. Kupferberg, R. Fullon, C. Lee, H. Jeong, H. Shin and M. Chhowalla, *Science*, 2017, **353**, 6306.
216. S. Bhunia and D. Bose, *J. Cryst. Growth*, 1998, **186**, 535.
217. R. Kerner, O. Palchik and A. Gedanken, *Chem. Mater.*, 2001, **13**, 1413.
218. A. Whittaker and D. Mingos, *J. Chem. Soc., Dalton Trans.*, 1992, 2751.
219. C. Landry and A. Barron, *Science*, 1993, **260**, 1653.
220. P. M. Ajayan and B. I. Yakobson, *Nat. Mater.*, 2011, **10**, 415.
221. M. Chhowalla, D. Voiry, J. Yang, H. S. Shin and K. P. Loh, *MRS Bull.*, 2015, **40**, 585.
222. A. K. Geim and I. V. Grigorieva, *Nature*, 2013, **499**, 419.

CHAPTER 5

Liquid Phase Synthesis of Two-dimensional Crystals: from Top-down to Bottom-up

HUI WANG, WEI SHAO, SHICHUAN CHEN AND XIAODONG ZHANG*

University of Science and Technology of China, Department of Chemistry, Hefei, Anhui 230026, P. R. China
*E-mail: zhxid@ustc.edu.cn

5.1 Introduction

Since the discovery and identification of graphene in 2004,[1] two-dimensional (2D) nanosheets with atomic thickness have attracted increasing research interest in a series of areas for their unique electronic structures. Benefiting from their ultra-large specific surface area and unique electronic structures, ultrathin 2D crystals exhibit promising applications in numerous fields, such as catalysis, sensing, energy storage and conversion, and so on.[2–30] Given the correlative surface effects, electronic effects and quantum confined effects, ultrathin 2D nanosheets are considered as a satisfactory junction that could combine the microscopic structural features with outstanding macroscopic properties, realizing the maximum functionality with a minimized size.[29,30] Moreover, ultrathin 2D nanosheets also present excellent mechanical flexibility and optical transparency associated with their large lateral size and

Smart Materials No. 27
Inorganic Two-dimensional Nanomaterials: Fundamental Understanding, Characterizations and Energy Applications
Edited by Changzheng Wu

Published by the Royal Society of Chemistry, www.rsc.org

atomic thickness, enabling them to be fascinating candidates for application as next-generation electronic devices with highly transparent and flexible characters.[26–35] For example, exfoliated metallic MoS_2 nanosheets could electrochemically intercalate ions with extraordinary efficiency and achieve high capacitance values in a variety of aqueous electrolytes,[36] ultrathin black phosphorus nanosheets show outstanding electron mobility which promise the applications in field effect transistors and phototransistors,[37–40] ultrathin g-C_3N_4 nanosheets show enhanced intrinsic photo-absorption and photoresponse with respect to the bulk counterpart, and can be used for bioimaging applications,[28] single layered SnS_2 has been found to show efficient visible light water splitting properties,[31] and so on.

Nevertheless, the difficulties in synthetic methods for the atomically thick 2D nanosheets undoubtedly restrict research interests. Although various methods such as chemical vapor deposition (CVD)[41,42] and mechanical cleavage[9,43,44] strategies have been proposed, most of them were restricted to layered materials with weak van der Waals forces between the layers and simple components, such as graphite,[45,46] hexagonal boron nitride (h-BN),[47,48] and layered transition metal dichalcogenides,[49,50] which surely limit their further explorations. As for the complicated non-layered materials and quasi-layer materials with relatively strong bonds between the layers, these techniques seem to be useless. Here, non-layered materials imply the materials possess isotropic structures such as Co_9Se_8,[25] VO_2(M),[51] ZnSe,[52] and so on. In detail, VO_2(M) can be taken as an example, whose monoclinic structure does not show obvious anisotropy as all octahedral [VO_6] are linked by sharing an edge along the *a*-axis to form chains, and the chains are then interlinked by sharing corner oxygen atoms to form a 3D framework. The quasi-layered materials are the materials which possess an obviously anisotropic layered structure but with relatively strong bonds between the layers. For example, layered ternary carbides or nitrides[53–56] (MAX phase, M is a transition metal, A is a group IIIA or IVA element, and X is carbon or nitrogen atom) consist of alternately arranged MX and A layers, where the bonds between the MX layer and A layer are much stronger than the van der Waals forces but weaker than the bonds of M–X. Moreover, the extremely low yields for ultrathin 2D crystals through these methods also remain problematic. Therefore, the exploration of novel and general synthetic methods to extend research objects to atomically thick 2D crystals with a non-layered structure or quasi-layered structure with relatively strong bonds between the layers is urgently demanded.

To address these problems, liquid phase synthesis[25,57–60] has been proposed for the preparation of 2D nanosheets, which exhibits intriguing potential for various advantages like facile control, economic saving, environmental protection, and mass production, which could greatly enrich the family of 2D crystals with atomic thickness and give rise to enormous properties. Generally, liquid phase synthesis could be mainly divided into 'top-down' and 'bottom-up' methods, based on the relationship between reactants and target products. The top-down liquid phase synthesis usually implies the liquid exfoliation process, during which layered materials with weak van der Waals forces between layers are expected to be exfoliatable into 2D nanosheets with atomic thickness with proper solvent under ultrasonication treatments.[57–59] However, the bottom-up strategy in liquid phase synthesis denotes controlled assembly of small building blocks to 2D crystals,[24,51,52,60] which is considered an exclusive method for the preparation of nanosheets with atomic thickness of non-layered structure materials due to the lack of intrinsic driving force for 2D anisotropic growth. Accompanied by the boom in graphene analogues, an increasing number of modified liquid phase synthetic methods have been proposed, giving rise to a wide range of 2D nanosheets with atomic thickness.

Herein, we give an overview of the recent advances in liquid phase synthesis of atomically thick 2D nanosheets, especially modified strategies for the synthesis of 2D materials with non-layered structures and quasi-layered structures with relatively strong weak van der Waals forces between layers. As for top-down liquid phase synthesis, various modified liquid exfoliation methods are summarized, while the self-assembly strategy, oriented attachment strategy and template-directed strategy are listed for bottom-up liquid phase synthesis.

5.2 Top-down Strategies

In 2008, a liquid exfoliation method was first proposed for the preparation of defect-free graphene by Coleman and co-workers,[57] in whose study pristine graphite was peeled into few-layered nanosheets in organic solvents by ultrasound treatment, establishing a new era for the preparation of 2D materials. According to several guiding works,[58,59] the key factor for liquid exfoliation strategy is the well matched surface energies between the as-exfoliated materials and the disperse solvents, which could reduce the driving energy and thus improve the exfoliation efficiency. Up to now, the liquid exfoliation method has been expanded to the ultrasonic exfoliation of various layered materials in proper solvents (such as water, ethanol, isopropanol, *N*-methylpyrrolidone, dimethylformamide and so on), resulting in numerous layered ultrathin 2D nanosheets. For example, Zhang *et al.* reported the fabrication of few-layered graphitic carbon nitride nanosheets through a facile liquid exfoliation strategy, in which water was demonstrated to be the most efficient solvent (Figure 5.1(a) and (b)), facilitating their further applications for bioimaging.[28]

Recently, Wang *et al.* successfully obtained ultrathin black phosphorus nanosheets with a thickness of only about 2 nm through liquid exfoliation of bulk counterpart in water (Figure 5.1(c) and (d)).[61] The simple liquid exfoliation method has many advantages for the preparation of ultrathin 2D nanosheet such as facile control, cost saving, environmental protection, and mass production. However, the co-existing disadvantages of liquid exfoliation method are also obvious: the relatively low exfoliation efficiency (Figure 5.1(e)) and uncontrollable thickness of the as-exfoliated nanosheets (Figure 5.1(f)); and most importantly, the application of direct liquid exfoliation

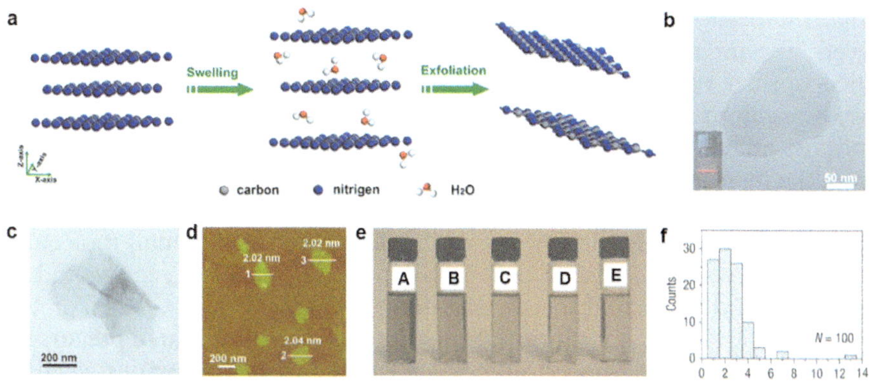

Figure 5.1 (a) Schematic illustration of liquid-exfoliation process for preparing ultrathin g-C_3N_4 nanosheets from corresponding bulk counterpart. (b) A typical TEM image of an ultrathin g-C_3N_4 nanosheet. Inset: photograph of the Tyndall effect of the ultrathin g-C_3N_4 nanosheet dispersed in water. (c) TEM image of ultrathin black phosphorus nanosheets obtained through liquid-exfoliation process with water as dispersant. (d) AFM image of as-exfoliated ultrathin black phosphorus nanosheets, showing a height of about 2.0 nm, which is consistent with about four individual black phosphorus layers. (e) Photograph of graphite flakes dispersed in *N*-methyl-2-pyrrolidone with a series of concentrations ranging from 6 to 4 μg mL^{-1} after centrifugation. (f) Histogram of the number of visual observations of as-exfoliated flakes as a function of the number of monolayers per flake for NMP dispersions. (a) and (b) Reproduced with permission from ref. 28. Copyright 2013 American Chemical Society. (c) and (d) Reproduced with permission from ref. 61, Copyright 2015 American Chemical Society. (e) Reprinted with permission from Macmillan Publishers Ltd; *Nature Nanotechnology* (ref. 57), Copyright 2008.

method for preparing two dimensional crystals is mainly confined to layered materials with weak van der Waals forces between the layers. As for the quasi-layered materials with strong van der Waals interactions between the layers and the non-layered materials, direct liquid exfoliation method seemed to be useless. Just recently, several modified liquid exfoliation strategies have been proposed to synthesize 2D crystals of quasi-layered materials with relatively strong van der Waals interactions or even the non-layered materials. Here, we review the advances in preparing atomically thick 2D crystals through 'top-down' liquid phase strategies.

5.2.1 Small Molecules Assisted Exfoliation Strategy

A small molecule assisted exfoliation strategy[30,62–65] is proposed for preparing 2D crystals of layered materials with strong van der Waals interactions that could not be exfoliated through direct liquid exfoliation method (or extremely low exfoliation efficiency). By designing layered precursors with small intercalated molecules (such as NH_3, H_2O, amine, *etc.*) between stacked layers, the interlayer distances would be enlarged, thus leading to weakened van der Waals interactions. In that case, the intercalated precursors can be easily exfoliated into single- or few-layer target nanosheets in proper solvents under sonication. The small intercalated molecules would escape from the interlayers during ultrasonication treatment, accelerating the exfoliation rate and improving the exfoliation efficiency towards simple liquid exfoliation process. In view of these, the design of proper small molecules intercalated precursors (such as $TiS_2·NH_3$, $VS_2·NH_3$, $VSe_2·NH_3$, *etc.*) would be critical towards a small molecule assisted exfoliation strategy. Based on the gas molecule assisted exfoliation strategy, numerous intercalated precursors with small molecules (including but not limited to gas molecules, such as H_2O, amine, ethylene glycol, and so on) are used for preparation of corresponding 2D crystals. For example, Wu *et al.* successfully achieved the scalable production of high-quality ultrathin $VOPO_4$ nanosheets by exfoliating a bulk $VOPO_4·2H_2O$ precursor. The hydrogen bonds between molecular H_2O and $VOPO_4$ layers could be broken during ultrasonication in 2-propanol for 15 min (Figure 5.2(a)).

During the exfoliation process, the intercalated H_2O molecules escaped, leading to the formation of freestanding ultrathin $VOPO_4$ nanosheets (transmission electron microscopy image, TEM, Figure 5.2(b)) with a thickness of only about 4.07 nm, which is consistent with about 5–6 single layers (atomic force microscopy image, AFM, Figure 5.2(c)).[62] They also reported obtaining atomically large-area VS_2 nanosheets through exfoliation of $VS_2·NH_3$ in water.[30] The layered $VS_2·NH_3$ precursor was ultrasonicated in water for about 1 h, during which the intercalated NH_3 molecules escaped, leading to the formation of ultrathin VS_2 nanosheets with a thickness of only about 2.5 nm (corresponding to 4 or 5 S–V–S single layers). The clean VS_2 nanosheets without residues of foreign ions could reflect the intrinsic properties of the 2D VS_2 crystals. Note that the exfoliation efficiency of $VS_2·NH_3$ is much higher than that of bulk VS_2 with a 1 h ultrasonication. Recently Jeong *et al.* have reported a tandem intercalation strategy for preparing single layer nanostructures of group IV (TiS_2, ZrS_2), group V (NbS_2) and VI (WSe_2, MoS_2) transition metal chalcogenides (TMCs).[63] The tandem intercalation requires tandem Lewis base intercalates, where short initial molecules are intercalated into TMCs firstly (Figure 5.2(d) and (e)), giving rise to an opened up interlayer gap, then the long primary molecules extend the gap to full width so that a random mixture of intercalates overcomes the interlayer force. This exfoliation method establishes a fast process for producing single layer TMCs under ambient conditions, avoiding the occasional degradation of single layer TMCs by avoiding harsh exfoliation processes, extending molecule assisted exfoliation strategies. Note that the gas molecule assisted exfoliation strategy is different from lithium ion intercalation (see details in **5.2.2**), as the latter involved partial reduction of layered materials.

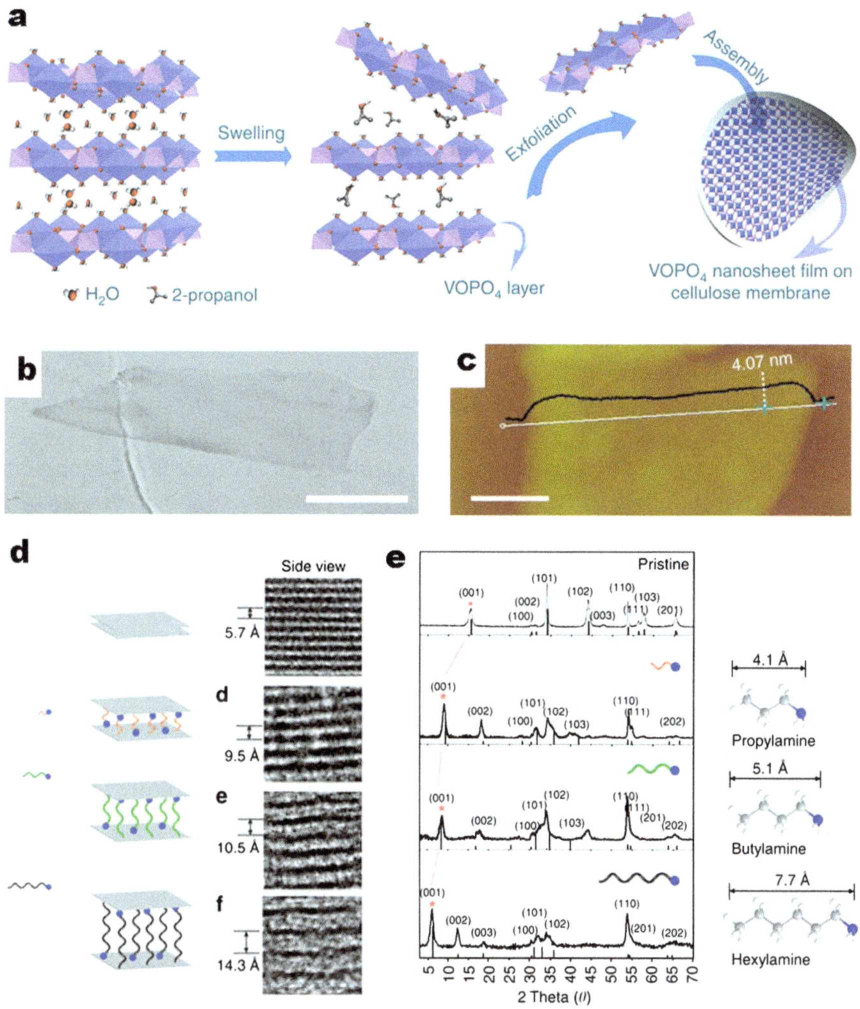

Figure 5.2 (a) Schematic illustration of small molecules assisted exfoliation strategy to fabricate $VOPO_4$ nanosheets, using layered $VOPO_4{\cdot}2H_2O$ as precursor. (b) TEM image of a $VOPO_4$ nanosheet, of which the high transparency implies its ultrathin feature. (c) AFM image of a $VOPO_4$ nanosheet, showing a thickness of 4.07 nm that corresponds to less than six atomic layers. (d) XRD patterns of pristine TiS2 and intercalated TiS2 with different amines (propylamine, butylamine and hexylamine, respectively), while the vertical lines indicate reference peaks of bulk and expanded TiS_2 (JCPDS #15-0853). (a)–(c) reprinted with permission from Macmillan Publishers Ltd: *Nature Communications* (ref. 62), Copyright 2013. (d) and (e) reprinted with permission from Macmillan Publishers Ltd; *Nature Communications* (ref. 63), Copyright 2015.

5.2.2 Ions Intercalation–deintercalation Assisted Exfoliation Strategy

For layered materials, enlarged interlayer distances and thus weakened van der Waals interactions between stacking layers could be realized by inserting ions (Li, Na, K, *etc.*) into the interlayer spaces. The resulting ion intercalated intermediates would be effectively exfoliated into atomically thick nanosheets in certain solvents (as schematically illustrated in Figure 5.3(a)).[5,36,66–69]

The ion intercalation–deintercalation assisted exfoliation strategy has been widely applied in preparing atomically thick 2D crystals of layered materials, especially for transitional metal chalcogenides. Recently, Lin *et al.* successfully obtained ultrathin TiS_2 nanosheets by treating the corresponding bulk materials in a solution of *n*-butyllithium in an inert atmosphere, followed by an ultrasonication treatment in water.[66] The heights of the as-exfoliated TiS_2 nanosheets measured from an AFM image ranged from 1.97 to 2.97 nm, which corresponded to three and five individual TiS_2 layers (Figure 5.3(b)). *n*-Butyllithium has been widely used as a typical intercalation reagent for preparing ultrathin transitional metal chalcogenides nanosheets, resulting in numerous atomic atomically thick 2D crystals such as MoS_2 (as AFM and HRTEM image shown in Figure 5.3(d)),[36] $MoSe_2$, WS_2, WSe_2, $NbSe_2$,[68] and so on. Apart from transitional metal chalcogenides, some other 2D crystals were also obtained in this way such as ZrNCl (Figure 5.3(c)),[67] and Cu_2WS_4 nanosheets.[69] It is worth noting that the reduction of metal atom (or, unfilled d orbital) is the key factor for the intercalation of *n*-butyllithium into layered materials. Given the partial reduction during lithium intercalation, this strategy should be distinguished from the molecule assisted exfoliation strategy, and the involved reduction and surface adsorption would greatly influence the properties of the materials.

Although the ion intercalation–deintercalation assisted exfoliation strategy has been proven to be effective for the preparation of atomically thick nanosheets, the drawback is equally clear: the involved organolithium are extremely reactive and sensitive to water and oxygen, and the experimental process is relatively complicated and must be carried out in an inert atmosphere. To solve this problem, Zhang and co-workers proposed an effective approach for preparing high-yield single-layer semiconducting nanosheets, based on a controllable electrochemical lithiation process. By using corresponding layered bulk materials as cathode, the lithium intercalation in the bulk material can be realized during the discharge process, giving rising to a series of 2D crystals with atomic thicknesses, including MoS_2, WS_2, TiS_2, TaS_2, ZrS_2, and graphene (as shown in Figure 5.3(e)).[70] The controllable electrochemical lithiation process established an intriguing potential for preparing high quality 2D nanosheets.

However, these strategies are mainly restricted to the design of ultrathin nanosheets of layered materials, and are inapplicable for non-layered materials. On the basis of crystal structure, Liu *et al.* developed a room-temperature lithium intercalation-deintercalation strategy to prepare atomically

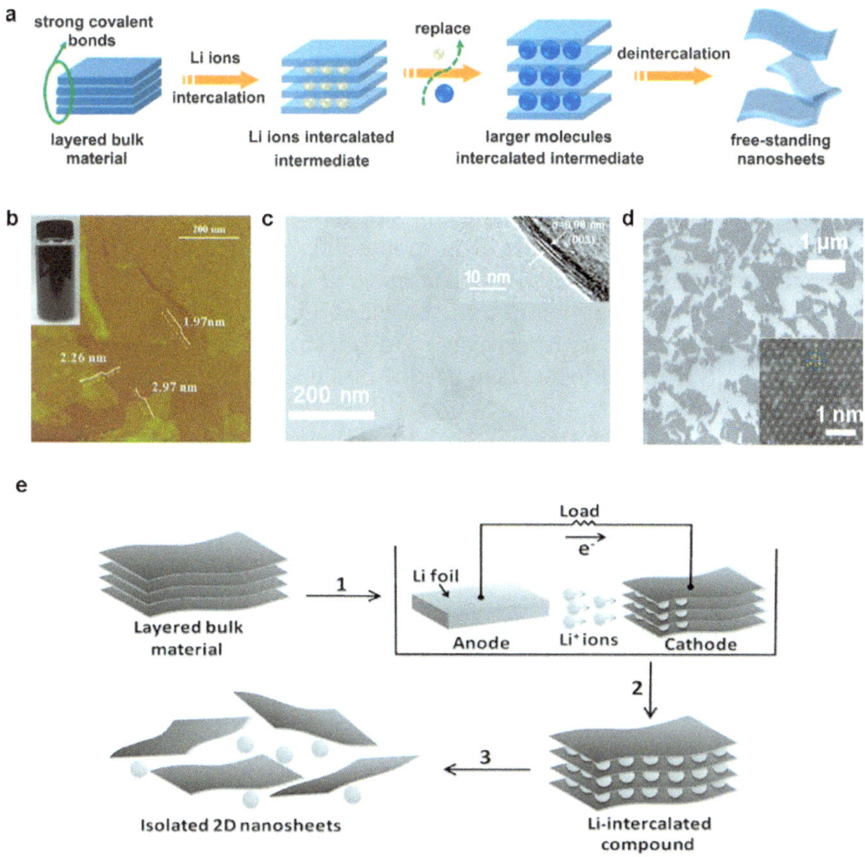

Figure 5.3 (a) Schematic illustration of the preparation of atomic thick 2D crystals through the lithium Ions intercalation-deintercalation assisted exfoliation strategy. (b) AFM image of the ultrathin TiS_2 nanosheets, exhibiting heights ranging from 1.97 to 2.97 nm, corresponding to three and five layers of TiS_2, respectively. Inset: photograph of the as-exfoliated TiS_2 nanosheets suspension. (c) TEM image of the exfoliated ZrNCl nanosheets. Inset: HRTEM image of the folded edge of an ultrathin nanosheet. (d) SEM image of monolayer 1T phase MoS_2 nanosheets prepared through lithium Ions intercalation-deintercalation assisted exfoliation strategy. (e) Illustration of the electrochemical lithiation process for the fabrication of atomic thick 2D crystals. (a) Reproduced from ref. 5 with permission of Royal Society of Chemistry (b) Reproduced with permission from ref. 66, Copyright 2013 American Chemical Society. (c) Reproduced with permission from ref. 67, Copyright 2015 American Chemical Society. (d) Reprinted with permission of Macmillan Publishers Ltd: *Nature Nanotechnology* (ref. 36), Copyright 2015. (e) Reproduced with permission from Z. Zeng *et al.*, Single-layer Semiconducting Nanoshets: High-Yield Preparation and Device Fabrication, *Angewandte Chemie International Edition*, John Wiley and Sons, Copyright © 2011 WILEY-VCH Verlag GmbH & Co. KGaA, Weinheim.[70]

thick nanosheets of valence-sensitive quasi-layered material with relatively strong interactions between the layers.[71] This method is relatively suitable for the preparation of atomically thick 2D crystals of valence-sensitive materials, due to the absence of any high temperature treatment during the lithium intercalation-deintercalation process. As a representative case, clean and free-standing single quasi-layered nanosheets of VO_2(B) were prepared. The structure of VO_2(B) is transformed from a bulk sample to corresponding single-layered nanosheets with a thickness of about 0.6 nm and an unchanged in-plane crystal structure by the lithium intercalation-deintercalation exfoliation strategy. Beyond the above preparation of atomically thick 2D crystals of quasi-layered materials with strong covalent bonds, it is worth noting that quasi-layered materials with weak van der Waals forces can also be exfoliated into atomically thick nanosheets by the lithium intercalation-deintercalation exfoliation method.

Recently, Yao *et al.* developed a structural phase transition (SPT) assisted lithium intercalation-deintercalation strategy for the exfoliation of non-layered VO_2(M), and there is a totally reversible SPT between VO_2(M) and VO_2(R) accompanied by lattice expansion-contraction at the critical temperature (T_c, ~ 68 °C).[51] The detailed steps of the SPT assisted lithium intercalation-deintercalation strategy are chemical lithiation of VO_2(M) at low temperatures (below T_c) which forms a intermediate with a lamellar morphology and maintains the monoclinic phase and deintercalation-exfoliation of lithiated VO_2(R) at a temperature higher than T_c, during which the lattice expansion and anisotropic strain induced by SPT could weaken the bond strength of V–O and promote the deintercalation-exfoliation process. The SPT assisted lithium intercalation-deintercalation strategy offers a practical way to design atomically thick 2D crystals of non-layered materials with reversible phase transition.

The lithium-intercalation strategy can provide an excellent opportunity for realizing phase transformation during the exfoliation process, which enables some unprecedented properties for promising applications.[36] Nevertheless, despite fact that the lithium-intercalation method can synthesize ultrathin 2D nanosheets with high production yields, it still has some drawbacks, and is not a universal strategy for preparing ultrathin nanosheets of non-layered materials.

5.2.3 Ion Exchange Based Exfoliation Strategy

The emphasis for the exfoliation of quasi-layered materials with relatively strong bonds between the layers into atomically thick nanosheets is on the establishment of effective methods to weaken the van der Waals forces between the layers. As one of the most universal methods, the ion exchange strategy has long been studied for its easy operation, and has been widely applied in the exfoliation of layered double hydroxide (LDH),[72,73] layered titanate,[74] metal oxides,[75,76] and so on. By selecting suitable intercalated ions,

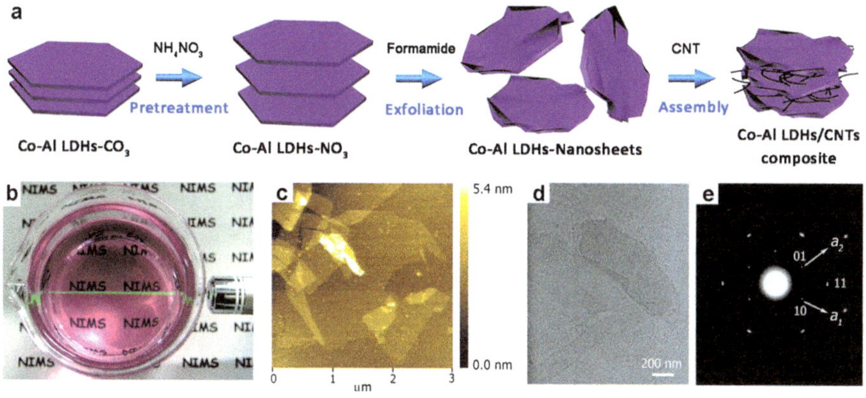

Figure 5.4 (a) The schematics of anion-change strategy for the exfoliation of Co-Al LDHs-CNTs composite. (b) Photograph of a colloidal suspension of exfoliated Co-Al LDH nanosheets. The light beam was incident from the side to demonstrate the Tyndall effect. (c) Tapping-mode AFM image of the exfoliated LDH nanosheets deposited on a Si wafer substrate. (d) TEM image and (e) SAED pattern of LDH nanosheets. (a) Reproduced from ref. 72 with permission of Royal Society of Chemistry. (b)–(e) Reproduced with permission from ref. 74, Copyright 2006 American Chemical Society.

the van der Waals forces could be significantly reduced, followed by a liquid exfoliation process to obtain atomically thick 2D nanosheets. The delamination of LDHs is the most typical example, where Sasaki's group[73,74] synthesized a wide range of LDH 2D nanosheets with atomic thickness. Taking the Co-Al LDH nanosheets as an example, the Co-Al LDH hexagonal platelets were firstly obtained through a hydrothermal method, then filled with CO_3^{2-} that endowed the precursor (as shown in Figure 5.4(a)) with a tough Van der Waals force between the layers, leading to the negligible exfoliation efficiency of Co-Al-CO_3 LDH platelets.

The as-prepared precursor was converted to Cl^--LDH by treating with a NaCl–HCl mixed solution, which could further be exchanged by a variety of anions (such as NO^{3-}, ClO^{4-}, acetate, lactate, dodecyl sulfate, and oleate) *via* an anion-exchange process employing corresponding salts, retaining its high crystallinity and hexagonal plate-like morphology. Exchanged products in various anion forms were found to show different delamination behaviors in amide form, obtaining Co-Al LDH nanosheets with an average thickness of ~0.80 nm (as illustrated in Figure 5.4(b)–(e)). Sasaki's group also reported the exfoliation process of layered titanate to obtain TiO_2, in which layered protonic titanate obtained through ion exchange has been exfoliated by the action of an aqueous solution of tetrabutylammonium (TBA) hydroxide, paving a general pathway for the preparation of atomically thick oxide 2D materials MnO_2, HNb_3O_8, $HNbMoO_6$, $Ca_2Nb_3O_{10}$, and so on.[75–83]

5.2.4 Lamellar Hybrid Intermediate Based Exfoliation Strategy

'Top-down' liquid phase strategies are usually considered to be feasible for the preparation of 2D crystals of layered materials with weak van der Waals interactions between stacking layers. In the case of quasi-layered materials with strong bonds or even covalent bonds between the layers, these exfoliation strategies seem to be useless, and it is necessary to find new synthesis strategies. To this end, a lamellar hybrid intermediate based exfoliation strategy was proposed for the preparation of atomically thick nanosheets of non-layered materials.[52,84–87] For the preparation of atomically thick 2D crystals of non-layered materials, organic molecule intercalated lamellar hybrid intermediates were first synthesized, followed by a typical liquid exfoliation process, and the as-obtained lamellar hybrid intermediates were peeled into hybrid single layers covered with organic molecules. Finally, organic species of the hybrid single layers were removed by heat or other treatments, which would give rise to clean and free-standing atomically thick nanosheets. Benefiting from the novel lamellar hybrid intermediate based exfoliation strategy, a series of 2D crystals of non-layered materials have been prepared, including ZnSe,[52] $CoSe_2$,[84,85] CdS,[86] and CdSe.[87] As shown in Figure 5.5(a), lamellar hybrid intermediate $CoSe_2$/DETA (DETA stands for diethylenetriamine) was prepared through a solvothermal method.

The intermediate was then dispersed into ethanol and ultrasonicated in ice water to obtain ultrathin $CoSe_2$ nanosheets with an average thickness of 1.4 nm (as shown in Figure 5.5(b) and (c)). Similarly, single layered ZnSe (zinc blende structure) nanosheets with a lateral size of more than 500 nm and thickness of only about 0.9 nm were obtained by exfoliating the lamellar hybrid (Zn_2Se_2)(pa) (pa stands for *n*-propylamine) intermediate, in which propylamine molecules were removed by further heat treatment (Figure 5.5(d)). As another typical case, the (Zn_2S_2)(pa) intermediate can also be exfoliated into atomically thick nanosheets. In principle, the strategy introduced above may be a universal method expected to prepare atomically thick 2D crystals of all the other non-layered structure materials. Xu *et al.* synthesized ultrathin CdS nanosheets *via* this strategy, with lamellar CdS–DETA hybrid nanosheets as starting materials and L-cysteine as a stabilizing agent (as shown in Figure 5.5(e)).[86] The key point is preparation of the 2D crystals with clean surfaces, the intercalated organic species should be easily removed. There is a report on CdSe-RNH_2 inorganic-organic hybrid intermediate,[87] which is not suitable for the preparation of 2D CdSe nanosheets with clean surfaces because RNH_2 means long-chain alkylamine and cannot be easily removed.

5.2.5 2D Precursor-based Topotactic Reaction Strategy

The topotactic reaction strategy refers to the preparation of target nanosheets through an *in-situ* conversion from an atomic thick intermediate, and has been an effective method for 2D crystal fabrication, especially for non-layered

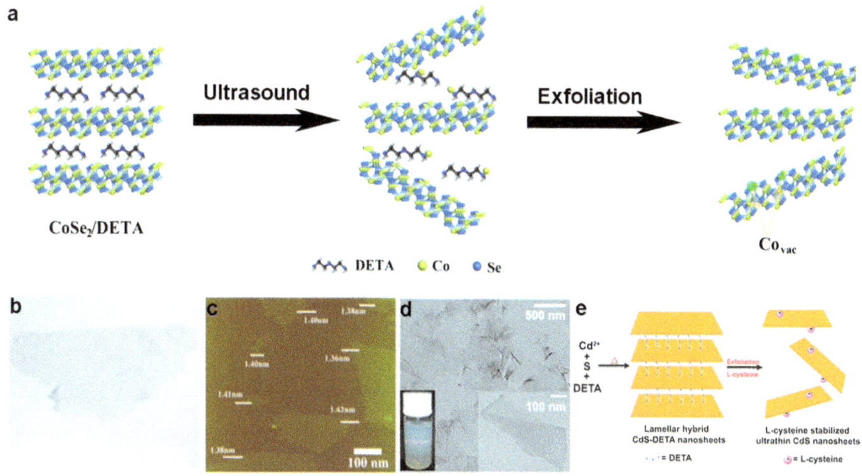

Figure 5.5 (a) Schematic illustration of the preparation of ultrathin $CoSe_2$ nanosheets from lamellar hybrid $CoSe_2$/DETA intermediate. (b) TEM image of an as-exfoliated $CoSe_2$ nanosheet. (c) AFM image of ultrathin $CoSe_2$ nanosheets, exhibiting an average thickness of ~1.4 nm. (d) TEM images of typical ZnSe single layered nanosheets, and the Tyndall effect of colloidal ethanol dispersion. (e) Schematic illustration of the preparation of ultrathin CdS nanosheets through the lamellar hybrid intermediate strategy with L-cysteine as assisted reagent. (a)–(c) Reproduced with permission from ref. 84, Copyright 2014 American Chemical Society. (d) Reprinted with permission from Macmillan Publishers Ltd: *Nature Communications*, ref. 52, Copyright 2012. (e) Reproduced from ref. 86 with permission of Royal Society of Chemistry.

materials. In detail, the as-obtained ultrathin 2D precursors need to undergo a certain treatment for topotactic conversion, such as heat, nitridation and hydrothermal reaction, to form target 2D nanosheets. Recently, Xie's group reported a range of metal oxide and nitrides nanosheets with non-layered structure (*e.g.*, CeO_2,[88] In_2O_3,[89,90] Co_3O_4,[91] $NiCo_2O_4$,[92] Co_4N, and Ni_3N[93]) topotactic conversion strategy. For instance, ultrathin CeO_2 nanosheets with numerous surface pits were successfully synthesized by a fast-heating topotactic conversion process. Briefly, the precursors of ultrathin $CeCO_3OH$ nanosheets were first fabricated through a hydrothermal reaction in the presence of oleate ions. Then, the resulting intermediate nanosheets were converted to atomically thick CeO_2 nanosheets with numerous surface pits *via* a fast annealing treatment in air, possessing an average thickness of only 0.6 nm (as illustrated in Figure 5.6(a)–(c)).[88]

Further experiments demonstrated that the heating temperature and time could greatly influence the formation of pits on their surface, enabling an unexpected fascinating property for further applications. Xu *et al.* reported that ultrathin nickel nitride nanosheets were prepared *via* a nitrogenizing reaction by topotactically converting the precursors of ultrathin Ni-based nanosheets

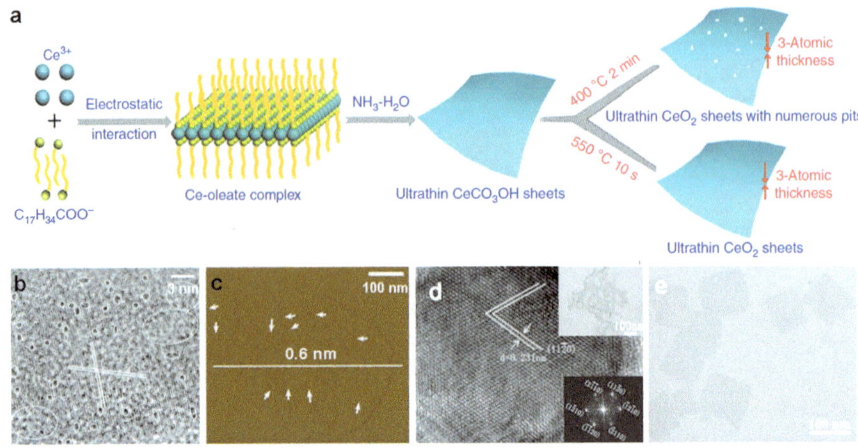

Figure 5.6 (a) Schematic illustration of the formation of ultrathin CeO_2 sheets with numerous surface pits and intact ultrathin CeO_2 sheets. (b) HRTEM image of an ultrathin CeO_2 sheet with numerous surface pits, from which one can figure out the presence of sub-nm pits. (c) AFM image of an ultrathin CeO_2 sheet with numerous surface pits, possessing a thickness of ~0.6 nm. (d) HRTEM image of a typical Ni_3N nanosheet. Inset: corresponding FFT pattern and TEM image. (e) TEM image of the single-unit-cell Bi_2WO_6 layers obtained through 2D precursor-based topotactic reaction strategy. (a)–(c) Reprinted with permission of Macmillan Publishers Ltd: *Nature Communications* (ref. 88), Copyright 2013. (d) Reproduced with permission from ref. 93, Copyright 2015 American Chemical Society. (e) Reproduced with permission from L. Liang *et al.*, Single unit cell bismuth tungstate layers realizing robust solar CO_2 reduction to methanol, *Angewandte Chemie Internaional Edition*, John Wiley and Sons, Copyright © 2015 WILEY-VCH Verlag GmbH & Co. KGaA, Weinheim.[94]

under heat treatment in NH_3 atmosphere (Figure 5.6(d)).[93] Moreover, the sulfurizing topotactic conversion strategy was carried out to successfully synthesize metallic ultrathin iron-nickel sulfide nanosheets from the Fe–Ni layered double hydroxide precursors through a solvothermal reaction. Note that layered materials could also be obtained *via* topotactic conversion strategy, such as Bi_2WO_6 nanosheets (Figure 5.6(e)), using layered Bi-oleate as precursor.[94]

Furthermore, the free-standing silicon (Si) nanosheets can also be obtained by 2D precursor-based topotactic reaction strategy.[95,96] Calcium disilicide ($CaSi_2$), a quasi-layered material consists of alternating arranged Ca and Si layers held together by strong van der Waals interactions, was selected as a precursor. A direct liquid exfoliation process could not break the bonds. However, the Ca layers can be extracted from $CaSi_2$ to form Si layers by treatment with hydrochloric acid (HCl) solution. Similarly, the surface of as-formed Si layers will be terminated by different terminating groups which are intercalated with each other by weakened bonds. Then, the silicon (Si) nanosheets with different terminated groups (H or OH) and thickness lower than 1 nm can be obtained by further ultrasonication treatment.

Another example of the topotactic conversion strategy is the well-known layered ternary carbides or nitrides (MAX phases). It has long been considered that the MAX phases formed by alternately arranged MX and A layers cannot be exfoliated into ultrathin nanosheets by direct exfoliation processes because there are relatively strong bonds between the MX and A layers. A topotactic conversion strategy by a chemical etching method was generally applied for the fabrication of 2D crystals of transition metal nitrides or carbides (that is, MX layers).[53–56] A schematic illustration of selective HF (hydrofluoric acid) etching of A layers of MAX phases is shown in Figure 5.7(a). After etching, the interaction between the residual MX layers should become weakened and can be successfully exfoliated into ultrathin nanosheets (named 'MXene'). As one of the representative cases, the Ti_3AlC_2 of the '312' MAX phase would be converted into $Ti_3C_2(OH)_2/Ti_3C_2F_2$ for the extraction of Al layers in HF solution *via* the following reactions:

$$Ti_3AlC_2 + 3HF \rightarrow AlF_3 + 3/2H_2 + Ti_3C_2 \tag{5.1}$$

$$Ti_3C_2 + 2H_2O/HF \rightarrow Ti_3C_2(OH)_2/Ti_3C_2F_2 + H_2 \tag{5.2}$$

As shown in eqn (5.1), individual Ti_3C_2 layers can be obtained through the treatment with HF, due to the weakened interaction caused by the removal of Al atoms from the middle layers. At once, the surface of as-formed Ti_3C_2 are then chained to OH or F groups, then the free-standing $Ti_3C_2(OH)_2/Ti_3C_2F_2$ nanosheets will be obtained by further ultrasonication treatment (Figure 5.7(b)). The as-exfoliated nanosheets named 'MXene' vary from a single layer (about 1 nm in thickness) to several layers. Note that it is crucial for the design of the target 2D crystals to choose appropriate corrosive reagents during the etching process. It is worth noting that the process of chemical etching can make the as-exfoliated nanosheets functionalized unavoidably by external groups, so the as-exfoliated nanosheets will show very different electronic structures compared with the pristine nanosheets.

In addition, a substitution-solid-solution based exfoliation strategy has also been proposed to exfoliate the quasi-layered MAX materials into ultrathin nanosheets by forming a Hume-Rothery based substitution-solid-solution in those materials to activate the A layers.[97] Compared to the MX layers, he A layers of MAX phases are more reactive, so A atoms of the A layers should be substituted partly by other elements to form a substitution solid solution, and the A layers would be further activated to make MAX phases exfoliable (as shown in Figure 5.7(c)). As a typical case, as-doped $Ti_3Si_{0.75}Al_{0.25}C_2$ (TSAC) that was obtained by substituting 1/4 of Si with Al atoms of Ti_3SiC_2 ('312' MAX phase) have been successfully exfoliated to nanosheets with a thickness of about 4 nm by the substitution-solid-solution based exfoliation strategy.[97] As shown in Figure 5.7(d), the as-exfoliated TSAC nanosheets with sizes ranging from 100 to 200 nm show a thickness of only about 4 nm, which is matched well with the about two to three layer thickness of TSAC

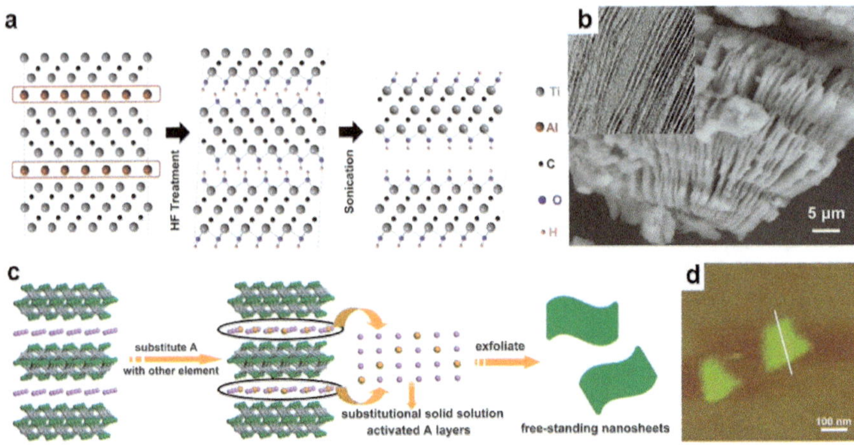

Figure 5.7 (a) Schematic illustration of selective HF etching of A layers of MAX phases for the preparation of MXene. (b) SEM image of bulk Ti3AlC2 sample after HF treatment. Inset: HRTEM image of the MXene nanosheets in side view. (c) Schematic illustration of the substitutional-solid-solution based exfoliation for the preparation of atomically thick 2D crystals in the case of MAX phases. (d) AFM image of the TSAC nanosheets, showing a thickness of ~4 nm. (a) and (b) Reproduced with permission from M. Naguib *et al.*, Two-Dimensional Nanocrystals Produced by Exfoliation of Ti3AlC2, *Advanced Materials*, John Wiley and Sons, Copyright © 2011 WILEY-VCH Verlag GmbH & Co. KGaA, Weinheim.[53] (c) and (d) Reproduced with permission from X Zhang *et al.*, Ultrathin Nanosheets of MAX Phases with Enhanced Thermal and Mechanical Properties in Polymeric Compositions: Ti3Si0.75Al0.25C2, *Angewandte Chemie International Edition*, John Wiley and Sons, Copyright © 20XX WILEY-VCH Verlag GmbH & Co. KGaA, Weinheim.[97]

crystal. In fact, there are various pairs of MAX phases that can form stable substitution-solid-solutions with each other, so long as they obey the basic Hume-Rothery rules, such as, Ti_2AlC and Ti_2GaC, Hf_2InC and Hf_2SnC, V_2GaC and V_2GeC, Ti_3AlC_2 and Ti_3SiC_2, Ti_4SiC_3 and Ti_4GeC_3, and so on, indicating that substitution-solid-solution based exfoliation strategy may be considered as a universal method to obtain atomically thick nanosheets of MAX phases. Therefore, the design of proper substitution-solid-solution activated layered materials obeying the basic Hume-Rothery rules becomes the key point for this strategy.

On the basis of these reports, one could expect the topotactic conversion strategy to be a universal method for the preparation of ultrathin 2D nanosheets with non-layered structures.

5.3 Bottom-up Strategies

Due to the hardship of strong in-plane bond breaking (*e.g.*, covalent, metallic and ionic bonding) and the lack of intrinsic driving force, it remains greatly challenging for anisotropic growth of atomic-thick 2D crystals of non-layered structures. In that case, the controlled assembly of small building blocks, or namely bottom-up strategies, is considered to be an exclusive way for the preparation of 2D crystals of non-layered structure materials. Benefiting from recent developments in chemical synthetic strategies, a series of new atomically thick 2D crystals with non-layered structures have been successfully obtained.[25,60] The bottom-up strategies are also known as one-pot syntheses, which have been comprehensively studied in the synthesis of one-dimensional (1D) and three dimensional (3D) nanostructures.[98,99] However, their application in 2D crystals with atomic thickness still lacks systematic study. In this section, we focus on three strategies for direct chemical synthesis for the preparation of 2D crystals with atomic thickness, especially for the non-layered structural materials.

5.3.1 Self-assembly Strategy

As a common phenomenon in nature, self-assembly is a process where disordered building blocks (varying from nano to mesoscopic structures) spontaneously organize with each other to form an ordered structure through relatively weak interactions (*e.g.* van der Waals, π–π, hydrogen bonds, *etc.*). The method has long been used by scientists for the design of a range of hierarchical structures. In recent times, the atomically thick 2D nanosheets of organic, inorganic, and inorganic–organic hybrid materials have been successfully obtained by self-assembly of corresponding building blocks.[100–104] The self-assembly of organic molecular fragments into superstructures has drawn much attention for their promising applications in device fabrication, nanoscale synthesis and imaging, membrane mimetic, sensors and separations, and the self-assembly strategy for 2D organic crystals is attracting more and more attention. For example, Nam *et al.* reported the self-assembly of two oppositely charged peptoids of a specific sequence in aqueous solution into ultrathin nanosheets with a thickness of 2.7 nm, driven by periodic amphiphilicity, electrostatic recognition and aromatic interactions (as illustrated in Figure 5.8(a) and (b)).[100]

Zheng *et al.* further reported a hierarchical self-assembly of the copolymer through poly(Ph2TPh-OEG) artificial polymer folding, giving rise to the formation of ultrathin thiophene nanosheet.[101]

As for inorganic aspects, Tang *et al.* firstly reported the template-free self-assembly of 0D CdTe nanoparticles into free-standing 2D CdTe nanosheets with a thickness of about 3.4 nm, during which there are small positive charges, dipole moment and directional hydro-phobic interaction of the CdTe nanoparticles in organic solvent.[102] Acharya *et al.* reported the controllable synthesis of ultrathin PbS nanosheets (~1.8 nm from AFM analysis) from the self-assembly of individual nanowires (Figure 5.8(c)–(e)).[103]

Moreover, Zhang *et al.* developed a universal in-plane co-assembly method for preparing transition-metal chalcogenide-alkylamine (M_xT_y–alkylamine) inorganic–organic hybrid nanosheets with atomic thickness, where M = Co, Ni, Fe; T = S, Se, and alkylamine is $C_nH_{2n+x}NH_2$ with $n \geq 12$ (Figure 5.8(f) and (g)).[104] The micrometer-level Co_9S_8-oleylamine (Co_9S_8-OA) hybrid nanosheet

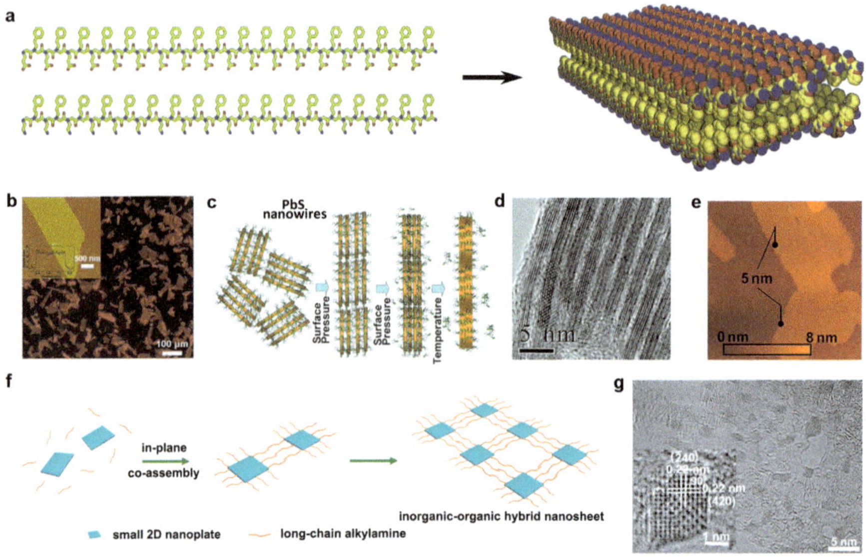

Figure 5.8 (a) Schematic illustration of 2D crystalline nanosheet formed from two oppositely charged peptoid polymers. Atomic color scheme: carbon, yellow; nitrogen, blue; oxygen, red. Left: the chemical structure of a negatively charged periodic amphiphilic peptoid and the positively charged periodic amphiphilic peptoid, respectively. Right: molecular model of the assembled nanosheet. (b) Fluorescent optical microscope image of the assembled nanosheets. Inset: AFM image of a typical assembled polymeric nanosheet. (c) Schematic illustration of the self-assembly strategy for preparing ultrathin PdS nanosheets from corresponding nanowires. (d) HRTEM images of PbS nanowires self-assembled into 2D supercrystalline arrays. (e) AFM image of the assembled PdS nanosheets, which exhibit an average thickness of 5 nm. (f) Schematic illustration of the in-plane co-assembly route to the inorganic–organic hybrid nanosheets with atomic thickness in the presence of long-chain alkylamine. (g) HRTEM image of a typical Co_9S_8-OA nanosheet obtained from self-assembly strategy. Inset: HRTEM image of a typical 2D Co_9S_8 nanoplate. (a) and (b) Reprinted with permission of Macmillan Publishers Ltd: *Nature Materials*, (ref. 100), Copyright 2010. (c)–(e) Reproduced with permission from ref. 103, Copyright 2013 American Chemical Society. (f) and (g) Reproduced with permission from ref. 104, Copyright 2013 American Chemical Society.

exhibits a thickness of ~0.5 nm, which is just in accordance with the half-unit-cell thickness of the Co_9S_8 crystal. The atomic thick inorganic–organic nanosheet is made up of randomly arranged small Co_9S_8 nanoplates which should be connected by oleylamine molecules. Theoretical simulations suggested that the oleylamine molecules energetically tend to adsorb on the corner sites and side surface sites of Co atoms rather than the top surface sites of Co atoms of the Co_9S_8 nanoplate, favoring the in-plane co-assembly strategy for the formation of the Co_9S_8-OA inorganic–organic hybrid nanosheets.

Generally speaking, in view of the spontaneous occurrence character of self-assembly process, there is still no universal law for the controllable design of atomically thick 2D crystals. Up to now, there are still very limited reports for the synthesis of atomically thick 2D crystals *via* self-assembly strategy, and further efforts are required to make the spontaneous process more 'controllable and universal'.

5.3.2 Oriented Attachment Strategy

Different from the self-assembly process discussed above, oriented attachment refers the small inorganic building blocks connect with each other in a confined 2D space by sharing a common crystallographic face, which could minimize their high surface energy and meet a thermodynamic balance.[105] Due to the lack of intrinsic driving force for 2D anisotropic growth, the oriented attachment of small building blocks during the reaction process is considered the most promising and controlling way to achieve the atomically thick 2D crystals of non-layered crystal structures. Weller's group firstly proposed two-dimensional oriented attachment of PbS nanocrystals into ultrathin single-crystal nanosheets with dimensions on the micrometer scale.[105] Zhang *et al.* reported atomically thick Co_9Se_8 nanosheets with unique lamellar stacking obtained *via* a 2D oriented attachment strategy (as shown in Figure 5.9(a)–(e)).[25]

The investigation of time-dependent morphology evolution suggests that the small 2D nanoplates with diameters of about 2 nm are randomly dispersed without obvious orientation in the initial stage. To minimize their high surface energy, the nanoplates could attach with each other to form larger nanoplates. After the reaction has been executed for about 5 h, there are still parts of the 2D nanoplates attached on the previously formed nanosheets. Thus, the 2D nanoplates could act as new nuclei for further two-dimensional attachment and formation of atomically thick lamellar stacking nanosheets. Wang *et al.* systematically investigated the growth mechanism and stress-driven synthesis of PbS nanoparticles, demonstrating that transition-catalyzed surface atoms played crucial roles in the inter-nanoparticles coalescing process and the formation of low-energy structure nanosheets.[106] Chen *et al.* reported the formation of WO_3 nanosheets *via* the laterally oriented attachment of tiny WO_3 nanocrystals using a solid–liquid phase arc discharge route in an aqueous solution (as schemed in Figure 5.9(f)).[107]

A variety of atomically thick 2D crystals of non-layered structural materials (SnO_2[108] and CeO_2[109]) have also been prepared by the 2D oriented attachment strategy (Figure 5.9(d) and (e)). Moreover, atomically thick 2D crystals of layered structural materials also can be successfully achieved by the oriented attachment strategy as expected, such as SnSe,[110] Bi_2Se_3,[111] and so on. By controlling the reaction conditions, oriented attachment strategy could provide a new way to synthesize 2D crystals of layered and non-layered structural materials. The 2D nanoplates could attach to each other to form larger nanoplates to minimize their high surface energy in the initial stage. It is

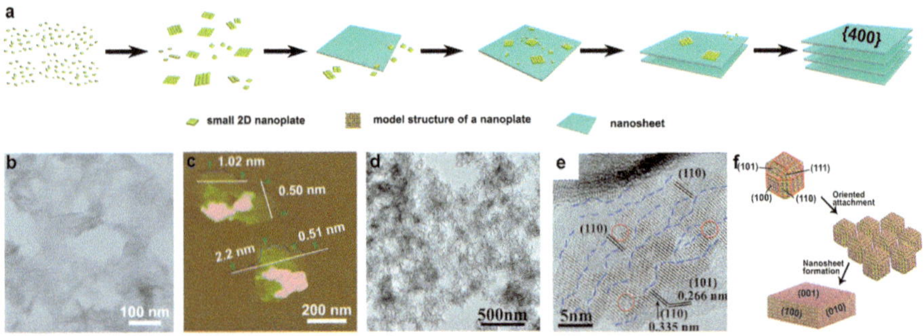

Figure 5.9 (a) Schematic illustration of the oriented attachment strategy for the preparation of atomically thick Co_9Se_8 nanosheets. (b) TEM image of ultrathin Co_9Se_8 nanosheets. (c) AFM image of Co_9Se_8 nanosheets, implying an average thickness of 0.52 nm that equals half of its unit cell parameter. (d) TEM image of SnO_2 nanosheets synthesized through oriented attachment strategy. (e) HRTEM image of the as-obtained SnO_2 nanosheets, in which dashed blue lines figure out the obvious crystal interfaces in SnO_2 nanostructures. (f) Schematic illustration of the WO_3 nanosheet formation from small nanocrystals, which implies the facet orientation during the attachment process. (a)–(c) Reproduced with permission from ref. 25, Copyright 2012 American Chemical Society. (d) and (e) Reproduced with permission from ref. 108, Copyright 2012 American Chemical Society. (f) Reproduced with permission from ref. 107, Copyright 2012 American Chemical Society.

worth mentioning that oriented attachment is very sensitive to the reaction conditions, such as the solvents, the concentration of materials, reaction temperature, *etc.* Therefore, the reaction parameters need to be adjusted for the oriental attachment of small building blocks in a confined 2D space.

5.3.3 Template-based Strategy

2D-templated based synthetic strategy is an effective approach to fabricate ultrathin 2D nanosheets with both layer and non-layered structures, in which the nanocrystal growth can be confined in specific dimensions by the templates used, where the prototype 2D templates could previously exist or be generated during the reaction procedures. For example, Cheng *et al.* have successfully synthesized the atomically thick α-Fe_2O_3 nanosheets with only half-unit-cell thickness by using CuO nanoplates as the 2D template (as shown in Figure 5.10(a)).[112]

In detail, at the initial phase, some iron hydroxide nanosheets seeds came out on the surface of CuO nanoplates as the Fe^{2+} hydrolysis reaction occurred. And then iron hydroxide nanosheets with uniform thicknesses were slowly formed due to the oriented growth of the adsorbed ferrous ions at the interface of the CuO template. As the CuO template etched out, the freestanding ultrathin iron hydroxide nanosheets with large lateral

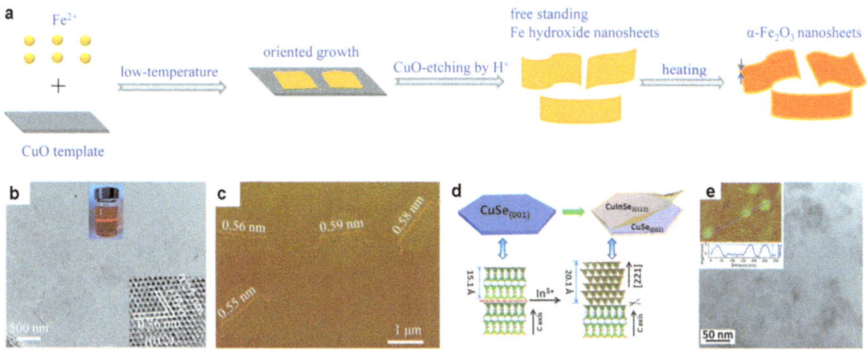

Figure 5.10 (a) Schematic illustration of template-based strategy for the preparation of ultrathin α-Fe_2O_3 nanosheets. (b) TEM image of the α-Fe_2O_3 nanosheets. Inset: HRTEM image of α-Fe_2O_3 nanosheet (upper) and Tyndall effect of the colloidal dispersion (down). (c) AFM image of 2D α-Fe_2O_3 crystals with atomic thickness. (d) Schematic structural transformation for the formation of ultrathin $CuInSe_2$ nanosheets. (e) TEM image of the $CuInSe_2$ nanosheets. Inset: AFM image of $CuInSe_2$ nanosheets, showing an average thickness of 2 nm. (a)–(c) Reproduced with permission from ref. 112, Copyright 2014 American Chemical Society. (d) and (e) Reproduced from ref. 113 with permission from Royal Society of Chemistry.

size were achieved, following which a heating treatment was performed to enable the transformation of iron hydroxide nanosheets into freestanding α-Fe_2O_3 nanosheets with thicknesses in the range from 0.55 to 0.59 nm and sizes of up to about 1 μm (illustrated in Figure 5.10(b) and (c)). Xie's group has synthesized non-layered chalcopyrite type $CuInSe_2$ nanosheets through a novel self-sacrificial template-directed strategy, with a thickness of about only 2 nm,[113] as schematically illustrated in Figure 5.10(d) and (e). Based on the similar structures between CuSe (001) and $CuInSe_2$ (112), by In^{3+} etching the as-formed CuSe template during the reaction process, atomically thick $CuInSe_2$ nanosheets were achieved. In detail, the discrepancies of lattice planes in CuSe {110} and $CuInSe_2$ {220}, CuSe {001} and $CuInSe_2$ {112} are 3.6% and 3.9%, respectively. Meanwhile, the Se atomic arrangement in CuSe (001) and $CuInSe_2$ (112) are similar to each other, which could facilitate the epitaxial growth of $CuInSe_2$ nanoplatelets with (112) facets. The matched crystal structures and similarity of Se atomic arrangements facilitate the epitaxial growth of $CuInSe_2$ nanoplatelets with (112) facets by etching the (001) facets of CuSe templates with In^{3+} ions. It is worth noting that the self-sacrificial template-directed strategy could be rigorous for further applications to synthesize other atomically thick 2D crystals, as strict lattice-matching between the as-prepared 2D crystals and the template is the prerequisite, which will undoubtedly lead to the extensive application of this method.

Note that the selection of 2D template is crucial, as the template could effectually restrain the growth of 2D nanosheets along the third dimension in the synthesis process. Usually, graphene and its derivatives as well as the previously reported 2D nanosheets could serve as potential templates for the synthesis of non-layered ultrathin 2D nanosheets. As a typical example, Zhang and co-workers developed a 2D-templated growth method to prepare the atomically thick Au nanosheets on the surface of graphene oxide, exhibiting a length range from 200 to 500 nm and thickness only of about 2.4 nm, which is in accordance with 16 Au atomic layers.[114] It is worth noting that this method could further prepare ultrathin 2D nanosheets with hetero-structures, such as Au@Ag, Au@Pt or Au@Pd nanosheets, through using the hcp Au nanosheets as seed and phase transformation in the secondary growth process.[115–117]

5.4 Conclusion

In view of their intriguing properties and promising applications, numerous efforts have been devoted to research into two-dimensional crystals with atomic thickness. Although a series of methods have been established, effective strategies for the scalable fabrication of 2D crystals remain lacking, especially for non-layered materials and quasi-layer materials with relatively strong bonds between the layers. Just recently, great progress has been achieved in the liquid phase synthesis, offering an alternative pathway for the preparation of ultrathin 2D crystals. Here we have reviewed these strategies in the categories of top-down and bottom-up synthesis, based on the relationship between reactants and target products. The 'top-down' strategies, including the small molecule assisted exfoliation strategy, ion intercalation-deintercalation assisted exfoliation strategy, and ion-exchange-based exfoliation strategy, are mainly proposed for the exfoliation of layer materials that involve intermediates with weaker van der Waals interaction, while the lamellar hybrid intermediate based exfoliation strategy and 2D precursor-based topotactic reaction strategy are feasible for preparing 2D crystals of non-layered materials. On the other hand, 'bottom-up' strategies such as self-assembly, oriented attachment, and template-based strategies, involving the construction of small building blocks, are known as to effectively conquer the hardship of strong in-plane bond breaking and the lack of intrinsic driving force, and thus more favor the preparation of 2D crystals of non-layered materials. All these liquid phase strategies significantly enrich the family of 2D crystals with atomic thicknesses.

Acknowledgements

The support from the National Natural Science Foundation of China (21401181, U1362149), Fundamental Research Funds for the Central Universities (WK2060190027, WK2340000063) is acknowledged.

References

1. K. S. Novoselov, A. K. Geim, S. V. Morozov, D. Jiang, Y. Zhang, S. V. Dubonos, I. V. Grigorieva and A. A. Firsov, *Science*, 2004, **306**, 666.
2. Y. F. Sun, S. Gao, F. C. Lei and Y. Xie, *Chem. Soc. Rev.*, 2015, **44**, 623.
3. Y. F. Sun, S. Gao, F. Lei, C. Xiao and Y. Xie, *Acc. Chem. Res.*, 2015, **48**, 3.
4. Y. F. Sun, S. Gao and Y. Xie, *Chem. Soc. Rev.*, 2014, **43**, 530.
5. X. Zhang and Y. Xie, *Chem. Soc. Rev.*, 2013, **42**, 8187.
6. X. Peng, L. Peng, C. Wu and Y. Xie, *Chem. Soc. Rev.*, 2014, **43**, 3303.
7. Y. Guo, K. Xu, C. Wu, J. Zhao and Y. Xie, *Chem. Soc. Rev.*, 2015, **44**, 637.
8. C. Tan and H. Zhang, *Chem. Soc. Rev.*, 2015, **44**, 2713.
9. H. Li, J. Wu, Z. Y. Yin and H. Zhang, *Acc. Chem. Res.*, 2014, **47**, 1067.
10. M. Chhowalla, Z. Liu and H. Zhang, *Chem. Soc. Rev.*, 2015, **44**, 2584.
11. H. Wang, H. Yuan, S. S. Hong, Y. Li and Y. Cui, *Chem. Soc. Rev.*, 2015, **44**, 2664.
12. D. Voiry, A. Mohite and M. Chhowalla, *Chem. Soc. Rev.*, 2015, **44**, 2702.
13. H. Liu, Y. Du, Y. Deng and P. D. Ye, *Chem. Soc. Rev.*, 2015, **44**, 2732.
14. G.-B. Liu, D. Xiao, Y. Yao, X. Xu and W. Yao, *Chem. Soc. Rev.*, 2015, **44**, 2643.
15. Y. Chen, C. Tan, H. Zhang and L. Wang, *Chem. Soc. Rev.*, 2015, **44**, 2681.
16. R. Raccichini, A. Varzi, S. Passerini and B. Scrosati, *Nat. Mater.*, 2015, **14**, 271.
17. Y. Shao, M. F. El-Kady, L. J. Wang, Q. Zhang, Y. Li, H. Wang, M. F. Mousavi and R. B. Kaner, *Chem. Soc. Rev.*, 2015, **44**, 3639.
18. C. Cervetti, A. Rettori, M. G. Pini, A. Cornia, A. Repollés, F. Luis, M. Dressel, S. Rauschenbach, K. Kern, M. Burghard and L. Bogani, *Nat. Mater.*, 2015, **15**, 164.
19. Y. Zhao, L. Peng, B. Liu and G. Yu, *Nano Lett.*, 2014, **14**(5), 2849.
20. X. Wang, A. M. Jones, K. L. Seyler, V. Tran, Y. Jia, H. Zhao, H. Wang, L. Yang, X. Xu and F. Xia, *Nat. Nanotechnol.*, 2015, **10**, 517.
21. A. K. Geim and K. S. Novoselov, *Nat. Mater.*, 2007, **6**, 183.
22. M. Chhowalla, H. S. Shin, G. Eda, L.-J. Li, K. P. Loh and H. Zhang, *Nat. Chem.*, 2013, **5**, 263.
23. S. Z. Butler, S. M. Hollen, L. Cao, Y. Cui, J. A. Gupta, H. R. Gutiérrez, T. F. Heinz, S. S. Hong, J. Huang, A. F. Ismach, E. Johnston-Halperin, M. Kuno, V. V. Plashnitsa, R. D. Robinson, R. S. Ruoff, S. Salahuddin, J. Shan, L. Shi, M. G. Spencer, M. Terrones, W. Windl and J. E. Goldberger, *ACS Nano*, 2013, 7, 2898.
24. Q. H. Wang, K. Kalantar-Zadeh, A. Kis, J. N. Coleman and M. S. Strano, *Nat. Nanotechnol.*, 2012, **7**, 699.
25. X. Zhang, J. Zhang, J. Zhao, B. Pan, M. Kong, J. Chen and Y. Xie, *J. Am. Chem. Soc.*, 2012, **134**, 11908.
26. Y. Sun, H. Cheng, S. Gao, Q. Liu, Z. Sun, C. Xiao, C. Wu, S. Wei and Y. Xie, *J. Am. Chem. Soc.*, 2012, **134**, 20294.
27. M. Xu, T. Liang, M. Shi and H. Chen, *Chem. Rev.*, 2013, **113**, 3766.

28. X. Zhang, X. Xie, H. Wang, J. Zhang, B. Pan and Y. Xie, *J. Am. Chem. Soc.*, 2012, **135**, 18.
29. J. Feng, L. Peng, C. Wu, X. Sun, S. Hu, C. Lin, J. Dai, J. Yang and Y. Xie, *Adv. Mater.*, 2012, **24**, 1917.
30. J. Feng, X. Sun, C. Wu, L. Peng, C. Lin, S. Hu, J. Yang and Y. Xie, *J. Am. Chem. Soc.*, 2011, **133**, 17832.
31. Y. Sun, H. Cheng, S. Gao, Z. Sun, Q. Liu, Q. Liu, F. Lei, T. Yao, J. He, S. Wei and Y. Xie, *Angew. Chem., Int. Ed.*, 2012, **51**, 8727.
32. B. Radisavljevic, A. Radenovic, J. Brivio, V. Giacometti and A. Kis, *Nat. Nanotechnol.*, 2011, **6**, 147.
33. Z. Yin, H. Li, H. Li, L. Jiang, Y. Shi, Y. Sun, G. Lu, Q. Zhang, X. Chen and H. Zhang, *ACS Nano*, 2012, **6**, 74.
34. Z. Zeng, T. Sun, J. Zhu, X. Huang, Z. Yin, G. Lu, Z. Fan, Q. Yan, H. H. Hng and H. Zhang, *Angew. Chem., Int. Ed.*, 2012, **51**, 9052.
35. H. Li, G. Lu, Y. Wang, Z. Yin, C. Cong, Q. He, L. Wang, F. Ding, T. Yu and H. Zhang, *Small*, 2013, **9**, 1974.
36. M. Acerce, D. Voiry and M. Chhowalla, *Nat. Nanotechnol.*, 2015, **10**, 313.
37. L. Li, Y. Yu, G. J. Ye, Q. Ge, X. Ou, H. Wu, D. Feng, X. H. Chen and Y. Zhang, *Nat. Nanotechnol.*, 2014, **9**, 372.
38. F. H. L. Koppens, T. Mueller, P. Avouris, A. C. Ferrari, M. S. Vitiello and M. Polini, *Nat. Nanotechnol.*, 2014, **9**, 780.
39. A. S. Rodin, A. Carvalho and A. H. Castro Neto, *Phys. Rev. Lett.*, 2014, **112**, 176801.
40. M. Buscema, D. J. Groenendijk, S. I. Blanter, G. A. Steele, H. S. J. van der Zant and A. Castellanos-Gomez, *Nano Lett.*, 2014, **14**, 3347.
41. X. Duan, C. Wang, J. C. Shaw, R. Cheng, Y. Chen, H. Li, X. Wu, Y. Tang, Q. Zhang, A. Pan, J. Jiang, R. Yu, Y. Huang and X. Duan, *Nat. Nanotechnol.*, 2014, **9**, 1024–1030.
42. J. Yang, D. Voiry, S. J. Ahn, D. Kang, A. Y. Kim, M. Chhowalla and H. S. Shin, *Angew. Chem., Int. Ed.*, 2013, **52**, 13751.
43. Y. Guo, J. Dai, J. Zhao, C. Wu, D. Li, L. Zhang, W. Ning, M. Tian, X. C. Zeng and Y. Xie, *Phys. Rev. Lett.*, 2014, **113**, 157202.
44. K. S. Novoselov, D. Jiang, F. Schedin, T. J. Booth, V. V. Khotkevich, S. V. Morozov and A. K. Geim, *Proc. Natl. Acad. Sci. U. S. A.*, 2005, **102**, 10451.
45. A. Reina, X. Jia, J. Ho, D. Nezich, H. Son, V. Bulovic, M. S. Dresselhaus and J. Kong, *Nano Lett.*, 2009, **9**, 30.
46. I.-Y. Jeon, Y.-R. Shin, G.-J. Sohn, H.-J. Choi, S.-Y. Bae, J. Mahmood, S.-M. Jung, J.-M. Seo, M.-J. Kim, D. Wook Chang, L. Dai and J.-B. Baek, *Proc. Natl. Acad. Sci. U. S. A.*, 2012, **109**, 5588.
47. K. H. Lee, H.-J. Shin, J. Lee, I.-Y. Lee, G.-H. Kim, J.-Y. Choi and S.-W. Kim, *Nano Lett.*, 2012, **12**, 714.
48. L. H. Li, Y. Chen, G. Behan, H. Z. Zhang, M. Petravic and A. M. Glushenkov, *J. Mater. Chem.*, 2011, **21**, 11862.
49. Y. H. Lee, X. Q. Zhang, W. Zhang, M. T. Chang, C. T. Lin, K. D. Chang, Y. C. Yu, J. T. W. Wang, C. S. Chang, L. J. Li and T. W. Lin, *Adv. Mater.*, 2012, **24**, 2320.

50. X. Huang, Z. Zeng and H. Zhang, *Chem. Soc. Rev.*, 2013, **42**, 1934.
51. Y. Tao, L. Liu, C. Xiao, X. Zhang, Q. Liu, S. Wei and Y. Xie, *Angew. Chem., Int. Ed.*, 2013, **52**, 7554.
52. Y. Sun, Z. Sun, S. Gao, H. Cheng, Q. Liu, J. Piao, T. Yao, C. Wu, S. Hu, S. Wei and Y. Xie, *Nat. Commun.*, 2012, **3**, 1057.
53. M. Naguib, M. Kurtoglu, V. Presser, J. Lu, J. Niu, M. Heon, L. Hultman, Y. Gogotsi and M. W. Barsoum, *Adv. Mater.*, 2011, **23**, 4248.
54. M. Naguib, O. Mashtalir, J. Carle, V. Presser, J. Lu, L. Hultman, Y. Gogotsi and M. W. Barsoum, *ACS Nano*, 2012, **6**, 1322.
55. M. Naguib, J. Halim, J. Lu, K. M. Cook, L. Hultman, Y. Gogotsi and M. W. Barsoum, *J. Am. Chem. Soc.*, 2013, **135**, 15966–15969.
56. M. R. Lukatskaya, O. Mashtalir, C. E. Ren, Y. Dall'Agnese, P. Rozier, P. L. Taberna, M. Naguib, P. Simon, M. W. Barsoum and Y. Gogotsi, *Science*, 2013, **341**, 1502.
57. Y. Hernandez, V. Nicolosi, M. Lotya, F. M. Blighe, Z. Y. Sun, S. De, I. T. McGovern, B. Holland, M. Byrne, Y. K. Gun'ko, J. J. Boland, P. Niraj, G. Duesberg, S. Krishnamurthy, R. Goodhue, J. Hutchison, V. Scardaci, A. C. Ferrari and J. N. Coleman, *Nat. Nanotechnol.*, 2008, **3**, 563.
58. J. N. Coleman, M. Lotya, A. O'Neill, S. D. Bergin, P. J. King, U. Khan, K. Young, A. Gaucher, S. De, R. J. Smith, I. V. Shvets, S. K. Arora, G. Stanton, H.-Y. Kim, K. Lee, G. T. Kim, G. S. Duesberg, T. Hallam, J. J. Boland, J. J. Wang, J. F. Donegan, J. C. Grunlan, G. Moriarty, A. Shmeliov, R. J. Nicholls, J. M. Perkins, E. M. Grieveson, K. Theuwissen, D. W. McComb, P. D. Nellist and V. Nicolosi, *Science*, 2011, **331**, 568.
59. V. Nicolosi, M. Chhowalla, M. G. Kanatzidis, M. S. Strano and J. N. Coleman, *Science*, 2013, **340**, 1226419.
60. C. Schliehe, B. H. Juarez, M. Pelletier, S. Jander, D. Greshnykh, M. Nagel, A. Meyer, S. Foerster, A. Kornowski, C. Klinke and H. Weller, *Science*, 2010, **329**, 550.
61. H. Wang, X. Yang, W. Shao, S. Chen, J. Xie, X. Zhang, J. Wang and Y. Xie, *J. Am. Chem. Soc.*, 2015, **137**, 11376.
62. C. Wu, X. Lu, L. Peng, K. Xu, X. Peng, J. Huang, G. Yu and Y. Xie, *Nat. Commun.*, 2013, **4**, 2431.
63. S. Jeong, D. Yoo, M. Ahn, P. Miró, T. Heine and J. Cheon, *Nat. Commun.*, 2015, **6**, 5763.
64. K. Xu, P. Chen, X. Li, C. Wu, Y. Guo, J. Zhao, X. Wu and Y. Xie, *Angew. Chem., Int. Ed.*, 2013, **52**, 10477.
65. L. Liang, K. Li, C. Xiao, S. J. Fan, J. Liu, W. S. Zhang, W. H. Xu, W. Tong, J. Y. Liao, Y. Y. Zhou, B. J. Ye and Y. Xie, *J. Am. Chem. Soc.*, 2015, **137**, 3102.
66. C. Lin, X. Zhu, J. Feng, C. Wu, S. Hu, J. Peng, Y. Guo, L. Peng, J. Zhao, J. Huang, J. Yang and Y. Xie, *J. Am. Chem. Soc.*, 2013, **135**, 5144.
67. F. Feng, H. Guo, D. Li, C. Wu, J. Wu, W. Zhang, S. Fan, Y. Yang, X. Wu, J. Yang, B. Ye and Y. Xie, *ACS Nano*, 2015, **9**, 1683.
68. X. Zhu, Y. Guo, H. Cheng, J. Dai, X. An, J. Zhao, K. Tian, S. Wei, X. Zeng, C. Wu and Y. Xie, *Nat. Commun.*, 2016, **7**, 11210.

69. X. Hu, W. Shao, X. Hang, X. Zhang, W. Zhu and Y. Xie, *Angew. Chem., Int. Ed.*, 2016, **55**, 5733.
70. Z. Zeng, Z. Yin, X. Huang, H. Li, Q. He, G. Lu, F. Boey and H. Zhang, *Angew. Chem., Int. Ed.*, 2011, **50**, 11093.
71. L. Liu, T. Yao, X. Tan, Q. Liu, Z. Wang, D. Shen, Z. Sun, S. Wei and Y. Xie, *Small*, 2012, **8**, 3752.
72. L. Yu, N. N. Shi, Q. Liu, J. Wang, B. Yang, B. Wang, H. J. Yan, Y. B. Sun and X. Y. Jing, *Phys. Chem. Chem. Phys.*, 2014, **16**, 17936.
73. T. Sasaki, M. Watanabe, H. Hashizume, H. Yamada and H. Nakazawa, *J. Am. Chem. Soc.*, 1996, **118**, 8329.
74. Z. Liu, R. Ma, M. Osada, N. Iyi, Y. Ebina, K. Takada and T. Sasaki, *J. Am. Chem. Soc.*, 2006, **128**, 4872.
75. M. Osada and T. Sasaki, *J. Mater. Chem.*, 2009, **19**, 2503.
76. L. Z. Wang, N. Sakai, Y. Ebina, K. Takada and T. Sasaki, *Chem. Mater.*, 2005, **17**, 1352.
77. R. Ma, Z. Liu, L. Li, N. Iyi and T. Sasaki, *J. Mater. Chem.*, 2006, **16**, 3809.
78. F.-F. Xu, Y. Ebina, Y. Bando and T. Sasaki, *J. Phys. Chem. B*, 2003, **107**, 9638.
79. A. Takagaki, D. Lu, J. N. Kondo, M. Hara, K. Domen and S. Hayashi, *Chem. Mater.*, 2005, **17**, 2487.
80. T. Sasaki and M. Watanabe, *J. Am. Chem. Soc.*, 1998, **120**, 4682.
81. K. Fukuda, K. Akatsuka, Y. Ebina, R. Ma, K. Takada, I. Nakai and T. Sasaki, *ACS Nano*, 2008, **2**, 1689.
82. L. Z. Wang, Y. Omomo, N. Sakai, K. Fukuda, I. Nakai, Y. Ebina, K. Takada, M. Watanabe and T. Sasaki, *Chem. Mater.*, 2003, **15**, 2873.
83. H. Wang, J. Zhang, X. Hang, X. Zhang, J. Xie, B. Pan and Y. Xie, *Angew. Chem., Int. Ed.*, 2015, **54**, 1195.
84. Y. Liu, H. Cheng, M. Lyu, S. Fan, Q. Liu, W. Zhang, Y. Zhi, C. Wang, C. Xiao, S. Wei, B. Ye and Y. Xie, *J. Am. Chem. Soc.*, 2014, **136**, 15670.
85. Y. Liu, X. Hua, C. Xiao, T. Zhou, P. Huang, Z. Guo, B. Pan and Y. Xie, *J. Am. Chem. Soc.*, 2016, **138**, 5087.
86. Y. Xu, W. Zhao, R. Xu, Y. Shi and B. Zhang, *Chem. Commun.*, 2013, **49**, 9803.
87. J. S. Son, X.-D. Wen, J. Joo, J. Chae, S.-i. Baek, K. Park, J. H. Kim, K. An, J. H. Yu, S. G. Kwon, S.-H. Choi, Z. Wang, Y.-W. Kim, Y. Kuk, R. Hoffmann and T. Hyeon, *Angew. Chem., Int. Ed.*, 2009, **48**, 6861.
88. Y. F. Sun, Q. H. Liu, S. Gao, H. Cheng, F. C. Lei, Z. H. Sun, Y. Jiang, H. B. Su, S. Q. Wei and Y. Xie, *Nat. Commun.*, 2013, **4**, 2899.
89. F. Lei, L. Zhang, Y. Sun, L. Liang, K. Liu, J. Xu, Q. Zhang, B. Pan, Y. Luo and Y. Xie, Atomic-Layer-Confined Doping for Atomic-Level Insights into Visible-Light Water Splitting, *Angew. Chem., Int. Ed.*, 2015, **54**, 9266.
90. F. Lei, Y. Sun, K. Liu, S. Gao, L. Liang, B. Pan and Y. Xie, *J. Am. Chem. Soc.*, 2014, **136**, 6826.
91. S. Gao, Y. Lin, X. Jiao, Y. Sun, Q. Luo, W. Zhang, D. Li, J. Yang and Y. Xie, *Nature*, 2016, **529**, 68.

92. J. Bao, X. Zhang, B. Fan, J. Zhang, M. Zhou, W. Yang, X. Hu, H. Wang, B. Pan and Y. Xie, *Angew. Chem., Int. Ed.*, 2015, **54**, 7399.
93. K. Xu, P. Chen, X. Li, Y. Tong, H. Ding, X. Wu, W. Chu, Z. Peng, C. Wu and Y. Xie, *J. Am. Chem. Soc.*, 2015, **137**, 4119.
94. L. Liang, F. Lei, S. Gao, Y. Sun, X. Jiao, J. Wu, S. Qamar and Y. Xie, *Angew. Chem., Int. Ed.*, 2015, **54**, 13971.
95. H. Nakano, M. Ishii and H. Nakamura, *Chem. Commun.*, 2005, **23**, 2945.
96. H. Nakano, T. Mitsuoka, M. Harada, K. Horibuchi, H. Nozaki, N. Takahashi, T. Nonaka, Y. Seno and H. Nakamura, *Angew. Chem., Int. Ed.*, 2006, **45**, 6303.
97. X. Zhang, J. Xu, H. Wang, J. Zhang, H. Yan, B. Pan, J. Zhou and Y. Xie, *Angew. Chem.*, 2013, **52**, 4361.
98. Y. Du, Z. Yin, J. Zhu, X. Huang, X.-J. Wu, Z. Zeng, Q. Yan and H. Zhang, *Nat. Commun.*, 2012, **3**, 1177.
99. S. Jeong, D. Yoo, J.-t. Jang, M. Kim and J. Cheon, *J. Am. Chem. Soc.*, 2012, **134**, 18233.
100. K. T. Nam, S. A. Shelby, P. H. Choi, A. B. Marciel, R. Chen, L. Tan, T. K. Chu, R. A. Mesch, B.-C. Lee, M. D. Connolly, C. Kisielowski and R. N. Zuckermann, *Nat. Mater.*, 2010, **9**, 454.
101. Y. Zheng, H. Zhou, D. Liu, G. Floudas, M. Wagner, K. Koynov, M. Mezger, H.-J. Butt and T. Ikeda, *Angew. Chem., Int. Ed.*, 2013, **52**, 4845.
102. Z. Tang, Z. Zhang, Y. Wang, S. C. Glotzer and N. A. Kotov, *Science*, 2006, **314**, 274.
103. S. Acharya, B. Das, U. Thupakula, K. Ariga, D. D. Sarma, J. Israelachvili and Y. Golan, *Nano Lett.*, 2013, **13**, 409.
104. X. Zhang, Q. Liu, L. Meng, H. Wang, W. Bi, Y. Peng, T. Yao, S. Wei and Y. Xie, *ACS Nano*, 2013, **7**, 1682.
105. C. Schliehe, B. H. Juarez, M. Pelletier, S. Jander, D. Greshnykh, M. Nagel, A. Meyer, S. Foerster, A. Kornowski, C. Klinke and H. Weller, *Science*, 2010, **329**, 550.
106. Z. Wang, C. Schliehe, T. Wang, Y. Nagaoka, Y. C. Cao, W. A. Bassett, H. Wu, H. Fan and H. Weller, *J. Am. Chem. Soc.*, 2011, **133**, 14484.
107. X. Chen, Y. Zhou, Q. Liu, Z. Li, J. Liu and Z. Zou, *ACS Appl. Mater. Interfaces*, 2012, **4**, 3372.
108. C. Wang, G. Du, K. Ståhl, H. Huang, Y. Zhong and J. Z. Jiang, *J. Phys. Chem. C*, 2012, **116**, 4000.
109. T. Yu, B. Lim and Y. Xia, *Angew. Chem., Int. Ed.*, 2010, **49**, 4484.
110. L. Li, Z. Chen, Y. Hu, X. Wang, T. Zhang, W. Chen and Q. Wang, *J. Am. Chem. Soc.*, 2013, **135**, 1213.
111. Y. Min, G. D. Moon, B. S. Kim, B. Lim, J.-S. Kim, C. Y. Kang and U. Jeong, *J. Am. Chem. Soc.*, 2012, **134**, 2872.
112. W. R. Cheng, J. F. He, T. Yao, Z. H. Sun, Y. Jiang, Q. H. Liu, S. Jiang, F. C. Hu, Z. Xie, B. He, W. S. Yan and S. Q. Wei, *J. Am. Chem. Soc.*, 2014, **136**, 10393.
113. W. Bi, M. Zhou, Z. Ma, H. Zhang, J. Yu and Y. Xie, *Chem. Commun.*, 2012, **48**, 9162.

114. X. Huang, S. Li, Y. Huang, S. Wu, X. Zhou, S. Li, C. L. Gan, F. Boey, C. A. Mirkin and H. Zhang, *Nat. Commun.*, 2011, **2**, 292.
115. X. Huang, H. Li, S. Z. Li, S. X. Wu, F. Boey, J. Ma and H. Zhang, *Angew. Chem., Int. Ed.*, 2011, **50**, 12245.
116. Z. X. Fan, X. Huang, Y. Han, M. Bosman, Q. X. Wang, Y. H. Zhu, Q. Liu, B. Li, Z. Y. Zeng, J. Wu, W. X. Shi, S. Z. Li, C. L. Gan and H. Zhang, *Nat. Commun.*, 2015, **6**, 6571.
117. Z. X. Fan, Y. H. Zhu, X. Huang, Y. Han, Q. X. Wang, Q. Liu, C. L. Gan and H. Zhang, *Angew. Chem., Int. Ed.*, 2015, **54**, 5672.

CHAPTER 6

Growth of Inorganic Two-dimensional Heterostructures Based on Transition Metal Dichalcogenides

NING LU[a], HONGYAN GUO[a] AND ZHENYU LI*[b]

[a]Department of Physics, Anhui Normal University, Wuhu, Anhui 241000, China; [b]Hefei National Laboratory for Physical Sciences at the Microscale, University of Science and Technology of China, Hefei, Anhui 230026, China
*E-mail: zyli@ustc.edu.cn

6.1 Introduction

Two-dimensional (2D) materials, such as graphene, boron nitride, and 2D transition metal dichalcogenides (TMDCs), have been widely studied recently due to their unique properties.[1–20] If suitable combinations are chosen, van der Waals heterostuctures of these 2D materials have great potential in electronic and optoelectronic applications.[12,21–42] As a representative of 2D TMDCs materials, molybdenum disulfide (MoS_2) has an electronic structure very sensitive to the number of layers due to the interlayer interaction. Heterostructures based on TMDCs have become a new research focus.[15,43–55]

Smart Materials No. 27
Inorganic Two-dimensional Nanomaterials: Fundamental Understanding, Characterizations and Energy Applications
Edited by Changzheng Wu

Published by the Royal Society of Chemistry, www.rsc.org

Mechanical exfoliation in combination with physical transferring process has already been widely used to construct 2D heterostructures.[56–65] However, there are some disadvantages for such a mechanical method. First, bringing different materials into contact *via* physical transferring process can bring impurities into the interface, such as hydrocarbons. Also, a mechanical method is difficult to scale up for massive production of large-area stacked layers.

However, progress has been made in chemical vapor deposition (CVD) growth of graphene, MoS_2, and other 2D materials. In CVD growth, typically precursor gases are fed to the furnace. Decomposition or other chemical reactions occur when they are heated. Deposition of target film material is then realized accompanied by these reactions. Compared to other expensive epitaxial growth techniques, CVD growth can be directly used in industry. By controlling the growth conditions, such as the temperature and pressure, samples obtained from CVD growth can have very high quality. Therefore, direct growth of van der Waals heterostructures by a CVD method becomes a promising way to get high-quality samples on a large scale.[66–74] Very good performance has been achieved in heterostructures grown with the CVD technique,[66,72,75–79] such as MoS_2/graphene, $MoSe_2$/graphene, MoS_2/h-BN *etc.* Here, we give a brief review on CVD growth of 2D heterostructures.

6.2 TMDCs/TMDCs

Both vertical and in-plane 2D heterostructures of different TMDCs have been synthesized. Mechanical exfoliation in combination with a physical transfer process is a popular method to achieve vertical heterostructures.[56–58] However, we will focus on the CVD method to construct TMDC heterostructures.

Intensive studies have already been done on CVD growth of TMDCs, and different precursors and substrates have been used. Generally speaking, mainly two routes have been used.[80,81] The first one is a two-step route.[82,83] In the first step, Mo-contained precursors are deposited on the substrate. Then, thermal annealing of the deposited thin layers is performed in a sulfur-vapor environment. The second route is a one-step route,[84,85] where metal source species (*e.g.* MoO_3) are evaporated and directly react with gaseous chalcogen precursors to deposit on the substrate.

To construct vertical heterostructures, two different TMDCs can be grown *via* the CVD method independently. Then, vertical heterostructures can be obtained by stacking these two different TMDCs together. Using such a protocol, a series of TMDC heterostructures has been produced.[67,86] A thermal treatment has proved to be useful to enhance the coupling between two monolayers of MoS_2/WeS_2.[67] Moreover, a type-II alignment between MoS_2 and WSe_2 bands is determined. As measured by microbeam X-ray photoelectron spectroscopy and scanning tunneling microscopy/spectroscopy,[86] the valence band offset is 0.83 eV and the conduction band offset is 0.76 eV.

To avoid the manual stacking process and the extra treatment needed to enhance interlayer interaction, more CVD methods have been proposed. These methods can also be classified into one-step and two-step routes.

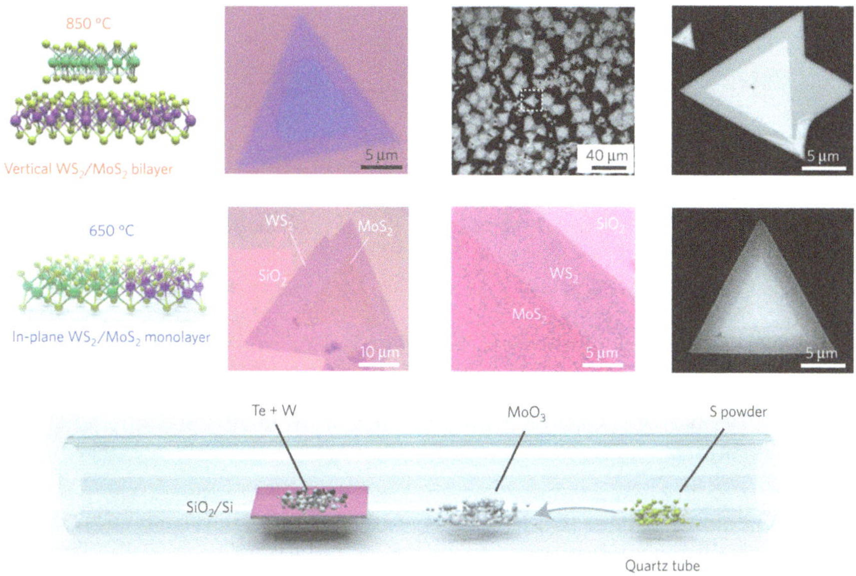

Figure 6.1 Schematic of the synthesis and the overall morphologies of the vertically stacked and in-plane WS_2/MoS_2 heterostructures. Reprinted with permission from Macmillan Publishers Ltd: *Nature Materials*, (ref. 87), Copyright 2014.

A one-step CVD method has been used to grow both in-plane and vertical heterostructures.[87] The scheme for the growth of WS_2/MoS_2 heterostructures is shown in Figure 6.1. MoO_3 powder, mixed powder of tungsten and tellurium, and S powder are used as the precursors. The existence of Te accelerates the melting of tungsten. If using WO_3 as a precursor, MoWS alloy will be obtained due to the similar sublimation rate of WO_3 and MoO_3. Different temperatures have been used to produce vertically stacked bilayers (850 °C) and in-plane lateral heterojunctions (650 °C) It is suggested that the in-plane heterostructure is a kinetic product while the vertical stacked bilayer is a thermodynamic product. To understand the detailed mechanism requires more research effort. It is found that the 2H stacking order dominates the production, which is also an advantage of the CVD method over the mechanical method. The obtained sample has a size of 20 μm. Due to the type II band alignment, strong interlayer excitonic transition is observed. At low temperature, where the in-plane heterostructure is created, a strong enhancement of localized photoluminescence is observed due to the seamless and atomically sharp in-plane boundary. Besides the one-step CVD method, a one-step physical vapor transport method using $MoSe_2$ and WSe_2 powders as the precursors has also been successfully used to synthesize the in-plane herostucture of $MoSe_2/WSe_2$. Enhanced photoluminescence has been observed.[88] WS_2 and WSe_2 powders are also directly used as the solid source to grow WS_2–WSe_2 lateral heterostructure. MoO_3 and elemental S and Se powders are

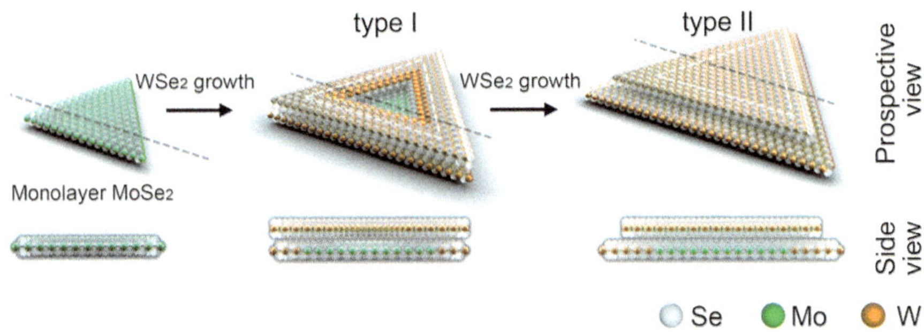

Figure 6.2 Schematics of the $MoSe_2$-templated WSe_2 growth. WSe_2 always grows from the edge regions of the $MoSe_2$ layer, forming both vertical bilayer heterostructures on $MoSe_2$ surface and seamless in-plane heterostructures. Reprinted with permission from ref. 70. Copyright 2015 American Chemical Society.

used as the solid source to grow MoS_2–$MoSe_2$ lateral heterostructures.[36] A single crystalline structure with opposite modulation of S and Se distributions across the heterostructure interface is revealed. Lateral p–n diodes and photodiodes are demonstrated for the WSe_2–WS_2 heterojunctions, which can be used to create complementary high voltage gain inverters.

A much larger (up to 169 μm) heterostructure has been obtained using a two-step CVD method as shown in Figure 6.2. MoO_3 powder and Se powder are used as the precursors to grow $MoSe_2$ on SiO_2/Si substrate using CVD as the first step. Then, the $MoSe_2$/SiO_2/Si system is transferred into another CVD setup, and WSe_2 is grown on it. Interestingly, the WSe_2 layer always prefers to grow from the edge regions of the monolayer $MoSe_2$. Two different types of heterostructures have been produced by controlling the growth time. In the type I heterostructure, the WSe_2 bilayer and WSe_2/$MoSe_2$ bilayer are at the edges with monolayer $MoSe_2$ at the center. In the type II heterostructure, WSe_2 monolayer, WSe_2 bilayer, and WSe_2/$MoSe_2$ bilayer are achieved from edge to the center. An in-plane heterostructure is not obtained with such a two-step CVD process, indicating that in-plane and vertical growth might proceed at the same time. The rectification characteristics of a p–n junction have been confirmed *via* optoelectronic tests in the type-I heterostructure, with an internal quantum efficiency as high as 91% when used as a photodetector.[70] A rotationally aligned MoS_2/WSe_2 van der Waals heterostructure has also been obtained by a two-step CVD method.[89] A WSe_2 monolayer was grown on the HOPG substrate first. Then the as-grown WSe_2 sample was put into a separate furnace as the substrate for MoS_2 growth. The finally obtained bilayer is still a direct gap semiconductor, and a Moiré pattern with a well-defined periodicity is observed due to the lattice mismatch.

A two-step CVD approach for epitaxial growth of WSe_2/MoS_2 lateral junction has also been reported.[38] WSe_2 single crystal monolayer is firstly synthesized by CVD method on sapphire substrate. The as-grown WSe_2 was then put into a separate furnace for the edge epitaxial growth of MoS_2. This epitaxial growth method provides a controllable method to obtain lateral heterojunction with an atomically sharp interface.

By using sulfurization of pre-deposited Mo films, large-area and uniform MoS_2 film is grown.[90–95] Though controlling the thickness of the pre-deposited Mo films, precise layer number controllability is also achieved. A small variation in electronic characteristics is observed with a positive dependence of drain currents with increasing layer numbers of the 1-, 3- and 5- layer MoS_2 transistors. A $WS_2/MoS_2/WS_2$ double hetero-structure is demonstrated by sequential transition metal deposition and sulfurization procedures.[90] Such a growth method shows excellent layer number controllability.

Besides the vertical and in-plane heterostructures with atomic layers parallel to the substrate, vertically aligned layers of TMDC heterostructures have also been synthesized *via* CVD. Two types of vertically aligned layers have been reported.

In Figure 6.3, W thin film is first sputtered on a SiO_2/Si substrate and selectively etched. Mo is subsequently sputtered onto the W-etched areas, resulting in the final pattern of Mo/W lines. The patterned Mo/W lines are heated in the furnace with S (Se) powders for the growth of heterostructure films. Such a one-step method has successfully grown MoS_2/WS_2 and $MoSe_2/WSe_2$ heterostructure films with vertically aligned layers over a large area (>1 cm^2).[68]

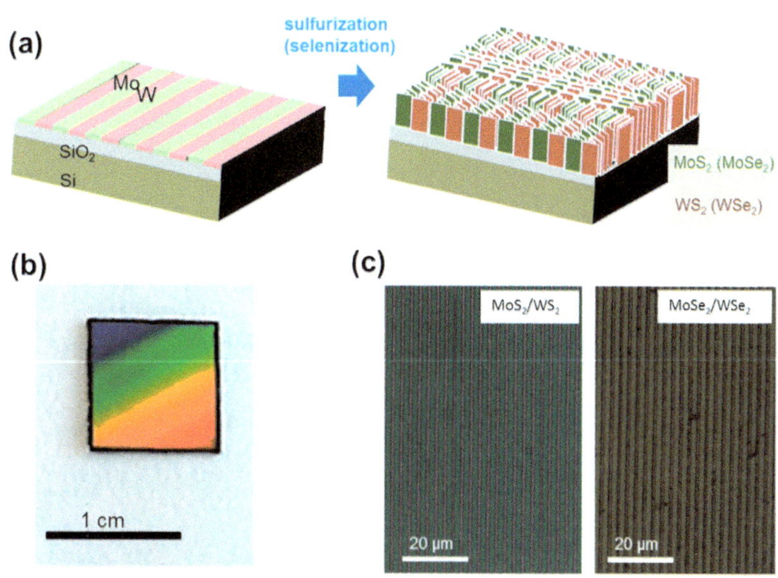

Figure 6.3 Synthesis of TMDC heterostructure films with vertically aligned 2D layers. (a) Schematic to depict the synthesis process of TMDC heterostructure films with vertically aligned 2D layers. (b) Photography image of a MoS_2/WS_2 heterostructure film grown on a SiO_2/Si substrate. (c) Zoomed-in optical microscopy images of MoS_2/WS_2 (left) and $MoSe_2/WSe_2$ (right) heterostructure films. (Reprinted with permission from ref. 68. Copyright (2014) by the American Chemical Society).

Limited carrier transport properties have been demonstrated for this heterostructure. However, a variety of potential applications are suggested for the novel structure of this 2D TMDC vertically aligned 2D layer, such as the inherent chemically active edge sites being useful in catalytic applications.

In Figure 6.4, another vertically aligned heterostructure with WSe_2 and MoS_2 layers is also synthesized. The films are grown through kinetically

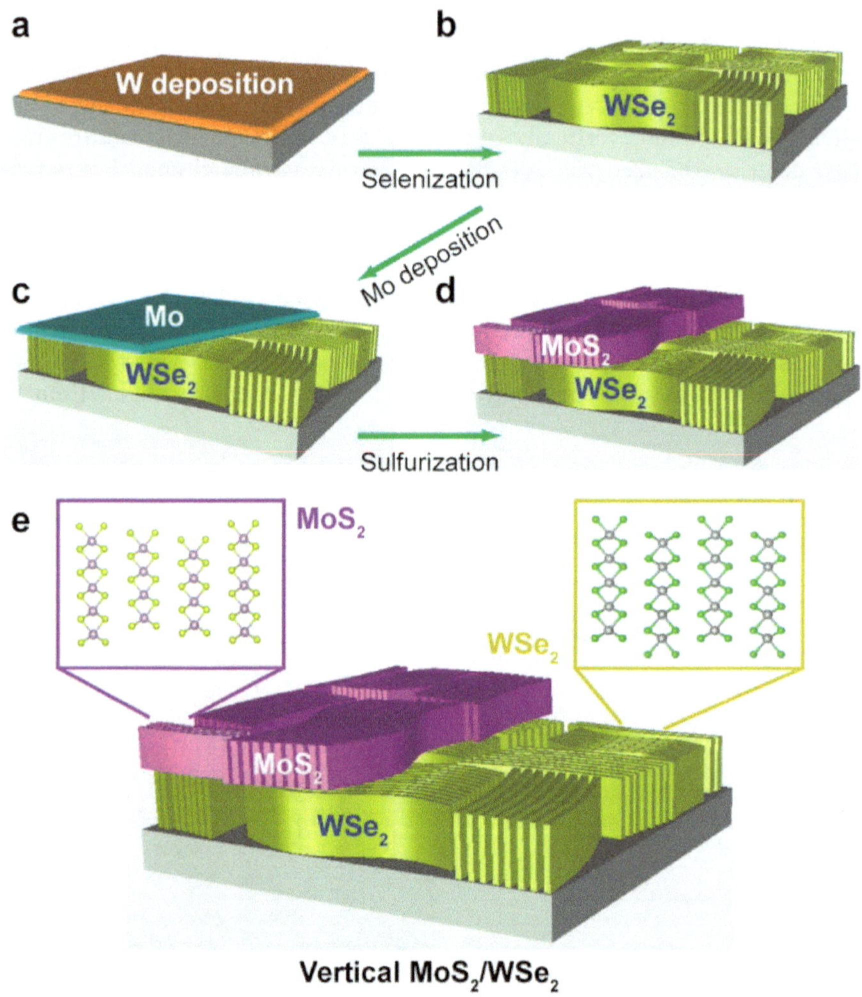

Figure 6.4 Schematics of MoS_2/WSe_2 vertical heterostructure synthesis. (a) Tungsten (W) is coated onto the substrate. (b) WSe_2 film with vertically aligned layer is formed *via* rapid selenization. (c) Molybdenum (Mo) is coated on top of the synthesized WSe_2 film. (d) MoS_2/WSe_2 vertical heterostructure is formed *via* rapid sulfurization. (e) The heterostructure consists of the MoS_2 and WSe_2, in which their van der Waals layers are aligned perpendicular to the substrate. (Reprinted with permission from ref. 96. Copyright (2015) by the American Chemical Society).

controlled rapid selenization and sulfurization, respectively. W films are rapidly selenized at 600 °C to produce edge-exposed WSe_2 film. A Mo thin film is then coated on top of the edge-exposed WSe_2 film by sputtering and rapidly sulfurized at 600 °C to produce MoS_2/WSe_2 vertically aligned layers. The p–n junction diode behavior has been confirmed for these vertically aligned layers.[96] It is noted that the synthesis of the heterostructure is scalable to a large area over 1 cm^2. The unique structure of this heterostructure is an excellent platform to exploit both science and engineering aspects of the 2D materials and opens a new way for a variety of electronic and optoelectronic devices.

6.3 TMDCs/Graphene

Van der Waals heterostructures between TMDCs and other 2D materials have also been widely investigated,[66,78,79,97] such as MoS_2/graphene, MoS_2/BN, and MoS_2/phosphorene. As the CVD synthesis process of these heterostructures is similar, here we focus on the MoS_2/graphene hybrid heterostructure.

The MoS_2/graphene hybrid heterostructure was first synthesized by Shi *et al.*[66] Graphene was first deposited on Cu foil *via* CVD. $(NH_4)_2MoS_4$ is then transported to the graphene face, and forms MoS_2 on the surface *via* thermal decomposition. The growth temperature is about 400 °C. A typical lateral size ranging from several hundred nanometers to several micrometers has been achieved for single crystalline hexagonal flakes of MoS_2. Furthermore, other hexagonal structured substrates, such as hexagonal boron nitride, are also suggested to be used as a growth substrate for the MoS_2 layers. They have provided a simple approach to synthesize MoS_2/graphene heterostructures. However, the obtained monolayer MoS_2 film is polycrystalline, and precise control on the number of MoS_2 layers is difficult.

Similar to the construction of TMDCs/TMDCs heterostructure, manual stacking of CVD grown MoS_2 and graphene is also possible.[76] By using the vapor-phase reaction of MoO_3 and S powders, Zhang *et al.* has directly grown MoS_2 monolayer crystal flakes on SiO_2 substrate with an average domain size around several microns (Figure 6.5). Then, by using PMMA as a transfer supporting layer, a graphene film was transferred on top of the grown MoS_2 film to construct the MoS_2/graphene heterostructure. High photogain greater than 10^8 is achieved for the photodetector based on this graphene/MoS_2 heterostructure. After light absorption, the electron–hole pairs are produced in the MoS_2 layer, and separated at the interface. This manually stacked heterostructure shows a good photoresponsivity higher than 10^7 A/W.

Ago *et al.* prepared monolayer MoS_2 on transferred graphene domains on SiO_2/Si by using MoO_3 and S precursors.[69] The graphene was grown by ambient pressure CVD at 1075 °C on a crystalline Cu(111) thin film. Then the hexagonal graphene domains are transferred onto a SiO_2/Si substrate. Using MoO_3 and S powder as precursors, MoS_2 crystals were grown on graphene by a second CVD process. Two preferred equivalent orientations are observed for the MoS_2 domains. An upshift of the Raman G and 2D bands of the graphene and significant photoluminescence quenching for the monolayer MoS_2 are

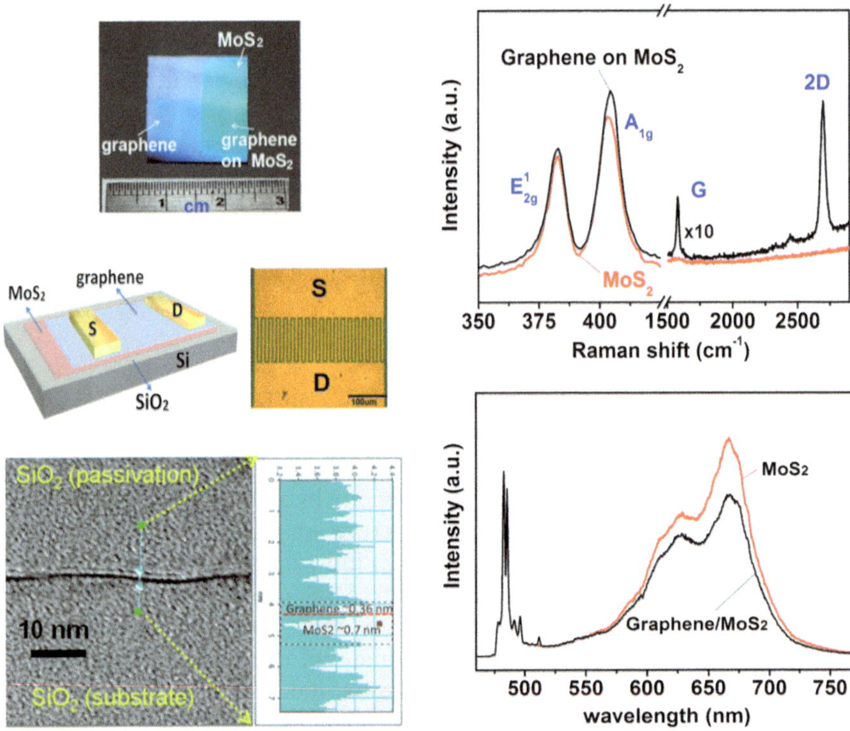

Figure 6.5 Spectroscopic characterizations of a graphene/MoS_2 bilayer. Reprinted with permission from Macmillan Publishers Ltd: *Scientific Reports* (ref. 76), Copyright 2014.

observed due to the interaction between MoS_2 and graphene. Photo-induced current modulation is also demonstrated.

To avoid the transfer process of graphene, Shi *et al.* have designed a direct all-CVD process for the construction of the MoS_2/graphene vertical heterostructures (Figure 6.6).[71] First, by using atmospheric-pressure CVD with CH_4 as precursor, a monolayer graphene film with nearly full coverage was grown on an Au substrate. Then, using MoO_3 and S precursors, monolayer MoS_2 were deposited on graphene/Au through low-pressure CVD. Spectroscopic characterizations reveal the high quality of MoS_2/fraphene vertical heterostructures with a clean interface, and the existence of graphene weakens the interface interaction between MoS_2 and Au. Weak *n*-doping effects of MoS_2 in the MoS_2/Gr/Au sandwich have been observed, and can be attributed to the intrinsic defects of MoS_2 and the electron donation from Au foils. Good transferability of this MoS_2/Gr heterostructure is also demonstrated, which will be helpful for the development of optoelectronic devices.

However, by sequentially growing two different TMD on epitaxial graphene layers, the heterostructure of MoS_2–WSe_2–graphene and WSe_2–MoS_2–graphene has been realized.[98] The individual TMD layers are grown *ex-situ via* powder vaporization or MOCVD methods. A sharp negative differential

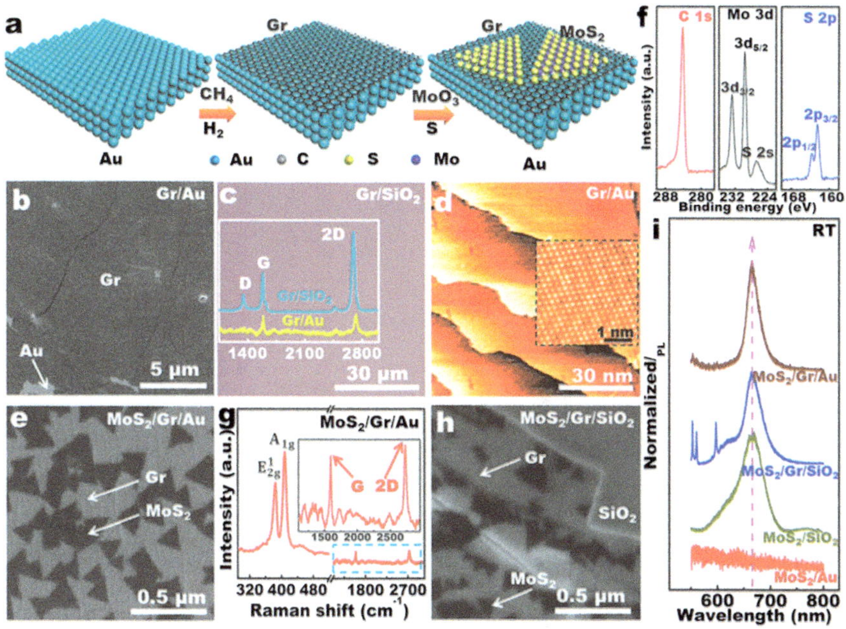

Figure 6.6 (a) Schematic illustration of the all-CVD growth process. (b) and (c) SEM and optical microscopy images of graphene on Au foils and on SiO_2/Si substrates before and after sample transfer, respectively. Inset of (c): Raman spectra of as-grown and transferred graphene. (d) STM image of as-grown graphene. Inset: an atomically resolved image. (e) SEM image of as-grown MoS_2/Graphene vertical heterostructures on Au foils. (f) XPS and (g) Raman spectra of as-grown MoS_2/Graphene/Au (h) SEM image of MoS_2/Graphene heterostructures transferred onto SiO_2/Si substrate. (i) Room-temperature PL spectra of MoS_2/Au, MoS_2/SiO_2, MoS_2/Gr/SiO_2, and MoS_2/Gr/Au, normalized to the intensity of the Raman peak (A_{1g}). Reproduced from J. Shi *et al.*, All Chemical Vapor Deposition Synthesis and Intrinsic Bandgap Observation of MoS_2/Graphene Heterostructures, *Advanced Materials*, John Wiley and Sons, © 2015 WILEY-VCH Verlag GmbH & Co. KGaA, Weinheim.[71]

resistance at room temperature is demonstrated, which arises from the special resonant tunneling of charge carriers at interface of the heterostructure. It is noted that these unique electronic transport properties are not typically found in exfoliated structures. The peak of the resonant tunneling is also demonstrated to be able to be tuned by changing the layer composition or stacking order, which may open a new way to utilize heterostructures for ultra-low power electronic devices.

Besides 2D layered material based heterostructures, a 1D heterostructure is also constructed, which has proved to be an interesting way to bring out new functionalities. In Figure 6.7, a 1D heterostructure of layered of MoS_2/graphene nanoribbon was constructed.[72] The authors offer a new way to synthesize heterostructures with controlled dimensionality. First, single-layer graphene nanoribbons (GNRs) are grown *via* CVD on an epitaxial Cu(100) film.

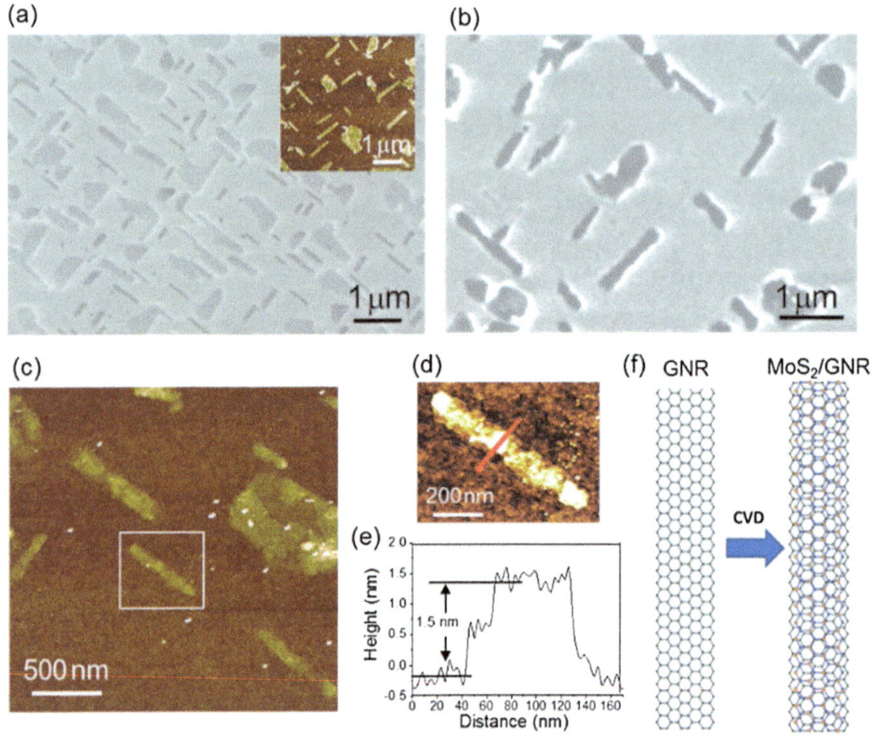

Figure 6.7 (a) SEM and AFM (inset) images of transferred GNRs on SiO_2. (b) and (c) SEM and AFM images of the MoS_2/GNR heterostructures obtained after the MoS_2 CVD on the GNR/SiO_2 substrate. (d) Enlarged AFM image of one of the MoS_2/GNRs, which is marked by white square in (c). (e) Height profile of the MoS_2/GNR measured along the red line in (d). (f) A schematic diagram of MoS_2 growth on GNRs by the second CVD. (Reproduced from ref. 72 with permission from the PCCP Owner Societies.)

Then, MoS_2 is grown on the GNRs by a second CVD process. By AFM, SEM, Raman, and TEM-EDX measurements, the formation of the MoS_2/GNR heterostructures was confirmed. Furthermore, heterostructures with different MoS_2 coverage are achieved through the control of the CVD growth conditions. A much higher photosensitivity to visible light is achieved for the obtained 1D MoS_2/graphene heterostructure than that of 2D heterostructures.

6.4 Conclusion and Perspectives

Van der Waals heterostructures between these 2D materials form a new field, and show a broad range of applications, such as in biosensors, light emitting diodes, light detectors, photovoltaic, energy storage devices and field effect/tunneling transistors. A series of techniques can be used to assemble them. However, advantages of the CVD method make it a very promising route, with the transfer process being totally avoided. Some typical CVD methods for growing TMDC based 2D heterostructures have been reviewed. However, some challenges like the control of surface reconstruction, charge transfer, and built-in electric field in these heterostructures are still ahead.

To address these challenges, the most important thing is understanding the atomistic mechanisms of the CVD growth. Encouraging progress has been made in the CVD growth of graphene and other 2D atomic layers. However, CVD growth of heterostructures is even more complicated. The first thing that needs to be clarified is which species exist on the surface. Since more types of element are involved in the CVD growth of heterostructures, there are more possibilities to find those species which play an important role in the growth. When surface species are known, it is then critical to understand the edge morphology of growing islands. Interaction between the edge and the substrate should be carefully investigated. With information about both surface species and edge morphology available, then it is possible to study the dynamics of attachment and detachment of different species at the edge. Combining these three parts together, we obtain an overall picture of CVD growth. Of course, there is another more complicated issue which is also important in optimizing the growth procedure. That is the nucleation process. Nucleation is driven by thermal fluctuation and it is sensitive to the perfection of the substrate. From the theoretical side, first principles calculation combined with kinetic Monte Carlo simulation or molecular simulation with enhanced sampling techniques provides a powerful approach to study growth mechanisms. From the experimental side, useful insight comes from kinetics studies.

Acknowledgement

This work is partially supported by NSFC (21403005, 21503002).

References

1. K. S. Novoselov, A. K. Geim, S. V. Morozov, D. Jiang, Y. Zhang, S. V. Dubonos, I. V. Grigorieva and A. A. Firsov, *Science*, 2004, **306**, 666–669.
2. N. Lu, Z. Li and J. Yang, *J. Phys. Chem. C*, 2009, **113**, 16741–16746.
3. H. Guo, Y. Zhao, N. Lu, E. Kan, X. C. Zeng, X. Wu and J. Yang, *J. Phys. Chem. C*, 2012, **116**, 11336–11342.
4. J. Dai, X. Wu, J. Yang and X. C. Zeng, *J. Phys. Chem. Lett.*, 2013, **4**, 3484–3488.
5. K. F. Mak, C. Lee, J. Hone, J. Shan and T. F. Heinz, *Phys. Rev. Lett.*, 2010, **105**, 136805.
6. A. Splendiani, L. Sun, Y. Zhang, T. Li, J. Kim, C.-Y. Chim, G. Galli and F. Wang, *Nano Lett.*, 2010, **10**, 1271–1275.
7. B. Radisavljevic, M. B. Whitwick and A. Kis, *ACS Nano*, 2011, **5**, 9934–9938.
8. Y. Yoon, K. Ganapathi and S. Salahuddin, *Nano Lett.*, 2011, **11**, 3768–3773.

9. Z. Zhang, W. Guo and B. I. Yakobson, *Nanoscale*, 2013, 6381–6387.
10. Y. Li, Z. Zhou, P. Shen and Z. Chen, *ACS Nano*, 2009, **3**, 1952–1958.
11. Y. Li, Z. Zhou, S. Zhang and Z. Chen, *J. Am. Chem. Soc.*, 2008, **130**, 16739–16744.
12. S. Bertolazzi, D. Krasnozhon and A. Kis, *ACS Nano*, 2013, **7**, 3246–3252.
13. H. L. Zhuang and R. G. Hennig, *J. Phys. Chem. C*, 2013, **117**, 20440–20445.
14. J. Kang, J. Li, S.-S. Li, J.-B. Xia and L.-W. Wang, *Nano Lett.*, 2013, **13**, 5485–5490.
15. N. Lu, H. Guo, L. Wang, X. Wu and X. C. Zeng, *Nanoscale*, 2014, **6**, 4566–4571.
16. N. Lu, D. Yin, Z. Li and J. Yang, *J. Phys. Chem. C*, 2011, **115**, 11991–11995.
17. A. H. Castro Neto, F. Guinea, N. M. R. Peres, K. S. Novoselov and A. K. Geim, *Rev. Mod. Phys.*, 2009, **81**, 109–162.
18. Y. Xu, J. Dai and X. C. Zeng, *J. Phys. Chem. Lett.*, 2015, **6**, 1996–2002.
19. M. Wu and X. C. Zeng, *Nano Lett.*, 2016, **16**, 3236–3241.
20. X. Wu, J. Dai, Y. Zhao, Z. Zhuo, J. Yang and X. C. Zeng, *ACS Nano*, 2012, **6**, 7443–7453.
21. L. Britnell, R. Gorbachev, R. Jalil, B. Belle, F. Schedin, A. Mishchenko, T. Georgiou, M. Katsnelson, L. Eaves and S. Morozov, *Science*, 2012, **335**, 947–950.
22. S. Haigh, A. Gholinia, R. Jalil, S. Romani, L. Britnell, D. Elias, K. Novoselov, L. Ponomarenko, A. Geim and R. Gorbachev, *Nat. Mater.*, 2012, **11**, 764–767.
23. W. J. Yu, Z. Li, H. Zhou, Y. Chen, Y. Wang, Y. Huang and X. Duan, *Nat. Mater.*, 2012, **12**, 246–252.
24. A. Anto Jeffery, C. Nethravathi and M. Rajamathi, *J. Phys. Chem. C*, 2013, **118**, 1386–1396.
25. T. Georgiou, R. Jalil, B. D. Belle, L. Britnell, R. V. Gorbachev, S. V. Morozov, Y.-J. Kim, A. Gholinia, S. J. Haigh and O. Makarovsky, *Nat. Nanotechnol.*, 2013, **8**, 100–103.
26. Y. Y. Hui, X. Liu, W. Jie, N. Y. Chan, J. Hao, Y.-T. Hsu, L.-J. Li, W. Guo and S. P. Lau, *ACS Nano*, 2013, **7**, 7126–7131.
27. A. K. Geim and I. V. Grigorieva, *Nature*, 2013, **499**, 419–425.
28. H. Wang, F. Liu, W. Fu, Z. Fang, W. Zhou and Z. Liu, *Nanoscale*, 2014, **6**, 12250–12272.
29. M. Buscema, J. O. Island, D. J. Groenendijk, S. I. Blanter, G. A. Steele, H. S. J. van der Zant and A. Castellanos-Gomez, *Chem. Soc. Rev.*, 2015, **44**, 3691–3718.
30. P. T. K. Loan, W. Zhang, C.-T. Lin, K.-H. Wei, L.-J. Li and C.-H. Chen, *Adv. Mater.*, 2014, **26**, 4838–4844.
31. T. Roy, M. Tosun, J. S. Kang, A. B. Sachid, S. B. Desai, M. Hettick, C. C. Hu and A. Javey, *ACS Nano*, 2014, **8**, 6259–6264.
32. M. Shanmugam, R. Jacobs-Gedrim, E. S. Song and B. Yu, *Nanoscale*, 2014, **6**, 12682–12689.
33. R. Moriya, T. Yamaguchi, Y. Inoue, S. Morikawa, Y. Sata, S. Masubuchi and T. Machida, *Appl. Phys. Lett.*, 2014, **105**, 083119.

34. L. Britnell, R. M. Ribeiro, A. Eckmann, R. Jalil, B. D. Belle, A. Mishchenko, Y. J. Kim, R. V. Gorbachev, T. Georgiou, S. V. Morozov, A. N. Grigorenko, A. K. Geim, C. Casiraghi, A. H. C. Neto and K. S. Novoselov, *Science*, 2013, **340**, 1311.
35. M.-Y. Li, C.-H. Chen, Y. Shi and L.-J. Li, *Mater. Today*, 2016, **19**, 322–335.
36. X. Duan, C. Wang, J. C. Shaw, R. Cheng, Y. Chen, H. Li, X. Wu, Y. Tang, Q. Zhang, A. Pan, J. Jiang, R. Yu, Y. Huang and X. Duan, *Nat. Nanotechnol.*, 2014, **9**, 1024–1030.
37. M. P. Levendorf, C.-J. Kim, L. Brown, P. Y. Huang, R. W. Havener, D. A. Muller and J. Park, *Nature*, 2012, **488**, 627–632.
38. M.-Y. Li, Y. Shi, C.-C. Cheng, L.-S. Lu, Y.-C. Lin, H.-L. Tang, M.-L. Tsai, C. -W. Chu, K.-H. Wei, J.-H. He, W.-H. Chang, K. Suenaga and L.-J. Li, *Science*, 2015, **349**, 524.
39. J. Lu, L. C. Gomes, R. W. Nunes, A. H. Castro Neto and K. P. Loh, *Nano Lett.*, 2014, **14**, 5133–5139.
40. C. Zhu, Z. Zeng, H. Li, F. Li, C. Fan and H. Zhang, *J. Am. Chem. Soc.*, 2013, **135**, 5998–6001.
41. A. H. Loo, A. Bonanni, A. Ambrosi and M. Pumera, *Nanoscale*, 2014, **6**, 11971–11975.
42. K. F. Mak, K. He, C. Lee, G. H. Lee, J. Hone, T. F. Heinz and J. Shan, *Nat. Mater.*, 2013, **12**, 207–211.
43. J. Kang, J. Li, S.-S. Li, J.-B. Xia and L.-W. Wang, *Nano Lett.*, 2013, **13**, 5485–5490.
44. H.-P. Komsa and A. V. Krasheninnikov, *Phys. Rev. B*, 2013, **88**, 085318.
45. K. Kośmider and J. Fernández-Rossier, *Phys. Rev. B*, 2013, **87**, 075451.
46. H. Terrones, F. López-Urías and M. Terrones, *Sci. Rep.*, 2013, **3**, 1549.
47. W. Zhao, R. M. Ribeiro, M. Toh, A. Carvalho, C. Kloc, A. H. Castro Neto and G. Eda, *Nano Lett.*, 2013, **13**, 5627–5634.
48. N. Lu, H. Guo, L. Li, J. Dai, L. Wang, W.-N. Mei, X. Wu and X. C. Zeng, *Nanoscale*, 2014, **6**, 2879–2886.
49. Y. Yu, S. Hu, L. Su, L. Huang, Y. Liu, Z. Jin, A. A. Purezky, D. B. Geohegan, K. W. Kim, Y. Zhang and L. Cao, *Nano Lett.*, 2015, **15**, 486–491.
50. D. Jariwala, V. K. Sangwan, L. J. Lauhon, T. J. Marks and M. C. Hersam, *ACS Nano*, 2014, **8**, 1102–1120.
51. G.-H. Lee, Y.-J. Yu, X. Cui, N. Petrone, C.-H. Lee, M. S. Choi, D.-Y. Lee, C. Lee, W. J. Yoo, K. Watanabe, T. Taniguchi, C. Nuckolls, P. Kim and J. Hone, *ACS Nano*, 2013, **7**, 7931–7936.
52. K. Roy, M. Padmanabhan, S. Goswami, T. P. Sai, G. Ramalingam, S. Raghavan and A. Ghosh, *Nat. Nanotechnol.*, 2013, **8**, 826–830.
53. E. Pop, V. Varshney and A. K. Roy, *MRS Bull.*, 2012, **37**, 1273–1281.
54. C.-C. Chen, Z. Li, L. Shi and S. B. Cronin, *Nano Res.*, 2015, **8**, 666–672.
55. X. Hong, J. Kim, S.-F. Shi, Y. Zhang, C. Jin, Y. Sun, S. Tongay, J. Wu, Y. Zhang and F. Wang, *Nat. Nanotechnol.*, 2014, **9**, 682–686.
56. M. M. Furchi, A. Pospischil, F. Libisch, J. Burgdörfer and T. Mueller, *Nano Lett.*, 2014, **14**, 4785–4791.

57. A. F. Rigosi, H. M. Hill, Y. Li, A. Chernikov and T. F. Heinz, *Nano Lett.*, 2015, **15**, 5033–5038.
58. P. Rivera, J. R. Schaibley, A. M. Jones, J. S. Ross, S. Wu, G. Aivazian, P. Klement, K. Seyler, G. Clark, N. J. Ghimire, J. Yan, D. G. Mandrus, W. Yao and X. Xu, *Nat. Commun.*, 2015, **6**, 6242.
59. H. Fang, C. Battaglia, C. Carraro, S. Nemsak, B. Ozdol, J. S. Kang, H. A. Bechtel, S. B. Desai, F. Kronast, A. A. Unal, G. Conti, C. Conlon, G. K. Palsson, M. C. Martin, A. M. Minor, C. S. Fadley, E. Yablonovitch, R. Maboudian and A. Javey, *Proc. Natl. Acad. Sci.*, 2014, **111**, 6198–6202.
60. F. Ceballos, M. Z. Bellus, H.-Y. Chiu and H. Zhao, *ACS Nano*, 2014, **8**, 12717–12724.
61. W. J. Yu, Y. Liu, H. Zhou, A. Yin, Z. Li, Y. Huang and X. Duan, *Nat. Nanotechnol.*, 2013, **8**, 952–958.
62. G. Gao, W. Gao, E. Cannuccia, J. Taha-Tijerina, L. Balicas, A. Mathkar, T. N. Narayanan, Z. Liu, B. K. Gupta, J. Peng, Y. Yin, A. Rubio and P. M. Ajayan, *Nano Lett.*, 2012, **12**, 3518–3525.
63. C. R. Dean, A. F. Young, I. Meric, C. Lee, L. Wang, S. Sorgenfrei, K. Watanabe, T. Taniguchi, P. Kim, K. L. Shepard and J. Hone, *Nat. Nanotechnol.*, 2010, **5**, 722–726.
64. W.-T. Hsu, Z.-A. Zhao, L.-J. Li, C.-H. Chen, M.-H. Chiu, P.-S. Chang, Y.-C. Chou and W.-H. Chang, *ACS Nano*, 2014, **8**, 2951–2958.
65. K. S. Kim, Y. Zhao, H. Jang, S. Y. Lee, J. M. Kim, K. S. Kim, J.-H. Ahn, P. Kim, J.-Y. Choi and B. H. Hong, *Nature*, 2009, **457**, 706–710.
66. Y. Shi, W. Zhou, A.-Y. Lu, W. Fang, Y.-H. Lee, A. L. Hsu, S. M. Kim, K. K. Kim, H. Y. Yang, L.-J. Li, J.-C. Idrobo and J. Kong, *Nano Lett.*, 2012, **12**, 2784–2791.
67. M.-H. Chiu, M.-Y. Li, W. Zhang, W.-T. Hsu, W.-H. Chang, M. Terrones, H. Terrones and L.-J. Li, *ACS Nano*, 2014, **8**, 9649–9656.
68. Y. Jung, J. Shen, Y. Sun and J. J. Cha, *ACS Nano*, 2014, **8**, 9550–9557.
69. H. Ago, H. Endo, P. Solís-Fernández, R. Takizawa, Y. Ohta, Y. Fujita, K. Yamamoto and M. Tsuji, *ACS Appl. Mater. Interfaces*, 2015, **7**, 5265–5273.
70. Y. Gong, S. Lei, G. Ye, B. Li, Y. He, K. Keyshar, X. Zhang, Q. Wang, J. Lou, Z. Liu, R. Vajtai, W. Zhou and P. M. Ajayan, *Nano Lett.*, 2015, **15**, 6135–6141.
71. J. Shi, M. Liu, J. Wen, X. Ren, X. Zhou, Q. Ji, D. Ma, Y. Zhang, C. Jin and H. Chen, *Adv. Mater.*, 2015, **27**, 7086–7092.
72. R. M. Yunus, H. Endo, M. Tsuji and H. Ago, *Phys. Chem. Chem. Phys.*, 2015, **17**, 25210–25215.
73. J.-K. Huang, J. Pu, C.-L. Hsu, M.-H. Chiu, Z.-Y. Juang, Y.-H. Chang, W.-H. Chang, Y. Iwasa, T. Takenobu and L.-J. Li, *ACS Nano*, 2014, **8**, 923–930.
74. Y. H. Lee, X. Q. Zhang, W. Zhang, M. T. Chang, C. T. Lin, K. D. Chang, Y. C. Yu, J. T. W. Wang, C. S. Chang and L. J. Li, *Adv. Mater.*, 2012, **24**, 2320–2325.

75. L. Yu, Y.-H. Lee, X. Ling, E. J. G. Santos, Y. C. Shin, Y. Lin, M. Dubey, E. Kaxiras, J. Kong, H. Wang and T. Palacios, *Nano Lett.*, 2014, **14**, 3055–3063.
76. W. Zhang, C.-P. Chuu, J.-K. Huang, C.-H. Chen, M.-L. Tsai, Y.-H. Chang, C.-T. Liang, Y.-Z. Chen, Y.-L. Chueh, J.-H. He, M.-Y. Chou and L.-J. Li, *Sci. Rep.*, 2014, **4**, 3826.
77. J. A. Miwa, M. Dendzik, S. S. Grønborg, M. Bianchi, J. V. Lauritsen, P. Hofmann and S. Ulstrup, *ACS Nano*, 2015, **9**, 6502–6510.
78. S. Wang, X. Wang and J. H. Warner, *ACS Nano*, 2015, **9**, 5246–5254.
79. L. Fu, Y. Sun, N. Wu, R. G. Mendes, L. Chen, Z. Xu, T. Zhang, M. H. Rümmeli, B. Rellinghaus, D. Pohl, L. Zhuang and L. Fu, *ACS Nano*, 2016, **10**, 2063–2070.
80. Q. Ji, Y. Zhang, Y. Zhang and Z. Liu, *Chem. Soc. Rev.*, 2015, **44**, 2587–2602.
81. Y. Shi, H. Li and L.-J. Li, *Chem. Soc. Rev.*, 2015, **44**, 2744–2756.
82. K.-K. Liu, W. Zhang, Y.-H. Lee, Y.-C. Lin, M.-T. Chang, C.-Y. Su, C.-S. Chang, H. Li, Y. Shi, H. Zhang, C.-S. Lai and L.-J. Li, *Nano Lett.*, 2012, **12**, 1538–1544.
83. X. Wang, H. Feng, Y. Wu and L. Jiao, *J. Am. Chem. Soc.*, 2013, **135**, 5304–5307.
84. Y.-H. Lee, L. Yu, H. Wang, W. Fang, X. Ling, Y. Shi, C.-T. Lin, J.-K. Huang, M.-T. Chang, C.-S. Chang, M. Dresselhaus, T. Palacios, L.-J. Li and J. Kong, *Nano Lett.*, 2013, **13**, 1852–1857.
85. X. Ling, Y.-H. Lee, Y. Lin, W. Fang, L. Yu, M. S. Dresselhaus and J. Kong, *Nano Lett.*, 2014, **14**, 464–472.
86. M.-H. Chiu, C. Zhang, H.-W. Shiu, C.-P. Chuu, C.-H. Chen, C.-Y. S. Chang, C.-H. Chen, M.-Y. Chou, C.-K. Shih and L.-J. Li, *Nat. Commun.*, 2015, **6**, 7666.
87. Y. Gong, J. Lin, X. Wang, G. Shi, S. Lei, Z. Lin, X. Zou, G. Ye, R. Vajtai, B. I. Yakobson, H. Terrones, M. Terrones, B. K. Tay, J. Lou, S. T. Pantelides, Z. Liu, W. Zhou and P. M. Ajayan, *Nat. Mater.*, 2014, **13**, 1135–1142.
88. C. Huang, S. Wu, A. M. Sanchez, J. J. P. Peters, R. Beanland, J. S. Ross, P. Rivera, W. Yao, D. H. Cobden and X. Xu, *Nat. Mater.*, 2014, **13**, 1096–1101.
89. C. Zhang, C.-P. Chuu, X. Ren, M.-Y. Li, L.-J. Li, C. Jin, M.-Y. Chou and C.-K. Shih, *Sci. Adv.*, 2017, **3**, e1601459.
90. C.-R. Wu, X.-R. Chang, C.-H. Wu and S.-Y. Lin, *Sci. Rep.*, 2017, **7**, 42146.
91. J. Tao, J. Chai, X. Lu, L. M. Wong, T. I. Wong, J. Pan, Q. Xiong, D. Chi and S. Wang, *Nanoscale*, 2015, **7**, 2497–2503.
92. L. Hongfei, K. K. A. Antwi, Y. Jifeng, C. Soojin and C. Dongzhi, *Nanotechnology*, 2014, **25**, 405702.
93. N. Choudhary, J. Park, J. Y. Hwang and W. Choi, *ACS Appl. Mater. Interfaces*, 2014, **6**, 21215–21222.
94. C. M. Orofeo, S. Suzuki, Y. Sekine and H. Hibino, *Appl. Phys. Lett.*, 2014, **105**, 083112.

95. Y. Jung, J. Shen, Y. Liu, J. M. Woods, Y. Sun and J. J. Cha, *Nano Lett.*, 2014, **14**, 6842–6849.
96. J. H. Yu, H. R. Lee, S. S. Hong, D. Kong, H.-W. Lee, H. Wang, F. Xiong, S. Wang and Y. Cui, *Nano Lett.*, 2015, **15**, 1031–1035.
97. Y. Deng, Z. Luo, N. J. Conrad, H. Liu, Y. Gong, S. Najmaei, P. M. Ajayan, J. Lou, X. Xu and P. D. Ye, *ACS Nano*, 2014, **8**, 8292–8299.
98. Y.-C. Lin, R. K. Ghosh, R. Addou, N. Lu, S. M. Eichfeld, H. Zhu, M.-Y. Li, X. Peng, M. J. Kim, L.-J. Li, R. M. Wallace, S. Datta and J. A. Robinson, *Nat. Commun.*, 2015, **6**, 7311.

Section II

Characterizations

CHAPTER 7

The Investigations of Mono-element Two Dimensional Materials by Scanning Tunneling Microscopy/Spectroscopy

LAN CHEN*[a,b] AND KEHUI WU*[a,b,c]

[a]Institute of Physics, Chinese Academy of Sciences, Beijing 100190, China; [b]University of Chinese Academy of Sciences, Beijing 100049, China; [c]Collaborative Innovation Center of Quantum Matter, Beijing 100871, China
*E-mail: lchen@iphy.ac.cn, khwu@iphy.ac.cn

7.1 Introduction

In 2004, a team at Manchester University, led by Andre Geim, demonstrated graphene, a one-atom thick two-dimensional (2D) crystal of carbon atoms arranged in a honeycomb lattice, which could be isolated from graphite by a simple mechanical exfoliation process.[1] The subsequent measurement on graphene revealed its extraordinary physical properties, such as half-integer quantum Hall effect,[2,3] extremely high carrier mobility,[4] high thermal conductivity,[5] high Young's modulus,[6] and so on. Because of the remarkable

Smart Materials No. 27
Inorganic Two-dimensional Nanomaterials: Fundamental Understanding, Characterizations and Energy Applications
Edited by Changzheng Wu

Published by the Royal Society of Chemistry, www.rsc.org

properties, graphene is proposed to be used in a wide variety of areas, including high-speed electronic and optical devices.[7]

Since the discovery of graphene, intensive interest has been focused on new discoveries towards graphene-analogous 2D materials, comprising single layer or few layers with compositions other than carbon. Due to the nature of sp^2 hybridization of bonding between the neighbor atoms in 2D materials, elements sitting around carbon in the periodic table of elements (PTE) have the biggest possibility to form 2D materials. For example, other group 14 elements such as silicon, germanium and tin, as well as carbon's nearest neighbors such as boron and phosphorus, are the candidates. 2D graphene-like materials such as silicene, germanene, and stanene have progressed from theoretical predictions[8–12] to experimental fabrication[13–21] in only a few years. They are found to exhibit electronic characteristics like graphene. Black phosphorus is a stable allotrope of phosphor in nature and layered materials similar to graphene.[22] But a single layer of black phosphorus has still not been realized or synthesized until now. Boron is the fifth element in the PTE, and possesses sp^2 hybridized orbitals similar to carbon, which revealed that 2D boron is also energetically stable by theoretical studies.[23,24] Very recently, the 2D boron sheets have been experimentally realized on metal substrate, and exhibit metallic property and chemical inertness against oxidization.[25,26]

These novel 2D materials have fantastic applications because of their distinct properties and high specific surface areas. The quasiparticles in graphene can be described as massless Dirac fermions.[27] Besides the Dirac fermion behaviors, other group 14 elements (silicene, germanene and stanene) further have larger spin-orbital coupling which results in gap opening at the Dirac point and leads to observable quantum spin Hall effects in experiments.[11,12] It is obviously that the properties of 2D materials have remarkable changes compared with their 3D bulk. Based on the innovative requirements, the new and unexpected properties of 2D materials need to be investigated more elaborately at the nanoscale.

Scanning tunneling microscopy (STM) and spectroscopy (STS) are powerful techniques of surface science to investigate the atomic structures and electronic properties of 2D materials at atomic scale.[28] In this chapter, we will give a brief review of the investigation of 2D materials by STM and STS in the past few years. Section 7.2 gives a brief introduction to the STM/STS technique including principle and setup. Section 7.3 focus on STM/STS measurements on graphene supported on different substrates, including twist bilayer graphene and Landau levels spectroscopy. Sections 7.4 and 7.5 cover other 2D materials made of group 14 elements and boron, respectively. Section 7.6 gives a summary for this chapter.

7.2 Scanning Tunneling Microscopy/Spectroscopy

The STM was invented by Binnig and Rohrer in 1981,[28] and brought a revolution to the characterization of surface science. It can resolve single atoms on the surface in real space. The construction of STM is quite simple, and

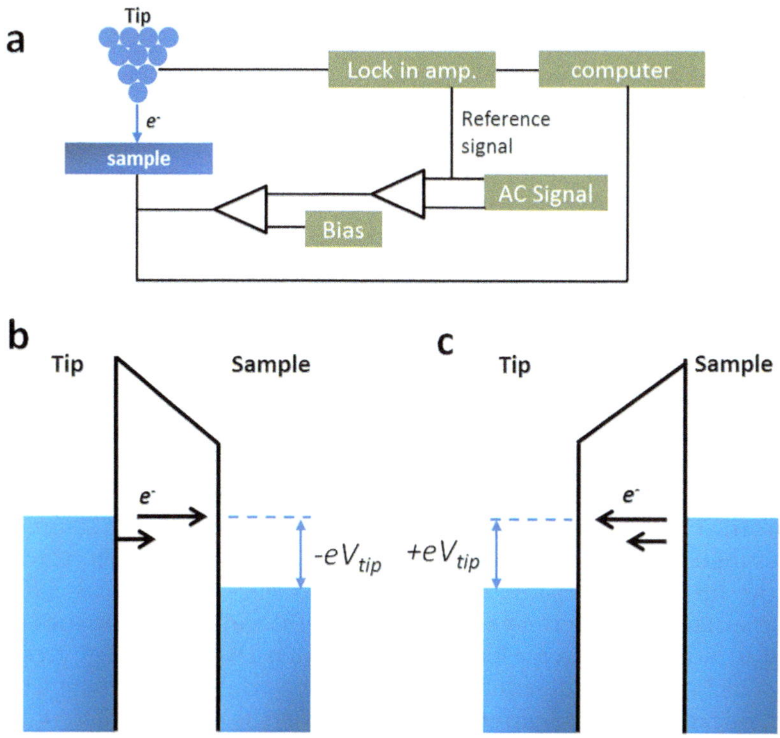

Figure 7.1 (a) The schematics of facility of scanning tunneling microscopy. (b) and (c) The schematics of tip and sample separated by a vacuum gap with a negative and positive bias voltage applied on the junction, respectively.

consists of an atomically sharp metallic tip and conducting sample with an atomically flat surface (Figure 7.1(a)).

Based on the tunneling mechanism in quantum mechanics, when the tip is brought within several angstroms of the surface, and a voltage bias is applied across the gap, electrons can tunnel through the potential barrier of the gap (Figure 7.1(b) and (c)). This induced tunneling current depends exponentially on the tip–surface separation, providing the high spatial resolution. In topography image mode of STM, tip motion is controlled by a feedback loop with a piezoelectric motor. When raster scanning the tip over the sample, the feedback loop instructs the tip to retract (or approach) from (to) the surface as the tunneling current is larger (smaller) than the current set point. The resultant variation of the tip height and current with the tip position are recorded as an STM topography image.

As per the Fermi golden rule with modified Barteen approximation,[29] the tunneling current of STM is expressed as the following equation:

$$I = \frac{4\pi e}{\hbar}\int_0^{eV} \rho_{\mathrm{S}}(E_{\mathrm{F}} - eV + \varepsilon)\rho_{\mathrm{T}}(E_{\mathrm{F}} + \varepsilon)|M|^2 \mathrm{d}\varepsilon \tag{7.1}$$

where ρ_S, ρ_T are the density of states (DOS) of surface and tip, respectively. M is the tunneling matrix, and E_F is the Fermi energy. Additionally, the sample DOS may contain energy dependent features that influence the topography measurement, so all the STM topography image must be interpreted carefully.

Simply, considering that the DOS of tip ρ_T and tunneling matrix M have no obvious change in the energy range eV, taking a derivative of the tunneling current with respect to the bias voltage yields:

$$\frac{dI}{dV} \propto \rho_S(E_F - eV) \tag{7.2}$$

That means the differential tunneling conductance (dI/dV) is directly proportional to the DOS of surface. In spectroscopy mode of STM (in other word, STS), the tip-surface distance is kept constant by turning off the feedback loop. A varied voltage is applied onto the gap of tip and surface, and then the measured derivative of tunneling current with respect to bias voltage is recorded. The dI/dV measured at negative tip bias (or positive sample bias) corresponds to unoccupied states of surface above Fermi energy, while at positive tip bias (or negative sample bias) it corresponds to occupied states of the surface below Fermi energy (Figure 7.1(b) and (c)). To enhance signal-to-noise ratio, an electronic lock-in method is used whereby a high-frequency sinusoidal voltage modulation is superimposed on the bias voltage. The first harmonic of modulated tunneling current gives the desired differential conductance at different bias voltage (dI/dV curves). Finite temperatures induce thermal broadening which leads to reduced energy resolution in STS. Correspondingly, the amplitude of ac voltage modulation should be comparable to this broadening in order to achieve the highest resolution. Of course, other surface analyzing techniques such as photoemission spectroscopy can measure the average DOS over a larger area of surface. But based on the ultrahigh imaging ability, STM can measures the local DOS at the atomic scale.

If a bias modulation is applied during scanning topography image, the modulated signal of dI/dV at certain bias voltage can be recorded at every pixel as a dI/dV map. Therefore, the spatial distribution of electronic states of surface with the energy specified by the applied bias voltage can be resolved. If the standing waves arising from quasiparticle scattering by impurities on surface are observed in the dI/dV maps, it is possible to directly image the Fermi surface contour of surface by performing the power spectrum (Fourier transform (FT)) of dI/dV maps. Obtaining and analyzing the FT of dI/dV maps of STM has been named the FT-STM (or FT-STS) technique.[30,31] The information on the band structure, such as the band dispersion and Fermi velocity, can be determined by this technique directly. The FT-STM technique can be compared with angle resolved photoemission spectroscopy, but the latter only obtains the band structure below Fermi level.

In practice, The STM can work in various ambient conditions from a liquid to ultrahigh vacuum (UHV). But for precise investigation on properties of two dimensional materials at atomic scale, the UHV condition is required due to the minimized contamination. In this chapter, we only focus on the STM in UHV.

7.3 Graphene

Graphene is a one-atom-thick planar sheet of carbon atoms arranged in a honeycomb crystal lattice, which can be described as a triangle lattice with a basis of two atoms per unit cell. In other word, there are two equivalent carbon sublattices in graphene (Figure 7.2(a)).

The carbon atom has four valence electrons, one in s and three in p orbitals. Each carbon atom in graphene is bound to three nearest neighbor carbon atoms. One s and two p orbitals are hybridized to form the sp^2 hybridization orbitals, which lead to strong σ bonds with 120° apart and the distance of 1.42 Å between neighbor carbon atoms. These strong σ bonds make graphene one of the strongest materials, and are responsible for its exceptionally high thermal conductivity. However, the fourth orbital, p_z orbital of carbon atom is perpendicular to the plane of graphene and forms a weak π bond by overlapping with other p_z orbitals.

The σ bonds forms the deep valence bands in electronic structure of graphene, while the low energy electronic bands are determined by the electrons occupied in p_z orbitals. The tight binding Hamiltonian is used to describe the hopping of electrons in graphene between nearest neighbor

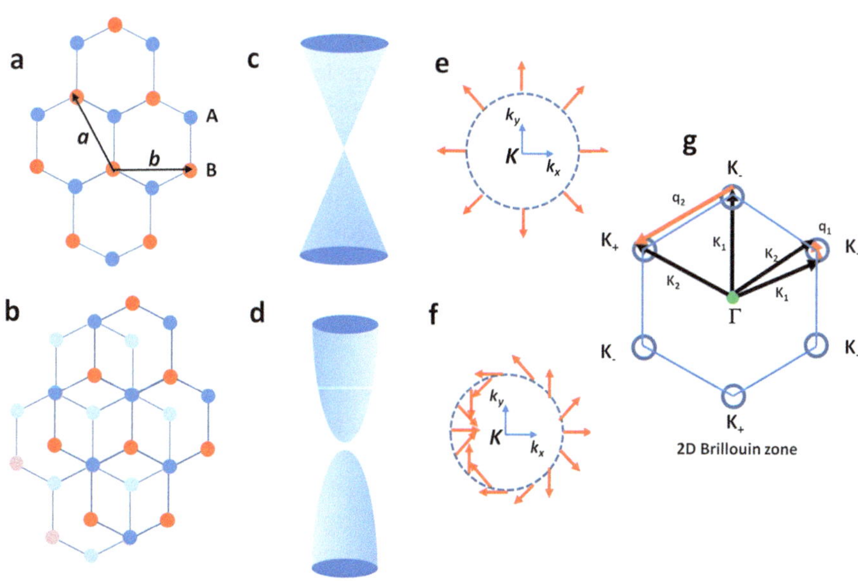

Figure 7.2 (a) Structure model of monolayer graphene. (b) Structure model of bilayer graphene with Bernal stacking. The red and blue balls are carbon atoms in the first layer graphene. The light blue and red balls are carbon atoms in the second layer graphene. (c) and (d) band structures at K points of Brillouin zone of monolayer and bilayer graphene, respectively. (e) and (f) Schematic of pseudospin texture of monolayer and bilayer graphene, respectively. (g) Schematic of constant energy contours (blue circles) in Brillouin zone of graphene. The scattering vectors of intravalley (q_1) and intervalley (q_2) are indicated by red arrows.

atoms, resulting in the valence π bands and conduction π* bands meeting at the corners (K and K' points) of the hexagonal Brillouine zone of graphene directly at Fermi energy (E_F).[27] The band dispersion of the bands close to E_F is linear and described as:

$$H_K = h v_F \vec{\sigma} \cdot \vec{p} \tag{7.3}$$

where $\vec{\sigma} = (\sigma_x, \sigma_y)$ are the Pauli matrices, which now operate on the sublattice degrees of freedom instead of spin, which is now called pseudospin. This eqn (7.3) is exactly the Dirac–Weyl equation in 2D, which is usually used to describe an ultra-relativistic (or massless) particle with spin 1/2 with velocity of light c. Hence, the low energy quasiparticles in graphene are often referred to as "massless Dirac fermions" with Fermi velocity $v_F = 1 \times 10^6$ m s^{-1}. The point where π bands intersect at K or K' point is called the Dirac point (DP), as shown in Figure 7.2(c). Due to the pseudospin in definition of wave functions of quasiparticles in graphene, it allows one to introduce chirality as the projection of pseudospin in the direction of the momentum ((Figure 7.2(e)). As a consequence, the backscattering of quasiparticle within a valley (same K point, intravalley) or between two valleys (K and K' points, intervalley) should be strongly suppressed because the backscattering will reverse the chirality of the quasiparticle.[27]

The structure of bilayer graphene is shown in Figure 7.2(b), which indicate the so-called 'Bernal stacking'. In the top layer, the atoms of sublattice A are located right above atoms of B in the bottom layer, while the atoms of B in the top layer are located above a hexagonal hollow site of the bottom layer. The inter-layer interaction breaks the equivalency between the two sublattices in both layers.[32] Therefore, the electronic structure of bilayer graphene in Bernal stacking is different from monolayer graphene. The bands have a parabolic dispersion at low energy[33] (Figure 7.2(d)), and the chirality of quasiparticles is different from the one in monolayer[34] (Figure 7.2(f)). As a consequence, backscattering is no longer forbidden in bilayer graphene.

Despite the amazing electronic properties described in pristine graphene, the experimental characterization and practice application of graphene need the sheet to be supported by a substrate. Because the carriers are at the surface and carrier density is low at the DP, the interfacial interactions between graphene and substrate will strongly influence the electronic properties of graphene. There are three methods to prepare the graphene: mechanical exfoliated graphene, epitaxial graphene on SiC and graphene grow on metal surface by chemical vapor deposition (CVD) and molecular beam epitaxy (MBE) method. In order to access the intrinsic properties of graphene, it is necessary to find out what the influence of substrate is on the properties of graphene and how to reduce it. The STM/STS can investigate the structure and electronic states of a surface at atomic scale. In the next section, we will review the works of STM/STS on graphene on different substrates.

7.3.1 Exfoliated Graphene

Mechanically exfoliated graphene results from layers peeled off from graphite by sticky tape.[1] The thickness of the layers can be determined from the contrast seen in an optical microscope[35] and through Raman spectroscopy.[36] When compared with other methods, exfoliated graphene has the lowest number of defects, and the devices have the highest mobility. Exfoliated graphene can be transferred to a variety of substrates. For the requirement of gating in device fabrication, the research focused on the graphene on SiO_2, graphite and *h*-BN substrates.

7.3.1.1 Graphene on SiO_2

The SiO_2 is insulating substrate, which is convenient for transport measurements on graphene device. But for STM measurements, the surface should be conductive. To overcome this, one can integrate the exfoliated graphene devices with STM, in which the gold electrode contact with graphene and back-gate voltage is applied to the underlying Si substrate.[37–40] The STM topography image (inset of Figure 7.3(a)) shows a honeycomb structure for single layer graphene because both the sublattices of carbon atoms have the electronic contribution,[37] while a triangle lattice is observed for multilayer graphene or in the case of 3D graphite.[27,39]

The electrons tunneling from a STM tip usually have $k = 0$ and cannot couple to the electrons at DP, which occurs at $K(K')$ point in the graphene band structure. But the STS measurements performed by Zhang *et al.* on graphene on SiO_2 show a local minimum at $V = -138$ mV and a large gap-like feature centered at E_F[38] (Figure 7.3(d)). The locale minimum varies with the change of gate voltage, indicating it should be DP. The gap-like feature pinned with E_F arises from a unique phonon-mediated inelastic tunneling process. It is indicated that the 67 meV out-of-plane acoustic graphene phonon mode enhances the probability of electrons tunneling to DP.

The SiO_2 is obtained by thermal growth and is amorphous. In STM image shown in Figure 7.3(a), random corrugations with a vertical dimension of ~1.5 Å are observed on larger length scales about several nanometers, probably due to the roughness in the underlying SiO_2 surface and/or intrinsic ripples of the graphene sheet. The position of the local minimum in the d*I*/d*V* curve (DP) has a spatial variation, reflecting the spatial profile of charge inhomogeneity in graphene, in other words, electron and hole puddles, which can be directly observed in d*I*/d*V* maps (Figure 7.3(c)) and spatial maps of the variations of the DP energy E_D (Figure 7.3(b)). Zhang *et al.* attribute the fluctuations in graphene charge density to the charge-donating impurities below the graphene.[37] That is why conductivity of graphene was finite at DP in transport measurement and probing the DP physics in graphene on SiO_2 fails. Samaddar *et al.* observed the puddles' growth, both in spatial extent and amplitude, as the Fermi level approaches to DP.[41] Additionally, STM/STS studies on graphene supported on SiO_2 substrates that were treated by chlorination reveal that the trapped charges are minimized.[42]

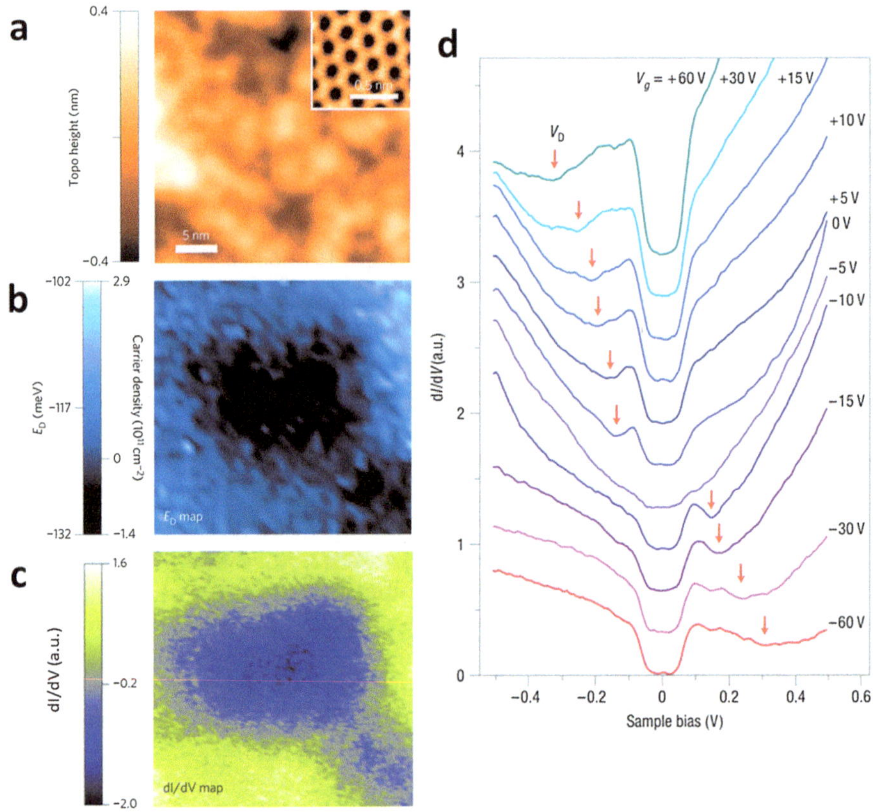

Figure 7.3 (a) STM topography image ($V_s = -0.25$ V, $I = 20$ pA) of a 30 × 30 nm^2 of graphene on SiO_2. The inset is high-resolution STM image of graphene honeycomb lattice. (b) Dirac point energy map of a single energy puddle lying in the same area as (a). (c) dI/dV map ($V_s = -0.25$ V, $I = 20$ pA) over the same area as (a) and (b). (a)–(c) reprinted with permission from Macmillan Publishers Ltd: *Nature Physics* (ref. 37), copyright (2009). (d) dI/dV curves taken on graphene on SiO_2 with different gate voltage. The red arrows indicate the positions of Dirac point. (d) reprinted with permission from Macmillan Publishers Ltd: *Nature Physics* (ref. 38), copyright (2008).

7.3.1.2 Graphene on h-BN

The problem of disorder potential in graphene on SiO_2 due to random impurities hampers performance of graphene device. One strategy to resolve the issue is to place graphene on h-BN. h-BN is a layer material much similar to graphite, and can also be exfoliated to flakes with atomic flat surfaces. The single layer of h-BN has a honeycomb structure and triangle lattice consisting of one boron and nitrogen atoms in one unit cell. Because boron and nitrogen are the neighbor elements to carbon on the PTE, the lattice of h-BN is very close to graphene (about 1.8% larger). But the inter-bonding between boron and nitrogen atoms in h-BN is an ionic bond, resulting in a large band gap of about 6 eV.[43] These properties make h-BN an ideal substrate for graphene.

Xue *et al.*[44] and Decker *et al.*[45] both showed that the graphene on h-BN has very low surface roughness due to the atomically flat nature of the h-BN surface, by the STM topography image and its histogram shown in Figure 7.4(a) and (b).

The surface roughness for graphene on h-BN is similar to that for it on graphite, suggesting it has reached its ultimate limit of flatness. The d*I*/d*V* curve has a minimum near zero bias and increases nearly linearly with energy (Figure 7.4(d)). This is in agreement with the linear DOS from linear band dispersion of pristine graphene. A significant reduction (about two orders of magnitude) in local charge inhomogeneity for graphene on h-BN compared to graphene on SiO_2 is observed on Dirac Point maps of Figure 7.4(c), and the local shift in the energy of DP is much smaller than graphene on SiO_2.[37] That is why the transport measurements on graphene/h-BN devices exhibit greatly enhanced charge carrier mobility compared to graphene/SiO_2 devices.[46]

Due to the relative orientations and small difference of lattice constant between graphene and h-BN, the graphene on h-BN will result in moiré pattern, which are displayed as long wavelength hexagonal superstructures superimposed on the hexagonal structure of the graphene atomic lattice in STM images shown in Figure 7.4(f)–(i). The wavelength of the Moiré pattern is determined by the orientation angle.[45] As shown in Figure 7.4(f)–(i), smaller orientation angles give a longer wavelength of the Moiré pattern. The hexagonal superstructure of graphene on h-BN results in periodic potential which connects the K and −K bands of graphene and creates six new Dirac points in the conduction and valence band respectively[47] (inset in Figure 7.4(c)). The STS shown in Figure 7.4(d) experimentally exhibits two dips symmetrically placed about the original DP, which prove the existence of superlattice DP. Figure 7.4(e) also shows the superlattice DP move to lower energy when the wavelength of moiré pattern increases. Furthermore, the presence of the periodic superlattice potential also leads to a spatial distribution of LDOS with a matching period.[47,48]

The cleaner charge environment and flat surface of h-BN allow the investigation of intrinsic electronic properties of graphene. For most orientation angles between graphene and h-BN, the graphene band structure influenced by h-BN can be negligible at low energy.

7.3.1.3 Graphene on Graphite

During exfoliation of the graphite, the area on the graphite adjacent to the removed layers has the possibility to obtain a decoupled region. Li *et al.*[49] found that the top layer separation of region A in the STM image of Figure 7.5(a) is about 0.44 nm, which is about 30% larger than interlayer spacing of graphite.

The high resolution STM image taken on region A (Figure 7.5(b)) also shows a honeycomb structure, not the triangle lattice observed on normal graphite (Figure 7.5(c)) which results from the sublattice asymmetry due to the Bernal stacking in graphite.[50] The Figure 7.5(d) shows STS measurement

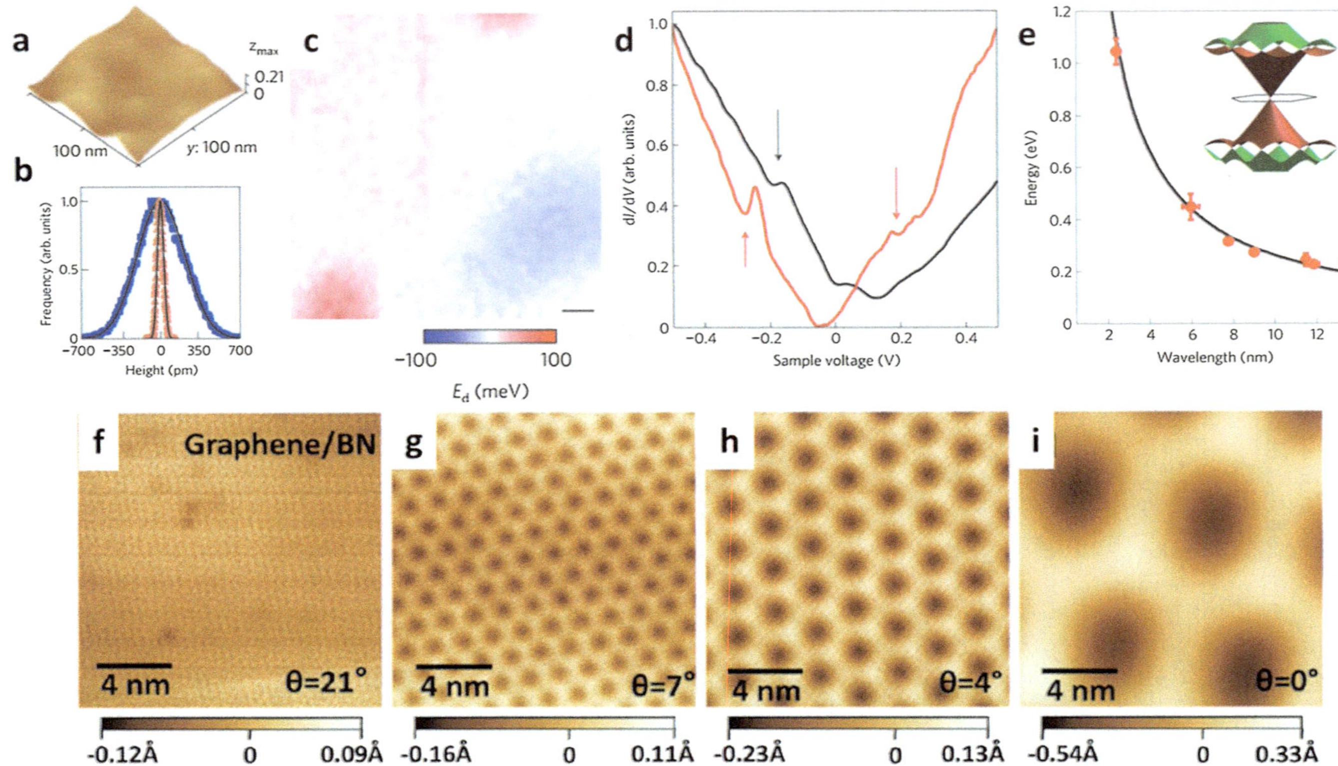

Figure 7.4 (a) STM topography image ($V_t = -0.3$ V, $I = 100$ pA) of monolayer graphene on h-BN. (b) Histogram of height distribution for graphene on SiO_2 (blue squares) and graphene on h-BN (red triangles) along with Gaussian fit. (c) Dirac point energy map of monolayer graphene on h-BN. (a)–(c) are reprinted with permission from Macmillan Publishers Ltd: *Nature Materials* (ref. 44), copyright (2011). (d) dI/dV curves of monolayer graphene on h-BN for two different Moiré wavelengths, 9.0 nm (black) and 13.4 nm (red). The dips are marked by arrows. (e) The dips away from Dirac point as a function of a moiré wavelengths. The red points are the experimentally measured values and the black line is the expected theoretical dependence. (d) and (e) are reprinted by permission from Macmillan Publishers Ltd: *Nature Physics* (ref. 47), copyright (2012). (f)–(i) STM topography images of 20 × 20 nm^2 of monolayer graphene on h-BN, showing Moiré patterns changes depending on the rotational orientation (θ) between graphene and h-BN lattices. (f)–(i) reprinted with permission from ref. 45. Copyright (2011) American Chemical Society.

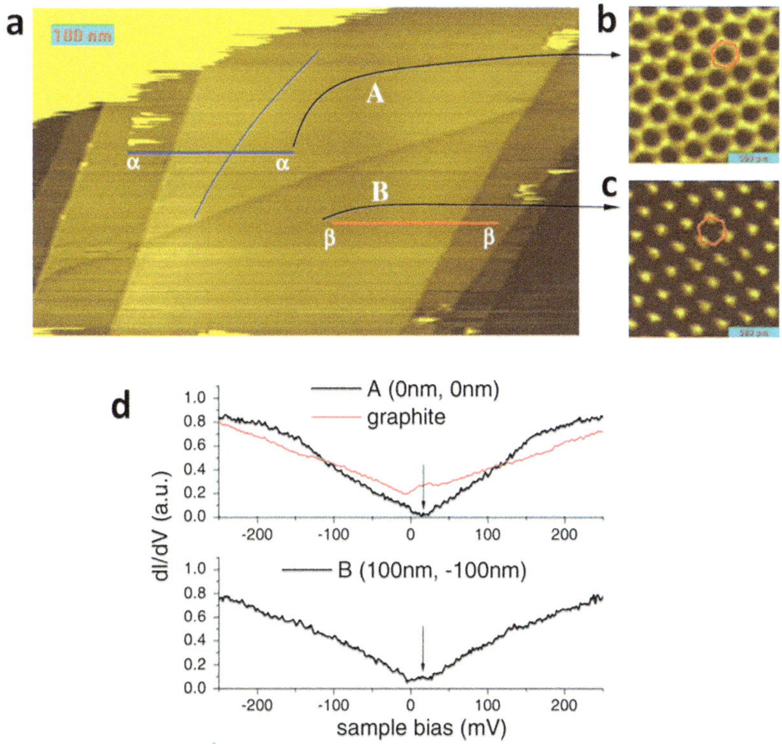

Figure 7.5 (a) STM Topography image of isolated graphene on a graphite surface. (b) and (c) High-resolution STM images taken on region A and B in (a), respectively. Scanning parameters: (a) V_s = 300 mV, I = 9 pA; (b) V_s = 300 mV, I = 55 pA; (c) V_s = 200 mV, I = 22 pA. (d) dI/dV curves taken on two layers with different coupling strength. A typical spectrum for graphite is included for comparison. The position of DP is marked by arrows. Reprinted with permission from G. Li, A. Luican and E. Y. Andrei, *Phys. Rev. Lett.*, **102**, 176804, 2009.[49] Copyright (2009) by the American Physical Society.

on the top layer of graphite with weak coupling, indicating the V-shape DOS and the differential conductance vanishes at DP. For comparison, the differential conductance of more strongly coupled layers does not vanish at DP. Therefore, the STM images and STS measurements suggest that the top layer of graphite in special regions is electronically decoupled from the substrate, and can be regarded as monolayer graphene.

7.3.1.4 Graphene on MoS_2

The h-BN and graphite substrates are both good candidates for the investigation of intrinsic properties of graphene. However, they may cause significant band structure reconstruction of graphene due to the interface interactions.[51] MoS_2 is a semiconductor with an indirect band gap of about 1.2 eV in the layered transition-metal-dichalcogenite family consisting of S–Mo–S

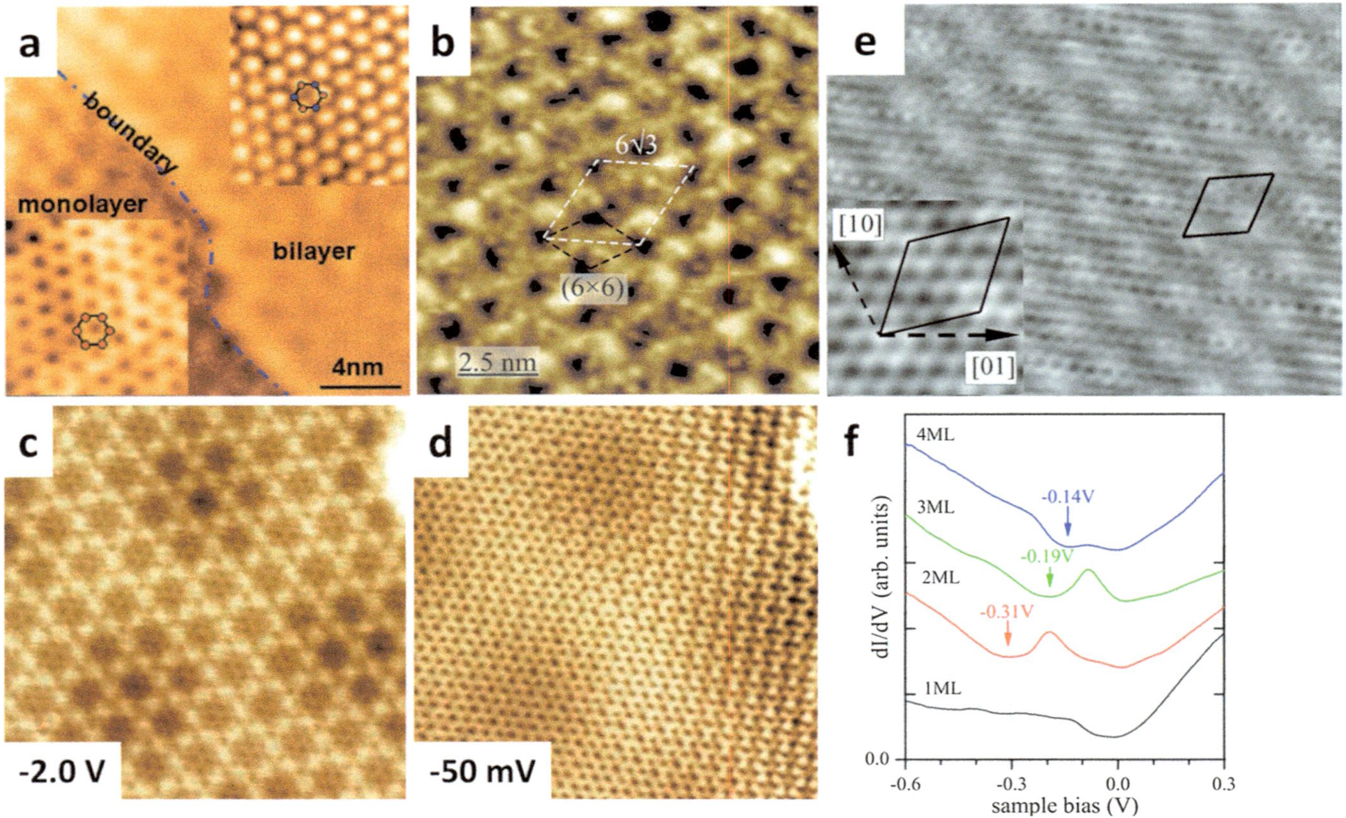

Figure 7.6 (a) STM topography image (20×20 nm^2, $V_t = 0.5$ V) of epitaxial graphene on 6H-SiC(0001), showing the coexistence of monolayer and bilayer. The insets shows the high-resolution STM images of monolayer and bilayer graphene, respectively. The hexagons are used to highlight the honeycomb lattice. (a) Reprinted with permission from ref. 62. Copyright (2008) American Chemical Society. (b) STM image ($V_t = -1.85$ V) of epitaxial monolayer graphene on SiC(0001). (c) and (d) STM images (7.5×7.5 nm^2) of epitaxial monolayer graphene on SiC(000-1) taken at bias voltages $V_s = -2$ V and $V_s = -50$ mV, respectively. (c) and (d) Reprinted with permission from F. Hiebel *et al.*, *Phys. Rev. B*, **78**, 153412, 2008.[61] Copyright (2008) by the American

sheets stabilized by the van der Waals interactions.[52] The chemical inertness and tunable screening of MoS_2, and the large lattice mismatch between MoS_2 and graphene render MoS_2 suitable for gated STM/STS studies of graphene. Lu *et al.* investigated the graphene on MoS_2 in a device configuration by STM/STS.[53] They found that graphene on MoS_2 is ultraflat through STM topography images and height histograms obtained from these images. The STS in the presence of a magnetic field and tuning of charge carriers by gate voltages reveals enhanced screening and substantial increase in the mean free path and quasiparticle lifetime. Their results indicate that MoS_2 substrates provide opportunities to access the intrinsic electronic properties of graphene.

7.3.1.5 Epitaxial Graphene on SiC(0001)

Graphitization of SiC(0001) substrates upon high temperature treatment has been investigated long before. The thin carbon films with different thickness can be formed depending on the temperature and duration of the thermal treatment.[54,55] The mechanism of the formation of carbon film is thought to be the preferential sublimation of silicon atoms from SiC substrate, then the excess surface C atoms accumulate on the surface to form graphitic film.[56] Early experiments by surface science techniques focus on the structural and electronic signatures of graphitic layers. After the boom of graphene, researchers realized this technique could yield high quality samples with physical properties like those expected for ideal graphene.[57,58] This has led to the renewal of interest in this system.

Graphitization is usually performed on the polar faces of hexagonal 4H or 6H-SiC wafers. Due to the polar nature of SiC material there are two different faces: Si face and C face. Below the graphene layers, it has been confirmed that there is a reconstruction of SiC surface: $(6\sqrt{3} \times 6\sqrt{3})R30°$ for the Si face[54,59] or (3×3) and (2×2) for the C face,[55,60,61] which consists of a graphitic layer on top of SiC substrate and is recognized as buffer layer between graphene and SiC substrate. The features of STM topography images taken at low and high bias voltages are different. At low bias (<0.4 V), electrons mainly tunnel from graphene because of the lack of electronic states in band gap of SiC substrate.[62,63] The monolayer graphene exhibits a honeycomb structure in the STM image in Figure 7.6(a), while the triangle lattice is observed on

Physical Society. (e) STM image of epitaxial multilayer graphene on SiC(000-1) showing a superlattice with a $\sqrt{13} \times \sqrt{13}$ cell. Inset: high-resolution STM image ($V_s = -0.8$ V) of the $\sqrt{13} \times \sqrt{13}R46.1°$ superlattice. (e) Reprinted with permission from J. Hass *et al.*, *Phys. Rev. Lett.*, **100**, 125504, 2008.[65] Copyright (2008) by the American Physical Society. (f) d*I*/d*V* curves of epitaxial graphene on SiC(0001) with different thickness. (b) and (f) Reprinted with permission from P. Lauffer *et al.*, *Phys. Rev. B*, 77, 155426, 2008.[64] Copyright (2008) by the American Physical Society.

bilayer graphene because the Bernal stacking in bilayer graphene leads the LDOS on the two sublattices of top layer graphene to be different.[62,63]

At high bias, the interfacial buffer layer can be directly observed with a reduced contrast compared with bared reconstruction. The contrast also decreases with the thickness of graphene layers, allowing one to distinguish the regions of epitaxial graphene with different thickness.[64] For Si face of SiC, the $(6\sqrt{3} \times 6\sqrt{3})R30°$ reconstruction of buffer layer commonly looks like a 6×6 superstructure of the surface in high bias voltage STM images of Figure 7.6(b).[62] Figure 7.6(c) and (d) are the STM images of monolayer graphene on C face of SiC taken at high (−2.0 V) and low bias voltages (−50 mV), indicating the 3×3 superstructure of the buffer layer. Because of bottom-up growth mode and the well-defined orientation of buffer layer, all of the grown graphene layers are Bernal stacked. However, the (3×3) and (2×2) reconstruction of buffer layer on C face of SiC show significant amount of rotational disorder, so the graphene layers are also rotationally disordered.[65–67] For example, the multilayer graphene in Figure 7.6(e) shows a $(\sqrt{13} \times \sqrt{13})R46.1°$ superlattice because of about 30° rotations between two adjacent graphene layers on C face of SiC.[65]

A characteristic STS obtained on few layer graphene on SiC(0001) is shown in Figure 7.6(f). It is very hard to find DP in the spectra taken on monolayer graphene. The absence of a clear signature of DP for monolayer graphene might be due to the large contribution of interfacial states of SiC to the tunneling current. For bilayer graphene on SiC(0001), the STS presents a conductance dip at −0.31 V, which is attributed to the DP, and a small band gap (<0.1 eV) around DP. This small gap is thought to result from asymmetry charge transfer from substrate to the adjacent graphene layers. The shift of DP to Fermi level with the increase of thickness is considered to result from the reduced charge transfer from the substrate.[64] Kim *et al.* observed an unexpected negative differential resistance (NDR) at DP of bilayer graphene, which spatial varies within the single unit cell.[68] The origin of NDR is explained by two-gap van Hove singularities emerging in the electronic states of bilayer graphene under a transverse electric field, which are strongly localized on two sublattices in different layers.

7.3.1.6 Graphene on Metal Substrates

Graphene can be grown on metal substrates by decomposition of hydrocarbons or solid carbon at high temperatures, or by segregation of the dissolved carbon from bulk metal. This was first realized by annealing an ethylene covered Pt(111) surface in UHV,[69] and then similar method has been used to generate graphene on Ir(111),[70,71] Ni(111),[72,73] Pd(111),[74,75] Pt(111),[76,77] and so on. The graphene can be grown on Au(111) and Ag(111) by direct evaporation of element carbon source,[78,79] or on Ni and Ru by the surface segregation of carbon atoms from bulk metal.[80,81] Growth of graphene on Cu has attracted great interest since the graphene can be readily formed in ambient pressure chemical vapor deposition (CVD) processes.[82] The advances in CVD growth have obtained large single grain graphene islands on polycrystalline Cu foil.[83] The STM studies of graphene on polycrystalline Cu foil show that the graphene sheet can grow over facets and substrate grain boundaries to continuously cover the whole substrate.[84]

The interfacial interactions may be the driving force for the formation of graphene on metal. This can explain why the monolayer formation can be easily accomplished while formation of a controlled number of graphene sheets is difficult. Generally, there is a lattice mismatch between graphene and metal surfaces which ranges from 1.3% for Ni(111) to 12.8% for Pt(111). For metal surfaces for which the lattice is close to graphene, such as Ni(111)[72] and Co(0001),[85] the graphene adopts the lattice constant of the metallic substrate. As a result, the STM topography image in Figure 7.7(a) shows that the graphene on Ni(111) has a 1×1 structural periodicity.

The carbon atoms in honeycomb alternatively occupy the top and three-fold hollow sites of the substrate. Therefore, the atoms belonging to two sub-lattices show obvious contrast in STM images. Other techniques combining theoretical calculations reveal the distance between graphene and Ni and Co substrates of about 2.1 Å and very small corrugations of 0.05 Å, indicate the strong interfacial interactions.[86]

If the lattice of graphene and that of substrate are incommensurate, growth of graphene on other metals will induce the formation of Moiré structures that can be observed by STM. The binding of graphene to the substrate depend strongly on the carbon arrangement relative to substrate. This give rise to a strong buckling of the graphene as the relative lattice positions within the Moiré pattern change. Here we introduce two cases of graphene on Ru(0001)[87,88] and Ir(111),[89] where strongly and weakly corrugated graphene forms. Figure 7.7(b) shows the STM topography images obtained on graphene/Ru(0001) indicating the Moiré pattern,[87] which is composed of three regions with different contrast: low, intermediate, and high areas. The widely accepted structure of graphene/Ru(0001) is (25×25)gr/(23×23)Ru, which means that 25 unit cells of graphene are aligned along 23 unit cells of Ru(0001) surface.[90] DFT calculations showed that the carbon atoms located on the top of Ru atoms interact strongly with the substrate, while carbon atoms in-between Ru atoms interact much weaker with the substrate. This difference in the interaction with the substrate gives rise to the strong corrugation of the graphene sheet (about 1.5 Å).[90] The dI/dV maps (Figure 7.7(c) and (d)) taken at different voltages indicate the different interactions between graphene and substrate also produce spatial charge redistribution in the graphene.[88] For graphene/Ir(111), there are domains of graphene with several specific rotation angles with respect to the substrate observed.[91] The prefer orientation is referred to 0°, which forms a Moiré structure with a periodicity of (10×10)gr/(9×9)Ir (Figure 7.7(e)). The STM topography images in Figure 7.7(f) taken at different bias voltages reveal the distinguished contrast of the Moiré pattern.[89] The highest places in the unit cell observed in STM images at low bias voltages change to dark areas at high bias voltages, indicating the bias dependence of corrugation. The DFT calculations and

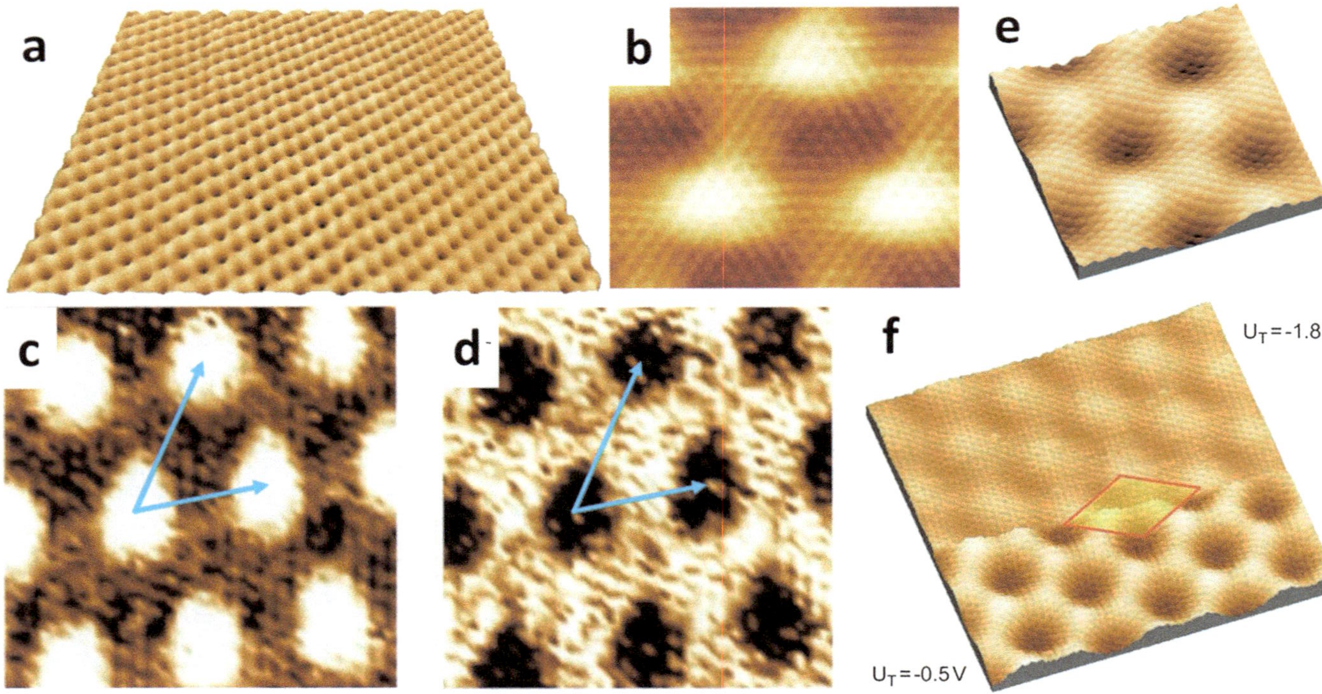

Figure 7.7 (a) High-resolution STM image (V_s = 2 mV) of monolayer graphene on Ni(111) surface. (a) Reproduced with permission from ref. 72. © IOP Publishing & Deutsche Physikalische Gesellschaft. CC BY-NC-SA. (b) High-resolution STM image (5 × 4 nm^2, V_s = −50 mV) of monolayer graphene on Ru(0001) surface. (b) Reprinted with permission from S. Marchini *et al.*, *Phys. Rev. B*, **76**, 075429, 2007.[87] Copyright (2007) by the American Physical Society. (c) and (d) d*I*/d*V* maps of graphene on Ru(0001) surface taken at −100 meV and 200 meV, respectively. (c) and (d) Reprinted with permission from A. L. Vázquez de Parga *et al.*, *Phys. Rev. Lett.*, **100**, 056807, 2008.[88] Copyright (2008) by the American Physical Society. (e) High-resolution STM image (5.2 × 5.2 nm^2, V_s = 550 mV) of monolayer graphene on Ir(111) surface. (f) STM image of graphene/Ir(111) obtained at V_s = −1.8 V (top) and V_s = −0.5 V (bottom). (e) and (f) reprinted by permission from Macmillan Publishers Ltd: *Scientific Reports* (ref. 89), copyright (2009).

AFM experiments reveal that the corrugation is about 0.3 Å, which is much smaller than that on Ru(0001).[89]

Adsorption of a graphene layer on metallic surface always leads the modification of the electronic structure of graphene. The STS measurements on graphene/metal can show strong alternation of electronic structure of graphene compared to free-standing graphene.[88] However, STS only reflects the LDOS in a finite range of energy, and the tunneling current is mainly contributed by interfacial states. Therefore, it is very hard to obtain other properties of electronic bands of graphene, such as dispersion and DP by STS. The angle resolved photoemission spectroscopy (ARPES) and DFT calculation studies reveal the graphene is strongly *n*-doped by the electron transfer from the substrate to graphene π^* states, leading to the overlap of metal d bands with graphene π states. Depending on the relative energy overlap of the π states and d bands, the Dirac cone of the graphene is fully destroyed or symmetry energy gap is opened directly at the DP.[92]

7.3.2 Quasiparticle Interferences

As we mentioned before, the electron backscattering in monolayer graphene should be suppressed because of the chirality of the massless Dirac fermion,[27] while it could not happen in bilayer graphene because of the different chirality.[34] This effect has been checked on epitaxial graphene on SiC(0001) by several research groups.[93–95] The STM topography images taken at low bias voltage and the d*I*/d*V* maps contain the spatial variations of LDOS. The standing waves induced by quasiparticles interferences can be resolved by these images,[30,31] and the FT of these images gives the reciprocal-space distribution of the scattering vectors, as shown in Figure 7.8(a) (monolayer) and Figure 7.8(b) (bilayer).[95]

For monolayer graphene, a circle at the center (Γ point) is absent, which is observed for bilayer graphene. This circle has a radius $2q_F$, where $|q_F|$ is the radius of Dirac cone. The quasiparticle interferences at wave vector $2q_F$ associate to the intravalley backscattering process (Figure 7.2(g)). Therefore, the absence of center circle means the strongly suppression of intravalley backscattering that occurs in monolayer graphene because the pseudospins at opposite ends of the constant energy contour of Dirac cones are the inverse of each other (Figure 7.2(c)). However, the intravalley backscattering of quasiparticles is allowed in the bilayer of graphene due to the different orientations of pseudospins (Figure 7.2(f)). At the six *K* points of Brillouine zone in FT-STM images for both monolayer and bilayer graphene (Figure 7.8(a) and (b)),[95] six rings with radius $2q_F$ are observed and could be associated with intervalley scattering (Figure 7.2(g)). The difference is that the intensity of such rings for monolayer graphene is suppressed in the direction perpendicular to ΓK or $\Gamma K'$ due to the orientation of pseudospin, while the $2q_F$ rings is complete for bilayer graphene. All the rings' radius at the center (intravalley scattering) or *K* points (intervalley scattering) increase with the bias voltages, so the dispersion relations for monolayer and bilayer graphene can

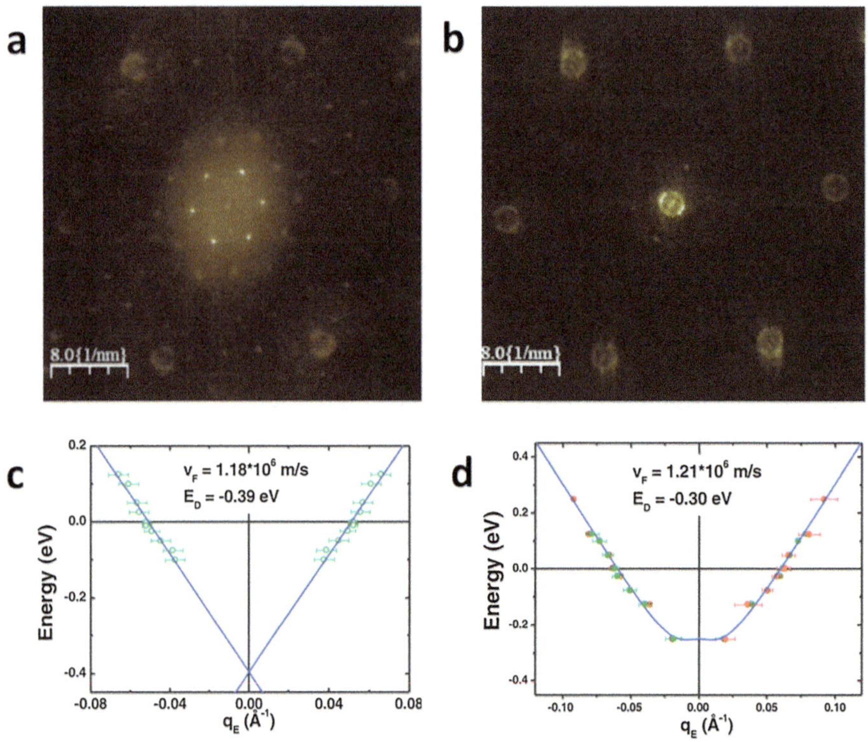

Figure 7.8 (a) and (b) Fourier transform STM images obtained from STM measurements on monolayer and bilayer graphene on SiC(0001), respectively. (c) and (d) Dispersion relation derived from the STM data on monolayer and bilayer graphene, respectively. Reprinted with permission from P. Mallet *et al.*, *Phys. Rev. B*, **86**, 045444, 2008.[95] Copyright (2008) by the American Physical Society.

be extracted respectively, and the results are shown in Figure 7.8(c) and (d). For monolayer graphene, the linear fit to the dispersion is perfect and gives a DP located at 0.4 eV below Fermi level (Figure 7.8(c)), which agrees with the small dip observed in dI/dV curve very well.[95,96] However, the fit to dispersion in bilayer graphene is parabolic and gives an energy band gap about 0.1 eV (Figure 7.8(d)), which is in accordance with theory[34,97] and dI/dV spectra measurements very well.[98]

Additionally, intervalley quasiparticle scattering can be directly observed in STM topography images Figure 7.9(a) near the edges of graphene.[99,100]

But only armchair edges are found to favor intervalley scattering (Figure 7.9(b) and (c)), leading to ($\sqrt{3} \times \sqrt{3}$)R30° superstructure near edges. The component of the wave vector of the incident state parallel to the step edge should be conserved upon scattering by an extended linear step edge, while the zigzag edge is rotated by 30° compared with the armchair edge. Therefore, intervalley scattering is kinematically allowed at the armchair edge, and forbidden at the zigzag edge.

The decay of standing waves from scattering centers, such as step edges, provides important information on the quasipartical scattering mechanism. In conventional 2D electron gas with parabolic momentum-energy dispersion,[101] the decay of standing waves from a straight step edge as a function of the distance obeys a power law with decay factor of −0.5, *i.e.*, $\delta\rho(x) \propto \cos(2kx + \varphi)x^{-0.5}$. This situation will be different for Dirac fermion systems with chirality. Xue *et al.* observed standing waves near the h-BN step in the STM d*I*/d*V* maps shown in Figure 7.9(d)–(f) on graphene on h-BN,[102] corresponding to the quasiparticle intravalley scattering. The wavelengths of the oscillations increase as the tip bias decreases, and are on the order of 10 nm, which is much larger than that on noble metal surfaces. The fitting to LDOS oscillation (Figure 7.9(g)) indicates that the amplitude decay as a function of distance from the step edge is much faster than on noble metal surfaces (the decay factor of $\alpha = -1.5$), suggesting the backscattering suppression due to the existence of pseudospin.

Zhang *et al.* observed standing wave patterns on monolayer graphene on SiO_2 in d*I*/d*V* maps corresponding to the backscattering of Dirac fermions by underlying impurities.[37] But the backscattering of Dirac fermions should be suppressed in single layer graphene due to the chirality of the quasiparticles. The broken sublattice symmetry induced by the topological corrugations may explain the observation of backscattering of Dirac fermions in monolayer graphene on SiO_2.

7.3.3 Twist Graphene Layers

Despite the Bernal stacking in bilayer graphene, the two graphene layers can rotate relative to each other by an angle θ, which can produce a commensuration superstructure known as a Moiré pattern.[103] The period L of Moiré pattern corresponding to a twist angle θ is given by:

$$L = \frac{a}{2\sin(\theta/2)} \tag{7.4}$$

where a is the lattice constant of graphene. The twist graphene layers are commonly found on different substrates, such as metals, the C face of SiC,[65] or graphite surfaces.[104] An example of a Moiré pattern on the surface of bilayer graphene on Rh foil is shown in Figure 7.10(e).[105]

The highly ordered triangular pattern has a period of about 4.9 nm, much larger than the lattice constant of graphene. The twist angle can be deduced by equation, and is 2.9°. The high resolution STM image Figure 7.10(f) shows the bright spots of patterns consist of triangle lattice indicating Bernal stacking. The honeycomb structure is observed in between bright spots, indicating lost registry between layers due to twisting. The evolution of the patterns with twist angles shown in Figure 7.10(a)–(d) illustrates the decrease in period with increasing twist angle.

The twist between layers has a profound effect on the electronic band structure of graphene.[106] When two graphene layers rotate against to each other,

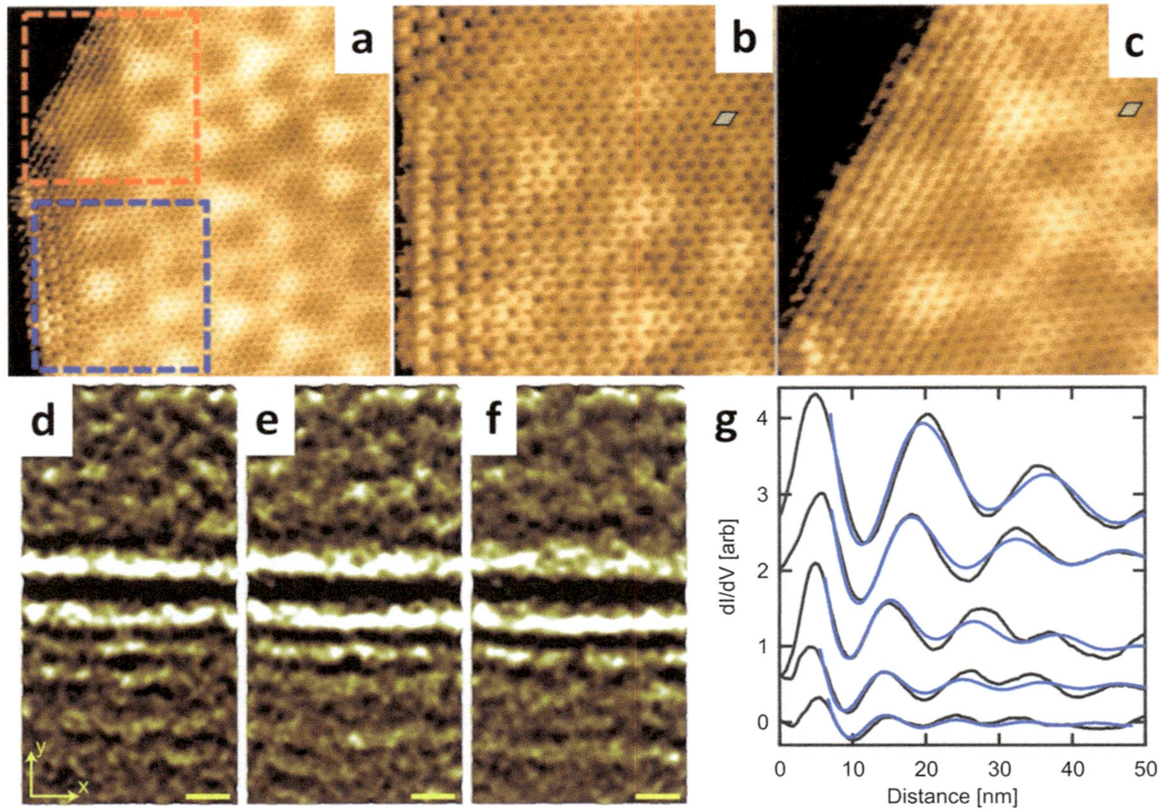

Figure 7.9 (a) 10×10 nm^2 STM image ($V_s = 100$ mV) of monolayer graphene on SiC(0001) with armchair and zigzag edges. (b) and (c) Zoom-in STM images (5.5×5.5 nm^2) of perturbations near armchair and zigzag edges, respectively. (a)–(c) Reprinted with permission from ref. 100. (d)–(f) dI/dV maps of graphene on h-BN across a step edge obtained at different bias voltage. The tip voltage are (d) 138 mV, (e) 118 mV, (f) 98 mV. (g) dI/dV as a function of position for several different tip voltages showing the decay with distance. The black curves are the experimental data and the blue curves are from the theoretical calculations. (d)–(g) Reprinted with permission from J. Xue *et al.*, *Phys. Rev. Lett.*, **108**, 016801, 2012.[102] Copyright (2012) by the American Physical Society.

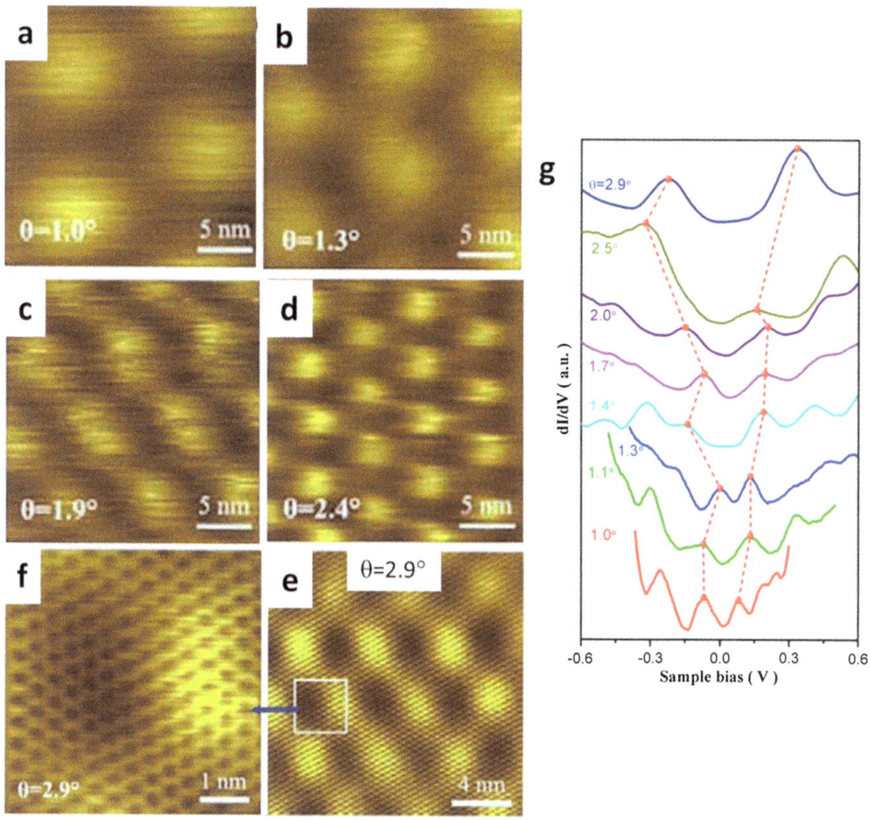

Figure 7.10 (a)–(e) STM images of bilayer graphene on Rh foil with different twist angles. (a) $\theta = 1.0°$, $V_s = 317$ mV. (b) $\theta = 1.3°$, $V_s = 560$ mV. (c) $\theta = 1.9°$, $V_s = -375$ mV. (d) $\theta = 2.4°$, $V_s = -304$ mV. (e) $\theta = 2.9°$, $V_s = -351$ mV. (f) High-resolution STM image ($V_s = -496$ mV) of the white square in (e). (g) dI/dV curves taken on bilayer graphene with different twist angles. Reprinted with permission from W. Yan *et al.*, *Phys. Rev. Lett.*, **109**, 126801, 2012.[105] Copyright (2012) by the American Physical Society.

the two hexagonal Brilouin zones also rotate around the Γ point. The Dirac cones of the two layers no longer coincide and the zero-energy states occur at $k = -\Delta K/2$ in one layer and $k = \Delta K/2$ in the other layer. The displaced Dirac cones cross at energies $\pm h v_F \Delta K/2$ and two saddle points are unavoidable along the intersection of the two cones where there is a finite interlayer hopping. Therefore, two low-energy van Hove singularities (vHSs), which originate from the two saddle points in the band, can be observed as two pronounced peaks in the LDOS measured by STS in the twist bilayer graphene.[105–107] Li *et al.* studied the vHSs of a bilayer graphene on graphite with three different twist angles by STS.[106] They demonstrated that the energy difference of the two vHSs shows angle dependence. Yan *et al.* showed the dI/dV curves taken on a graphene bilayer on Rh foil with different twist angles (Figure 7.10(g)),[105] and the energy difference between two peaks (vHSs) flanking a zero bias follows $\Delta E_{vHS} \sim hv_F\Delta K$ between 1.0° and 3.0°. Their results indicate the twist angle between graphene layers does not result in a significant reduction of Fermi velocity. An abrupt reduction of ΔE_{vHS} at $\theta \sim 1.3°$ was also found, which might originate from reduction of Fermi velocity by flat bands in twist graphene layers with 1.3°. The vHSs in twist graphene layers on SiC were also observed by Brihuega *et al.*[108] They found that vHSs are ubiquitously present in a broad range of twist angles, and the Fermi velocity of the graphene as well as the strength of the interlayer interaction can be recovered from the variation of the energy difference of the vHS with twist angles.

7.3.4 Landau Levels

In the presence of a perpendicular magnetic field, charge carriers in graphene circulate in cyclotron orbits with quantized energies referred to as Landau levels (LLs). Due to the relativistic behavior of quasiparticles in graphene, the LL energy can be expressed as: $E_N = E_D + \mathrm{sgn}(N)\sqrt{2ehV_F^2NB}$.[109] The LL sequence contains a level exactly at zero energy, and is unequally spaced in energy and varies with both the square root of the magnetic field and the LL index. The spectroscopy of LL quantization in a magnetic field is another way to probe quasiparticle properties of graphene.

The LL spectroscopy by STM measurements were firstly carried out on graphene layers grown on SiC(000-1).[110] The dI/dV curve obtained at a magnetic field of 5T exhibit the zero energy state (the $N = 0$ LL), and non-equally-spaced energy level spectrum of LLs are observed. This also shows that the top layer of epitaxial graphene is effectively decoupled from other layers. Song *et al.*[111] examined the fine structure with in a LL by performing high-resolution STS at temperature as low as 10 mK on epitaxial graphene on SiC (Figure 7.11(a)). They found the LL_1 level is composed of four separate peaks, indicating that the valley and spin degenerate is lifted.

Li *et al.* gave the magnetic field dependence of STS spectra on a special surface region of graphite.[49] The dI/dV spectra shown in Figure 7.11(b) exhibits spaced sequence of peaks flanking symmetrically a peak at the DP, which is identified with the $N = 0$ LLs. The peak positions as a function of the $(|N|B)^{1/2}$ can run along a straight line. The field-independent state at the DP together with a square-root dependence on both field and level index are the strong evidence of massless Dirac fermion, indicating that the top layer of graphite in special region is electronically decoupled from substrate. The degree of decoupling to the substrate can be quantified by the LLs spectroscopy.

Jung *et al.* performed STS measurements on exfoliated graphene on SiO_2.[112] In the presence of a perpendicular magnetic field, the STS shown in Figure 7.11(c) exhibit resonances corresponding to LLs. The LL linewidths are significantly broadened compared with graphene on SiC and graphite, reflecting the shorter lifetimes and lower carrier mobility due to the substrate induced disorder potential. Furthermore, in high magnetic field the 2D electron gas

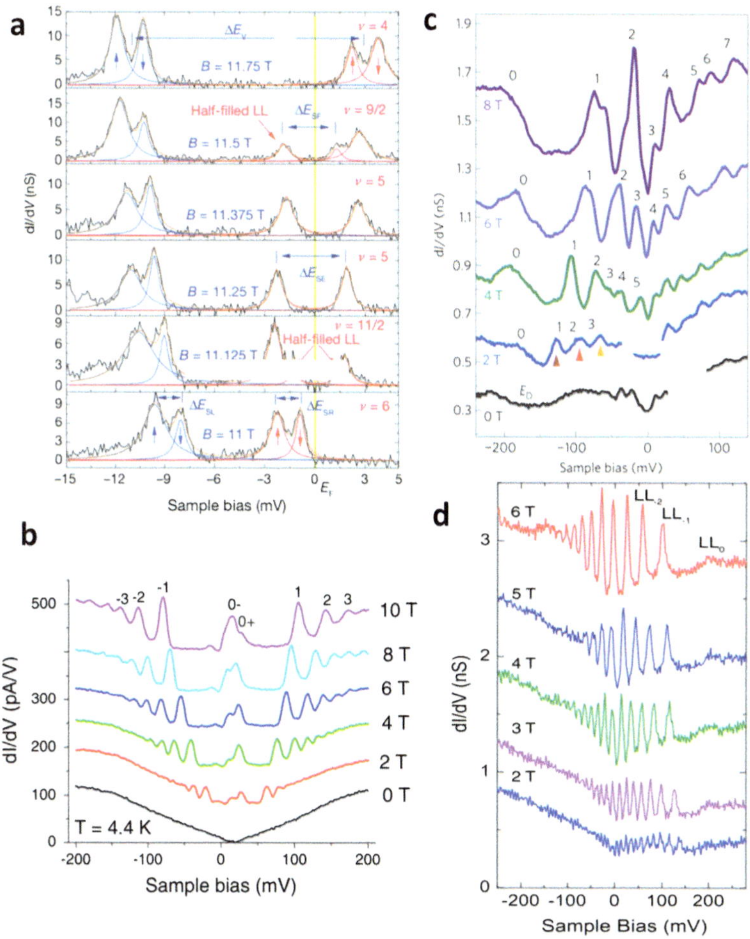

Figure 7.11 (a) High-resolution STS revealing fourfold states that make up $n = 1$ LL of epitaxial graphene on SiC at different magnetic fields. Reprinted by permission from Macmillan Publishers Ltd: *Nature* (ref. 111), copyright (2010). (b) dI/dV spectra of graphene on graphite at different magnetic fields. The peaks are labeled with LL index. Reprinted with permission from G. Li, A. Luican and E. Y. Andrei, *Phys. Rev. Lett.*, **102**, 176804, 2009.[49] Copyright (2009) by the American Physical Society. (c) dI/dV spectra of graphene on SiO_2 at different magnetic fields. The peaks are labelled with LL index. Reprinted by permission from Macmillan Publishers Ltd: *Nature* (ref. 112), Copyright (2010). (d) dI/dV spectra of graphene on h-BN at different magnetic fields. Reprinted with permission from J. Chae *et al.*, *Phys. Rev. Lett.*, **109**, 116802, 2012.[113] Copyright (2012) by the American Physical Society.

in graphene breaks into a network of interacting quantum dots formed at the potential highs and valleys of the disorder potential. Thus, a quartet of peaks in d*I*/d*V* spectra at high magnetic field is observed, which is identified as single-electron resonant tunneling through quantized energy levels inside the graphene quantum dots.

Chae *et al.* investigated the electronic behavior of graphene on h-BN in magnetic field.[113] The LLs in Figure 7.11(d) are extremely sharp and are well developed at field as small as 2T. Compared with graphene on SiO_2,[112] more LLs were observed over a wide energy range. For massless Dirac fermions, the LL energy varies with $\sqrt{B}$ and with the level index as $\sqrt{n}$, and the Fermi velocity can be deduced from the slope. They found the Fermi velocity changes as function of carrier density, suggesting that the Dirac cone is squeezed due to interaction effect at low carrier density.

The LL spectroscopy was also used to investigate the low energy electronic properties of twist graphene layers on Ni film. Luican *et al.* found for twist angles greater than 20° the LL spectra had the characteristics of massless Dirac fermions.[114] That is to say, the twist layers behave like single layer graphene. For the small twist angle the spectra are dominated by vHSs and the Dirac fermion behavior breaks down.

A small perturbation in the form of mechanical strain can deform the energy bands and shifts the Dirac points in both the energy and the momentum space, which is analogue to the effect induced by an external magnetic field applied perpendicular to the graphene plane.[115] Strained graphene nanobubbles were created on Pt(111) surface to test the distortion of graphene lattice give rises to a pseudo magnetic field.[116] The d*I*/d*V* spectra taken on the nanobubble showed peaks separated by more than 100 meV. These peaks are attributed to the LLs arising from a strain induced magnetic field, and follow the expected scaling behavior for massless Dirac fermions in monolayer graphene. The value of magnetic field is obtained from the peak intervals and varies from 300 T to 400 T. The similar effect is also observed on the bilayer graphene grown on Rh foil.[117]

7.4 Other 2D Materials of Group 14 Elements

7.4.1 Silicene

Theoretical investigations on silicene actually dates to 1994, well before the discovery of graphene, when Takeda and Shiraishi proposed that silicon can exist in a two-dimensional sheet structure.[8] In this pioneering work they also pointed out the key difference between silicene and graphene: in contrast to graphene where all C atoms are in exactly the same plane, in silicene the two Si sublattices are relatively shifted in the direction perpendicular to the atomic plane, forming a so-called low buckled structure (see Figure 7.12(a) and (b)).

However, this work was not much noticed until silicene received renewed interest inspired by the great impact brought by graphene. As a new 2D material that is most closely related to graphene, the stability of low-buckled

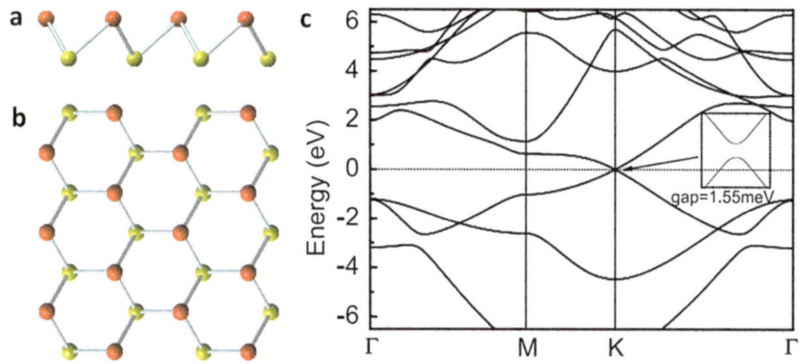

Figure 7.12 (a) and (b) Side and top view of lattice geometry of low-buckled silicene, respectively. The two sublattice are not co-planar. (c) Calculated relativistic band structure of silicene. Inset: zooming in the energy dispersion near the K point and the gap induced by SOC. Reprinted with permission from C. C. Liu *et al.*, *Phys. Rev. Lett.*, **107**, 076802, 2011.[11] Copyright (2011) by the American Physical Society.

structure of silicene was confirmed theoretically,[9,10] and the name 'silicene' was also given.[9] Theoretical works also proposed various intriguing properties of silicene, including the existence of Dirac cone (Figure 7.12(c)),[10,11] and a larger spin-orbit coupling gap which can result in detectable quantum spin Hall effect (QSHE),[11] and unconventional properties tuned by external electric and magnetic fields.[118]

The successful fabrication of silicene took place in 2012, when several groups reported in parallel the growth of silicene on Ag(111),[13,14,16,119] as well as on ZrB_2 film.[15] The preparation of silicene is usually perform by evaporation of pure silicon on to a clean substrate surface in ultrahigh vacuum (UHV). Increasing the Si coverage multilayer silicene films can also be obtained.[120,121] Following this strategy silicene was also reported to be grown on a number of other substrates, such as Ir(111),[17] MoS_2,[122] *etc.* Even a FET device based on silicene films grown on Ag(111), which shows ambipolar transport behavior, has been reported recently.[123]

STM/STS has been the major technique to investigate the structure and electronic properties of silicene. For example, on Ag(111), silicene forms a variety of phases with different superstructures, including the 4×4, $\sqrt{13} \times \sqrt{13}$, $2\sqrt{3} \times 2\sqrt{3}$ and $\sqrt{3} \times \sqrt{3}$ phases,[13,14,16,119] the atomic structures of which have been revealed in detail by STM in combination with theoretical modeling. The electronic structure of silicene has also been investigated by STS. For examples, on the $\sqrt{3} \times \sqrt{3}$ silicene phases STS reveals pronounced quasiparticle interference patterns and linear dispersion was derived,[14] together with the intriguing observation of suppressed backscattering,[124] low temperature phase transition,[125] and an superconducting-like energy gap.[126] These results demonstrated that silicene is a promising playing ground for exotic electronic properties as well as potential applications.

7.4.1.1 Monolayer Silicene Superstructures on Ag(111)

Depending on the Si coverage and substrate temperature, silicene can form a variety of superstructures on Ag(111) surface, including 4 × 4,[13,16,119] (√13 × √13)R ± 13.9°,[13,119] 2√3 × 2√3,[13] and finally to a √3 × √3 phases.[14] The origination of silicene reconstructions is the low-buckled structure of silicene. Although silicene has a honeycomb lattice like graphene, the silicon atoms are not exactly located on the same plane, but instead form a buckled structure. In free standing silicene, the Si atoms in sublattice A is upper buckled whereas those in B are lower buckled, forming a 1 × 1 structure. While the silicene sheet adsorbs on a substrate, the interaction between silicene and substrate results in rearrangement of the buckling degree, forming a variety of superstructures corresponding to different buckling patterns. It should be note that overall the honeycomb lattice of silicene is still preserved regardless of the different patterns.

In literature, the notation of these reconstruction phases is referred to both the pristine silicene-1 × 1 lattice and the Ag(111)-1 × 1 lattice. Therefore, the 4 × 4 phase is also cited as 3 × 3, and the 2√3 × 2√3 phase as √7 × √7 with respective to the silicene-1 × 1. To avoid possible confusion, we will occasionally use Si(3 × 3)/Ag(4 × 4) and Si(√7 × √7)/Ag(2√3 × 2√3) in the following discussions. The term of silicene 1 × 1 as reference is sometimes useful if the film is not commensurate with the substrate.

At relatively low substrate temperature, the first perfectly ordered phase observed in experiments is Si(3 × 3)/Ag(4 × 4) phase. As displayed in Figure 7.13(a), the STM image of Si(3 × 3)/Ag(4 × 4) phase exhibits triangular half unit cells (HUC) each consisting of three bright spots, which is quite similar to the well-known Si(111)-5 × 5 surface structure. The STM image does not show much bias dependence, indicating that the observed protrusions in the image correspond to geometric protrusions.

The Si(3 × 3)/Ag(4 × 4) structure can be easily understood by the so called 'magic mismatch' between the lattice constant of Ag (a_{Ag} = 2.88 Å) and Si (a_{Si} = 3.84 Å), *i.e.*, 4 × a_{Ag} = 3 × a_{Si} within 0.5% of error. Since both Ag(111) surface and silicene are hexagonal lattices, a commensurate Si(3 × 3)/Ag(4 × 4) superstructure can be constructed by simply overlapping a low-buckled, freestanding silicene 1 × 1 lattice on top of a Ag(111)-1 × 1 lattice in the same orientation.[127] Such a magic commensuration results in minimal interface strain and thus stabilizes the silicene structure.

In the case of graphene on metal surfaces (Ru(0001)),[87,88] overlapping two periodic lattices results Moiré patterns. But in the Si(3 × 3)/Ag(4 × 4) phase, STM image shows six protrusions in each unit, not corresponding to a Moiré pattern. Each protrusion originates from a real Si atom with higher position. In each Si(3 × 3)/Ag(4 × 4) unit cell, there are 18 Si atoms. Originally, half of them are upper buckled and half are lower buckled. Once the silicene-1 × 1 sheet is placed on a Ag(111) substrate, the buckling configuration will spontaneously rearrange to lower the surface energy. Therefore, there will be six upper-buckled Si atoms per unit cell, while the other twelve Si atoms are all

Figure 7.13 (a) A STM topography image (16×16 nm^2) of silicene Si(3×3)/Ag(4×4) phase. In the upper-right part, there is a small area consisting of a metastable 4×4-β phase. The white rhombus and the red rhombus mark the 4×4-α and 4×4-β unit cells, respectively. (b) and (c) Structural model of 4×4-α and 4×4-β, respectively. Reprinted with permission from Z.-X. Guo *et al.*, *Phys. Rev. B*, **87**, 235435, 2013.[127] Copyright (2013) by the American Physical Society.

lower-buckled, as depicted in Figure 7.13(b). Moreover, the three upper-buckled Si atoms in one HUC do not belong to the same sublattice with the three upper-buckled Si atoms in the other HUC. This contrasts with the case of freestanding silicene where all upper-buckled (or lower-buckled) Si atoms constitute a sublattice.[10,11] That the protrusions observed in STM correspond to upper-buckled Si atoms that are indeed geometrically higher can be further confirmed by atomic force microscopy (AFM) measurements.[128]

In addition to the normal Si(3×3)/Ag(4×4) phase (denoted as 4×4-α), one can often observe another phase with a 4×4 periodicity, namely the 4×4-β phase.[129] It emerges in the domain boundaries between 4×4-α phases probably because that this phase is less stable than the normal 4×4-α phase and that the strain in domain boundary helps to stabilize it. The STM images in Figure 7.13(a) reveal that the two HUCs of the 4×4-β phase are different: one with six spots, another with only one spot in the center. The structural model of the 4×4-β phases is shown in Figure 7.13(c).

The $\sqrt{13} \times \sqrt{13}$R13.9° ($\sqrt{13} \times \sqrt{13}$ for short) phase usually coexists with the 4×4 phase.[13,119,129] With appropriate Si coverage, such a mix-phase monolayer film can fully cover the substrate surface. This fact suggests that 4×4 phase and $\sqrt{13} \times \sqrt{13}$ have comparable thermal stability. The orientation of $\sqrt{13} \times \sqrt{13}$ phase is rotated 13.9° regarding the 4×4 phase. Note that in sufficiently large $\sqrt{13} \times \sqrt{13}$ domains, long-range Moiré patterns are commonly observed.[129]

The high-resolution STM image shown in Figure 7.14(a) reveals the atomic structure of $\sqrt{13} \times \sqrt{13}$ phase as well as its Moiré pattern. Within a $\sqrt{13} \times \sqrt{13}$ unit cell (marked by rhombus), the most prominent feature is a trimer in the center of one HUC, and a protrusion at the corner.

The large 'Moiré pattern' with periodic vortex structure is indeed a domain wall structure, with each vortex surrounded by six triangular domains. The unit cell directions in neighboring domains are rotated by an angle of 60°. It should be emphasized again that the whole surface is a complete and continuous honeycomb silicene lattice. The appearance of these vortex patterns is solely due to the rearrangement of the buckling configuration of silicon atoms.

There have been several theoretical models proposed for the $\sqrt{13} \times \sqrt{13}$ phase.[129–131] The most reasonable one is depicted in Figure 7.14(d). The simulated STM image well reproduced the bright protrusion in the corner of each unit cell and the trimer feature inside the unit cell. Note that the model only reproduce the local structure $\sqrt{13} \times \sqrt{13}$ phase, while the important feature of the rotational triangular domains and the boundaries are not reflected in this model. The origination of the domain and boundary structures remains an open question to be explored.

In contrast to the coexistence of 4×4 and $\sqrt{13} \times \sqrt{13}$ phases, the Si($\sqrt{7} \times \sqrt{7}$)/Ag($2\sqrt{3} \times 2\sqrt{3}$) phase, which was reported first by Feng *et al.*[13] and subsequently observed by some other groups,[132] is the only superstructure that can be prepared with single phase spreading over the entire Ag(111) surface. By slightly annealing a 4×4 and $\sqrt{13} \times \sqrt{13}$ mix-phase surface, a complete Si($\sqrt{7} \times \sqrt{7}$)/Ag($2\sqrt{3} \times 2\sqrt{3}$) phase can be derived without apparent change of the area of the film. This indicates that the densities of Si atoms in the 4×4, $\sqrt{13} \times \sqrt{13}$ and $2\sqrt{3} \times 2\sqrt{3}$ structures are comparable, while the Si($\sqrt{7} \times \sqrt{7}$)/Ag($2\sqrt{3} \times 2\sqrt{3}$) phase has a higher thermal stability.

Figure 7.15 shows typical large-area and zoom-in STM images of the Si($\sqrt{7} \times \sqrt{7}$)/Ag($2\sqrt{3} \times 2\sqrt{3}$) structure.

Although the prepared monolayer silicon film is entirely uniform in a large scale, the atomically resolved structure appears rather defective and disordered. The large-size image shows a periodic pattern of brighter areas, while the zoomed-in image shows that the brighter region consists of a few perfectly arranged hexagonal rings. However, the areas around the brighter area are rather disordered. With the aid of first-principles calculations, the atomic structure model of the perfect area can be established (Figure 7.15(c)),[13] which involves the Si-$\sqrt{7} \times \sqrt{7}$ super cell matching to the Ag-$2\sqrt{3} \times 2\sqrt{3}$ super cell. Within each Si-$\sqrt{7} \times \sqrt{7}$ super cell, there is one Si atom sitting exactly on top of a Ag atom that corresponds to the brighter spot. These spots form a honeycomb lattice with period of $0.38 \times \sqrt{7} = 1.0$ nm, in line with experimental observation. The 30° angle between the lattice direction of the $\sqrt{7} \times \sqrt{7}$ superstructure and <1-10> direction of Ag(111) is also confirmed by experiments.

Careful examination of the disordered areas between the ordered areas reveals a hidden order (Figure 7.15(b)). Around a hole site there are always

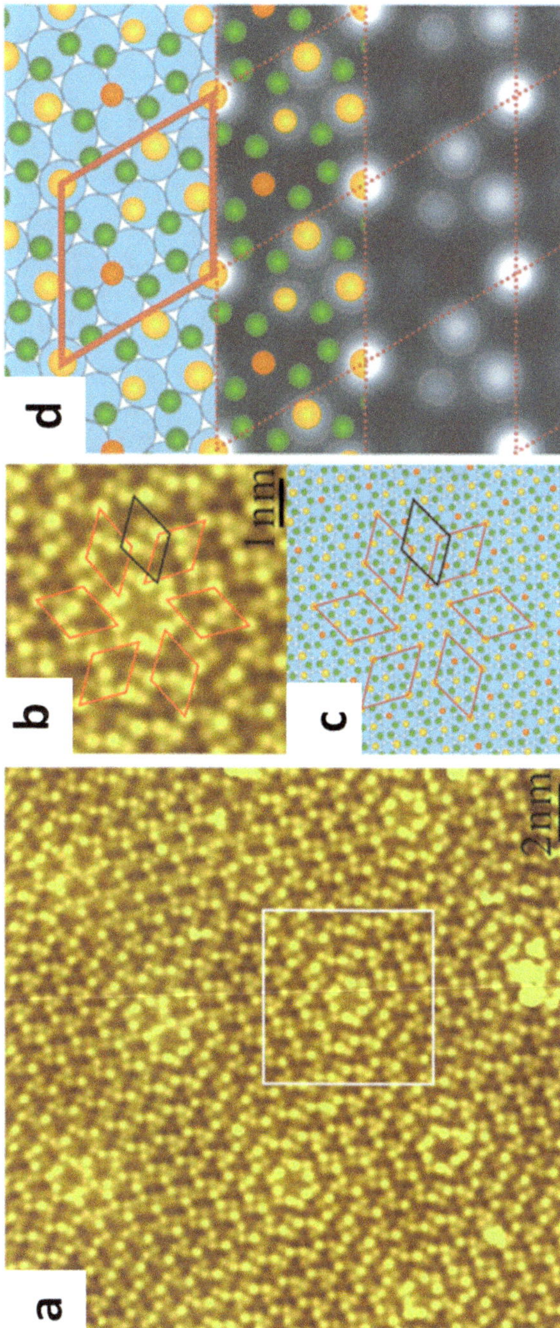

Figure 7.14 (a) High-resolution topography STM image (U = 3 mV and I = 0.2 nA) taken on a $\sqrt{13} \times \sqrt{13}$ phase. (b) Enlarged STM image of the area marked by the white square indicated in (a). (c) Ball structural model for the vortex pattern shown in (b). (d) Atomic structural model for the $\sqrt{13} \times \sqrt{13}$ phase on Ag(111). Below ball model is its corresponding STM simulation result with partial superimposition. Reprinted with permission from ref. 129.

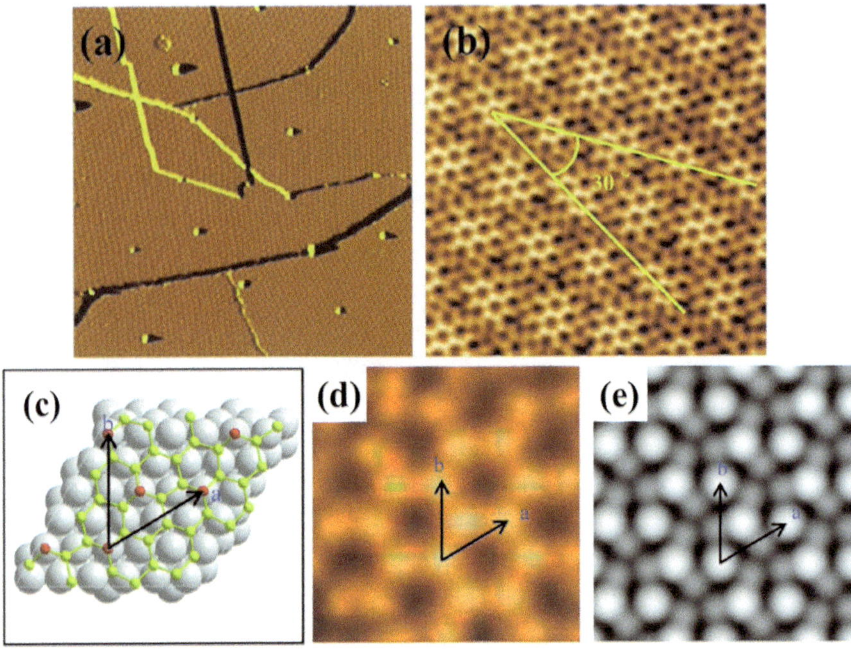

Figure 7.15 (a) A derivative STM image (200×200 nm^2, $V_t = 1.43$ V) of a surface fully covered by the $2\sqrt{3} \times 2\sqrt{3}$ phase. (b) High-resolution STM image (15×15 nm^2, $V_t = -1.0$ V) showing the atomic structure. (c) Calculated atomic model of $2\sqrt{3} \times 2\sqrt{3}$ superstructure of silicene. The grey, yellow, and red balls represent the silver, lower silicon, and higher silicon atoms, respectively. (d) and (e) Experimental and simulated STM images (1.0 eV above Fermi energy) showing the similar structure features and unit cell of lattice. Reprinted with permission from ref. 13. Copyright (2012) American Chemical Society.

six spots, and connecting these six spots would result in a hexagon. Some of these six spots appear darker, which is why the disordered area appears darker than the perfect area. Therefore, the large period pattern can be probably attributed to different electronic states in the ordered and disordered areas rather than the Moiré pattern. Recently, Jamgotchian *et al.*[133] proposed a structural model for $2\sqrt{3} \times 2\sqrt{3}$ phase: surrounding the perfect area are distorted hexagonal rings that correspond to disordered area. Again, we emphasize that the seemingly disordered area is not completely disordered. Most likely, the low-buckled honeycomb structure of silicene is distorted to accommodate the strain, but the basic honeycomb network is still preserved.

As substrate temperature reaches 500 K, silicene islands with $(\sqrt{3} \times \sqrt{3})$R30° periodicity (with respect to silicene-1×1 lattice) are observed on the surface. This phase is of particular interest for the following two reasons: (i) it exhibits a metallic surface state which generates quasi-particle interference (QPI) patterns (see Section 7.4.1.2); (ii) the surface of the multilayer silicene film exhibits exactly the same $(\sqrt{3} \times \sqrt{3})$R30° structure.[14,120,125–127]

The STM image in Figure 7.16(a) shows an island of silicene $(\sqrt{3} \times \sqrt{3})R30^\circ$ phase across the step edges of the underlying Ag(111) surface without losing continuity.

The high-resolution image in Figure 7.16(b) displays the honeycomb structure of silicene terrace. The lattice period of the honeycomb structure is about 0.64 nm, corresponding to a $(\sqrt{3} \times \sqrt{3})R30^\circ$ honeycomb superstructure with respect to the 1×1 lattice.[14]

To explain the atomic structure of $(\sqrt{3} \times \sqrt{3})R30^\circ$ phase, we should mention the characteristic phase transition at low temperature. When the silicene sample on substrate is cooled to liquid helium temperature (5 K), a dramatic phase transition occurs.[125] As depicted in Figure 7.16(c) and (d), one can see the emergence of atomic chains forming interconnected triangles, which are boundaries separating two symmetric domains. At 77 K, the two neighboring protrusions in each honeycomb unit cell are equally bright. After the phase transition ($T_c = \sim 40$ K), one of them becomes much brighter than the other. Since there are two possible configurations, the silicene film is phase separated into triangular domains with the two symmetric configurations divided by narrow domain boundaries.

The $(\sqrt{3} \times \sqrt{3})R30^\circ$ superstructure can be then explained by double symmetric $(\sqrt{3} \times \sqrt{3})R30^\circ$ reconstruction structures with dynamic flip-flop motion at high temperature,[125] as shown in Figure 7.16(e) and (f). If ignoring the substrate, these two $(\sqrt{3} \times \sqrt{3})R30^\circ$ superstructures have identical geometry and share the same central substrate atom for each hexagon unit. In every $(\sqrt{3} \times \sqrt{3})R30^\circ$ unit cell, only one Si atom is buckled upward, whereas the remaining five Si atoms have nearly identical lower heights, resulting in a rhombic $(\sqrt{3} \times \sqrt{3})R30^\circ$ superstructure. The two types of rhombic $(\sqrt{3} \times \sqrt{3})R30^\circ$ superstructures coincide well with the two mirror-symmetric phases observed in the low temperature experiments. At higher temperature, the flip-flop motion between the two phases leads to an averaged, symmetric $(\sqrt{3} \times \sqrt{3})R30^\circ$ phase.

In an early paper, Chen *et al.* claimed that this $(\sqrt{3} \times \sqrt{3})R30^\circ$ structure exists in both monolayer and multilayer silicene films.[13] In their experiments, only the $(\sqrt{3} \times \sqrt{3})R30^\circ$ phase was observed after growth of silicene on Ag(111) at relatively high temperature. Successive experiments by several other groups[134,135] obtained the same $(\sqrt{3} \times \sqrt{3})R30^\circ$ phase by continuing to grow Si on top of a mixed-phases surface at a relatively low temperature, and islands of $(\sqrt{3} \times \sqrt{3})R30^\circ$ phase would form on top of the 4×4 film. These experimental results indicate that 2 ML and multilayer silicene films all exhibit a $\sqrt{3} \times \sqrt{3}$ surface structure, and that the previously reported $(\sqrt{3} \times \sqrt{3})R30^\circ$ phase was likely be a 2 ML film. However, one cannot explicitly exclude the possibility that monolayer $(\sqrt{3} \times \sqrt{3})R30^\circ$ may still exist, but rarely be observed, since the monolayer $(\sqrt{3} \times \sqrt{3})R30^\circ$ structure is quite stable from theoretical calculations[125] and was actually observed on other substrates like Ir(111)[17] and ZrB_2.[15] In a theoretical study by Cahangirov *et al.*, it was suggested that the $(\sqrt{3} \times \sqrt{3})R30^\circ$ structure is constituted by dumbbell units of Si atoms arranged in a honeycomb pattern.[136]

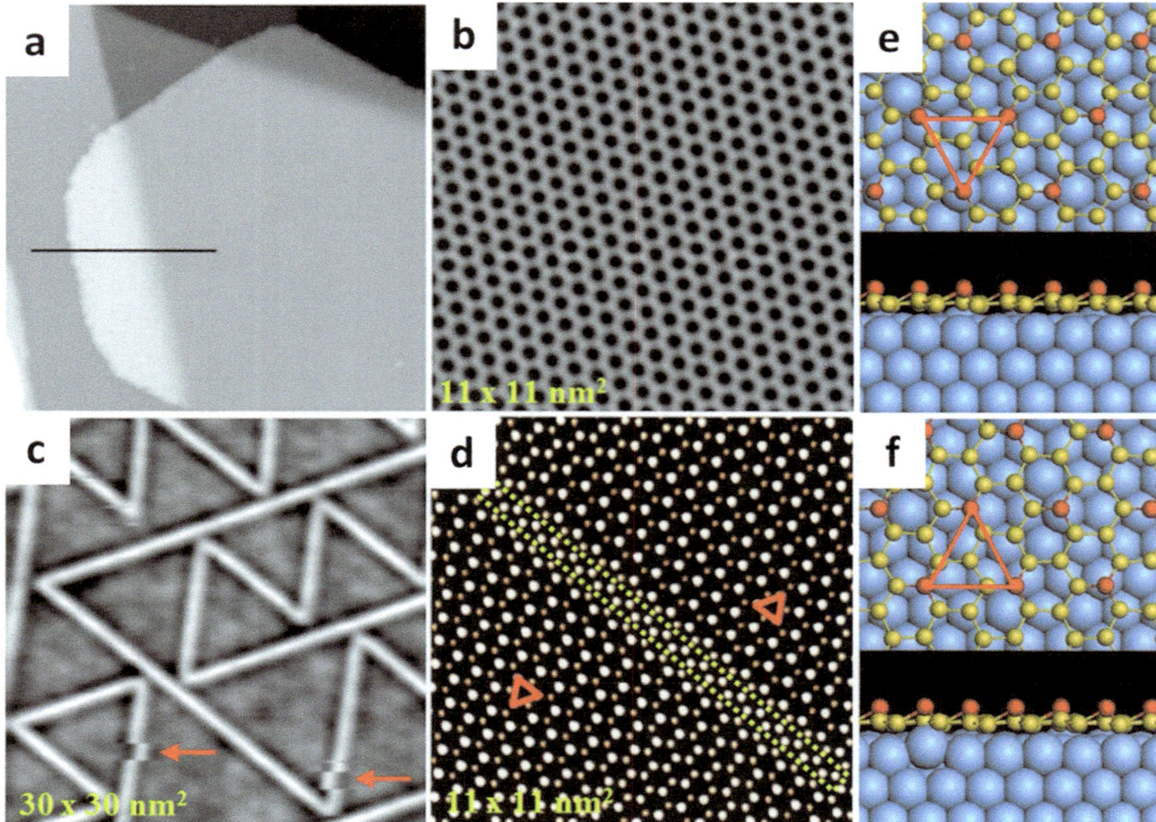

Figure 7.16 (a) The STM topography image of a large area (65×65 nm^2) consisting of a sheet of silicene on Ag(111) crossing two substrate steps. Reprinted with permission from L. Chen *et al.*, *Phys. Rev. Lett.*, **109**, 056804, 2012.[14] Copyright (2012) by the American Physical Society. (b) High-resolution STM image (V_t = 1.0 V) of the surface of silicene $\sqrt{3} \times \sqrt{3}$ phase taken at 77 K. The honeycomb structure is clearly observed. (c) STM image (V_t = −2.0 V) of silicene $\sqrt{3} \times \sqrt{3}$ phase taken at 5 K.

7.4.1.2 Scanning Tunneling Spectroscopy of Silicene

Following the successful preparation of silicene, the focusing issue is whether silicene indeed exhibits the exotic electronic properties (especially, the Dirac electronic state) predicted in theory. So far, ARPES and STM/STS are the two major spectroscopic techniques to characterize the electronic structures of silicene. P. Vogt *et al.* reported, together with their silicene structure, possible existence of the Dirac cone and linear dispersion in the Si(3 × 3)/Ag(4 × 4) phase of silicene from ARPES measurements.[16] Soon after the first ARPES experiment reported for Si(3 × 3)/Ag(4 × 4) silicene on Ag(111), there was a fierce debate about whether the Dirac cone of monolayer silicene can be preserved on Ag(111) surface.[137,138] The key issue is that, once a monolayer silicene sheet is placed on a clean Ag(111) substrate, the strong interaction between the silicene film and the Ag substrate may significantly influence the surface electronic band structure. Later, Avila *et al.*[138] revisited the electronic structures of the silicene Si(3 × 3)/Ag(4 × 4) phase. Analysis of the momentum distribution curves (MDC) taken at different photon energies of the bands around the $\Gamma_{3\times3}$ points further confirmed that the linear dispersion comes from silicene rather than Ag substrate.

Combining spectroscopic signatures of the quantum Hall effect by the Landau quantization under a magnetic field and DFT calculations, Lin *et al.*[137] proposed that silicene on the Ag(111) surface would no longer be a 2D Dirac fermion system due to substrate-induced symmetry breaking. They obtained STS spectra to measure the Landau levels of Si(3 × 3)/Ag(4 × 4) silicene in the presence of the magnetic field perpendicular to the sample between zero and 7 T, and did not find characteristic structures due to the Landau quantization. Therefore, they thought that the silicene on Ag(111) surface loses its characteristic of Dirac fermion system. In contrast, Landau levels appear in the highly-oriented pyrolytic graphite (HOPG), which can be reasonably rationalized by a combination of massless and massive Dirac fermions.

In contrast to the controversy about the electronic states of silicene Si(3 × 3)/Ag(4 × 4) phase, low-temperature electronic state measurements by STS conducted by Chen *et al.* have revealed that the √3 × √3 phase exhibits a metallic surface state, which can host Dirac fermions.[14,126] The typical d*I*/d*V* curves over wide energy range (from −1.5 V to +1.5 V) taken at 5 K and 77 K are shown in Figure 7.17(a) and reveal similar electronic structures. Besides the pronounced peak at 0.9 V, there is a small dip located at about 0.5 V, which is attributed to the position of DP of silicene.

(d) High-resolution STM image (V_t = 0.1 V) of silicene √3 × √3 phase taken at 5 K. (e) and (f) Models of two energy-degenerated √3 reconstructed structures of silicene sheet on Ag(111). Color code: Blue, yellow, and red spheres denote Ag atoms, Si atoms in lower layer, and Si atoms in higher layer, respectively. The red triangles denote the units of √3 silicene structures. (b)–(f) Reprinted with permission from L. Chen *et al.*, *Phys. Rev. Lett.*, **110**, 085504, 2013.[125] Copyright (2013) by the American Physical Society.

The dI/dV curve over a narrow energy range around Fermi energy (from −120 mV to 120 mV) shown in Figure 7.17(b) indicates a characteristic gap about 70 meV centered exactly at E_F and two significant shoulder peaks at both sides.[126] It is possible that the gap may be a superconducting one and the two shoulders are coherence peaks. The temperature-dependent dI/dV curves (Figure 7.17(c)) show that the gap disappears at 40 K, indicating the superconducting critical temperature of silicene is about 35–40 K. Additionally, when the STM tip approaches the surface of silicene the tunneling conductance in the middle of the gap region is lifted, consistent with the Andreev reflection in this superconductor-metal junction. Furthermore, a

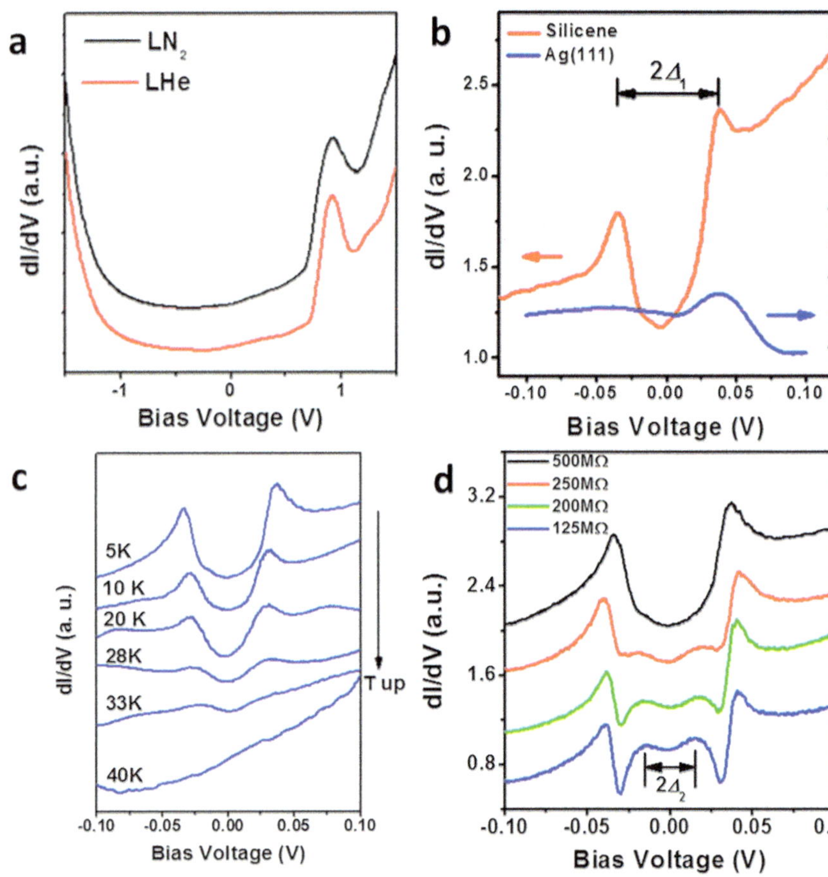

Figure 7.17 (a) dI/dV curves of silicene $\sqrt{3} \times \sqrt{3}$ phase with large energy range taken at 77 K and 5K. (b) dI/dV curves of silicene $\sqrt{3} \times \sqrt{3}$ phase and Ag(111) with narrow energy range taken at 5 K. (c) Temperature dependence of dI/dV curves at same silicene film. (d) dI/dV curves of silicene $\sqrt{3} \times \sqrt{3}$ phase taken at different tunneling resistance of tunneling junction between tip and surface. Reprinted from ref. 126, with permission of AIP Publishing.

smaller gap about 30 meV around E_F with is observed (Figure 7.17(d)), suggesting the silicene may be a two-band superconductor.

Similar to graphene, the Dirac cone of silicene is located at K points in the 2D Brillouin zone.[10,11] The constant energy contours in reciprocal space are small circles centered at the K points (Figure 7.2(g)). Quasiparticles can be scattered within the small circles (intravalley scattering) or between circles (intervalley scattering), resulting in standing waves (quasiparticle interference patterns) observed in d*I*/d*V* maps.[14] The patterns induced by intervalley scattering exhibit closed packed protrusions (shown in Figure 7.18(a)–(c)) in areas near the step edge and point defect, in contrast to the honeycomb structure observed in terrace of silicene island, because the wave vectors of intervalley scattering are close in length to the wave vectors $K_{1(2)}$. The standing waves with long wavelength corresponding to intravalley scattering are observed near the edges of the island of second layer silicene in d*I*/d*V* maps shown in Figure 7.18(d) and (f). As the bias increases from −0.4 V to −1.1 V, the wavelength decreases correspondingly from 2.8 nm to 1.6 nm. Since the value of wave vectors q_1 of intravalley scattering equal to the diameter of the constant energy contour of Dirac cone, the energy-momentum dispersion can be deduced from the $E(\kappa)$ curves in Figure 7.18(g), where $2\kappa = q_1$. The fitting curve give the linear dispersion, with Fermi velocity $V_F = (1.2 \pm 0.1) \times 10^6$ m s^{-1}. The linear E-κ dispersion proves the existence of the Dirac cone in electronic band structures of silicone.[14]

As we mentioned in Section 7.3.4, the behaviors of decay of standing waves from a straight step edge for conventional 2D electron gas and Dirac fermion systems are different. For 2D electron gas, the decay factor of standing waves is −0.5,[101] while in graphene or the surface of a topological insulator, the chirality of quasiparticles plays important role in scatting. This results in suppression of backscattering and a faster decay. A decay factor of α = −1.5 had been reported for graphene,[102] and α = −1.5 and −1 had been reported for 3D Tis.[140]

The decaying behavior of QPI patterns from the step edges of a √3 × √3 silicene film was reported by Feng *et al.*[124] The d*I*/d*V* intensity as a function of the distance from the armchair and zigzag step edges at various energies are shown in Figure 7.19(a)–(f).

The data were fitted using an exponential decaying equation $\delta\rho(x) \propto \cos(2kx + \varphi)x^{\alpha}$, where k is the wave vector of the standing wave, φ is a phase shift associated with the scattering potential, and α is the decay factor. The best fitting parameter of $\alpha = -1.5$ for armchair edge (Γ-K direction), and $\alpha = -1.0$ for zigzag edge (Γ-M direction) were obtained. The observation of strong decay of LDOS oscillation in silicene suggests that the quasiparticle in silicene is chiral. In topological insulator of Bi_2Te_3, the decay factors are −1.5 in Γ-K direction and −1.0 in Γ-M direction,[139] which dramatically coincides with the silicene. It was also noticed that in multilayer silicene, similar and persistent QPI patterns and decaying behavior have been observed on terraces with different thicknesses, indicating that the observed QPI indeed originates from the metallic surface state instead of from the bulk or substrate.

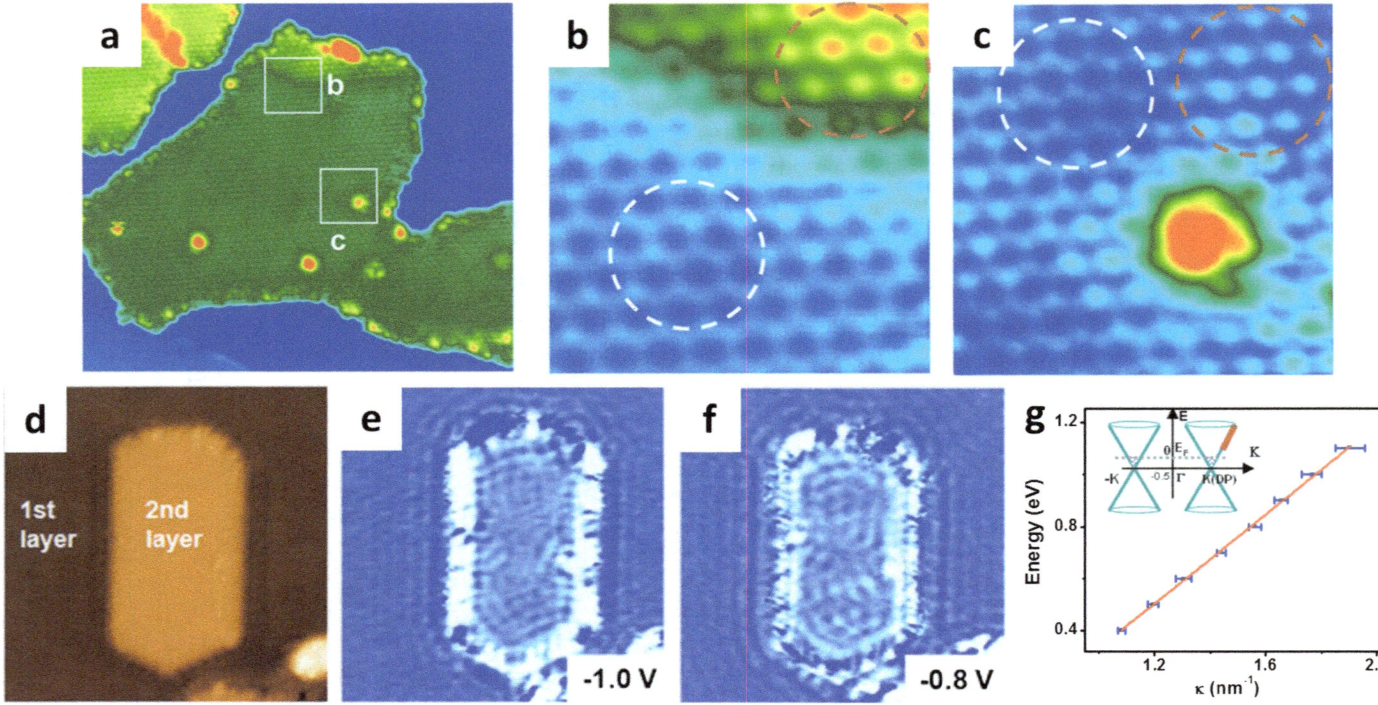

Figure 7.18 (a) STM image ($V_t = -1.1$ V) of a 50×45 nm^2 area of silicene $\sqrt{3} \times \sqrt{3}$ island. (b) and (c) Enlarged STM images of the areas marked by the white squares in (a), respectively. The white and red circles represent the normal honeycomb structure and hexagonal closed packed patterns induced by intervalley scattering of quasiparticles, respectively. (d) The STM image ($V_t = -1.0$ V) of 40×40 nm^2 of one monolayer silicene $\sqrt{3} \times \sqrt{3}$ phase containing an island of second layer. (e) and (f) dI/dV maps of the same area as (d) taken at $V_t = -1.0$ V and $V_t = -0.8$ V, respectively. (g) Energy dispersion as a function of κ for silicene $\sqrt{3} \times \sqrt{3}$ phase determined from the wavelength of standing waves in dI/dV maps. Reprinted with permission from L. Chen *et al.*, *Phys. Rev. Lett.*, **109**, 056804, 2012.[14] Copyright (2012) by the American Physical Society.

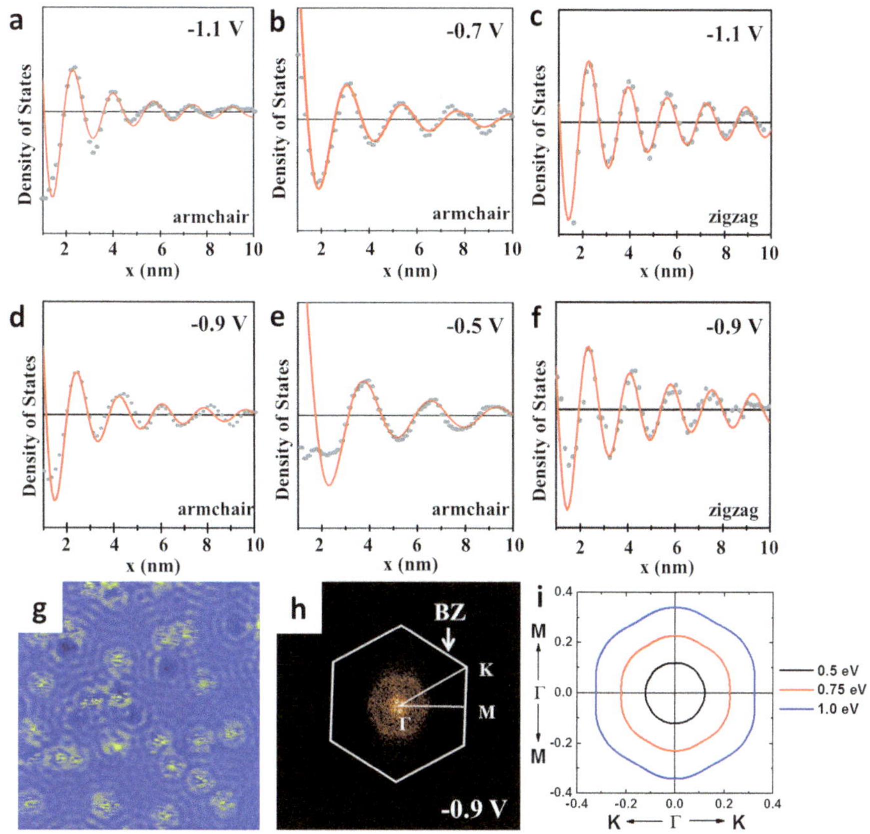

Figure 7.19 (a)–(f) LDOS on monolayer silicene as function of distance from the step edges of silicene √3 × √3 phase at different energies. The data were acquired near armchair edge (a)–(c) and zigzag edge (d)–(f). Grey dots: experimental values; red lines: fits to the data. (g) d*I*/d*V* map ($V_t = -0.9$ V) of surface of hydrogenated silicene showing standing wave patterns induced by hydrogen sites. (h) Fourier transform of (g). A hexagon of 1 × 1 Brillouin zone is superimposed. (i) Constant energy contour of Dirac cone of silicene √3 × √3 phase obtained from DFT calculations. Reprinted with permission from ref. 124. Copyright (2013) American Chemical Society.

Furthermore, other properties of Dirac cone of silicene can be revealed by the standing waves scattering from point defects. Feng *et al.*[124] performed the Fourier transform on the d*I*/d*V* maps containing the standing waves scattering from the hydrogen atoms adsorbed on silicene (√3 × √3)R30° phase, shown in Figure 7.19(g) and (h). A hexagon is found in the center with the edges in Γ-K direction and vertices in Γ-M direction. It means the Dirac cone of silicene is hexagonally warped, which is different to the triangle warping in graphene. Because of the (√3 × √3)R30° superstructure of silicene, the *K* and *K*′ points in the Brillouin zone of (1 × 1) phase are folded onto the Γ point of ($\sqrt{3} \times \sqrt{3}$) phase. Therefore, the six Dirac cones with different orientations of warping at K and K′ points of (1×1) phase are folded onto the Γ point of ($\sqrt{3} \times \sqrt{3}$) phase and give rise to one Dirac cone, which should be hexagonally warped (Figure 7.19(i)).

7.4.2 Germanene

Germanium is also group 14 element, of which electron configurations are very similar with carbon since there are four electrons in their outermost s and p orbitals. In bulk of germanium, the covalent bonds between the atoms are all equivalent and have a hybridized s, p_x, p_y and p_z character (sp^3). 2D graphene-like allotrope of germanium (germanene), in which the atoms have sp^2 character, does not exist in nature, but is theoretically predicted to be energetically stable.[8–10] Though the germanium has a similar honeycomb structure to graphene, the lattice of germanene is buckled, similar to silicene. That is to say, the two sublattices of the honeycomb lattice are slightly displaced with respect to each other in a direction normal to the sheet. The small buckling, and large atomic number of germanium that brings to a large spin-orbit coupling, facilitates the opening of considerable band gap at DP of the order of the thermal energy at room temperature (23.9 meV).[11]

The synthesis of gernamene and gernamene-related materials was initiated by Bianco *et al.* on the exfoliation of germanane (hydrogen terminated germanene) *via* topochemical deintercalation of $CaGe_2$.[140] But the growth of real germanene was only reported on a few substrates by the MBE technique.[141–144] Due to the similarities between silicene and germanene, it is reasonable to assume that germanene behaves more like silicene than graphene when it is adsorbed on metal substrate. In other word, the structural and electronic properties of germanene can be significantly modified by interactions with the underlying substrate. Li *et al.*[141] demonstrated the growth of germanene on Pt(111) surface by depositing germanium on substrate and then annealing sample up to 600 K. The STM topography images reveal the presence of continuous and well-ordered $\sqrt{19} \times \sqrt{19}$ superstructure with a corrugation of about 0.6 Å, which was interpreted as a monolayer germanene on Pt(111). DFT calculations reveal that the $\sqrt{19} \times \sqrt{19}$ superstructure accord with a germanene layer with $\sqrt{3} \times \sqrt{3}$ periodicity. However, a research suggested that the $\sqrt{19} \times \sqrt{19}$ reconstruction on Pt(111) is actually not germanene, but a surface alloy composed of Ge_3Pt tetramers. The synthesis of germanene on Au(111) was reported by Dávila *et al.*,[142] who found three different phases by STM: a $\sqrt{7} \times \sqrt{7}R19.1°$ phase, 5×5 phase and $\sqrt{19} \times \sqrt{19}R23.4°$ phase. Among them, the $\sqrt{7} \times \sqrt{7}$ phase indicate a nearly flat honeycomb structure as shown in STM images Figure 7.20(a), and was considered as germanene (Figure 7.20(b)).

The growth of germanium on Al(111) results in large continuous layer covering the surface with a domain size as large as 100 nm.[143] The high-resolution STM image Figure 7.20(c) taken on the layer exhibits a 3×3 reconstruction, which is also ascribed to germanene. The DFT calculations

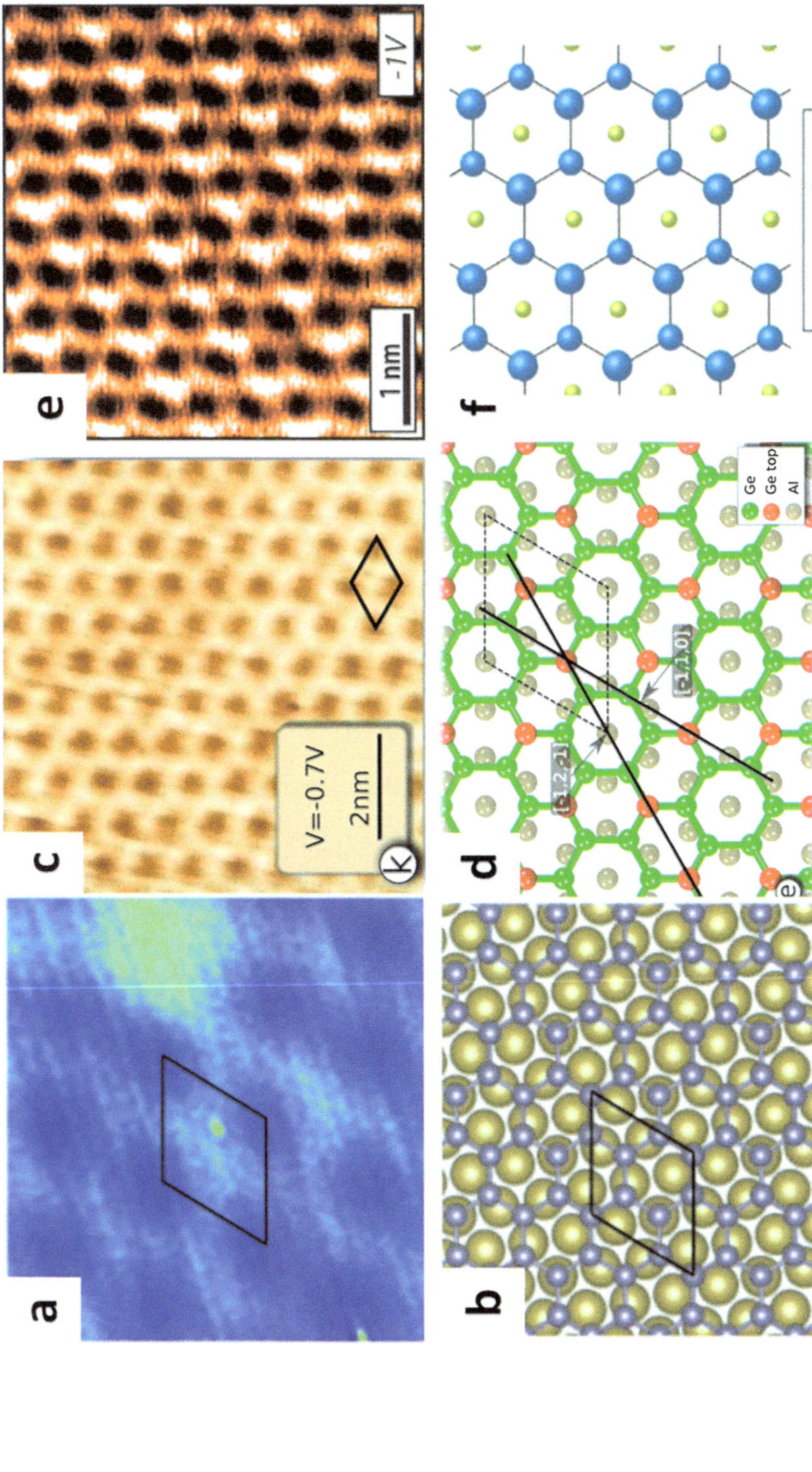

Figure 7.20 (a) STM image ($V_s = -1.2$ V) of germanene on Au(111) with $\sqrt{7} \times \sqrt{7}$ superstructure. (b) Structure model of $\sqrt{3} \times \sqrt{3}$ germanene on $\sqrt{7} \times \sqrt{7}$ Au(111) surface. (a) and (b) Reprinted with permission from ref. 142. (c) STM image ($V_s = -1.12$ V) of germanene on Al(111) with 3 × 3 superstructure. (d) Top view of atomic model of germanene on Al(111). The germanene unit cell is shown as a rhombus. (c) and (d) Reprinted with permission from ref. 143. Copyright (2015) American Chemical Society. (e) STM image ($V_s = -1.0$ V) of germanene on surface of Ge_2Pt cluster. (f) Schematic atomic model of germanene. (e) and (f) Reprinted with permission from ref. 144. © IOP Publishing. All rights reserved.

showed there are eight Ge atoms in one unit cell and two of them are upper buckled due to the interfacial interactions (Figure 7.20(d)), resulting in the protrusions observed in STM images. Bampoulis *et al.* deposited Pt on Ge(110) surface to form Ge_2Pt clusters.[144] After solidification by thermal treatment, the eutectic phases in Ge_2Pt clusters separate into a pure Ge phase and Ge_2Pt phase. The Ge phase has the lowest surface free energy and segregates toward the surface of the clusters. The STM image of Figure 7.20(e) indicates the buckled honeycomb lattice with a nearest-neighbor distance of 2.5 Å on surface of Ge_2Pt clusters, which is ascribed to germanene (Figure 7.20(f)).

Although several experimental researches demonstrated the germanene can be synthesized on metal surfaces, the exotic electronic properties of germanene, such as Dirac fermion and big spin-orbit coupling gap, do not be revealed. Actually, the germanene adsorbed on metallic substrate may destroy the Dirac character of the germanene, because the electronic states of germanene should be hybridized with electronic states of substrate, just like graphene and silicene. Therefore, the growth of germanene on a wide band gap material will also be solved primarily.

7.4.3 Stanene

Stanene is a 2D allotrope of Sn in a graphene-like structure. Similar to silicene and germanene, the two sublattices in honeycomb lattice are shifted along the direction normal to the sheet, forming a buckled honeycomb structure.[12] Due to the much stronger spin-orbit coupling than germanene and silicene, stanene and its derivatives could support a large-gap 2D quantum spin Hall state and thus enable the dissipationless electric conduction at room temperature.[12] Moreover, stanene also provides the near-room-temperature quantum anomalous Hall effect.[145] The first experimental realization was revealed by Zhu *et al.*, who reported the successful growth of stanene films on $Bi_2Se_3(111)$ surface by MBE.[146] After deposition of Sn atoms on $Bi_2Se_3(111)$ substrate, several layers of Sn are observed (Figure 7.21(a)).

The atomic-resolved STM image shown in Figure 7.21(b) revealed the triangular lattice of the stanene without any reconstruction. The STM image of Figure 7.21(c) also indicates the relative position of the topmost Sn layer and underlying Sn layer, suggesting these two atomic layers are A-B stacking like. Combined with the height difference of the two layers of about 0.12 nm, the biatomic-layer structure of stanene can be confirmed (Figure 7.21(d)). However, due to ultrathin thickness of stanene and the interface interaction, the electronic band of stanene will hybridize with the ones of underlying Bi_2Se_3 substrate. Though the ARPES technique combined with theoretical calculations can figure out the contribution from stanene, it is also hard to obtain the intrinsic properties of stanene on metallic surface. Therefore, the realization of the stanene on insulator substrate will be the next stage of investigation.

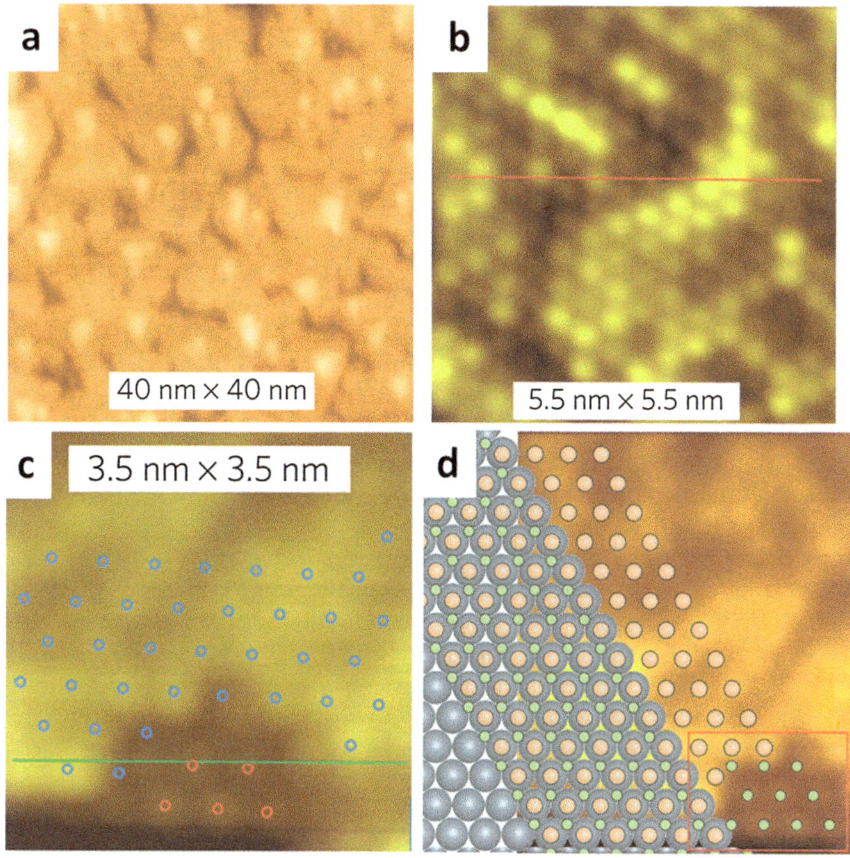

Figure 7.21 (a) STM image of stanene film on $Bi_2S_3(111)$ substrate. (b) High-resolution STM image of stanene surface. (c) High-resolution STM image of top and bottom atomic layers of stanene. Blue dots mark the lattice of the top Sn layer. Red dots mark the lattice of the bottom Sn atoms. (d) Top view atomic structure model for stanene on $Bi_2S_3(111)$ surface. The model is superimposed on the STM image. The grey balls, big green balls, and small green balls represent to Se, topmost Sn, and lower Sn atoms, respectively. Reprinted with permission from Macmillan Publishers Ltd: *Nature Materials* (ref. 146), copyright (2015).

7.5 Borophene

Boron, the fifth element on PTE, is positioned just one to the left of carbon in the PTE, and combines properties of both metal and non-metals. A rich variety of bonding configurations can form between boron atoms, ranging from the two-center two-electron bonds to three-center two-electron bonds. This leads to extreme bonding complexity in boron, boranes and boron-rich solids, which include a large number of allotropes. In past a few years, the theoretical calculations found that 2D boron is energetically stable. In contrast to the well-defined lattices of graphene, the 2D boron has been predicted to be polymorphic. That is to say, there are numerous variety of 2D boron lattices with competitive cohesive energy, which are set apart by the pattern of hexagon holes in a reference triangle grid.[147,148] Theoretically, the electronic properties of 2D boron sheets are related to their structures, exhibiting metallic or semiconducting.[147]

Different from graphene, h-BN or black phosphorus for which the layer form is the natural ground state, and the 3D bulk of boron is lower in energy than any of its 2D allotropes. To realize the 2D boron, choosing a suitable substrate, which is used to suppress the 2D boron nuclear barrier, is necessary. It was theoretically suggested that 2D boron sheets can form on metal substrates, such as Cu(111),[149] Ag(111) and Au(111),[150] due to the stabilization of sp^2 hybridization by metal passivation. In experiments, we directly evaporated a pure boron source on a clean single crystal Ag(111) surface.[151] When the substrate temperature is lower than 500 K, only clusters or disordered structures on the surface were found. At high substrate temperatures of about 570 K and 680 K, two kinds of sheets named S1 and S2 with one-atom thickness form (Figure 7.22(a) and (b)), respectively.

The S2 phase can also been obtained by annealing the S1 phase above 650 K. The STM topography image of Figure 7.22(c) shows the S1 phase consists of rows of parallel protrusions with a rectangle lattice 5.0 Å × 3.0 Å. While the S2 phase is also composed of parallel rows of protrusions with rhombus lattice 4.3 Å × 3.0 Å (Figure 7.22(d)), in which the protrusions along the rows are divided into sections with alternative brighter and darker parts. Combined with theoretical calculations and calibration of coverages, the S1 and S2 phases are determined to be β_{12}-sheets (Figure 7.22(e)) and χ_3-sheets (Figure 7.22(f)) of 2D boron described by a coordinate number of neighbor boron atoms. The inhomogeneous contrast of two phases observed in STM images is attributed to the modulation of the electron density on the boron sheets due to the incommensuration between the lattices of 2D boron sheets and Ag(111) surface. Increasing the boron coverage, the 2D boron sheets can extend their size until they percolate to cover almost the entire surface. When the boron coverage is close to 1 ML, many 3D clusters form on the surface, because the interface interaction has saturated. Mannix *et al.*[152] reported a stripe phase which forms at very high substrate temperature (1000 K), which is never observed by us. However, only 3D clusters form on surface. This stripe phase was explained as closed packed triangle lattice with buckling, which is less stable than β_{12}-sheet and χ_3-sheet in theoretical calculations.

The STS on 2D boron surface (the STS of S1 and S2 are similar) at lower bias voltage range shows a significant LDOS around Fermi level, proving that the 2D boron sheets are metallic. The dI/dV curves at wide voltage range shows a prominent peak at −2.5 V, which originate from p band of 2D boron. Additionally, the dI/dV curves on edges of 2D boron show additional shoulders around −3.0 eV and −0.2 eV, indicating the existence of edge states. DFT calculation reveal that all the edges of S1 and S2 ribbons are bonded to the Ag(111) substrate, as the distances between the B atoms at edge and

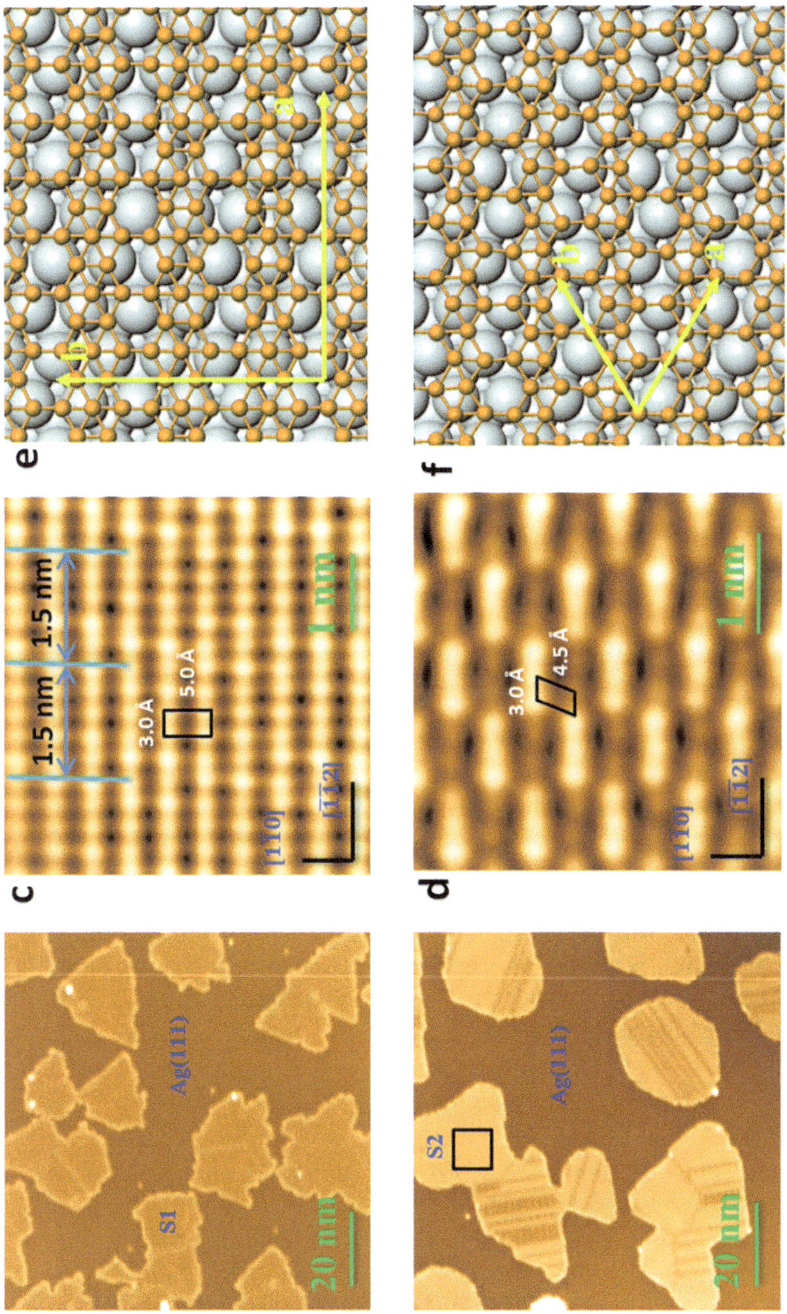

Figure 7.22 (a) STM image (V_t = −4.0 V) of 2D boron on Ag(111) surface with substrate temperature about 570 K during growth. (b) STM image (V_t = −4.0 V) of 2D boron on Ag(111) after annealing the surface in (a) to 650 K. (c) and (d) High-resolution STM images (V_t = 1.0 V) of S1 and S2 phases, respectively. (e) and (f) Top view of atomic model of S1 and S2 phases on Ag(111) surface, which correspond β_{12} and χ_3 sheets, respectively. The orange and grey balls represent boron and silver atoms, respectively. Reprinted with permission from Macmillan Publishers Ltd: *Nature Chemistry* (ref. 151), copyright (2016).

the Ag(111) substrate are significantly shorter than the interlayer distance between 2D boron and substrate, which can result in B-Ag hybrid states in the vicinity of the Fermi level.

The 2D boron sheets appear to be rather inert to oxidization by XPS experiments on the sample exposed to air. The *in-situ* STM experiments on 2D boron sheet exposed to oxygen gas with different dosing show that the oxidization only occurs at the edge of sheets and the lattice remains intact. Mannix *et al.* demonstrated that the oxidization of 2D boron was greatly impeded by a silicon/silicon oxidization capping layer.[152]

The 2D boron sheets are made of strong covalent bonds, leading to high in-plane stiffness. Combined with properties such as weak interfacial interactions and inertness to oxidization, it may be possible to obtain freestanding 2D boron sheets by exfoliation from substrate. Both 2D boron sheets are metallic, suggesting a possibility of applications in nano-electromechanical devices and flexible electronics. Metallicity and light atomic weight are typical prerequisites for phonon-mediated superconducting behavior, possible rendering boron an attractive system for investigating the 2D conventional superconductivity.

7.6 Summary

We have described the demonstrations of mono-element 2D materials, such as graphene, silicene, germanene, stanene and borophene by the STM and STS. These techniques can provide an atomically precise and local scale approach to elucidate the topological and electronic properties of these 2D materials. The atomic scale imaging, local probing of DOS and spatially resolved dI/dV measurements do not only allow us to understand the atomic and electronic structures of 2D materials, but also lead us to illuminate the quasiparticles scattering mechanism in 2D materials. The information provided by these measurements has an advantage over the transport measurements which give the position of DP and the mobility over the entire samples. Because of space limits, we have not been able to cover some interesting topics, for example, nanoribbons, defects, 2D materials heterojunctions, which are also widely investigated by STM/STS.

References

1. K. S. Novoselov, A. K. Geim, S. V. Morozov, D. Jiang, Y. Zhang, S. V. Dubonos, I. V. Grigorieva and A. A. Firosov, *Science*, 2004, **306**, 666.
2. K. S. Novoselov, A. K. Geim, S. V. Morozov, D. Jiang, M. I. Katsnelson, I. V. Grigorieva, S. V. Dubonos and A. A. Firsov, *Nature*, 2005, **438**, 197.
3. Y. Zhang, Y.-W. Tan, H. L. Stormer and P. Kim, *Nature*, 2005, **438**, 201.
4. S. V. Morozov, K. S. Novoselov, M. I. Katsnelson, F. Schedin, D. C. Elias, J. A. Jaszczak and A. K. Geim, *Phys. Rev. Lett.*, 2008, **100**, 016602.
5. A. A. Balandin, S. Ghosh, W. Bao, I. Calizo, D. Teweldebrhan, F. Miao and C. N. Lau, *Nano Lett.*, 2008, **8**, 902.

6. C. Lee, X. Wei, J. W. Kysar and J. Hone, *Science*, 2008, **321**, 385.
7. K. S. Kim, Y. Zhao, H. Jang, S. Y. Lee, J. M. Kim, K. S. Kim, J. H. Ahn, P. Kim, J. Y. Choi and B. H. Hong, *Nature*, 2009, **457**, 706.
8. K. Takeda and K. Shiraishi, *Phys. Rev. B*, 1994, **50**, 14916.
9. G. G. Guzman-Verri and L. C. Lew Yan Voon, *Phys. Rev. B*, 2007, **76**, 075131.
10. S. Cahangirov, M. Topsakal, E. Akturk, H. Sahin and S. Ciraci, *Phys. Rev. Lett.*, 2009, **102**, 236804.
11. C. C. Liu, W. Feng and Y. Yao, *Phys. Rev. Lett.*, 2011, **107**, 076802.
12. Y. Xu, B. Yan, H. J. Zhang, J. Wang, G. Xu, P. Tang, W. Duan and S. C. Zhang, *Phys. Rev. Lett.*, 2013, **111**, 136804.
13. B. J. Feng, Z. J. Ding, S. Meng, Y. G. Yao, X. Y. He, P. Cheng, L. Chen and K. H. Wu, *Nano Lett.*, 2012, **12**, 3507.
14. L. Chen, C. C. Liu, B. J. Feng, X. Y. He, P. Cheng, Z. J. Ding, S. Meng, Y. G. Yao and K. H. Wu, *Phys. Rev. Lett.*, 2012, **109**, 056804.
15. A. Fleurence, R. Friedlein, T. Ozaki, H. Kawai, Y. Wang and Y. Yamada-Takamura, *Phys. Rev. Lett.*, 2012, **108**, 245501.
16. P. Vogt, P. De Padova, C. Quaresima, J. Avila, E. Frantzeskakis, M. C. Asensia, A. Resta, B. Ealet and G. Le Lay, *Phys. Rev. Lett.*, 2012, **108**, 155501.
17. L. Meng, Y. Wang, L. Zhang, S. Du, R. Wu, L. Li, Y. Zhang, G. Li, H. Zhou, W. A. Hofer and H. J. Gao, *Nano Lett.*, 2013, **13**, 685.
18. M. E. Davila, L. Xian, S. Cahangirov, A. Rubio and G. Le Lay, *New J. Phys.*, 2014, **16**, 095002.
19. L. Li, S. Z. Lu, J. Pan, Z. Qin, Y. Q. Wang, Y. Wang, G. Y. Cao, S. Du and H. J. Gao, *Adv. Mater.*, 2014, **26**, 4820.
20. M. Derivaz, D. Dentel, R. Stephan, M. C. Hanf, A. Mehdaoui, P. Sonnet and C. Pirri, *Nano Lett.*, 2015, **15**, 2510.
21. F. F. Zhu, W. J. Chen, Y. Xu, C. L. Gao, D. D. Guan, C. H. Liu, D. Qian, S. C. Zhang and J. F. Jia, *Nat. Mater.*, 2015, **14**, 1020.
22. L. Li, Y. Yu, G. J. Ye, Q. Ge, X. Ou, H. Wu, D. Feng, X. H. Chen and Y. Zhang, *Nat. Nanotechnol.*, 2014, **9**, 372.
23. X. Wu, J. Dai, Y. Zhao, Z. Zhuo, J. Yang and X. C. Zeng, *ACS Nano*, 2012, **6**, 7443.
24. E. S. Penev, S. Bhowmick, A. Sadrzadeh and B. I. Yakobson, *Nano Lett.*, 2012, **12**, 2441.
25. B. Feng, J. Zhang, Q. Zhong, W. Li, S. Li, H. Li, P. Cheng, S. Meng, L. Chen and K. Wu, *Nat. Chem.*, 2016, **8**, 563.
26. A. J. Mannix, X. Zhou, B. Kiraly, J. D. Wood, D. Alducin, B. D. Myers, X. Liu, B. L. Fisher, U. Santiago, J. R. Guest, M. J. Yacaman, A. Ponce, A. R. Oganov, M. C. Hersam and N. P. Guisinger, *Science*, 2015, **350**, 1513.
27. A. H. Castro Neto, F. Guinea, N. M. R. Peres, K. S. Novoselov and A. K. Geim, *Rev. Mod. Phys.*, 2009, **81**, 109.
28. G. Binnig, H. Rohrer, C. Gerber and E. Weibel, *Phys. Rev. Lett.*, 1982, **49**, 57.
29. C. J. Chen, *Introduction to Scanning Tunneling Microscopy*, Oxford University Press, 1993.

30. L. Petersen, Ph. Hofmann, E. W. Plummer and F. Besenbacher, *J. Electron Spectrosc. Relat. Phenom.*, 2000, **109**, 97.
31. L. Simon, C. Bena, F. Vonau, M. Cranney and D. Aubel, *J. Phys. D: Appl. Phys.*, 2011, **44**, 464010.
32. Z. F. Wang, Q. Li, H. Su, X. Wang, Q. W. Shi, J. Chen, J. L. Yang and J. G. Hou, *Phys. Rev. B*, 2007, **75**, 085424.
33. J. Nilsson, A. Castro Neto, F. Guinea and N. M. R. Peres, *Phys. Rev. B*, 2008, **78**, 045405.
34. E. McCann and V. I. Fal'ko, *Phys. Rev. Lett.*, 2006, **96**, 086805.
35. P. Blake, E. W. Hill, A. Castro Neto, K. S. Novoselov, D. Jiang, R. Yang, T. J. Booth and A. K. Geim, *Appl. Phys. Lett.*, 2007, **91**, 063124.
36. A. C. Ferrari, J. C. Meyer, V. Scardaci, C. Casiraghi, M. Lazzeri, F. Mauri, S. Piscanec, D. Jiang, K. S. Novoselov, S. Roth and A. K. Geim, *Phys. Rev. Lett.*, 2006, **97**, 187401.
37. Y. Zhang, V. W. Brar, C. Girit, A. Zettl and M. F. Crommie, *Nat. Phys.*, 2009, **5**, 722.
38. Y. Zhang, V. W. Brar, F. Wang, C. Girit, Y. Yayon, M. Panlasigui, A. Zettl and M. F. Crommie, *Nat. Phys.*, 2008, **4**, 627.
39. E. Stolyarova, K. T. Rim, S. Ryu, J. Maultzsch, P. Kim, L. E. Brus, T. F. Heinz, M. K. Hybertsen and G. W. Flynn, *Proc. Natl. Acad. Sci. U. S. A.*, 2007, **104**, 9209.
40. M. Ishigami, J. H. Chen, W. G. Cullen, M. S. Fuhrer and E. D. Williams, *Nano Lett.*, 2007, **7**, 1643.
41. S. Samaddar, I. Yudhistira, S. Adam, H. Courtois and C. B. Winkelmann, *Phys. Rev. Lett.*, 2016, **116**, 126804.
42. A. Luican, G. Li and E. Y. Andrei, *Phys. Rev. B*, 2011, **83**, 041405(R).
43. K. Watanabe, T. Taniguchi and H. Kanda, *Nat. Mater.*, 2004, **3**, 404.
44. J. Xue, J. Sanchez-Yamagishi, D. Bulmash, P. Jacquod, A. Deshpande, K. Watanabe, T. Taniguchi, P. Jarillo-Herrero and B. J. LeRoy, *Nat. Mater.*, 2011, **10**, 282.
45. R. Decker, Y. Wang, V. W. Brar, W. Regan, H.-Z. Tsai, Q. Wu, W. Gannett, A. Zettl and M. F. Crommie, *Nano Lett.*, 2011, **11**, 2291.
46. C. R. Dean, A. F. Young, I. Meric, C. Lee, C. Wang, S. Sorgenfrei, K. Watanabe, T. Taniguchi, P. Kim, K. L. Shepard and J. Hone, *Nat. Nanotechnol.*, 2010, **5**, 722.
47. M. Yankowitz, J. Xue, D. Cormode, J. D. Sanchez-Yamagishi, K. Watanabe, T. Taniguchi, P. Jarillo-Herrero, P. Jacquod and B. J. LeRoy, *Nat. Phys.*, 2012, **8**, 382.
48. M. Neek-Amal and F. M. Peeters, *Appl. Phys. Lett.*, 2014, **104**, 041909.
49. G. Li, A. Luican and E. Y. Andrei, *Phys. Rev. Lett.*, 2009, **102**, 176804.
50. D. Tomanek, S. G. Louie, H. J. Mamin, D. W. Abraham, R. E. Thomson, E. Ganz and J. Clarke, *Phys. Rev. B*, 1987, **35**, 7790.
51. C. R. Woods, L. Britnell, A. Eckmann, R. S. Ma, J. C. Lu, H. M. Guo, X. Lin, G. L. Yu, Y. Cao, R. V. Gorbachev, A. V. Kretinin, J. Park, L. A. Ponomarenko, M. I. Katsnelson, Yu. N. Gornostyrev, K. Watanabe, T. Taniguchi, C. Casiraghi, H. J. Gao, A. K. Geim and K. S. Novoselov, *Nat. Phys.*, 2014, **10**, 451.

52. K. K. Kam and B. A. Parkinson, *J. Phys. Chem.*, 1982, **86**, 463.
53. C.-P. Lu, G. Li, K. Watanabe, T. Taniguchi and E. Y. Andrei, *Phys. Rev. Lett.*, 2014, **113**, 156804.
54. I. Forbeaux, J.-M. Themlin and J.-M. Debever, *Phys. Rev. B*, 1998, **58**, 16396.
55. I. Forbeaux, J.-M. Themlin and J.-M. Debever, *Surf. Sci.*, 1999, **442**, 9.
56. R. M. Tromp and J. B. Hannon, *Phys. Rev. Lett.*, 2009, **102**, 106104.
57. C. Berger, Z. Song, T. Li, X. Li, A. Y. Ogbazghi, R. Feng, Z. Dai, A. N. Marchenkov, E. H. Conrad, P. N. First and W. A. de Heer, *J. Phys. Chem. B*, 2004, **108**, 19912.
58. C. Berger, Z. Song, X. Li, X. Wu, N. Brown, C. Naud, D. Mayou, T. Li, J. Hass, A. N. Marchenkov, E. H. Conrad, P. N. First and W. A. de Heer, *Science*, 2006, **312**, 1191.
59. C. Riedl, U. Starke, J. Bernhardt, M. Franke and K. Heinz, *Phys. Rev. B*, 2007, **76**, 245406.
60. K. V. Emtsev, F. Speck, T. Seyller, L. Ley and J. D. Riley, *Phys. Rev. B*, 2008, **77**, 155303.
61. F. Hiebel, P. Mallet, L. Magaud and J.-Y. Veuillen, *Phys. Rev. B*, 2008, **78**, 153412.
62. H. Huang, W. Chen, S. Chen and A. T. S. Wee, *ACS Nano*, 2008, **2**, 2513.
63. P. Mallet, F. Varchon, C. Naud, L. Magaud, C. Berger and J.-Y. Veuillen, *Phys. Rev. B*, 2007, **76**, 041403(R).
64. P. Lauffer, K. V. Emtsev, R. Graupner, T. Seyller, L. Ley, S. A. Reshanov and H. B. Weber, *Phys. Rev. B*, 2008, 77, 155426.
65. J. Hass, F. Varchon, J. E. Millán-Otoya, M. Sprinkle, N. Sharma, W. A. de Heer, C. Berger, P. N. First, L. Magaud and E. H. Conrad, *Phys. Rev. Lett.*, 2008, **100**, 125504.
66. M. Sprinkle, D. Siegel, Y. Hu, J. Hicks, A. Tejeda, A. Taleb-Ibrahimi, P. Le Fèvre, F. Bertran, S. Vizzini, H. Enriquez, S. Chiang, P. Soukiassian, C. Berger, W. A. de Heer, A. Lanzara and E. H. Conrad, *Phys. Rev. Lett.*, 2009, **103**, 226803.
67. D. A. Siegel, C. G. Hwang, A. V. Fedorov, and A. Lanzara, *Phys. Rev. B*, 2008, 81, 241417(R).
68. K. S. Kim, T.-H. Kim, A. L. Walter, T. Seyller, H. W. Yeom, E. Rotenberg and A. Bostwick, *Phys. Rev. Lett.*, 2013, **110**, 036804.
69. T. A. Land, T. Michely, R. J. Behm, J. C. Hemminger and G. Comsa, *Surf. Sci.*, 1991, **264**, 261.
70. J. Coraux, A. T. N'Diaye, C. Busse and T. Michely, *Nano Lett.*, 2008, **8**, 565.
71. A. T. N'Diaye, J. Coraux, T. N. Plasa, C. Busse and T. Michely, *New J. Phys.*, 2008, **10**, 043033.
72. Y. S. Dedkov and M. Fonin, Electronic and magnetic properties of the graphene-ferromagnet interface, *New J. Phys.*, 2010, **12**, 125004.
73. J. Lahiri, T. Miller, L. Adamska, I. I. Oleynik and M. Batzill, *Nano Lett.*, 2011, **11**, 518.
74. B. Wang, X. Ma, M. Caffio, R. Schaub and W. X. Li, *Nano Lett.*, 2011, **11**, 424.

75. Y. Murata, E. Starodub, B. B. Kappes, C. V. Ciobanu, N. C. Bartelt, K. F. McCarty and S. Kodambaka, *Appl. Phys. Lett.*, 2010, **97**, 143114.
76. P. Sutter, J. T. Sadowski and E. Sutter, *Phys. Rev. B*, 2009, **80**, 245411.
77. P. Merino, M. Svec, A. L. Pinnardi, G. Otero and J. A. Martin-Gago, *ACS Nano*, 2011, **5**, 5627.
78. S. Nie, N. C. Bartelt, J. M. Wofford, O. D. Dubon, K. F. McCarty and K. Thürmer, *Phys. Rev. B*, 2012, **85**, 205406.
79. J. M. Wofford, E. Starodub, A. L. Walter, S. Nie, A. Bostwick, N. C. Bartelt, K. Thürmer, E. Rotenberg, K. F. McCarty and O. D. Dubon, *New J. Phys.*, 2012, **14**, 053008.
80. G. Odahara, S. Otani, C. Oshima, M. Suzuki, T. Yasue and T. Koshikawa, *Surf. Sci.*, 2011, **605**, 1095.
81. P. W. Sutter, J. I. Flege and E. A. Sutter, *Nat. Mater.*, 2008, **7**, 406.
82. K. S. Kim, Y. Zhao, H. Jang, S. Y. Lee, J. M. Kim, K. S. Kim, J. H. Ahn, P. Kim, J. Y. Choi and B. H. Hong, *Nature*, 2009, **457**, 706.
83. X. Li, C. W. Magnuson, A. Venugopal, R. M. Tromp, J. B. Hannon, E. M. Vogel, L. Colombo and R. S. Ruoff, *J. Am. Chem. Soc.*, 2011, **133**, 2816.
84. H. I. Rasool, E. B. Song, M. J. Allen, J. K. Wassei, R. B. Kaner, K. L. Wang, B. H. Weiller and J. K. Gimzewski, *Nano Lett.*, 2011, **11**, 251.
85. D. Eom, D. Prezzi, K. T. Rim, H. Zhou, M. Lefenfeld, S. Xiao, C. Nuckolls, M. S. Hybertsen, T. F. Heinz and G. W. Flynn, *Nano Lett.*, 2009, **9**, 2844.
86. D. E. Parreiras, E. A. Soares, G. J. P. Abreu, T. E. P. Bueno, W. P. Fernandes, V. E. de Carvalho, S. S. Carara, H. Chacham and R. Paniago, *Phys. Rev. B*, 2014, **90**, 155454.
87. S. Marchini, S. Günther and J. Wintterlin, *Phys. Rev. B*, 2007, **76**, 075429.
88. A. L. Vázquez de Parga, F. Calleja, B. Borca, M. C. G. Passeggi Jr, J. J. Hinarejos, F. Guinea and R. Miranda, *Phys. Rev. Lett.*, 2008, **100**, 056807.
89. E. N. Voloshina, E. Fertitta, A. Garhofer, F. Mittendorfer, M. Fonin, A. Thissen and Yu. S. Dedkov, *Sci. Rep.*, 2013, **3**, 1072.
90. W. Moritz, B. Wang, M. L. Bocquet, T. Brugger, T. Greber, J. Wintterlin and S. Guenther, *Phys. Rev. Lett.*, 2010, **104**, 136102.
91. E. Loginova, S. Nie, K. Thürmer, N. C. Bartelt and K. F. McCarty, *Phys. Rev. B*, 2009, **80**, 085430.
92. E. Starodub, A. Bostwick, L. Moreschini, S. Nie, F. Gabaly, K. McCarty and E. Rotenberg, *Phys. Rev. B*, 2011, **83**, 125428.
93. G. M. Rutter, J. N. Crain, N. P. Guisinger, T. Li, P. N. First and J. A. Stroscio, *Science*, 2007, **317**, 219.
94. I. Brihuega, P. Mallet, C. Bena, S. Bose, C. Michaelis, L. Vitali, F. Varchon, L. Magaud, K. Kern and J. Y. Veuillen, *Phys. Rev. Lett.*, 2008, **101**, 206802.
95. P. Mallet, I. Brihuega, S. Bose, M. M. Ugeda, J. M. Gomez-Rodriguez, K. Kern and J. Y. Veuillen, *Phys. Rev. B*, 2008, **86**, 045444.
96. L. Vitali, C. Riedl, R. Ohmann, I. Brihuega, U. Starke and K. Kern, *Surf. Sci.*, 2008, **602**, L127.

97. Z. F. Wang, Q. Li, H. Su, X. Wang, Q. W. Shi, J. Chen, J. L. Yang and J. G. Hou, *Phys. Rev.B*, 2007, **75**, 085424.
98. P. Lauffer, K. V. Emtsev, R. Graupner, Th. Seyller, L. Ley, S. A. Reshanov and H. B. Weber, *Phys. Rev. B*, 2008, **77**, 155426.
99. H. Yang, A. J. Mayne, M. Boucherit, G. Comtet, G. Dujardin and Y. Kuk, *Nano Lett.*, 2010, **10**, 943.
100. A. Mahmood, P. Mallet and J. Y. Veuillen, Quasiparticle scattering off phase boundaries in epitaxial graphene, *Nanotechnology*, 2012, **23**, 055706.
101. M. F. Crommie, C. P. Lutz and D. M. Eigler, *Nature*, 1993, **363**, 524.
102. J. Xue, J. Sanchez-Yamagishi, K. Watanabe, T. Taniguchi, P. Jarillo-Herrero and B. J. LeRoy, *Phys. Rev. Lett.*, 2012, **108**, 016801.
103. J. M. B. Lopes dos Santos, N. M. R. Peres and A. H. Castro Neto, *Phys. Rev. Lett.*, 2007, **99**, 256802.
104. S. Gwo and C. K. Shih, *Phys. Rev. B*, 1993, **47**, 13059.
105. W. Yan, M. Liu, R.-F. Dou, L. Meng, L. Feng, Z.-D. Chu, Y. Zhang, Z. Liu, J.-C. Nie and L. He, *Phys. Rev. Lett.*, 2012, **109**, 126801.
106. G. Li, A. Luican, J. M. B. Lopes dos Santos, A. H. Castro Neto, A. Reina, J. Kong and E. Y. Andrei, *Nat. Phys.*, 2009, **6**, 109.
107. L. Meng, Z.-D. Chu, Y. Zhang, J.-Y. Yang, R.-F. Dou, J.-C. Nie and L. He, *Phys. Rev. B*, 2012, **85**, 235453.
108. I. Brihuega, P. Mallet, H. Gonzalez-Herrero, G. Trambly de Laissardiere, M. M. Ugeda, L. Magaud, J. M. Gomez-Rodiguez, F. Yndurain and J.-Y. Veuillen, *Phys. Rev. Lett.*, 2012, **109**, 196802.
109. J. W. McClure, *Phys. Rev.*, 1957, **108**, 612.
110. D. L. Miller, K. D. Kubista, G. M. Rutter, M. Ruan, W. A. de Heer, P. N. First and J. A. Stroscio, *Science*, 2009, **324**, 924.
111. Y. J. Song, A. F. Otte, Y. Kuk, Y. Hu, D. B. Torrance, P. N. First, W. A. Heer, H. Min, S. Adam, M. D. Stiles, A. H. MacDonald and J. A. Stroscio, *Nature*, 2010, **467**, 185.
112. S. Jung, G. M. Rutter, N. N. Klimov, D. B. Newell, I. Calizo, A. R. Hight-Walker, N. B. Zhitenev and J. A. Stroscio, *Nat. Phys.*, 2011, **7**, 245.
113. J. Chae, S. Jung, A. F. Young, C. R. Dean, L. Wang, Y. Gao, K. Watanabe, T. Taniguchi, J. Hone, K. L. Shepard, P. Kim, N. B. Zhitenev and J. A. Stroscio, *Phys. Rev. Lett.*, 2012, **109**, 116802.
114. A. Luican, G. Li, A. Reina, J. Kong, R. R. Nair, K. S. Novoselov, A. K. Geim and E. Y. Andrei, *Phys. Rev. Lett.*, 2011, **106**, 126802.
115. F. Guinea, M. I. Katsnelson and A. K. Geim, *Nat. Phys.*, 2010, **6**, 30.
116. N. Levy, S. A. Burke, K. L. Meaker, M. Panlasigui, A. Zettl, F. Guinea, A. H. Castro Neto and M. F. Crommie, *Science*, 2010, **329**, 544.
117. W. Yan, W.-Y. He, Z.-D. Chu, M. Liu, L. Meng, R.-F. Dou, Y. Zhang, Z. Liu, J.-C. Nie and L. He, *Nat. Commun.*, 2013, **4**, 2159.
118. M. Ezawa, *Phys. Rev. Lett.*, 2011, **109**, 055502.
119. C. L. Lin, R. Arafune, K. Kawahara, N. Tsukahara, E. Minamitani, Y. Kim, N. Takagi and M. Kawai, *Appl. Phys. Express*, 2012, **5**, 045802.

120. J. Chen, Y. Du, Z. Li, W. B. Li, B. J. Feng, J. L. Qiu, P. Cheng, S. X. Dou, L. Chen and K. H. Wu, *Sci. Rep.*, 2015, **5**, 13590.
121. P. De Padova, C. Ottaviani, C. Quaresima, B. Olivieri, P. Imperatori, E. Salomon, T. Angot, L. Quagliano, C. Romano, A. Vona, M. Muniz-Miranda, A. Generosi, B. Paci and G. Le Lay, *2D Mater.*, 2014, **1**, 021003.
122. D. Chiappe, E. Scalise, E. Cinquanta, C. Grazianetti, B. van den Broek, M. Fanciulli, M. Houssa and A. Molle, *Adv. Mater.*, 2014, **26**, 2096.
123. L. Tao, E. Cinquanta, D. Chiappe, C. Grazianetti, M. Fanciulli, M. Dubey, A. Molle and D. Akinwande, *Nat. Nanotechnol.*, 2015, **10**, 227.
124. B. J. Feng, H. Li, C. C. Liu, T. N. Shao, P. Cheng, Y. Yao, S. Meng, L. Chen and K. H. Wu, *ACS Nano*, 2013, 7, 9049.
125. L. Chen, H. Li, B. J. Feng, Z. J. Ding, J. L. Qiu, P. Cheng, K. H. Wu and S. Meng, *Phys. Rev. Lett.*, 2013, **110**, 085504.
126. L. Chen, B. J. Feng and K. H. Wu, *Appl. Phys. Lett.*, 2013, **102**, 081602.
127. Z.-X. Guo, S. Furuya, J. Iwata and A. Oshiyama, *Phys. Rev. B*, 2013, **87**, 235435.
128. A. Resta, T. Leoni, C. Barth, A. Ranguis, C. Becker, T. Bruhn, P. Vogt and G. Le Lay, *Sci. Rep.*, 2013, **3**, 2399.
129. Z. L. Liu, M. X. Wang, J. P. Xu, J. F. Ge, G. Le Lay, P. Vogt, D. Qian, C. L. Gao, C. Liu and J. F. Jia, *New J. Phys.*, 2014, **16**, 075006.
130. H. Jamgotchian, Y. Colignon, N. Hamzaoui, B. Ealet, J. Y. Hoarau, B. Aufray and J. P. Bibérian, *J. Phys.: Condens. Matter*, 2012, **24**, 172001.
131. H. Enriquez, S. Vizzini, A. Kara, B. Lalmi and H. Oughaddou, *J. Phys.: Condens. Matter*, 2012, **24**, 314211.
132. D. Chiappe, C. Grazianetti, G. Tallarida, M. Fanciulli and A. Molle, *Adv. Mater.*, 2012, **24**, 5088.
133. H. Jamgotchian, B. Ealet, H. Maradj, J.-Y. Hoarau, J.-P. Biberian and B. Aufray, *J. Phys.: Condens. Matter*, 2016, **28**, 196002.
134. P. Vogt, P. Capiod, M. Berthe, A. Resta, P. De Padova, T. Bruhn, G. Le Lay and B. Grandidier, *Appl. Phys. Lett.*, 2014, **104**, 021602.
135. J. Zhuang, X. Xu, Y. Du, K. H. Wu, L. Chen, W. Hao, J. Wang, W. K. Yeoh, X. L. Wang and S. X. Dou, *Phys. Rev. B*, 2015, **91**, 161609(R).
136. S. Cahangirov, V. O. Özçelik, L. Xian, J. Avila, S. Cho, M. C. Asensio, S. Ciraci and A. Rubio, *Phys. Rev. B*, 2015, **90**, 035448.
137. C.-L. Lin, R. Arafune, K. Kawahara, M. Kanno, N. Tsukahara, E. Minamitani, Y. Kim, M. Kawai and N. Takagi, *Phys. Rev. Lett.*, 2013, **110**, 076801.
138. J. Avila, P. De Padova, S. Cho, I. Colambo, S. Lorcy, C. Quaresima, P. Vogt, A. Resta, G. Le Lay and M. C. Asensio, *J. Phys.: Condens. Matter*, 2013, **25**, 262001.
139. J. Wang, W. Li, P. Cheng, C. Song, T. Zhang, P. Deng, X. Chen, X. C. Ma, K. He, J. F. Jia, Q. K. Xue and B. F. Zhu, *Phys. Rev. B*, 2011, **84**, 235447.
140. E. Bianco, S. Butler, S. Jiang, O. D. Restrepo, W. Windl and J. E. Golberger, *ACS Nano*, 2013, **7**, 4414.
141. L. Li, S.-Z. Lu, J. Pan, Z. Qin, Y.-Q. Wang, Y. Wang, G. Cao, S. Du and H.-J. Gao, *Adv. Mater.*, 2014, **26**, 4820.

142. M. E. Dávila, L. Xian, S. Cahangirov, A. Rubio and G. Le Lay, *New J. Phys.*, 2014, **16**, 095002.
143. M. Derivaz, D. Dentel, R. Stephan, M.-C. Hanf, A. Mehdaoui, P. Sonnet and C. Pirri, *Nano Lett.*, 2015, **15**, 2510.
144. P. Bampoulis, L. Zhang, A. Safaei, R. van Gastel, B. Poelsema and H. J. W. Zandvliet, *J. Phys.: Condens. Matter*, 2014, **26**, 442001.
145. S.-C. Wu, G. Shan and B. Yan, *Phys. Rev. Lett.*, 2014, **113**, 256401.
146. F. Zhu, W. Chen, Y. Xu, C. Gao, D. Guan, C. Liu, D. Qian, S.-C. Zhang and J. F. Jia, *Nat. Mater.*, 2015, **14**, 1020.
147. X. Wu, J. Dai, Y. Zhao, Z. Zhuo, J. Yang and X. Zeng, *ACS Nano*, 2012, **6**, 7443.
148. E. S. Penev, S. Bhowmick, A. Sadrzadeh and B. I. Yakobson, *Nano Lett.*, 2012, **12**, 2441.
149. H. Liu, J. Gao and J. Zhao, *Sci. Rep.*, 2013, **3**, 3238.
150. Y. Liu, E. S. Penev and B. I. Yakobson, *Angew. Chem., Int. Ed.*, 2013, **52**, 3156.
151. B. J. Feng, J. Zhang, Q. Zhong, W. Li, S. Li, H. Li, P. Cheng, S. Meng, L. Chen and K. Wu, *Nat. Chem.*, 2016, **8**, 563.
152. A. J. Mannix, X. Zhou, B. Kiraly, J. D. Wood, D. Alducin, B. D. Myers, X. Liu, B. L. Fisher, U. Santiago, J. R. Guest, M. J. Yacaman, A. Ponce, A. R. Oganov, M. C. Hersam and N. P. Guisinger, *Science*, 2015, **350**, 1513.

CHAPTER 8

Synchrotron Radiation Spectroscopic Techniques for Two-dimensional Materials

TAO YAO*, WEI LIU, YUANJIE CAO AND LINLIN CAO

National Synchrotron Radiation Laboratory, University of Science and Technology of China, Hefei 230029, Anhui, P. R. China
*E-mail: yaot@ustc.edu.cn

8.1 Synchrotron Radiation X-ray Absorption Spectroscopy

Among multiple synchrotron radiation spectroscopic techniques, X-ray absorption fine structure (XAFS) spectroscopy is one of the most widely employed for investigating the structure of materials. XAFS has experienced a rapid development in the last four decades and has proved to be a powerful structure characterization technique in the study of local environments in condensed matter. XAFS spectroscopy has become a powerful probing tool for identifying the surface structure of two-dimensional (2D) materials, as proved by numerous structure studies that will be covered in this chapter.

Smart Materials No. 27
Inorganic Two-dimensional Nanomaterials: Fundamental Understanding, Characterizations and Energy Applications
Edited by Changzheng Wu

Published by the Royal Society of Chemistry, www.rsc.org

8.1.1 Basic XAFS Concepts

In the X-ray spectral region, matter interacts with electromagnetic radiation through dominant photoelectric absorption. For a specific element, there are characteristic energies called absorption edges, at which excitation of core-level electrons are energetically allowed. The X-ray absorption coefficient $\mu(E)$ exhibits a jump across the absorption edge, followed by oscillatory structure at higher energies.[1] The XAFS spectrum describes the details of X-ray absorption at energies near and above the absorption edge. Figure 8.1 displays the Cu K-edge XAFS spectrum of copper foil as an example. The spectral region covering the energy range from the absorption edge to typically 30–50 eV above it, is commonly denoted as XANES. The oscillation over a wide energy range extending 1000 eV or more above the edge is generally referred to as EXAFS. The distinction between the spectral region of XANES and EXAFS is somewhat arbitrary and there is no fundamental difference in the physics. However, the complexity and the analysis method for the EXAFS and XANES spectra are quite different.

XAFS possesses striking advantages such as element-specificity, and sensitivity to the short-range order (typically several Å) and chemical state. It can provide quantitative structural information at an atomic scale about the local environment around the absorbing atom, *i.e.*, near-neighbor species and distance, symmetry, coordination number and fluctuation in bond distance.[2–9] Specifically, EXAFS is sensitive to the distance, coordination number and species of the atoms immediately surrounding the absorbing atoms.[10,11] XANES can distinguish the electronic state and the three-dimensional coordination geometry (including symmetry, atomic types, bond angles and distances).[10] The EXAFS distance determination can be accurate to about 0.01 Å or better. The oscillatory part of $\mu(E)$, also known as the EXAFS signal, or

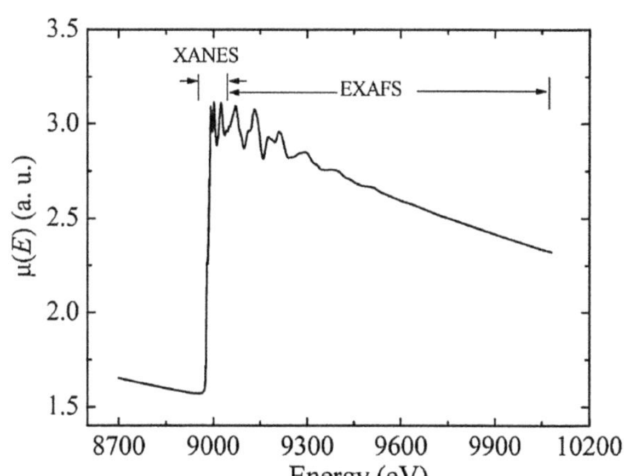

Figure 8.1 A typical Cu K-edge XAFS spectrum for Cu foil highlighting the EXAFS and XANES spectral region.

$\chi(k)$, is obtained by normalizing the $\mu(E)$ by the edge step and subtracting the smooth background function. These oscillations, which arise from the interference patterns of photoelectrons due to their scattering from neighboring atoms, contain quantitative information about the atomic environment in the proximity of the absorbing atom. The standard EXAFS formula is often written as the following eqn (8.1):

$$\chi(k) = \sum_j \frac{N_j f_j(k) S_0^2(k)}{kR_j^2} e^{-2k^2\sigma_j^2} e^{-2R_j/\lambda(k)} \sin\left[2kR_j + \delta_j(k) + 2\phi_C(k)\right] \tag{8.1}$$

where S_0^2 is the amplitude reduction factor, R is the distance between the absorbing and scattering atoms, $\lambda(k)$ is the mean free path of the excited photoelectron, $f(k)$ and $\delta(k)$ are the backscattering amplitude and scattering phase shift of the scattering atom, respectively, $\varphi_C(k)$ is phase-shift of the absorbing atom. The σ^2 term, known as the Debye–Waller factor, is the mean square variation in the interatomic distance R. The $e^{-2k\sigma_j^2}$ term accounts for the effects of dynamic (thermal) vibration and configuration (structural) disorder that smear out the EXAFS oscillations at high k region.

Another decisive factor contributing to the development of XAFS is the availability of brilliant synchrotron radiation light sources allowing for acquisition of better resolved experimental spectra.[3,12,13] In recent years, more new XAFS techniques have been developed, such as time-resolved XAFS,[14,15] micro-XAFS, X-ray magnetic circular dichroism (XMCD), to meet more specific and critical demands in structural investigations.

8.1.2 Structural Parameters Determined from XAFS Data Analysis

The aim of EXAFS data processing and analysis is to extract structural parameters such as interatomic distances, coordination numbers, disorder degrees and sometimes, types of atoms, in various coordination shells for the 'unknown' sample.[16] The general data processing of the raw data consists of the following steps: deglitching, pre-edge background removal; normalization; conversion to k-space; Fourier transform to R-space and inverse Fourier transform. The order of these steps is not critical and not all the steps are necessary.

Once the single-shell oscillation has been isolated, the quantitative structural parameters can be obtained through a fitting procedure according to the EXAFS standard formula eqn (8.1). Before doing so the amplitude and phase-shift functions should be known. These functions can be determined experimentally from a model compound through either fitting or ratio method, or from theoretical calculations or tabulated values. Thanks to the development of the accurate *ab-initio* calculations of FEFF code,[17] nowadays the theoretical amplitude and phase-shift are more routinely used than the experimental ones. The number of adjustable parameters should not exceed the number of relevant independent data points in the analysis: $N_{idp} = 2\Delta k \Delta r/\pi$.

There are various algorithms for fitting the filtered experimental EXAFS function, among which the least-squares algorithm is most commonly used. However, the least-squares algorithm suffers from some drawbacks, for example, many local minima in the sum of squares function may exist, and the obtained structural parameters are often physically impossible. To overcome this problem, some new algorithms such as simulated annealing method have been developed. For an EXAFS spectrum with a good signal–noise ratio, the accuracy of the structural parameter determination can be: interatomic distance R, ±0.01 Å or better for the first shell; coordination number N, typically within ±10%; and Debye–Waller factor σ^2 in the Gaussian approximation, typically within ±0.0005 Å^2. However, the coordination number and the Debye–Waller factor are strongly correlated, and in practice if both are treated as free parameters, the accuracy would be decreased. The analysis can be performed by several software packages that both perform non-linear least-square fitting and report uncertainties in the results. Among the most popular programs based on *ab-initio* calculations of X-ray absorption spectra are IFEFFIT, GNXAS, WIEN2k, and USTCXAFS3.0.[17,18]

The sensitivity to the short-range order makes XAFS capable of probing the local structure for a diversity of systems, no matter whether the studied sample is in solid (crystalline or amorphous), liquid or even gas state. Using high-brilliance synchrotron radiation light sources, many kinds of measurement methods have been developed to obtain high quality XAFS data, such as transmission XAFS for usual bulk material, fluorescence XAFS for dilute impurities and thin films, grazing-incidence XAFS for surface research,[19] magnetic XAFS for spin study,[20] *in-situ* (high-temperature, high-pressure) XAFS for state or stress transformation investigation, space-resolved XAFS and even time-resolved XAFS for kinetics studies.[21–29]

8.2 Advantages of Using XAFS Spectroscopy for Fine Structural Characterization of 2D Nanomaterials

2D nanomaterials, with atomic or molecular thickness and infinite planar lengths, are a conceptually-new class of materials, which provide promising opportunities to satisfy the requirements of next-generation flexible and transparent nanodevices.[30–37] The atomically-thick 2D crystals could serve as an ideal material platform for introducing surface defects to fully optimize the electronic properties, thanks to their atomically flat facet, atomic thickness and ultrahigh specific surface area.[38–47] Up to now, the studies toward the atomically-thick two dimensional crystals have mainly focused on the synthetic methodologies, while the characterization of these low-dimensional structures and the understanding of their clear structure–property relationship have been very limited.[48] Although X-ray diffraction (XRD) has been demonstrated to be a well-established technique in the field of structural investigations for decades, it is still not sufficient to precisely characterize the low-dimensional materials, especially for the 2D crystals with atomic thickness. This is mainly ascribed to the fact that the XRD technique is based on the long-range order of atomic positions in a crystal, while the lack of long-range order in the third dimension means it is unable to identify the spatial atomic distribution in the ultrathin two-dimensional crystals.[49] Intriguingly, thanks to the deep penetration of X-rays and the ability to identify atomic species, XAFS is a powerful element-specific tool to investigate the local atomic geometry and the chemical state of the atoms of one specific element in almost any type of substance. It is particularly suitable for studying the materials that lack long-range order such as nanomaterials, amorphous materials, highly disordered solids or even liquids.[14] In effect, systematic XAFS studies could indeed disclose the atomic structural variations in ultrathin 2D crystals compared with their three-dimensional bulk counterparts.[50–56] In other words, upon reducing the dimensionality from 3D bulk counterparts to 2D crystals with atomic thickness, synchrotron radiation XAFS studies show that their surface coordination numbers are decreased and bond lengths are elongated/contracted, while their disorder degrees are increased, which verify the presence of surface structural distortion. Obviously, the different local atomic arrangements could result in distinct electronic structures, which hence lead to many exciting applications by reason that the electronic structures are considered as the most effective bridge between the microscopic structure and macroscopic properties.[57] For example, the exotic electronic structure usually enables the atomically-thick 2D crystals to possess a variety of novel properties such as superconductivity, half-metallicity and topological insulator.[58–60] Therefore, it is very important for understanding and manipulating the electronic structure of the ultrathin two-dimensional crystals, with efforts to guide their controllable synthesis and bring a wealth of unprecedented physicochemical properties.

8.3 Recent Research Progress of XAFS Spectroscopy in 2D Materials

In the past decade, synchrotron radiation based XAFS analysis has increasingly been applied to investigate the local structures of considerable atomically-thick 2D materials. The XAFS spectroscopy can provide a wealth of information regarding the structural distortion, vacancy defects, and doping for the 2D materials, because it is sensitive to the local atomic environment of the probed atoms. In this section, several research progresses in discerning surface structure of 2D materials by using XAFS technique will be presented.

8.3.1 Structural Distortion of Ultrathin 2D Materials

In ultrathin 2D materials, more surface atoms are exposed. Thus, a very important question arises: how can the atomically-thick 2D crystals with many dangling bonds be stable in ambient conditions? The XAFS technique had been employed to reveal an appearently disordered structure that lowers the surface energy and hence endows two-dimensional materials with better stability. Taking a non-layered ZnSe with a zinc blend structure as an example, a clean 2D ZnSe crystal with a thickness of 4 atoms was obtained by an exfoliation method.[61] To interpret the stability of clean atomically-thick 2D ZnSe crystals with non-layered structures, synchrotron radiation XAFS measurements at both Zn and Se K-edges were performed (Figure 8.2(a) and (b)). The XAFS curve-fitting results showed that the coordination between Zn atoms and *n*-propylamine (pa) molecules resulted in the slight expansion of surface Zn–Se bond lengths, while the next nearest Zn–Zn distances were eminently contracted and the Se–Se distances were obviously elongated, compared with the values of bulk ZnSe (Figure 8.2(c)). This implied an apparent

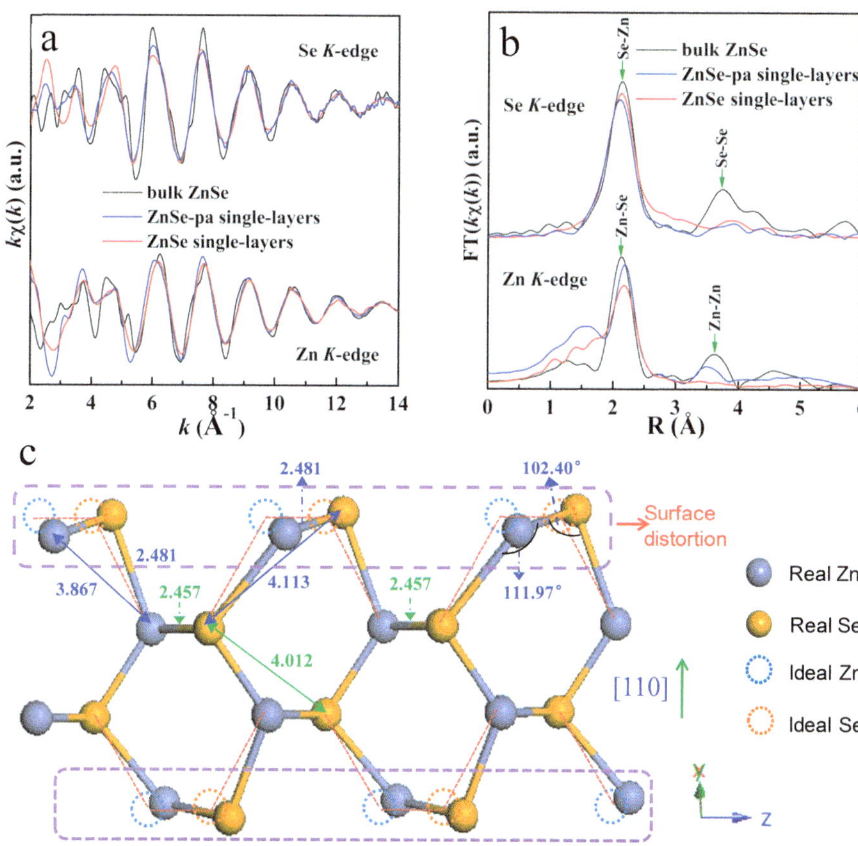

Figure 8.2 (a) Zn and Se K-edge extended XAFS oscillation function $k\chi(k)$ and (b) the corresponding Fourier transforms; the red, blue and black lines correspond to ZnSe single layers, ZnSe-pa single layers and bulk ZnSe, respectively; (c) structural model of ZnSe single layers viewed along the (110) plane. Reprinted with permission from Macmillan Publishers Ltd: *Nature Communications* (ref. 61), Copyright 2012.

surface distortion in the exfoliated two-dimensional ZnSe crystals. Notably, the finally obtained clean two-dimensional ZnSe crystals still preserve the surface distortion, which in turn reduces their surface energy and hence definitely determines their superior structural stability.

The surface atoms of 2D materials can easily adsorb molecules that would change their surface structure. The XAFS measurements were used to determine the local atomic and electronic structures of hydrazine-treated single-layer $NbSe_2$ nanosheets.[62] The pristine single-layer $NbSe_2$ nanosheet is shown in Figure 8.3(a). The $k^3\chi(k)$ oscillation curve presented in Figure 8.3(b) for the hydrazine-treated single-layer $NbSe_2$ nanosheets exhibits a spectral shape like that of the pure nanosheets; however, a noticeable reduction in the oscillation amplitude is observed. In Figure 8.3(c), the FT curve of the pure $NbSe_2$nanosheets displays two peaks at ~2.26 and 3.09 Å, which can be attributed to Se–Nb and Se–Se correlations, respectively. For the hydrazine-treated $NbSe_2$ nanosheets, the intensities of these two peaks are significantly decreased, accompanied by a noticeable shift in the position of the Se–Nb peak towards the high-*R* side and a shift of the Se–Se peak in the opposite direction. Based on these structural parameters, determined from the

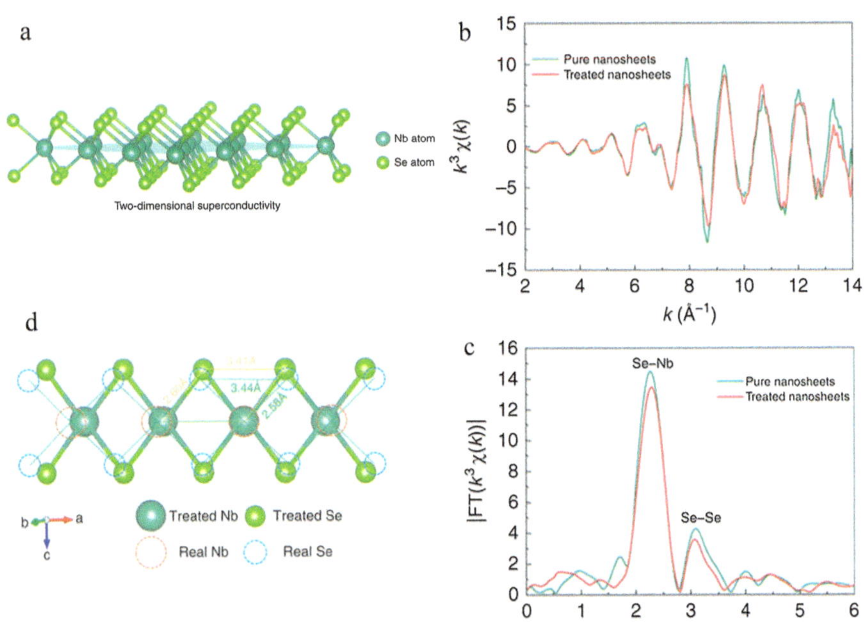

Figure 8.3 (a) Side view of the 2D superconducting $NbSe_2$. (b) Se K-edge extended XAFS oscillation function $k^3\chi(k)$ and (c) the corresponding Fourier transforms FT($k^3\chi(k)$), where k = wave vector and $\chi(k)$ = oscillation as a function of the photoelectron wavenumber; (d) 2D structural models showing the surface distortion. Reprinted with permission from Macmillan Publishers Ltd: *Nature Communications* (ref. 62), Copyright 2016.

least-squares fittings, the structural models for hydrazine-treated $NbSe_2$ nanosheets are displayed in Figure 8.3(d), which clearly illustrates the distorted bond lengths and bond angles of the Nb–Se, Nb–Nb and Nb–Se–Nb coordination geometries. It can be found that the inter-atomic distance of the Se–Nb coordination in the hydrazine-treated $NbSe_2$ nanosheets was elongated by 0.02 Å, whereas that of the Se–Se coordination was shortened by 0.03 Å, when compared with the pure $NbSe_2$ nanosheets. It is apparent that when compared with the pure $NbSe_2$ nanosheets, the adsorption of the polar hydrazine molecules resulted in local modulations in the hydrazine-treated $NbSe_2$ structure.

The surface structure distortion in 2D materials would lead to the distinct electron exchange and magnetic properties. Taking a half-unit-cell thick α-Fe_2O_3 nanosheets as an example, in Fe K-edge XANES spectra (Figure 8.4(a)), the pre-edge peak intensity of α-Fe_2O_3 nanosheets is 40% higher than that of bulk α-Fe_2O_3, while it is similar to that of the low-coordinated Fe_3O_4 reference compound, suggesting that this spectral characteristic is attributed to the inherent five-coordinated Fe (noted as 'Fe5-co' hereafter) of nanosheets.[63]

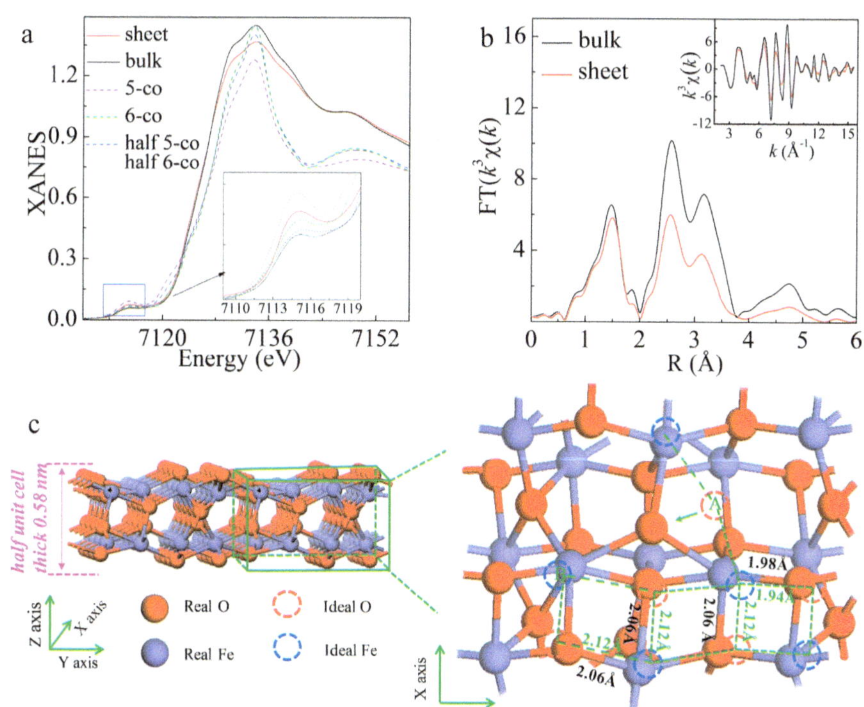

Figure 8.4 (a) Fe K-edge XANES spectra of nanosheet and bulk α-Fe_2O_3 and Fe_3O_4 for reference. (b) FT curves of Fe K-edge EXAFS $k^3\chi(k)$ functions of α-Fe_2O_3 nanosheets and bulk α-Fe_2O_3 (inset: $k^3\chi(k)$ curves). (c) Side and top views of the distortion model. Reprinted with permission from (ref. 63), Copyright 2014 American Chemical Society.

However, the white line peak shape for α-Fe_2O_3 nanosheets is close to that of six-coordinated Fe (noted as 'Fe6-co' hereafter) in bulk α-Fe_2O_3. Thus, the XANES results reveal that the non-degenerate Fe5-co and Fe6-co coordinations probably coexist in the nanosheets. Furthermore, for the FT function of nanosheets in Figure 8.4b, the peaks at 2.5, 3.3, and 4.7 Å corresponding to the second, third Fe–Fe, and the fourth Fe–O/Fe shells are preserved as those in the bulk, suggesting the retained integrity of hexagonal α-Fe_2O_3 structure in the nanosheets. The Fe–O first shell peak is slightly lowered in intensity, and the intensities of the Fe–Fe peaks are significantly reduced by 40%, compared to those of bulk. This indicates that the Fe–Fe second shell is greatly distorted and its structural homogeneity is reduced. The fitted coordination number 5.3 for the Fe–O first shell is slightly larger than that of the ideal nanosheets, further confirming the $Fe_{5\text{-co}} + Fe_{6\text{-co}}$ mixed coordination. Hence, it can be approximated that about 30% Fe ions take the six-coordinated sites, and about 70% of Fe ions take the five-coordinated sites. Based on the ideal half-unit-cell thickness α-Fe_2O_3 structure, it can be found that surface bridge O ions (noted as 'Obri' hereafter) may wedge into the nanosheet to bind with another $Fe_{5\text{-co}}$, resulting in the formation of six-coordinated Fe ions (Figure 8.4(c)). The unique $Fe_{5\text{-co}}$–O–$Fe_{6\text{-co}}$ ion chains endow the ultrathin α-Fe_2O_3 nanosheets robust intrinsic ferromagnetism of 0.6 μ_B/atom at 100 K and remain ferromagnetic at room temperature.

8.3.2 Low-coordinated Surface Atoms of Ultrathin 2D Materials

As is well known, catalysis is a surface reaction process, in which the coordination number of surface atoms plays a crucial role in affecting their catalytic activities. In this regard, the coordination number of the absorbing atoms can be immediately identified by the element-specific tool of XAFS. Taking the three theoretical models of bulk CeO_2, intact and pits-confined atomically-thin CeO_2 sheets as example, their coordination numbers and other structural information are accurately determined by the synchrotron radiation XAFS.[64] The Ce L_3-edge $k^2\chi(k)$ oscillation curve in Figure 8.5(a) for the ultrathin CeO_2 sheets with numerous surface pits displays obvious differences compared with the ultrathin CeO_2 sheets and bulk CeO_2. Furthermore, their corresponding FT curves show that the two main peaks gradually shift to the low R direction along with decreased intensities, qualitatively revealing their distinct local atomic arrangement (Figure 8.5(b)). The coordination numbers for Ce–O, Ce–Ce, Ce–O–O, Ce–Ce–O coordinations in the ultrathin CeO_2 sheets reduce as shown by the EXAFS data fitting results in Figure 8.5(c), indicating the presence of abundant dangling bonds on their surface. In addition, the pit-surrounding Ce–O distances, confined in the ultrathin CeO_2 sheets with numerous surface pits, are contracted from 2.33 to 2.30 Å, and their coordination numbers further decrease from 6.5 to 4.6, associated with increased disorder degrees as compared to the ultrathin CeO_2 sheets (Figure 8.5(c)). As a result, the lower coordination number,

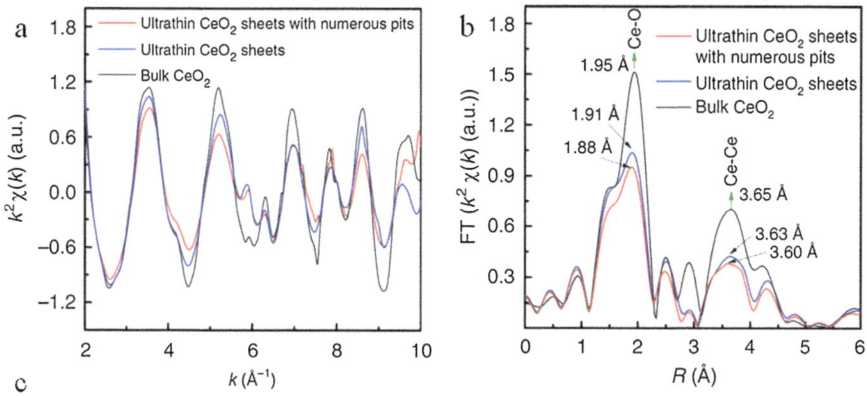

c

Sample	Path	R (Å)	N	σ^2 (10^{-3} Å^2)	ΔE_0 (eV)
Bulk CeO_2	Ce-O	2.344	8.0	4.7	2.9
	Ce-Ce	3.827	12.0	6.3	4.2
	Ce-O-O	3.697	24.0	5.9	3.3
	Ce-Ce-O	4.257	48.0	6.8	3.7
Ultrathin CeO_2 sheets	Ce-O	2.327	6.5 (6.7)	5.0	2.9
	Ce-Ce	3.812	7.7 (8.0)	6.7	4.5
	Ce-O-O	3.667	16.8 (17.3)	7.4	3.2
	Ce-Ce-O	4.234	29.7 (32.0)	9.0	3.4
Ultrathin CeO_2 sheets with numerous pits	Ce-O(80% remain)	2.327	6.5 (6.7)	5.6	2.7
	Ce-O(20% shrink)	2.302	4.6 (4.7)	8.3	3.0
	Ce-Ce	3.802	7.4 (7.8)	8.3	4.6
	Ce-O-O	3.644	15.1 (15.5)	8.7	3.3
	Ce-Ce-O	4.212	28.5 (30.3)	9.2	3.4

Structural parameters around Ce atoms extracted from EXAFS curve-fitting for the ultrathin CeO_2 sheets with numerous surface pits, ultrathin CeO_2 sheets and bulk CeO_2, respectively. The coordination numbers in brackets correspond to the ideal average coordination numbers of exterior and interior atoms for the ultrathin CeO_2 sheets with three-atomic-layer thicknesses.

Figure 8.5 (a) Ce L_3-edge extended XAFS oscillation function $k^2\chi(k)$, (b) the corresponding Fourier transforms FT($k^2\chi(k)$), (c) structural parameters around Ce atoms extracted from EXAFS curve fitting for the ultrathin CeO_2 sheets with numerous surface pits, ultrathin CeO_2 sheets and bulk CeO_2, respectively. Reprinted with permission from Macmillan Publishers Ltd: *Nature Communications* (ref. 64), Copyright 2013.

more dangling bonds and more obvious structural distortion enable the pit-surrounding Ce sites to act as the most favorable active sites for enhancing CO catalytic oxidation properties.

For metal 2D materials, the XAFS technique can also elucidate the local coordination environment of metal atoms and to determine the thickness of metal nanosheets on a macroscopic level. Taking single-layered Rh nanosheets for example, the XAFS spectra and k-space spectra of the PVP-capped Rh nanoparticles and PVP-capped Rh nanosheets are similar and closely resemble that of Rh foil, but with reduced oscillations in the case of the PVP-capped Rh nanosheets (Figure 8.6(a) and (b)).[65] Indeed, in the Fourier-transformed EXAFS spectra (Figure 8.6(c)), the intensity of the peak at 2.37 Å, assigned to the first shell Rh–Rh scattering, decreases in the order Rh foil > PVP-capped Rh nanoparticle > PVP-capped Rh nanosheet. The Rh–Rh coordination number is sensitive to both particle size and morphology at the nanoscale. The coordination environments in structures with different number of Rh layers are illustrated schematically in Figure 8.6(d). For a

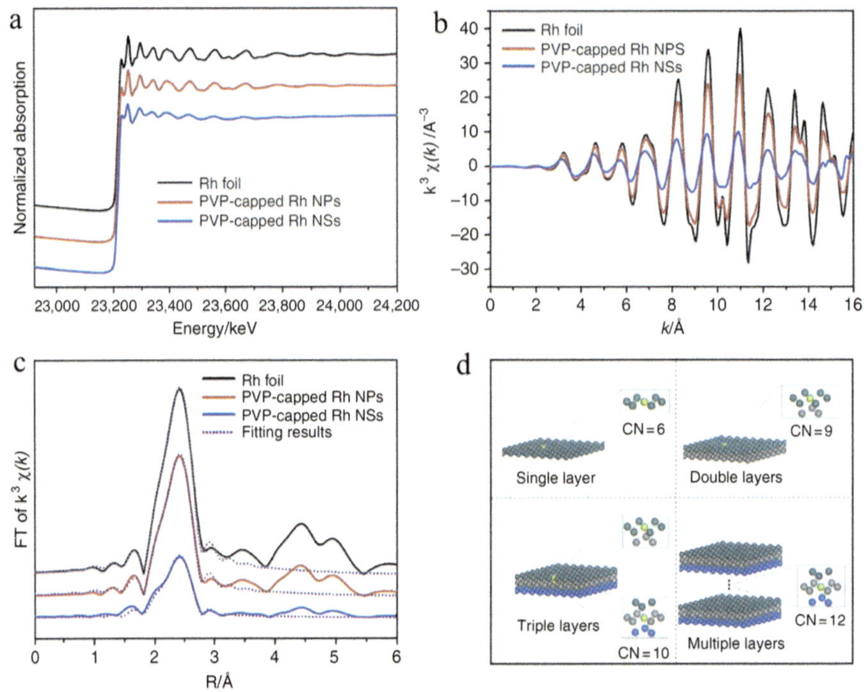

Figure 8.6 (a) Normalized Rh K-edge EXAFS spectra of Rh foil, PVP-capped Rh NPs, and PVP-capped Rh NSs. (b) k^3-weighted Rh K-edge k-space oscillations of Rh foil, PVP-capped Rh NPs and PVP-capped Rh NSs. (c) Fourier-transformed spectra of k^3-weighted Rh K-edge EXAFS of Rh foil (solid black line), PVP-capped Rh NPs (solid red line) and PVP-capped Rh NSs (solid blue line) and corresponding curve-fitting results (dotted purple lines). (d) Schematic illustration of the coordination states with CN of Rh NS with single, double, triple and multiple layers with an exposed Rh (111) lattice plane (the yellow balls indicate the 'core' Rh atoms). Reprinted with permission from Macmillan Publishers Ltd: *Nature Communications* (ref. 65), Copyright 2014.

single-layered, close-packed, infinite Rh structure, the first shell CN is 6, as each Rh atom is surrounded by 6 other Rh atoms. Double- and triple-layered Rh structures have higher CNs (9 and 10, respectively) due to the additional inter-layer Rh–Rh coordination. A CN of only 6.4 ± 0.7 is obtained for the PVP-capped Rh nanosheets, fitting well with the CN of a single-layered system. In addition, the average Rh–Rh bond length of 2.692 ± 0.004 Å in PVP-capped Rh nanosheets is similar to that of Rh foil (2.687 ± 0.003 Å), indicating the Rh nanosheet is composed of close-packed Rh atoms.

The XAFS technique was also used to reveal surface coordination loss in a half-unit-cell thick γ-CoOOH nanosheet.[66] As shown in Figure 8.7(a), the Co K-edge EXAFS $k^3\chi(k)$ oscillation curve for the ultrathin nanosheet displays a

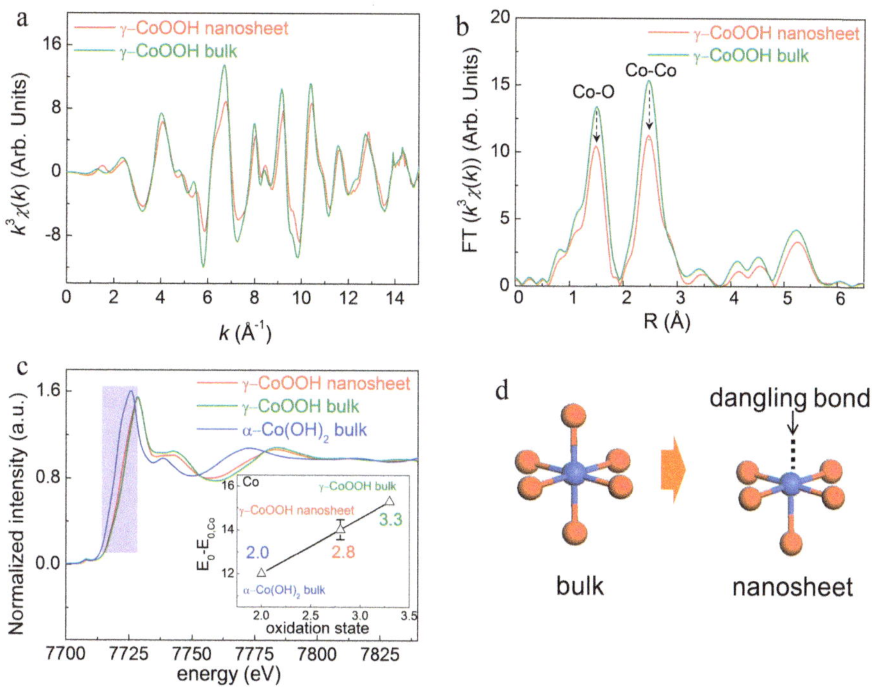

Figure 8.7 (a) Co K-edge EXAFS $k^3\chi(k)$ oscillation functions and (b) the corresponding FT curves of ultrathin γ-CoOOH nanosheets and bulk γ-CoOOH. (c) XANES spectra for nanosheet and bulk γ-CoOOH and bulk α-$Co(OH)_2$ (inset: fitted oxidation states of Co). (d) Structural model of local structural distortion of CoO_{6-x} octahedrons in nanosheets. Reprinted with permission from J. Huang, CoOOH Nanosheets with High Mass Activity for Water Oxidation, *Angewandte Chemie International Edition*, John Wiley and Sons, Copyright © 2015 WILEY-VCH Verlag GmbH & Co. KGaA, Weinheim.[66]

noticeable difference in 2–8 Å^{-1} in comparison with the bulk material, implying the different cobalt local atomic arrangements. Moreover, the FTs curves of bulk and nanosheet γ-CoOOH are characterized by two main peaks at 1.48 Å and 2.50 Å (Figure 8.7(b)), corresponding to the nearest Co–O and next nearest Co–Co coordination, respectively. Strikingly, the peak intensities of the γ-CoOOH nanosheet show a significant decrease. The Co–O coordination number decreased from 6.0 for the bulk to 5.1 for the nanosheet, suggesting that the ultrathin nature of the nanosheet resulted in the presence of a large number of dangling bonds in the surface Co octahedron (CoO_{6-x}) with a structural distortion. Correspondingly, the electronic structure for ultrathin nanosheet is changed by the local structural distortion of CoO_{6-x} octahedrons (Figure 8.7(d)). As shown in the XANES spectra (Figure 8.7(c)), the absorption edge of ultrathin γ-CoOOH nanosheet shifted 1.3 eV towards lower energy relative to that of bulk, corresponding to a change in the Co valence state from +3.2 for the bulk to +2.8 for nanosheet γ-CoOOH. Consequently, the surface distorted CoO_{6-x} octahedron greatly enhances the electrophilicity of H_2O and facilitates the interfacial electron transfer between Co ions and adsorbed-OOH species to form O_2, resulting in the high electrocatalytic activity of layered γ-CoOOH for water oxidation.

8.3.3 Vacancy and Doping of Ultrathin 2D Materials

The XAFS technique was also used to reveal the surface defects in ultrathin 2D materials. For example, the XAFS at Zn K-edge was performed to reveal the local atomic structures of the as-obtained O-doped $ZnIn_2S_4$ (ZIS) sample (Figure 8.8(a)).[67] It can be found that the Zn K-edge oscillation curve for the O-doped ZIS nanosheets displays remarkable differences in comparison with the calculated spectra for bulk ZnO counterpart, implying the different local atomic arrangements of the nanosheets (Figure 8.8(b)). Moreover, the Fourier transformed curve of O-doped ZIS nanosheets shows not only the nearest Zn–S coordination with a main peak at 1.90 Å, but also a new weak peak at about 1.39 Å. Comparing with the FT curve of ZnO, the 1.39 Å peak can only be assigned to the Zn–O coordination (Figure 8.8(c)). This provides strong evidences for the substitution of oxygen atoms for sulfur at the surface of ZIS. The absence of Zn–Zn coordination peak at 3.54 Å for the O-doped ZIS nanosheets can be attributed to the high structural distortion by the oxygen doping. The above results clearly indicate that the substitutional doping of oxygen in ZIS nanosheets with ultrathin thickness and structural distortions were obtained.

8.4 Outlook of Time- and Spatial-resolved Synchrotron Radiation XAFS Techniques for 2D Nanomaterials

In this section, the advantages of using the synchrotron radiation XAFS technique for the structural characterization of 2D nanomaterials, and some representative recent research progress in this field, are addressed. Atomically-thick 2D sheets are proposed as an ideal platform for providing atomic level insights into the active sites–catalysis relationships. The atomically flat facet helps to obtain a relatively simple type of active site with nearly the same coordination number, while the atomic thickness and ultrahigh specific surface area make the number of their active sites comparable with the total atoms. XAFS that is complementary to the XRD technique is a powerful and versatile probe for studying the local atomic geometry because it does not require crystallinity of the samples. Therefore, XAFS is particularly suitable for investigating the atomic coordination number, structural disorder and other structural defects in the ultrathin 2D sheets.

Despite atomically-thick 2D nanomaterials being intensively investigated by conventional XAFS to evaluate a clear structure–property relationship,

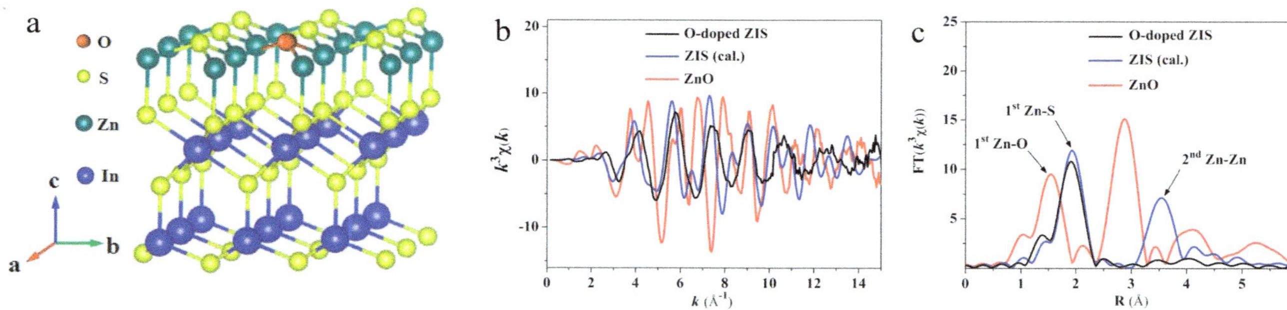

Figure 8.8 (a) The structural model of O-doped ZIS nanosheets (Zn atoms: green; In atoms: blue; S atoms: Yellow; O atoms: red). (b) Zn K-edge extended EXAFS oscillation function and (c) the corresponding Fourier transforms of the O-doped ZIS nanosheets, the calculated ZIS and ZnO. Reprinted with permission from W. Yang *et al.*, Enhanced Photoexcited Carrier Separation in Oxygen-Doped ZnIn2S4 Nanosheets for Hydrogen Evolution, *Angewandte Chemie International Edition*, John Wiley and Sons, © 2016 WILEY-VCH Verlag GmbH & Co. KGaA, Weinheim.

the number of reports about time- and spatial-resolved studies of 2D nanomaterials is still small. The time- and spatial-resolved studies on the growth or catalytic reaction of 2D nanomaterials could significantly enhance our abilities to design and fabricate new 2D nanomaterials with high performance. The atomically-thick 2D crystals always bring about a wealth of innovative properties and fascinating applications. There is a wish for scientists to directly observe the catalytic reaction process occurring at the surface of 2D nanomaterials. Hence, new spectroscopy techniques combining time- and space- resolved methods are required. For example, the time resolution should be improved to millisecond, microsecond, and even to nanosecond to capture some rapidly dynamic processes such as the formation and transition of supercritical nucleus and structural self-assembly.[68–70] Meanwhile, to observe the microstructure and evolution inside an individual 2D material, the probe beam should be focused to reach nanometer spatial resolution.

Recently 'operando' synchrotron radiation research has attracted attention for *in-situ* observation of functional materials such as catalysts or energy materials under operation. This concept arose from needs to understand the function as it operates by *in-situ* spectroscopic probe using synchrotron light sources. So far, *in-situ* observation of the catalysts placed under realistic environments (high temperature/pressure) or batteries during charging and discharging processes were widely studied. As operando experiments are inevitably *in-situ*, the future experimental configuration in general is expected to shift to techniques that could microscopically probe the structure–function relationship and the electronic structure, XAFS will play an important role to deepen our understanding, using high brilliance synchrotron beam. High energy and spatial resolution beamlines available at the newly built high brilliance synchrotron radiation facilities in the world would provide vital information on microscopic reaction mechanism and the intermediates (monomers) in the pathways leading to 2D nanomaterials.

Acknowledgements

We gratefully acknowledge the financial support of the National Natural Science Foundation of China (Grants No. 11422547, 21471143), and the Fundamental Research Funds for the Central Universities (WK2310000050, KY2310000019), and Youth Innovation Promotion Association CAS (CX2310000054).

References

1. D. C. Koningsberger and R. Prins, *X-ray Absorption: Principles: Applications, Techniques of EXAFS, SEXAFS and XANES*, Wiley, New York, 1988.
2. J. J. Rehr and R. C. Albers, *Rev. Mod. Phys.*, 2000, **72**, 621.
3. P. Lee, P. Citrin, P. t. Eisenberger and B. Kincaid, *Rev. Mod. Phys.*, 1981, **53**, 769.

4. A. E. Russell and A. Rose, *Chem. Rev.*, 2004, **104**, 4613–4636.
5. A. Ankudinov, J. Rehr, J. J. Low and S. R. Bare, *J. Chem. Phys.*, 2002, **116**, 1911–1919.
6. R. F. Pettifer, O. Mathon, S. Pascarelli, M. D. Cooke and M. R. Gibbs, *Nature*, 2005, **435**, 78–81.
7. Z. Sun, S. Wei, A. Kolobov, H. Oyanagi and K. Brunner, *Phys. Rev. B*, 2005, **71**, 245334.
8. S. Wei, H. Oyanagi, K. Sakamoto, Y. Takeda and T. Pearsall, *Phys. Rev. B*, 2000, **62**, 1883.
9. D. Chen, C. Li, Z. Zhu, J. Fan and S. Wei, *Phys. Rev. B*, 2005, **72**, 075341.
10. D. E. Sayers, E. A. Stern and F. W. Lytle, *Phys. Rev. Lett.*, 1971, **27**, 1204.
11. A. V. Kolobov, H. Oyanagi, S. Wei, K. Brunner, G. Abstreiter and K. Tanaka, *Phys. Rev. B*, 2002, **66**, 075319.
12. D. Norman, J. Stöhr, R. Jaeger, P. Durham and J. Pendry, *Phys. Rev. Lett.*, 1983, **51**, 2052.
13. J. Stöhr and K. Bauchspiess, *Phys. Rev. Lett.*, 1991, **67**, 3376.
14. T. Yao, Z. Sun, Y. Li, Z. Pan, H. Wei, Y. Xie, M. Nomura, Y. Niwa, W. Yan and Z. Wu, *J. Am. Chem. Soc.*, 2010, **132**, 7696–7701.
15. T. Yao, S. Liu, Z. Sun, Y. Li, S. He, H. Cheng, Y. Xie, Q. Liu, Y. Jiang and Z. Wu, *J. Am. Chem. Soc.*, 2012, **134**, 9410–9416.
16. C. Liu, F. Yun and H. Morkoc, *J. Mater. Sci.: Mater. Electron.*, 2005, **16**, 555.
17. J. Rehr, J. Mustre de Leon, S. Zabinsky and R. Albers, *J. Am. Chem. Soc.*, 1991, **113**, 5135–5140.
18. A. Filipponi, A. Di Cicco and C. R. Natoli, *Phys. Rev. B*, 1995, **52**, 15122.
19. H. Oyanagi, K. Sakamoto, R. Shioda, Y. Kuwahara and K. Haga, *Phys. Rev. B*, 1995, **52**, 5824.
20. G. Schütz, W. Wagner, W. Wilhelm, P. Kienle, R. Zeller, R. Frahm and G. Materlik, *Phys. Rev. Lett.*, 1987, **58**, 737.
21. A. Filipponi and A. Di Cicco, *Phys. Rev. B*, 1995, **51**, 12322.
22. A. J. Dent, *Top. Catal.*, 2002, **18**, 27–35.
23. T. Yao, X. Zhang, Z. Sun, S. Liu, Y. Huang, Y. Xie, C. Wu, X. Yuan, W. Zhang and Z. Wu, *Phys. Rev. Lett.*, 2010, **105**, 226405.
24. N. Kornienko, J. Resasco, N. Becknell, C.-M. Jiang, Y.-S. Liu, K. Nie, X. Sun, J. Guo, S. R. Leone and P. Yang, *J. Am. Chem. Soc.*, 2015, **137**, 7448–7455.
25. B. J. Trześniewski, O. Diaz-Morales, D. A. Vermaas, A. Longo, W. Bras, M. T. Koper and W. A. Smith, *J. Am. Chem. Soc.*, 2015, **137**, 15112–15121.
26. A. Kowal, M. Li, M. Shao, K. Sasaki, M. Vukmirovic, J. h. Zhang, N. Marinkovic, P. Liu, A. Frenkel and R. Adzic, *Nat. Mater.*, 2009, **8**, 325–330.
27. L. Nagarajan, R. A. De Souza, D. Samuelis, I. Valov, A. Börger, J. Janek, K.-D. Becker, P. C. Schmidt and M. Martin, *Nat. Mater.*, 2008, **7**, 391–398.
28. X. Kuang, X.-Y. Wu, J. Zhang and C.-Z. Lu, *Chem. Commun.*, 2011, **47**, 4150–4152.
29. M. Lutolf and J. Hubbell, *Nat. Biotechnol.*, 2005, **23**, 47–55.
30. B. Radisavljevic, A. Radenovic, J. Brivio, V. Giacometti and A. Kis, *Nat. Nanotechnol.*, 2011, **6**, 147–150.

31. Y. Zhu, S. Murali, M. D. Stoller, K. Ganesh, W. Cai, P. J. Ferreira, A. Pirkle, R. M. Wallace, K. A. Cychosz and M. Thommes, *Science*, 2011, **332**, 1537–1541.
32. Y. Zhang, Y.-W. Tan, H. L. Stormer and P. Kim, *Nature*, 2005, **438**, 201–204.
33. J. Feng, X. Sun, C. Wu, L. Peng, C. Lin, S. Hu, J. Yang and Y. Xie, *J. Am. Chem. Soc.*, 2011, **133**, 17832–17838.
34. J. Feng, L. Peng, C. Wu, X. Sun, S. Hu, C. Lin, J. Dai, J. Yang and Y. Xie, *Adv. Mater.*, 2012, **24**, 1969–1974.
35. Y. Hernandez, V. Nicolosi, M. Lotya, F. M. Blighe, Z. Sun, S. De, I. McGovern, B. Holland, M. Byrne and Y. K. Gun'Ko, *Nat. Nanotechnol.*, 2008, **3**, 563–568.
36. D. Teweldebrhan, V. Goyal and A. A. Balandin, *Nano Lett.*, 2010, **10**, 1209–1218.
37. C. Schliehe, B. H. Juarez, M. Pelletier, S. Jander, D. Greshnykh, M. Nagel, A. Meyer, S. Foerster, A. Kornowski and C. Klinke, *Science*, 2010, **329**, 550–553.
38. V. Nicolosi, M. Chhowalla, M. G. Kanatzidis, M. S. Strano and J. N. Coleman, *Science*, 2013, **340**, 1226419.
39. M. Katsnelson, *Eur. Phys. J. B*, 2006, **51**, 157–160.
40. J. N. Coleman, M. Lotya, A. O'Neill, S. D. Bergin, P. J. King, U. Khan, K. Young, A. Gaucher, S. De and R. J. Smith, *Science*, 2011, **331**, 568–571.
41. J. S. Son, J. H. Yu, S. G. Kwon, J. Lee, J. Joo and T. Hyeon, *Adv. Mater.*, 2011, **23**, 3214–3219.
42. S. Z. Butler, S. M. Hollen, L. Cao, Y. Cui, J. A. Gupta, H. R. Gutiérrez, T. F. Heinz, S. S. Hong, J. Huang and A. F. Ismach, *ACS Nano*, 2013, **7**, 2898–2926.
43. C. Rao, H. Ramakrishna Matte and U. Maitra, *Angew. Chem., Int. Ed.*, 2013, **52**, 13162–13185.
44. M. Xu, T. Liang, M. Shi and H. Chen, *Chem. Rev.*, 2013, **113**, 3766–3798.
45. M. Chhowalla, H. S. Shin, G. Eda, L.-J. Li, K. P. Loh and H. Zhang, *Nat. Chem.*, 2013, **5**, 263–275.
46. Q. H. Wang, K. Kalantar-Zadeh, A. Kis, J. N. Coleman and M. S. Strano, *Nat. Nanotechnol.*, 2012, **7**, 699–712.
47. G. D. Scholes, *Nat. Mater.*, 2011, **10**, 906.
48. M. Osada and T. Sasaki, *Adv. Mater.*, 2012, **24**, 210–228.
49. Y. Sun, S. Gao and Y. Xie, *Chem. Soc. Rev.*, 2014, **43**, 530–546.
50. T. Yao, L. Liu, C. Xiao, X. Zhang, Q. Liu, S. Wei and Y. Xie, *Angew. Chem., Int. Ed.*, 2013, **52**, 7554–7558.
51. Y. Sun, H. Cheng, S. Gao, Q. Liu, Z. Sun, C. Xiao, C. Wu, S. Wei and Y. Xie, *J. Am. Chem. Soc.*, 2012, **134**, 20294–20297.
52. X. Zhang, Q. Liu, L. Meng, H. Wang, W. Bi, Y. Peng, T. Yao, S. Wei and Y. Xie, *ACS Nano*, 2013, **7**, 1682–1688.
53. J. Huang, Q. Shang, Y. Huang, F. Tang, Q. Zhang, Q. Liu, S. Jiang, F. Hu, W. Liu and Y. Luo, *Angew. Chem.*, 2016, **128**, 2177–2181.
54. Y. Liu, H. Cheng, M. Lyu, S. Fan, Q. Liu, W. Zhang, Y. Zhi, C. Wang, C. Xiao and S. Wei, *J. Am. Chem. Soc.*, 2014, **136**, 15670–15675.

55. K. Fukuda, I. Nakai, C. Oishi, M. Nomura, M. Harada, Y. Ebina and T. Sasaki, *J. Phys. Chem. B*, 2004, **108**, 13088–13092.
56. Y. C. Chen, Y. G. Lin, Y. K. Hsu, S. C. Yen, K. H. Chen and L. C. Chen, *Small*, 2014, **10**, 3803–3810.
57. C. Wu, F. Feng and Y. Xie, *Chem. Soc. Rev.*, 2013, **42**, 5157–5183.
58. X. Zhang, J. Zhang, J. Zhao, B. Pan, M. Kong, J. Chen and Y. Xie, *J. Am. Chem. Soc.*, 2012, **134**, 11908–11911.
59. S. S. Hong, W. Kundhikanjana, J. J. Cha, K. Lai, D. Kong, S. Meister, M. A. Kelly, Z.-X. Shen and Y. Cui, *Nano Lett.*, 2010, **10**, 3118–3122.
60. N. E. Staley, J. Wu, P. Eklund, Y. Liu, L. Li and Z. Xu, *Phys. Rev. B*, 2009, **80**, 184505.
61. Y. Sun, Z. Sun, S. Gao, H. Cheng, Q. Liu, J. Piao, T. Yao, C. Wu, S. Hu and S. Wei, *Nat. Commun.*, 2012, **3**, 1057.
62. X. Zhu, Y. Guo, H. Cheng, J. Dai, X. An, J. Zhao, K. Tian, S. Wei, X. C. Zeng and C. Wu, *Nat. Commun.*, 2016, **7**, 11210.
63. W. Cheng, J. He, T. Yao, Z. Sun, Y. Jiang, Q. Liu, S. Jiang, F. Hu, Z. Xie and B. He, *J. Am. Chem. Soc.*, 2014, **136**, 10393–10398.
64. Y. Sun, Q. Liu, S. Gao, H. Cheng, F. Lei, Z. Sun, Y. Jiang, H. Su, S. Wei and Y. Xie, *Nat. Commun.*, 2013, **4**, 2899.
65. H. Duan, N. Yan, R. Yu, C.-R. Chang, G. Zhou, H.-S. Hu, H. Rong, Z. Niu, J. Mao and H. Asakura, *Nat. Commun.*, 2014, **5**, 3093.
66. J. Huang, J. Chen, T. Yao, J. He, S. Jiang, Z. Sun, Q. Liu, W. Cheng, F. Hu and Y. Jiang, *Angew. Chem.*, 2015, **127**, 8846–8851.
67. W. Yang, L. Zhang, J. Xie, X. Zhang, Q. Liu, T. Yao, S. Wei, Q. Zhang and Y. Xie, *Angew. Chem., Int. Ed.*, 2016, **55**, 6716–6720.
68. C. Bressler and M. Chergui, *Chem. Rev.*, 2004, **104**, 1781–1812.
69. C. Bressler, C. Milne, V.-T. Pham, A. ElNahhas, R. M. van der Veen, W. Gawelda, S. Johnson, P. Beaud, D. Grolimund and M. Kaiser, *Science*, 2009, **323**, 489–492.
70. A. Klopper, *Nat. Phys.*, 2015, **11**, 523.

Section III

Energy Applications

CHAPTER 9

Inorganic Two-dimensional Nanomaterials for Electrocatalysis

XIANGKAI KONG*[a] AND ZHENMENG PENG*[b]

[a]School of Physics and Electronic Information, Huaibei Normal University, Huaibei, Anhui 235000, P. R. China; [b]Department of Chemical and Biomolecular Engineering, University of Akron, Akron, OH 44325, United States
*E-mail: kxk@chnu.edu.cn, zpeng@uakron.edu

9.1 Introduction

Inorganic two-dimensional (2D) nanomaterials have received considerable and yet increasing research interest for studying their electrocatalytic properties in recent years. Benefiting from dimensional confinement in a unique 2D structure, they often exhibit abnormal physiochemical behaviors when compared with bulk counterparts. One example is graphene, which has a typical 2D structure and along the basal plane shows three to five orders of magnitude higher electrical and thermal conductivities than graphite, its 3D counterpart. Also, a 2D structure offers a large specific surface area and efficient utilization of materials because most of the atoms are exposed on the surface. These features make inorganic 2D nanomaterials promising candidates as catalysts for various applications, including energy generation and

Smart Materials No. 27
Inorganic Two-dimensional Nanomaterials: Fundamental Understanding, Characterizations and Energy Applications
Edited by Changzheng Wu

Published by the Royal Society of Chemistry, www.rsc.org

storage, chemicals and fuels production, environmental control, *etc.* A broad range of 2D nanomaterials, for instance graphene, metal dichalcogenides, metal carbides, metal nitrides, layered oxides and hydroxides, and their 2D derivatives, have been prepared and evaluated for their catalytic properties. Specifically, in the electrocatalysis field, 2D nanomaterials represent one new, but important, group of materials for many electrochemical reactions. A few examples include the oxygen reduction reaction (ORR), the oxygen evolution reaction (OER), the hydrogen evolution reaction (OER), and the CO_2 electrochemical reduction reaction, together with many new 2D materials being developed and demonstrated for these reactions.

In this chapter, we will briefly introduce the fundamentals of electrocatalysis, which has been addressed in detail in many review articles and book chapters. We will then discuss the advantages of using inorganic 2D nanomaterials as electrocatalysts. The focus of this chapter will be on summarizing the recent advances of 2D nanomaterials for catalyzing important electrochemical reactions, with an emphasis on water and CO_2 involved energy generation and fuel production applications.

9.2 Electrocatalysis Fundamentals

Electrocatalysis is typically used for promoting the kinetics of an electrochemical reaction, which involves electron transfer between a reacting species and an electrode. It can be considered as one special type of the catalysis process, during which the reactant species adsorbs on catalyst active sites with an imposed potential and undergoes surface reaction at the catalyst electrode–electrolyte interface.

For an electrochemical reaction, $O + ne \underset{k_b}{\overset{k_f}{\Leftrightarrow}} R$, the relationship between the reaction rate, which is usually represented using current density (i) and the electrode potential, E, can be described using the Butler–Volmer equation:

$$i = i_0[e^{-\alpha nF\eta/RT} - e^{(1-\alpha)nF\eta/RT}] \tag{9.1}$$

$$i_0 = nFAk_f^{1-\alpha}k_b^{\alpha}[O]^{1-\alpha}[R]^{\alpha} \tag{9.2}$$

where i_0 is the exchange current density, α is the charge transfer coefficient, η is the overpotential and equal to $E - E_{O/R}$, $E_{O/R}$ is the redox potential, F is Faraday constant, A is a pre-exponential constant in the Arrhenius equation, R is the gas constant, T is the reaction temperature, and k_f and k_b are the rate constant for the forward and reverse half-cell reaction. Electrochemical reduction occurs when η becomes negative, whereas a positive η drives the electrochemical process towards oxidation.

Electrochemical reactions can be grouped into reversible and irreversible processes based on their i_0 values. For a reversible process, the i_0 value is relatively large and a small η results in a significant change in i. When the i_0 value is small, a significantly bigger η is needed to obtain the same i (Figure 9.1).

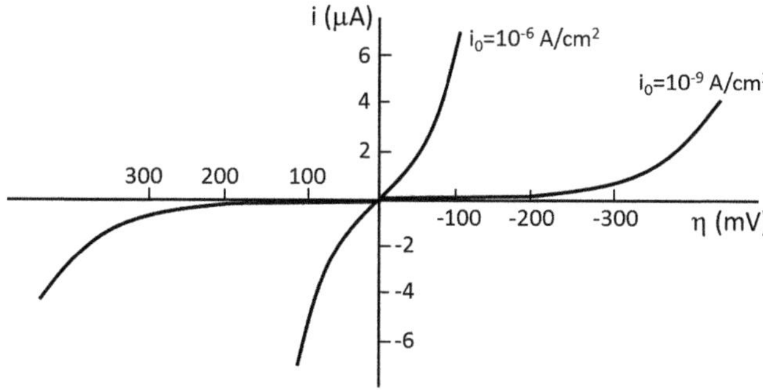

Figure 9.1 Current density (i)-overpotential (μ) curves for the reaction $O + ne \underset{k_b}{\overset{k_f}{\Leftrightarrow}} R$ with different i_0.

In practice, many processes are considered irreversible and have more sluggish reaction kinetics when compared with the reversible ones. In other words, i_0 serves as a descriptor for the kinetics of electrochemical reactions.

To promote an irreversible process, i_0 needs to be improved. From eqn (9.2), i_0 is a function of several parameters, among which rate constants, *i.e.* k_f and k_b, can be readily improved by employing electrocatalysis. A good electrocatalyst will enable the modification of the reaction pathways, decrease the activation energy, and increase the rate constant values and ultimately i_0.

9.3 Unique Features and Advantages of 2D Nanomaterials for Electrocatalysis

Electro-reduction and electro-oxidation are the two main research topics in electrocatalysis, through which chemicals can be converted into other species or be utilized to generate clean and alternative energy for the replacement of fossil feedstocks, overcoming the energy crisis and severe environmental problems. In addition to some common requirements as for conventional catalysts, electrocatalysts need to have good electric conductivity to facilitate electron transport and charge transfer. Thus, the research scope is largely limited to metals and some semiconductors. Meanwhile, electrocatalysis takes place on the active sites of a catalyst, which often are the low coordinated steps, edges, terraces, kinks and/or corner atoms on the surface. Inorganic 2D materials, which are composed of one or few atomic layers with a sheet structure, have attracted considerable interest among the public because of their unique electronic, optical, thermal and mechanical properties. With respect to electrocatalysis, their high specific surface area can provide many low coordinated atoms as the active sites for reactant adsorption and reaction. Their distinctive electronic structures can offer low reaction energy barriers as well as sufficient charge conducting abilities, differing from their bulk counterparts.

Besides, inorganic 2D nanomaterials could influence an electrochemical process from three different aspects. First, electrochemical reactions require electron transport between electrode and catalyst, and within catalyst materials. A big surface area of inorganic 2D electrocatalyst confers easy access to accept/give electrons from/to the electrode, which are driven by the applied electric fields. The 2D structure facilitates the transfer of electrons to/from the surface painlessly since the process can undergo short and efficient transportation pathways. Second, electrochemical reactions are accompanied by migration of charge compensating ions in the electrolyte. The unique layer structure of 2D nanomaterials makes it easier to reach charge balance at the catalyst–electrolyte interface, which in turn contributes to the charge transfer in electrolyte and that between electrode and catalyst. Last, transport of reaction species in electrolyte to the active sites for reaction can influence significantly the concentration of reaction species around the active sites and ultimately the reaction rate. When compared with their bulk counterparts, 2D inorganic nanocatalysts can offer substantially permeable channels for ion adsorption and transportation, allowing more atoms to be exposed and easy to interact with reaction molecules, which is beneficial for improving the whole reaction rate. Keeping these in mind, it is obviously noted that the performance of an electrocatalyst can be largely influenced by its nanostructure. Therefore, rational design and fabrication of new inorganic 2D nanomaterials for electrocatalysis is highly desired.

As is well known, a lower coordination number of surface atoms leads to more dangling bonds and makes them more energetic and reactive. It is often these surface atoms serve as active sites and possess higher catalytic activity compared to better coordinated ones, indicating the decisive role of the coordination number in determining the activity of reactive sites. With their atomic thickness nature, inorganic 2D nanomaterials can provide a greater portion of low coordinated atoms on their surface. In practice, it seems almost inevitable that structural defects and disorders would be generated during the fabrication of inorganic 2D nanomaterials, in which they help to decrease the coordination number of their neighboring sites and could help to improve the electrocatalytic activity. In addition, the confined dimension could confer novel and interesting changes on the electronic states, which could further have big effects on the electrocatalytic property: on one hand, it could modify the adsorption and desorption behavior of reacting species at the active sites; on the other hand, the charge transfer at the interface and within the electrocatalyst would be influenced.

Carrying out chemical modifications on inorganic 2D electrocatalysts as per the specific reaction mechanisms is also of great importance for improving the catalytic performance. Although the atomic-thickness nature has offered extremely high specific surface areas and important physical properties, artificial designer of chemical modifications of the surface still holds fundamental advances for improving the reactive activities, as well as further widening their applications in electrocatalysis. The applications of inorganic 2D nanomaterials in electrocatalysis are mainly focused on the following aspects: HER, OER, ORR, and CO_2 reduction reaction. Different mechanisms are behind these reactions, which lead to varying requirement on the electrocatalyst physiochemical parameters and subsequently materials development. Until now, many methods, for instance incorporation, defect engineering and surface modulations, have been implemented to tune and modify inorganic 2D nanomaterials to influence their electrocatalytic performance.

Here, we will discuss the latest advances of inorganic 2D nanomaterials for electrocatalysis applications. We will mainly focus on several most studied reactions including HER, OER, ORR, and CO_2 electrochemical reduction, and hope through the book chapter to increase the awareness of unique 2D electrocatalysts and stimulate further exploration of their application in more reactions.

9.4 Recent Research Progress

9.4.1 Hydrogen Evolution Reaction (HER)

Hydrogen is considered as a clean and efficient energy carrier, which could provide a solution for the energy and environmental problems we face. HER is the cathodic half-cell reaction in electrochemical water splitting, through which either protons in acid medium or water molecules in alkaline medium are reduced into H_2.[1] Two different reaction pathways, *i.e.* Volmer–Tafel and Volmer–Heyrovsky processes, have been identified for HER on variant catalysts. It is discovered that the first step of HER involves electrochemical reduction of H_3O^+ or H_2O on catalyst surface into H species, which adsorbs to an active site (the Volmer mechanism). It is then followed by either an electrochemical desorption step (the Heyrovsky mechanism) or a recombination step (the Tafel mechanism), as described in Table 9.1.[2]

9.4.1.1 Molybdenum Disulfide (MoS_2)

MoS_2, possessing an intrinsically layered structure and of high abundance, has been considered as one promising HER electrocatalyst to replace the noble metal-based ones. The past decade has witnessed intensive investigation of this material, with the HER performance keeping being improved by finely engineering the active sites. The use of MoS_2 nanomaterial for HER electrocatalysis was first reported by Tributsch and Bennett in 1977,[3] which

Table 9.1 Established reaction pathways for HER electrocatalysis.

Electrolyte	Acid media	Alkaline media
Step 1: Volmer pathway	$H_3O^+ + e^- \rightarrow H_{ads} + H_2O$	$H_2O + e^- \rightarrow H_{ads} + OH^-$
Step 2(a): Heyrovsky pathway	$H_3O^+ + e^- + H_{ads} \rightarrow H_2 + H_2O$	$H_2O^+ + e^- + H_{ads} \rightarrow H_2 + OH^-$
Step 2(b): Tafel pathway	$H_{ads} + H_{ads} \rightarrow H_2$	$H_{ads} + H_{ads} \rightarrow H_2$

soon attracted more studies on this material. These following studies have discovered a strong relationship between the MoS_2 parameters and the HER property. More specifically, the HER activity was found to be inversely proportional to the layer thickness of MoS_2 (Figure 9.2(a)). This interesting finding suggested the use of 2D MoS_2 for this reaction.[4]

There have been some debates on the active sites on MoS_2 surface, identification of which holds the importance to advance the research of this material for catalyzing HER. Recent studies conducted by Chorkendorff and his co-workers suggested the edge atoms on MoS_2 serve as active sites for HER. In their experiments, they grew MoS_2 nanosheets on a Au (111) substrate and characterized the working catalyst under HER condition with scanning tunneling microscopy (STM). They had two major observations: (1) the MoS_2 nanosheet perimeter region exhibited brighter contrast on the collected STM images, an indication of stronger interaction with reacting species; and (2) the measured HER activity showed a linear relationship with the number of edge sites.[5]

Low electrical conductivity is one issue for the pristine MoS_2 structure, which causes a significant ohmic drop and decreased electrocatalysis efficiency. As reported, MoS_2 can exhibit strong photoluminescence, a typical characteristic of a semiconductor. The band gap of MoS_2 is about 1.9 eV when the sheet thickness decreases to monolayer.[6,7] With such a relatively large band gap, MoS_2 shows poor electron conducting ability and thus high resistivity. Moreover, the resistivity varies along the transport direction within the material, with the values being measured to be 2200 times bigger normal to the basal planes compared with parallel to the planes ($0.1 \sim 1\ \Omega^{-1}cm^{-1}$).[8] The anisotropic property is originated from the weak van der Waals bond

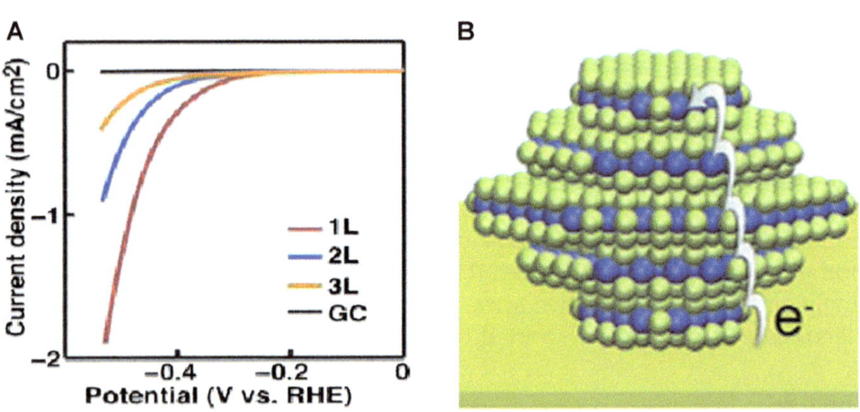

Figure 9.2 (a) Polarization curves of the synthesized monolayer, bilayer, and trilayer MoS_2 films and the curve of bare glass carbon substrates was also given. Reprinted with permission from ref. 4. Copyright (2014) American Chemical Society. (b) Schematic of the electron transfer between layers to the active sites. Adapted from Ref. 14 with permission from The Royal Society of Chemistry.

interaction between two adjacent S–Mo–S layers. With such weak electronic interactions, it is difficult for electrons to transfer through the stacking layer interfaces to reach all the active edge sites, as shown in Figure 9.2(b).

To resolve the low electric conductivity problem, many approaches have been explored in the past years. Three of them, including converting MoS_2 into a more conductive phase, vertical alignment of MoS_2 nanosheets, and using conductive support materials, are most promising and are briefly summarized below.

MoS_2 is most commonly in the 2H phase, which can be described as two S–Mo–S layers built from edge sharing MoS_6 trigonal prisms and is semiconducting with low electrical conductivity. The 2H phase could be converted to the metallic 1T polymorph, which can be described by a single S–Mo–S layer composed of edge sharing MoS_6 octahedra (Figure 9.3(a)),[9] after the lattices undergo subtle structural changes. Jin *et al.* used an exfoliation method to get 2D 1T-MoS_2 nanocrystals, which involved lithium intercalation of semiconducting 2H–MoS_2 nanostructures. The phase transition was confirmed from the changes on the X-ray diffractions (XRD) and Raman spectra. It was proposed that the electron transfer from intercalated Li would favor Mo atoms in an octahedral coordination. The obtained metallic nanosheets exhibited facile electrode kinetics and low electrical transport loss, with dramatically enhanced HER rates. As shown in Figure 9.3(b), its overpotential decreased to −187 mV *vs.* RHE at a current density of 10 mA cm^{-2} and the Tafel slope was measured to be 43 mV $decade^{-1}$ after iR correction.[10] Chhowalla's group also exfoliated MoS_2 nanosheets and obtained the excellent electrocatalytic performance in HER. In addition, they found the 2H nanosheets reduced dramatically in HER activity after the edge electrochemical oxidation treatment. In contrast, the 1T phase nanosheets were unaffected. The finding suggested that the active sites for HER were positioned primarily on the basal plane and the good charge transfer kinetics in metallic MoS_2 was pivotal for improving its electrocatalysis performance.[11]

Considering the difference between the in-plane and out-plane conductivity of 2H MoS_2, vertically aligned thin films have been designed and constructed to reduce the ohmic loss from electron transport along the direction perpendicular to MoS_2 basal planes recently. For instance, Cui *et al.* exposed a 5-nm electron beam-evaporated Mo film to 100 mTorr elemental sulfur vapor for 10 min at 550 °C for sulfurization, and finally obtained the vertically oriented MoS_2 nanosheets as shown in Figure 9.3(c).[12] The interesting edge-terminated vertical 2D nanostructures not only exposed primarily edge sites to the electrolyte but also allowed fast electron transport through the in-plane direction of a single MoS_2 layer to the conducting substrate. The measured HER exchange current per site was 4.1×10^{-21} A, with a high turnover frequency (TOF) value of 0.013 s^{-1} at 0 V.

A third approach for improving the electron transfer ability of MoS_2 is introducing conductive support, which help to facilitate more edge sites to be accessed. Two examples demonstrating the positive effect of employing conducting media were recently published. As shown in Figure 9.3(d),

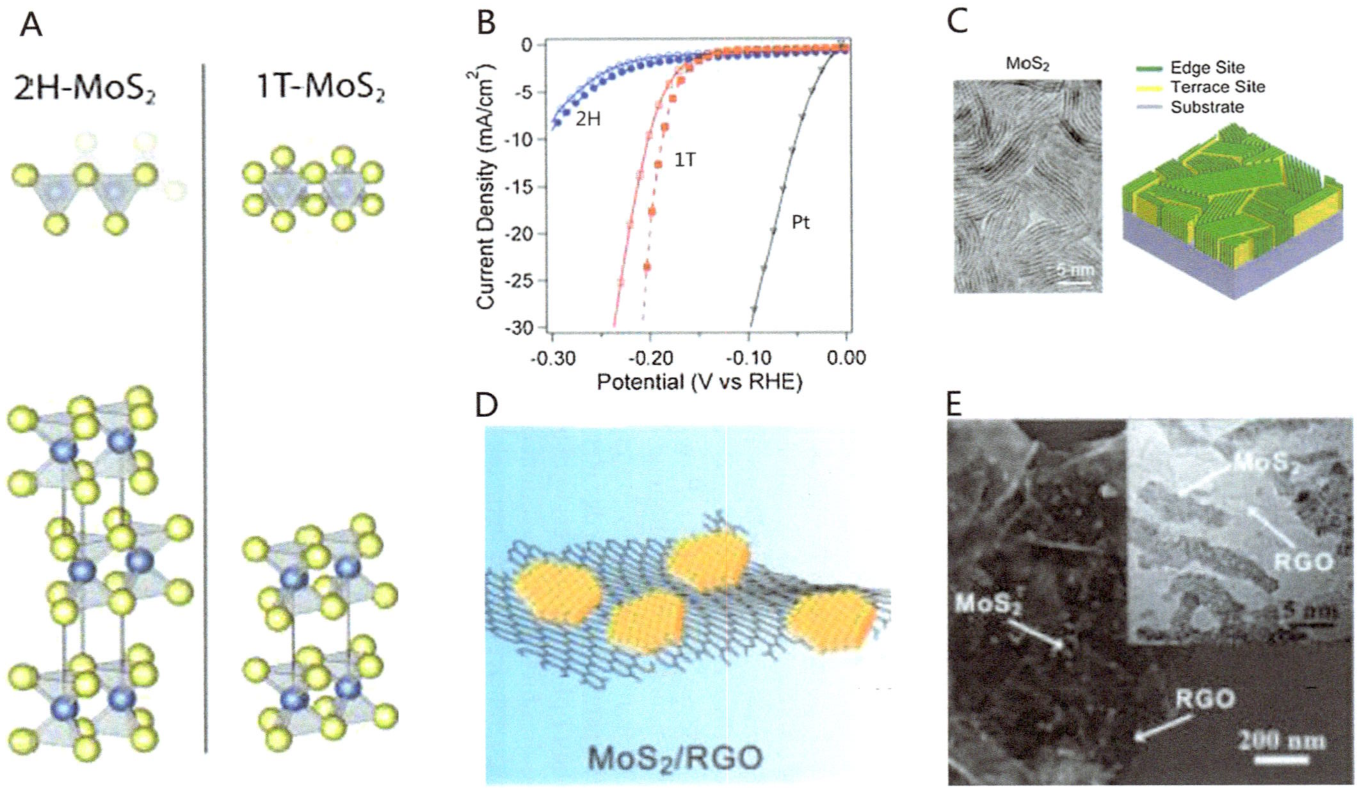

Figure 9.3 (a) Structure of the 2H and 1T polytypes of MoS_2. Reprinted with permission from ref. 9. Copyright (2014) American Chemical Society. (b) Measured polarization curves of 2H and 1T MoS_2 nanosheets, filled symbols with dotted lines showed iR-corrected data. Adapted with permission from ref. 10. Copyright (2013) American Chemical Society. (c) TEM images and schematic structure of edge-terminated 2D MoS_2 with the layers aligned perpendicular to the substrate. Reprinted with permission from ref. 12. Copyright (2013) American Chemical Society. (d) and (e) the schematic and TEM images for the prepared 2D MoS_2/reduced graphene oxide hybrids. Reprinted with permission from ref. 2. Copyright (2011) American Chemical Society.

Dai *et al.* synthesized MoS_2 nanoplates on reduced graphene oxide (RGO) through a one-step solvothermal reaction, in which $(NH_4)_2MoS_4$ and hydrazine were reacted at 200 °C in *N*,*N*-dimethylformamide (DMF) solution containing mildly oxidized graphene oxide.[2] The large specific surface area and the attached oxygen-containing groups on graphene oxide ensured localized growth of MoS_2 nanocrstals on graphene substrates. The 2D MoS_2/RGO yielded a Tafel slope of about 41 mV decade^{-1}, which was significantly smaller than the 94 mV decade^{-1} obtained using pure MoS_2. In the work by Chen's group, MoS_2 nanosheets supported on porous metallic MoO_2 (MoS_2/MoO_2) were produced by sulfurization treatment of porous and highly conductive MoO_2. Linear sweep voltammetry (LSV) and electrochemical impedance spectroscopy (EIS) measurements showed their clearly improved HER kinetics. The Nyquist plots determined the charge transfer resistance of the MoS_2/MoO_2 hybrid electrode to be 148.5 Ω, which was markedly lower than the 661.2 Ω measured for the pure MoS_2. In addition, at a potential of −200 mV *vs.* RHE, the series resistance was reduced from 7.3 Ω for MoS_2 to 6.2 Ω for MoS_2/MoO_2.[13]

As discussed above, the catalytic activity is proportional to the number of active sites. The active sites of intrinsic 2D MoS_2 for HER are limited and mainly located at the sheet edges, which often leads to a relatively poor activity and hampers its practical utilizations. Hence, creating more active sites on 2D MoS_2 becomes one effective strategy to achieve more effective HER electrocatalyst.[14,15] Based on this concept, Xie *et al.* put forward an interesting strategy for realizing controllable defect generation on MoS_2 ultrathin nanosheets.[15] They dissolved hexaamonium heptamolybdate tetrahydrate and thiourea in deionized water, and then used the precursor solution to prepare defect-rich 2D MoS_2 nanosheets *via* one-pot hydrothermal method. As illustrated in Figure 9.4(a), thiourea played a crucial role by adsorbing to the surface of primary MoS_2 nanocrystallites, which could hinder the oriented crystal growth and stabilize the ultrathin nanosheet morphology, resulting in the final defect-rich structure with quasiperiodic configuration. The obtained samples exhibited excellent HER activity with a small onset overpotential of 120 mV, a large cathodic current density, and a small Tafel slope of 50 mV decade^{-1}. The experimental data demonstrated a good performance of the defected 2D MoS_2 electrocatalyst. Jaramillo *et al.* have successfully designed and prepared an extended 3D mesoporous MoS_2 constructed by 2D nanostructures with double-gyroid morphology (Figure 9.4(b)). With virtue of the high surface curvature and high surface area, abundant edge sites were exposed, leading to the excellent HER activity and a low Tafel slope of about 50 mV decade^{-1}.[16]

Inspired by the many successes with heteroatom-doped graphene in electrocatalysis,[17] doping engineering has also been considered and investigated for stimulating the catalytic activity of inert in-plane atoms of 2D MoS_2, in which the electronic properties neighboring the doped heteroatoms could be significantly modulated, thereby modifying its original catalytic activity. The studies in this field can be divided into two parts based on the

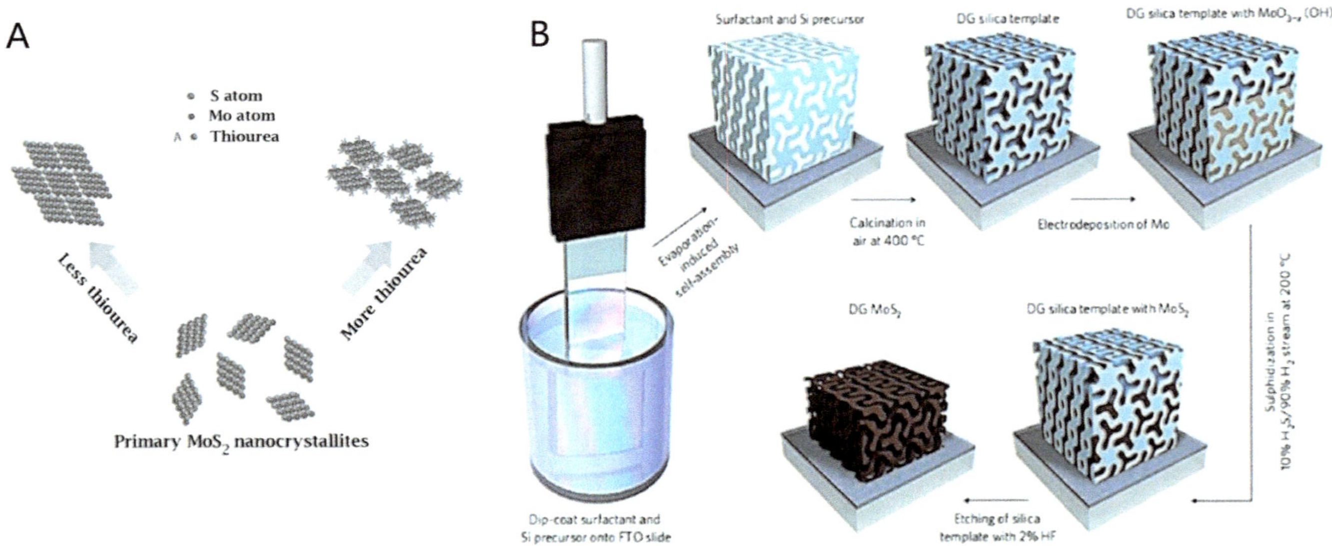

Figure 9.4 (a) As-designed synthetic pathways to obtain the two 2D MoS_2 nanosheet structures, adapted with permission from ref. 15, Copyright © 2013 WILEY-VCH Verlag GmbH & Co. KGaA, Weinheim. (b) Synthesis procedure and structural model for mesoporous MoS_2 with double gyroid morphology. Reprinted by permission from Macmillan Publishers Ltd: *Nature Materials* (ref. 16), copyright (2012).

dopant type: cation and/or anion doping. For cation doping, Chorkendorff and co-workers reported the significant effect of Co on MoS_2, as shown in Figure 9.5(a).[18] DFT calculations suggest that the addition of Co dopant could decrease the free energy of hydrogen adsorption from 0.22 eV to 0.07 eV on S edges, and consequently increase activity of the active sites. Another example is the Pt-doped 2D MoS_2 electrocatalyst reported by Bao *et al.*,[19] in which isolated Pt atoms were doped onto MoS_2 nanosheets using H_2PtCl_6 as precursor (Figure 9.5(b)). The Pt-doped MoS_2 showed significantly enhanced HER activity when compared with pure MoS_2. As per the DFT calculations, the catalytic enhancement was originated from the triggering of in-plane S sites, where H adsorption behavior has been tuned on the in-plane S sites neighboring the doped Pt atoms. They further studied the doping effect of Co and Ni on 2D MoS_2, and found improvement in the HER activity as well. Xie *et al.* doped V atoms into MoS_2 nanosheets and reported promoted HER, as determined by the LSV measurements in Figure 9.5(c).[20]

Anion doping has also attracted much attention and has been used to promote 2D MoS_2's HER performance. Jin *et al.* reported the preparation of Cl-doped 2D MoS_2 using chemical vapor deposition method and the observation of improved HER activity compared with pure MoS_2.[21] Furthermore, sulfur could be partially substituted by other chalcogens, such as Se, leading to a library of promising compounds for HER. As claimed by Li's group, they demonstrated a solution method using oleylamine and 1-octadecene as the solvent for synthesizing Se-doped 2D MoS_2 alloy nanoflakes, named $MoS_{2(1-x)}Se_{2x}$.[22] The HER tests showed that all the prepared $MoS_{2(1-x)}Se_{2x}$ composites exhibited improved activity compared to the pure MoS_2. The best performance was achieved using MoSSe, which gave an overpotential improvement of about 60 mV and suggested sufficient significance in electrocatalysis. Additionally, anion O doping has also been studied and exhibited a positive influence on the HER activity.[23]

9.4.1.2 *Other Chalcogenides*

Besides MoS_2, it is reasonable to expect that some other chalcogenides might also be active HER electrocatalysts because of their similarity in structure. Chhowalla *et al.* prepared monolayered WS_2 nanosheets through chemical exfoliation.[24] The AFM images, shown in Figure 9.6(a), reveal their 2D morphology and ultrathin nature (thickness of about 1 nm), with lateral dimensions in the 100–800 nm range. Figure 9.6(b) shows the corresponding LSV data. The reaction began at exceptionally low cathodic voltages ranging from 30–60 mV for the as-exfoliated WS_2 thin film electrode, above which the current increased rapidly. In addition, Qiao and co-workers have constructed a 3D vdW heterostructure by incorporating 2D WS_2 nanolayers into doped graphene film, as shown in Figure 9.6(c), which has been directly utilized as a catalyst electrode without extra binders, exhibiting extraordinary HER performance, a low overpotential to achieve significant hydrogen generation, a small Tafel slop, a high exchange current density, and a robust durability.[25]

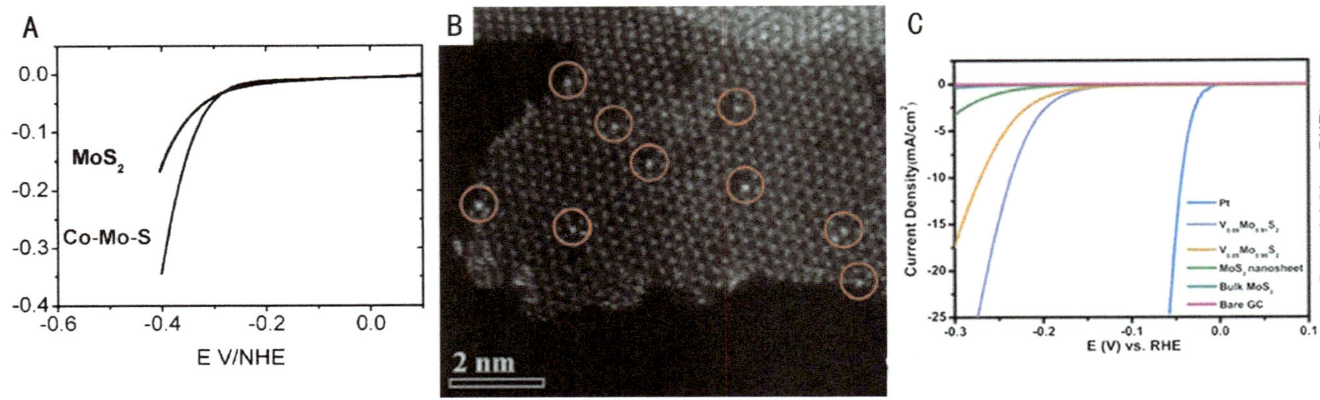

Figure 9.5 (a) Measured polarization curves of the HER on MoS_2 and cobalt promoted MoS_2. Reproduced from Ref. 18 with permission from The Royal Society of Chemistry. (b) HAADF-STEM image of Pt doped MoS_2 showing that single Pt atoms (marked by circles) uniformly disperse in the 2D MoS_2 plane. Reproduced from Ref. 19 with permission from The Royal Society of Chemistry. (c) Polarization curves of the V-doped MoS_2 in HER test. Adapted from Ref. 20 with permission from The Royal Society of Chemistry.

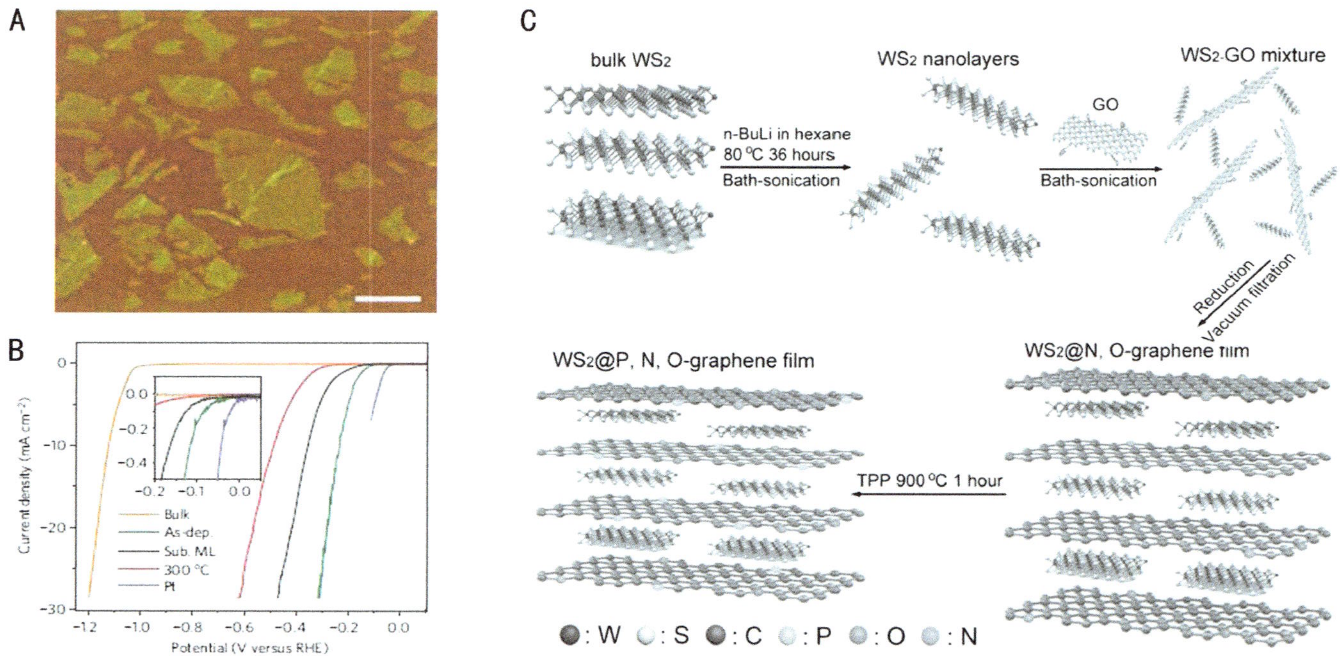

Figure 9.6 (a) Typical AFM image of individual exfoliated nanosheets of WS_2, with the scale bar of 500 nm, and (b) HER polarization curves of bulk and as-exfoliated WS_2 along with those corresponding to Pt nanoparticles and bulk WS_2 powder for comparison. With permission from ref. 25, Copyright © 2013 WILEY-VCH Verlag GmbH & Co. KGaA, Weinheim. (c) Schematic preparation process of 2D WS_2-P, N-, O-doped graphene hybrid. Reprinted by permission from Macmillan Publishers Ltd: Nature Materials (ref. 24), copyright (2015).

Apart from these, other 2D inorganic dichalcogenides have also been explored, such as $Fe_{1-x}Co_xS_2$[26], Ni_3S_2,[27] and CaS[28] nanosheets. All of them display good HER activity, confirming the importance of 2D dichalcogenide nanostructures in electrocatalysis.

9.4.1.3 *Other 2D HER Electrocatalysts*

Some other inorganic 2D nanomaterials also show good electrocatalytic properties for HER. Recently, our group reported the synthesis of free-standing 2D Ru nanosheets using a facile solvothermal method, in which Ru(III) was reduced *via* self-decomposition of the $Ru(acac)_3$ precursor and grew into ultrathin nanosheets with the aid of isopropanol and urea.[29] Figure 9.7(a) demonstrated its 2D growth mechanism, with the corresponding HER activity recorded in Figure 9.7(b). The 2D Ru exhibited excellent HER activity when compared with Ru powder. Sun *et al.* prepared single crystalline Ni nanosheets arrays using *in-situ* topotactic reduction by controlling the synthetic temperature.[30] As depicted in Figure 9.7(c) and (d), the freshly obtained Ni nanosheets showed good activity towards HER. Moreover, these Ni nanosheets not only retained the 2D nanostructure but also exhibited better HER performance, with an onset potential of −34 mV and excellent durability with 94.9% current density could remain after 10 h stability testing, after being partially oxidized by exposure in ambient atmosphere for designated time. Sasaki *et al.* reported the synthesis of 2D $NiMoN_x/C$ with low overpotential and a small Tafel slope for HER.[31] An interesting structural transformation from NiMo particles to $NiMoN_x$ nanosheets was observed during their thermal nitridation using ammonia. The improved HER activity of the 2D $NiMoN_x$ was attributed to the introduction of Ni, which produced a higher density of states (DOS) near the Fermi level, downshifted the d-band center of Mo atoms neighboring Ni, and finally led to relatively more moderate Mo–H binding strength.

9.4.2 Oxygen Evolution Reaction (OER)

OER represents one challenging electrochemical process that involves four proton-coupled electrons transferring and oxygen–oxygen bond formation. As an important component in water splitting and rechargeable metal-air batteries, it has attracted growing attention for the purposes of clean energy generation and storage.[32] Owing to the intrinsically sluggish OER kinetics, this reaction typically requires large overpotentials as the driving force.[33] Depending on the electrolyte environment, the reacting species as well as the overall reaction for OER can be different. In an acidic electrolyte, the reaction involves oxidation of two water molecules into four protons and one oxygen molecule and release of four electrons, whereas oxidation of hydroxide ions into water and oxygen occurs in an alkaline condition, as described below:

$$2H_2O \rightarrow 4H^+ + O_2 + 4e^- \text{ (acidic condition)} \quad (9.3)$$

$$4OH^- \rightarrow 2H_2O + O_2 + 4e^- \text{ (alkaline condition)} \quad (9.4)$$

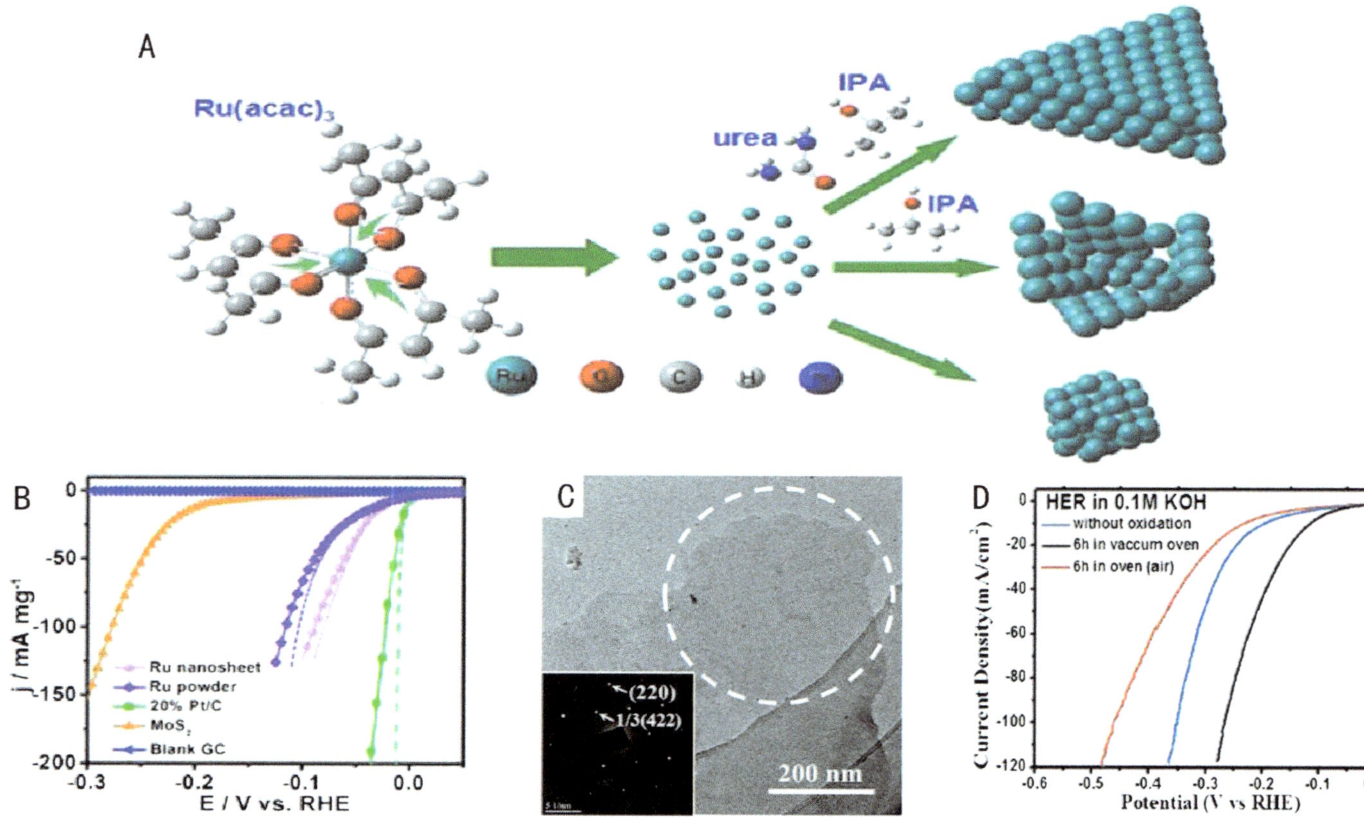

Figure 9.7 (a) Schematic illustration of the growing process of Ru nanosheets (with arrows in Ru(acac)$_3$ molecule standing for the electron-transfer direction) and (b) LSV characterizations of Ru nanosheets in 0.5 M H_2SO_4 electrolyte. Reprinted with permission from ref. 29. Copyright (2016) American Chemical Society. (c) TEM image of the as-synthesized single layered Ni nanosheet and (d) HER performance of 2D Ni with different oxidation degree. Adapted with permission from ref. 30, © 2015 Wiley-VCH Verlag GmbH & Co. KGaA, Weinheim.

RuO_2 and IrO_2 are the most active OER electrocatalysts. However, their high cost and limited reserves have prevented wide use and has stimulated the search for new catalysts.

9.4.2.1 Layered Double Hydroxides

Layered double hydroxides, which consist of positively charged cations with charge balancing anions within layers, represent one important group of 2D inorganic nanomaterials for OER electrocatalysis. One most representative catalyst is $FeNi(OH)_2$, the layered double hydroxide nature of which confers its nanosheet structure with hydroxyl groups terminating the sheet surfaces. Dai *et al.* used nickel acetate and iron nitrate to synthesize 2D ultrathin NiFe hydroxides *via* hydrolysis. The obtained 2D $NiFe(OH)_2$ was loaded on an electron conductive carbon support and found to exhibit both superior activity and durability compared with commercial IrO_2 catalysts for OER in alkaline solutions.[34] Shi's group reported a facile electrodeposition process to immobilize 2D $NiFe(OH)_2$ nanoplates on reduced graphene oxide (RGO), which exhibited a small onset overpotential of 0.259 V and a low Tafel slope of 39 mV decade^{-1}, outperforming IrO_2 in activity and stability.[35]

9.4.2.2 Other 2D OER Electrocatalysts

Xie and co-workers synthesized a series of 2D inorganic nanomaterials, such as Co_3O_4[36] and $NiCo_2O_4$,[37] and reported their robust and much enhanced OER properties when compared with bulk counterparts. DFT calculations and atomic micro binding structure characterizations attributed the improvement in the property to the modifications on the density of states and special electronic structures as resultant of 2D confinement. Recent studies have reported correlations between catalyst structure and property. Taking $CoSe_2$ for example: its metallic behavior and $t_{2g}{}^6e_g{}^1$ electronic configuration near the optimal e_g filling predict it to be a terrific OER electrocatalyst. However, the experimentally measured activity of pure $CoSe_2$ was far below the expectation, primarily caused by its poor active site exposure. Xie *et al.* overcame this problem confronted with $CoSe_2$ by reducing its thickness into the atomic scale, *i.e.* making 2D nanomaterials. After ultra-sonication in ice-water for 12 h, ultrathin $CoSe_2$ nanosheets were obtained and collected by centrifugation, with the schematic process shown in Figure 9.8(a). The positron annihilation spectrometry and Co K-edge X-ray absorption fine structure spectroscopy (Figure 9.8(b)) suggested there were many Co vacancies being formed in the ultrathin nanosheets, which served as the active site role as revealed by DFT calculation. The 2D $CoSe_2$ had an OER overpotential of as low as 0.32 V at 10 mA cm^{-2} in pH 13 medium (Figure 9.8(c)), which made it superior to its bulk counterpart as well as many other Co-based catalysts.[38] Metallic single-unit-cell $CoSe_2$ nanosheets in the orthorhombic phase were also synthesized by thermally exfoliating a lamellar $CoSe_2$-DETA hybrid (Figure 9.8(d)).[39] The measured Co K-edge $k^3\chi(k)$ oscillation curves showed clear differences

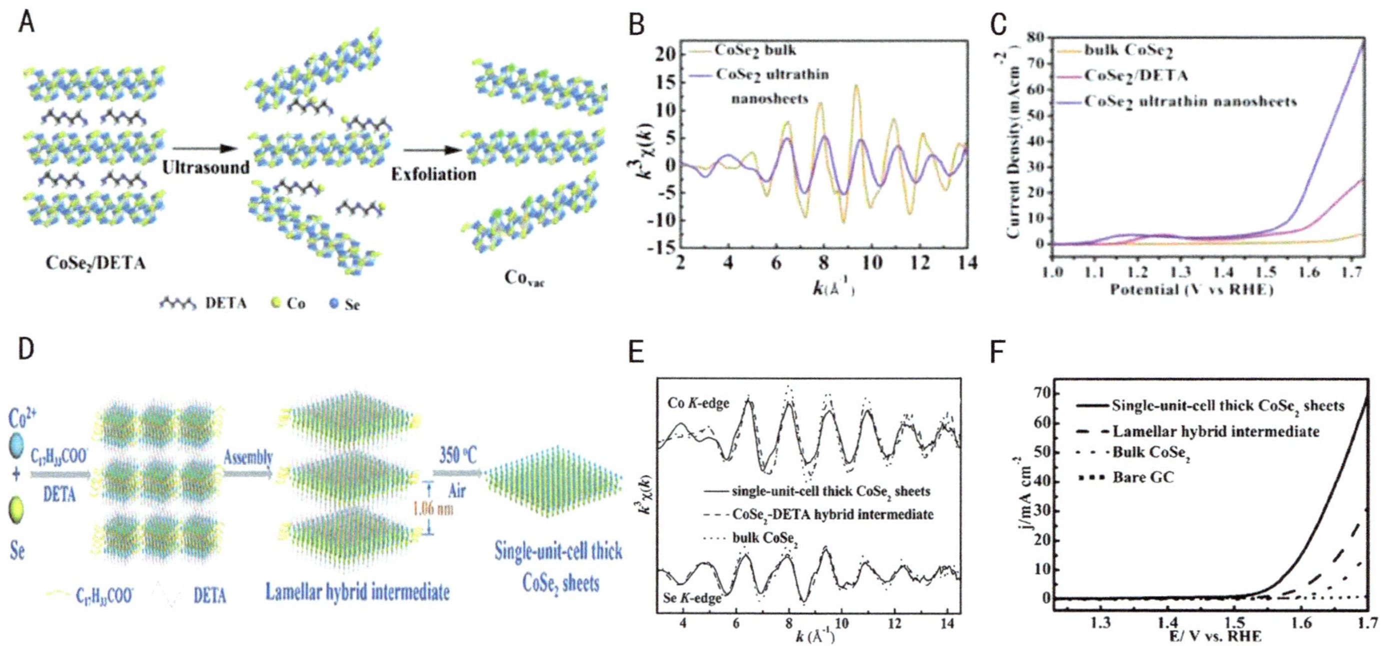

Figure 9.8 (a) Schematic of the formation of 2D $CoSe_2$ ultrathin nanosheets with Co vacancies; (b) Co K-edge extended XAFS oscillation function $k^3[\chi(k)]$; and (c) LSV curves for the water oxidation properties tests in 0.1 M KOH medium with ultrathin $CoSe_2$ nanosheets as the electrocatalyst. Reprinted with permission from ref. 38. Copyright (2014) American Chemical Society. (d) Schematic illustration for fabricating single-unit-cell thick $CoSe_2$ nanosheets; (e) Co K-edge and Se K-edge extended XAFS oscillation function $k^3[\chi(k)]$; and (f) Polarization curves obtained from electrochemical tests for single-unit-cell thick $CoSe_2$. Reproduced with permission from ref. 39, © 2015 Wiley-VCH Verlag GmbH & Co. KGaA, Weinheim.

among 2D $CoSe_2$ nanosheets, hybrid intermediates, and bulk $CoSe_2$, confirming their distinct local atomic arrangement (Figure 9.8(e)). DFT calculations and temperature-dependent resistances indicated its metallic characteristic, which allowed fast OER kinetics with a decreased onset overpotential of 0.27 V (Figure 9.8(f)). Our group also prepared 2D RuO_2 nanocrystals from mildly oxidation of 2D Ru nanosheets, which showed good OER property driven by the applied potential.[29]

9.4.3 Oxygen Reduction Reaction (ORR)

ORR takes place in several important electrochemical energy technologies, such as polymer electrolyte membrane fuel cells and metal-air batteries. Platinum has long been known as the most effective catalyst, but the applications are limited by the scarcity and prohibitive cost of this precious metal. Recently, cation-deficient transition metal sulfides have attracted increasing attention as alternative electrocatalyst due to their unique properties that arise from degenerate hole doping. Swihart *et al.* prepared Cu-deficient 2D $Cu_{2-x}S$ nanoplates at moderate temperature (below 100 °C) and studied the electrocatalytic property in ORR.[40] The resulting $Cu_{2-x}S$ nanoplates were shown to be active for ORR in alkaline electrolyte (Figure 9.9). In addition, through a reductant-free and *in-situ* solvothermal route, atomically thin 2D $MoSe_2$/RGO and WSe_2/RGO nanosheets were synthesized by Ma's group. Benefitting from the advantages of both inorganic graphene ultrathin analogues and RGO, the hybrid 2D nanomaterials showed excellent ORR activity associated with several exchanged electrons close to four, which corresponded to the complete reduction of oxygen into water. More interestingly, the two electrocatalysts exhibited a very small onset potential of −0.02 V and a high

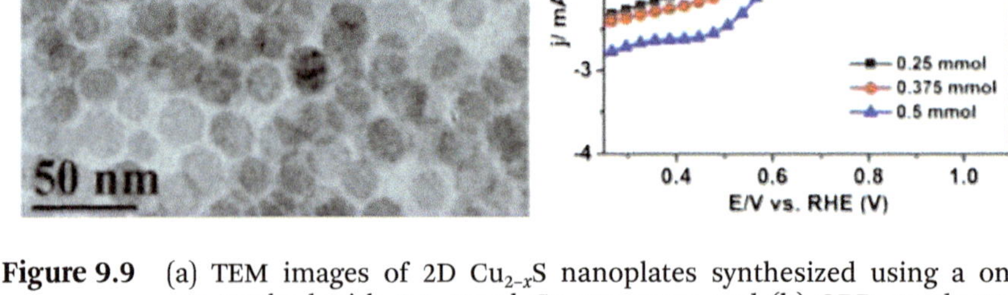

Figure 9.9 (a) TEM images of 2D $Cu_{2-x}S$ nanoplates synthesized using a one-pot method with 0.5 mmol Cu precursor and (b) ORR steady-state polarization curves of $Cu_{2-x}S$ nanocrystals prepared with varying Cu precursor amount. Reprinted with permission from ref. 40. Copyright (2015) American Chemical Society.

kinetic current density of 10.22 and 10.77 mA cm^{-2} for 2D $MoSe_2$ and WSe_2, respectively, which even outperformed Pt catalyst.[41]

9.4.4 CO_2 Electrochemical Reduction

The conversion of CO_2 into fuels *via* electrochemical reduction is also an important reaction, which can help to reduce CO_2 emission and mitigate energy shortage. Although some oxide-derived metal nanostructures have been shown to enable CO_2 reduction at low overpotential, it remains unclear how the electrocatalytic activity of these metals are influenced by their native oxides. Recently, Xie *et al.* prepared two kinds of 2D inorganic ultrathin nanosheets, metallic Co and Co compound containing domains of both metallic Co and Co oxide, which both were four atomic layers in thickness.[42] The fabrication process was depicted in Figure 9.10(a), with *n*-butylamine for tuning the growth direction of Co and dimethylformamide for gradually reducing the Co ions. XRD, elemental mapping and Raman spectroscopy were combined to characterize the obtained nanosheets. HRTEM and AFM images in Figure 9.10(b) and (c) demonstrate the [001]-oriented hexagonal Co nature with average sheet thickness of 0.84 nm. The fabricated 2D electrocatalysts were loaded onto a glassy carbon electrode and tested in CO_2-saturated 0.1 M Na_2SO_4 electrolyte. As can be seen from Figure 9.10(d), the partially oxidized Co nanosheets generated a current density of 10.59 mA cm^{-2} at −0.85 V *vs.* a saturated calomel electrode, roughly 10, 40 and 260 times larger than the current densities obtained with the pure Co nanosheets, partially oxidized bulk Co, and bulk Co, respectively. The authors also reported a good CO_2 electrocatalytic reduction property of Co_3O_4 nanosheets with different thicknesses.[43] It was reported that 1.72 nm thick Co_3O_4 nanosheets had over 1.5 and 20 times higher activity than 3.51 nm thick Co_3O_4 layers and the bulk counterpart, respectively, illustrating the advantages of using 2D Co_3O_4 for CO_2 reduction electrocatalysis.

9.5 Concluding Remarks and Outlook

This chapter has covered some of the basics of electrocatalysis and the recent research progress for 2D inorganic nanomaterial electrocatalysts. More specifically, the discussions have encompassed the electrocatalysis mechanism, reaction kinetics, unique features and advantages of inorganic 2D nanomaterials, and 2D electrocatalyst development for several important electrochemical reactions. It is obvious that 2D inorganic nanostructures could provide high surface area, many low coordinated atoms, substantially permeable channels for ion adsorption and transportation, short and portable transporting pathways, and easy access to exchange electrons with the electrodes, all of which are important features in electrocatalysis. It should be noted that most of the experimental successes in 2D inorganic nanomaterial synthesis and demonstration of their excellent electrocatalytic property have

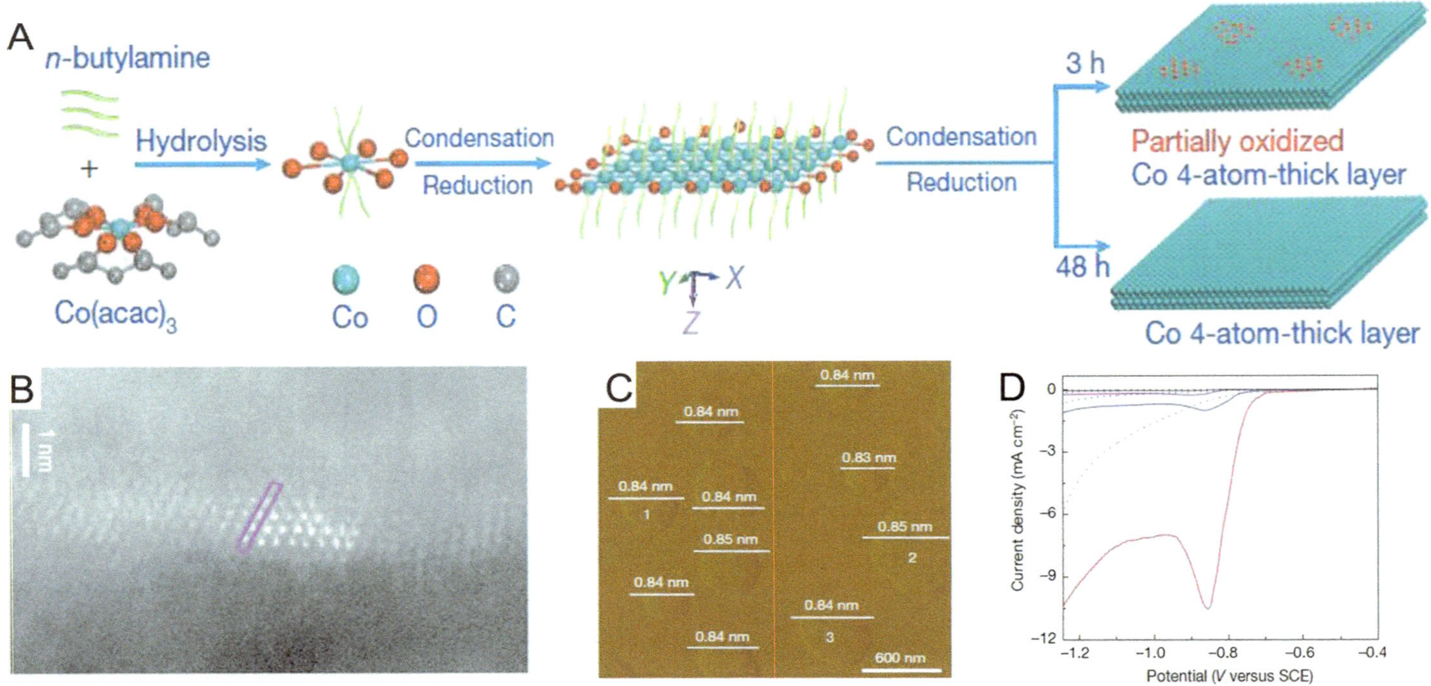

Figure 9.10 (a) Schematic formation process of the partially oxidized and pure Co 4-atomic-layer; (b) and (c) lateral HAADF-STEM and AFM images of Co 4-atomic-layer, respectively; and (d) LSV curves in a CO_2-saturated and N_2-saturated 0.1 M Na_2SO_4 aqueous solution (from down to up were successively partially oxidized Co 4-atom-thick layers, Co 4-atom-thick layers, partially oxidized bulk Co and bulk Co). Reprinted by permission from Macmillan Publishers Ltd: Nature (ref. 42), copyright (2016).

been achieved in the past few years. The great potential of 2D inorganic electrocatalysts have yet to be explored in depth, to study more reactions, with more thorough investigations. Some fundamental challenges, for instance realizing large scale fabrication of 2D nanomaterials and their easy recovery, understanding the underlying electrocatalysis mechanism, and controlling their physiochemical parameters to optimize their properties, need to be addressed in the future to allow unique 2D inorganic electrocatalysts to become real applications.

References

1. M. Gong, D. Y. Wang, C. C. Chen, B. J. Hwang and H. Dai, *Nano Res.*, 2015, **9**, 28.
2. Y. Li, H. Wang, L. Xie, Y. Liang, G. Hong and H. Dai, *J. Am. Chem. Soc.*, 2011, **133**, 7296.
3. H. Tributsch and J. C. Bennett, *J. Electroanal. Chem. Interfacial Electrochem.*, 1977, **81**, 97.
4. Y. Yu, S. Y. Huang, Y. Li, S. N. Steinmann, W. Yang and L. Cao, *Nano Lett.*, 2014, **14**, 553.
5. T. F. Jaramillo, K. P. Jørgensen, J. Bonde, J. H. Nielsen, S. Horch and I. Chorkendorff, *Science*, 2007, **317**, 100.
6. M. Acerce, D. Voiry and M. Chhowalla, *Nat. Nanotechnol.*, 2015, **10**, 313.
7. L. Yang, H. Hong, Q. Fu, Y. Huang, J. Zhang, X. Cui, Z. Fan, K. Liu and B. Xiang, *ACS Nano*, 2015, **9**, 6478.
8. H. Tributsch, *Ber. Bunsenges. Phys. Chem.*, 1977, **81**, 361.
9. J. D. Benck, T. R. Hellstern, J. Kibsgaard, P. Chakthranont and T. F. Jaramillo, *ACS Catal.*, 2014, **4**, 3957.
10. M. A. Lukowski, A. S. Daniel, F. Meng, A. Forticaux, L. Li and S. Jin, *J. Am. Chem. Soc.*, 2013, **135**, 10274.
11. D. Voiry, M. Salehi, R. Silva, T. Fujita, M. Chen, T. Asefa, V. B. Shenoy, G. Eda and M. Chhowalla, *Nano Lett.*, 2013, **13**, 6222.
12. D. Kong, H. Wang, J. J. Cha, M. Pasta, K. J. Koski, J. Yao and Y. Cui, *Nano Lett.*, 2013, **13**, 1341.
13. L. Yang, W. Zhou, D. Hou, K. Zhou, G. Li, Z. Tang, L. Li and S. Chen, *Nanoscale*, 2015, **7**, 5203.
14. A. B. Laursen, S. Kegnaes, S. Dahl and I. Chorkendorff, *Energy Environ. Sci.*, 2012, **5**, 5577.
15. J. Xie, H. Zhang, S. Li, R. Wang, X. Sun, M. Zhou, J. Zhou, X. W. Lou and Y. Xie, *Adv. Mater.*, 2013, **25**, 5807.
16. J. Kibsgaard, Z. Chen, B. N. Reinecke and T. F. Jaramillo, *Nat. Mater.*, 2012, **11**, 963.
17. X. K. Kong, C. L. Chen and Q. W. Chen, *Chem. Soc. Rev.*, 2014, **43**, 2841.
18. J. Bonde, P. G. Moses, T. F. Jaramillo, J. K. Norskov and I. Chorkendorff, *Faraday Discuss.*, 2009, **140**, 219.
19. J. Deng, H. Li, J. Xiao, Y. Tu, D. Deng, H. Yang, H. Tian, J. Li, P. Ren and X. Bao, *Energy Environ. Sci.*, 2015, **8**, 1594.

20. X. Sun, J. Dai, Y. Guo, C. Wu, F. Hu, J. Zhao, X. Zeng and Y. Xie, *Nanoscale*, 2014, **6**, 8359.
21. X. Zhang, F. Meng, S. Mao, Q. Ding, M. J. Shearer, M. S. Faber, J. Chen, R. J. Hamers and S. Jin, *Energy Environ. Sci.*, 2015, **8**, 862.
22. Q. Gong, L. Cheng, C. Liu, M. Zhang, Q. Feng, H. Ye, M. Zeng, L. Xie, Z. Liu and Y. Li, *ACS Catal.*, 2015, **5**, 2213.
23. J. Xie, J. Zhang, S. Li, F. Grote, X. Zhang, H. Zhang, R. Wang, Y. Lei, B. Pan and Y. Xie, *J. Am. Chem. Soc.*, 2013, **135**, 17881.
24. D. Voiry, H. Yamaguchi, J. Li, R. Silva, D. C. B. Alves, T. Fujita, M. Chen, T. Asefa, V. B. Shenoy, G. Eda and M. Chhowalla, *Nat. Mater.*, 2013, **12**, 850.
25. J. Duan, S. Chen, B. A. Chambers, G. G. Andersson and S. Z. Qiao, 3D WS2 Nanolayers@Heteroatom-Doped Graphene Films as Hydrogen Evolution Catalyst Electrodes, *Adv. Mater.*, 2015, **27**, 4234.
26. D. Y. Wang, M. Gong, H. L. Chou, C. J. Pan, H. A. Chen, Y. Wu, M. C. Lin, M. Guan, J. Yang, C. W. Chen, Y. L. Wang, B. J. Hwang, C. C. Chen and H. Dai, *J. Am. Chem. Soc.*, 2015, **137**, 1587.
27. L. L. Feng, G. Yu, Y. Wu, G. D. Li, H. Li, Y. Sun, T. Asefa, W. Chen and X. Zou, *J. Am. Chem. Soc.*, 2015, **137**, 14023.
28. A. Harvey, C. Backes, Z. Gholamvand, D. Hanlon, D. McAteer, H. C. Nerl, E. McGuire, A. Seral-Ascaso, Q. M. Ramasse, N. McEvoy, S. Winters, N. C. Berner, D. McCloskey, J. F. Donegan, G. S. Duesberg, V. Nicolosi and J. N. Coleman, *Chem. Mater.*, 2015, **27**, 3483.
29. X. Kong, K. Xu, C. Zhang, J. Dai, S. Norooz Oliaee, L. Li, X. Zeng, C. Wu and Z. Peng, *ACS Catal.*, 2016, **6**, 1487.
30. Y. Kuang, G. Feng, P. Li, Y. Bi, Y. Li and X. Sun, Single-crystalline ultrathin nickel nanosheets array from *in situ* topotactic reduction for active and stable electrocatalysis, *Angew. Chem.*, 2015, **128**, 703.
31. W.-F. Chen, K. Sasaki, C. Ma, A. I. Frenkel, N. Marinkovic, J. T. Muckerman, Y. Zhu and R. R. Adzic, *Angew. Chem., Int. Ed.*, 2012, **51**, 6131.
32. M. Gong and H. Dai, *Nano Res.*, 2014, **8**, 23.
33. M. T. M. Koper, *J. Electroanal. Chem.*, 2011, **660**, 254.
34. M. Gong, Y. Li, H. Wang, Y. Liang, J. Z. Wu, J. Zhou, J. Wang, T. Regier, F. Wei and H. Dai, *J. Am. Chem. Soc.*, 2013, **135**, 8452.
35. X. Yu, M. Zhang, W. Yuan and G. Shi, *J. Mater. Chem. A*, 2015, **3**, 6921.
36. Y. Sun, S. Gao, F. Lei, J. Liu, L. Liang and Y. Xie, *Chem. Sci.*, 2014, **5**, 3976.
37. J. Bao, X. Zhang, B. Fan, J. Zhang, M. Zhou, W. Yang, X. Hu, H. Wang, B. Pan and Y. Xie, *Angew. Chem., Int. Ed.*, 2015, **54**, 7399.
38. Y. Liu, H. Cheng, M. Lyu, S. Fan, Q. Liu, W. Zhang, Y. Zhi, C. Wang, C. Xiao, S. Wei, B. Ye and Y. Xie, *J. Am. Chem. Soc.*, 2014, **136**, 15670.
39. L. Liang, H. Cheng, F. Lei, J. Han, S. Gao, C. Wang, Y. Sun, S. Qamar, S. Wei and Y. Xie, Metallic single-unit-cell orthorhomic cobalt diselenide atomic layers: robust water-electrolysis catalysts, *Angew. Chem., Int. Ed.*, 2015, **54**, 12004.

40. X. Wang, Y. Ke, H. Pan, K. Ma, Q. Xiao, D. Yin, G. Wu and M. T. Swihart, *ACS Catal.*, 2015, **5**, 2534.
41. J. Guo, Y. Shi, X. Bai, X. Wang and T. Ma, *J. Mater. Chem. A*, 2015, **3**, 24397.
42. S. Gao, Y. Lin, X. Jiao, Y. Sun, Q. Luo, W. Zhang, D. Li, J. Yang and Y. Xie, *Nature*, 2016, **529**, 68.
43. S. Gao, X. Jiao, Z. Sun, W. Zhang, Y. Sun, C. Wang, Q. Hu, X. Zu, F. Yang, S. Yang, L. Liang, J. Wu and Y. Xie, *Angew. Chem.*, 2016, **128**, 708.

CHAPTER 10

Two-dimensional Nanomaterials for Applications in Flexible Supercapacitors

XU PENG AND CHANGZHENG WU*

Department of Chemistry & Hefei National Laboratory for Physical Sciences at the Microscale & Collaborative Innovation Center of Chemistry for Energy Materials, University of Science and Technology of China, No.96, JinZhai Road, Hefei, Anhui, 230026, P. R. China
*E-mail: czwu@ustc.edu.cn

10.1 Introduction

Since the discovery of the versatile properties of two-dimensional (2D) graphene,[1] 2D layered nanomaterials such as transition metal dichalcogenides (TMDs),[2] transition metal oxides (TMO),[3–11] MXenes,[12,13] and other 2D compounds have attracted renewed interest for application as electrode materials in energy storage areas.[14] Their 2D nature have brought great advantages in high flexibility achieved by layer-by-layer stacking of a single sheet to form flexible thin film, in which the overlapping areas would facilitate the mechanical stability and robust flexibility. Meanwhile, the high specific surface areas of 2D nanomaterials introduce lots of active sites for electrochemical reactions and electron transport, which is of great advantage in enhancing the capacitive performance of supercapacitors. Moreover,

Smart Materials No. 27
Inorganic Two-dimensional Nanomaterials: Fundamental Understanding, Characterizations and Energy Applications
Edited by Changzheng Wu

Published by the Royal Society of Chemistry, www.rsc.org

with a variety of alternative elements in 2D inorganic lattices for inorganic 2D nanomaterials, we can even stretch to higher capacitive performance as their pseudocapacitive properties come from the physicochemical properties of various inorganic elements. Because the planar geometries of 2D nanosheets are directly compatible with planar device design, the extensibility of 2D nanosheet materials also caters to achievement of an ultrathin electrode, finally constructing a 2D ion migration channel that is beneficial for electrochemical performance. In summary, because of their high specific surface areas and distinct properties with 2D dimensionality, 2D nanomaterials are of significant importance in the application of flexible supercapacitors as promising electrode materials.

To improve the flexibility and compact degree of flexible supercapacitors to a compelling level, further efforts are urgently needed to redesign the structure of the supercapacitor itself, rather than solely focus on the electrode materials. Planar supercapacitors, as an emerging branch of supercapacitors, provide the entire device with ultrathin thickness and ultra-flexibility owing to their architecture design of the integration of the three required components (electrode, electrolyte and current collector) into a 2D configuration in the same horizontal plane, thus making compact device design possible.[11,15–18] Meanwhile, the electrolyte ions are transported two-dimensionally in a planar supercapacitor that is compatible for 2D nanomaterials with a 2D migration channel, further eliminating the necessity of a separator as well as shortening the ion travel distance, which would increase the integration level. Moreover, 2D nanomaterials possess intrinsic ultra-flexibility and high mechanical properties. Ultrathin nanosheets can be stacked layer-by-layer, according to the demand for planar supercapacitors with a 2D configuration. In addition, the inorganic lattices of 2D nanomaterials could act as more active sites for enhancing the pseudocapacitance of planar supercapacitors. Therefore, high-performing, flexible planar configuration devices have been realized on account of ultrathin 2D nanomaterials as the portable stretchable/flexible electronic devices.

The electrolyte is also one of the most important elements in supercapacitors. Numerous studies have focused on the aqueous electrolyte for its high electrochemical potential window of water (1.23 V) and abundance.[19] Among them, gel electrolytes have attracted increasing attention because the thin film flexible supercapacitors in all-solid-state offer the synergic benefits of flexibility, thinness and transparency. The integration of supercapacitors is significantly enhanced when gel electrolytes are employed as separator and binder for fabricating solid-state supercapacitors. But the mechanisms of gel electrolytes that affect the performance of flexible supercapacitors are not fully understood. We hope that further understanding on how gel electrolytes help to improve the capacitance, flexibility, and so on, will be thoroughly investigated.

In this section, we outline a different class of 2D nanomaterials for flexible supercapacitors. In addition, further discussion and expectation are included to highlight the intrinsic merits of 2D nanomaterials for fabricating planar supercapacitors. Finally, the application of polymer gel electrolytes in 2D nanomaterials based flexible supercapacitors are summarized systematically.

10.2 Graphene-based Flexible Supercapacitors

Because of their high specific surface areas and high mechanical stability, 2D nanomaterials have been widely employed as the electrode materials of high performance flexible supercapacitors. Among them, 2D graphene, as one of the prototype 2D nanomaterials, was experimentally reported in 2004 for the first time,[1] consisting of a flat monolayer of sp^2 bonded carbon atoms in a 2D honeycomb lattice. Graphene possesses a high theoretical specific surface area of about 2630 $m^2\ g^{-1}$ along with outstanding mechanical strength, thus graphene has been extensively studied as electrodes of flexible energy devices since its discovery.[20] However, it is important to note that the values of specific capacitance, energy density, and power density of graphene-based supercapacitors have remained lower than expected due to the restacking of graphene sheets. From this, the regulation of the strong van der Waals interaction among sheet-to-sheet is required. To overcome this limitation, El-Kady *et al.* used a simple method for avoiding the restacking of graphene sheets. They used a direct laser from a standard LightScribe DVD optical drive to reduce graphite oxide films to graphene on a flexible substrate.[21] Consequently, the irradiation of the film with an infrared laser reduced the GO to laser-scribed graphene (LSG) with robust mechanical strength, high electrical conductivity (1738 S m^{-1}) and specific surface area (1520 $m^2\ g^{-1}$) (Figure 10.1). The cross section of the film, investigated with scanning electron microscopy (SEM), showed that the initially stacked GO sheets were converted into well-exfoliated LSG sheets without restacking. Based on this, the as-obtained supercapacitors with $EMIMBF_4$ ionic liquid showed outstanding electrochemical performance (276 F g^{-1} (LSG electrode)) at a wider potential window (about 4 V) and excellent cycling stability (97%) of the initial capacitance after 10 000 cycles. Similarly, Yang *et al.* reported a chemically converted graphene (CCG) hydrogel film with a metastable and adaptive pore structure. When the packing density of the electrode is 1.33 g cm^{-3}, the liquid electrolyte-mediated CCG (EM-CCG) film has a highly efficient ion transport channel, creating a continuous ion transport network that leads to exceptionally high volume capacitance at 261.3 F cm^{-3} at 0.1 A g^{-1} in organic electrolyte and a volume energy density at 59.9 Wh L^{-1}. In addition, these EM-CCG films are very stable under cycling tests, realizing 95% of the initial capacitance after 300 h in organic $EMIMBF_4$ electrolyte at a constant voltage of 3.5 V. All these attractive characters make 2D graphene a promising electrode material for large-scale real-world applications of energy storage devices.

Apart from the pure 2D graphene nanosheet, Lin *et al.* also synthesized a nitrogen-doped ordered mesoporous few-layer carbon with an outstanding 855 F g^{-1} at 1 A g^{-1} in 0.5 M H_2SO_4 aqueous electrolyte, much higher than the present pure carbon/graphene-based supercapacitors.[22] The large improvement of capacitance is mainly due to robust redox reactions at

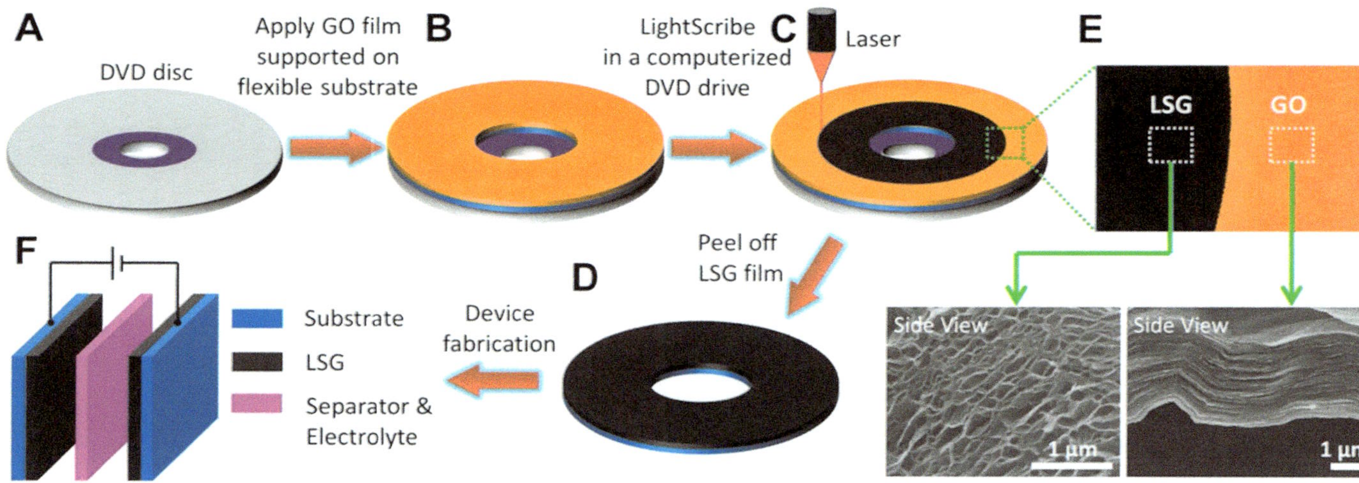

Figure 10.1 Schematic illustration of the fabrication of laser-scribed graphene-based flexible supercapacitor. (a) to (d) A GO film supported on a flexible substrate is placed on top of a LightScribe-enabled DVD media disc, and then a computer image is laser-irradiated on the GO film in a computerized LightScribe DVD drive. Finally, LSG film was peeled off for device fabrication. (e) As shown in the photograph, the color of the GO film changes from golden brown to black since it was reduced to laser-scribed graphene. As shown in the cross-sectional SEM images, the low-power infrared laser changes the stacked GO sheets (left) immediately into well-exfoliated few-layered LSG film (right). (f) A symmetric supercapacitor is fabricated from two LSG film electrodes and flexible substrates, ion-porous separator, and electrolyte. From ref. 21. Reprinted with permission from AAAS.

nitrogen-associated defects, which transform inert graphene-like layered carbon into an electrochemically active substance without affecting its electric conductivity. Moreover, the N-doped carbon layer is more hydrophilic than a non-N-doped carbon layer, showing improvement of the wetting behavior, which is beneficial for contact between the electrode and electrolyte (Figure 10.2). In addition, the symmetric electrochemical supercapacitor based on N-doped carbon layer exhibit remarkable 82% capacitance retention after 50 000 cycles in 0.5 M H_2SO_4 electrolyte from 0 to 1.2 V. This general method of doping other elements into graphene-like carbon

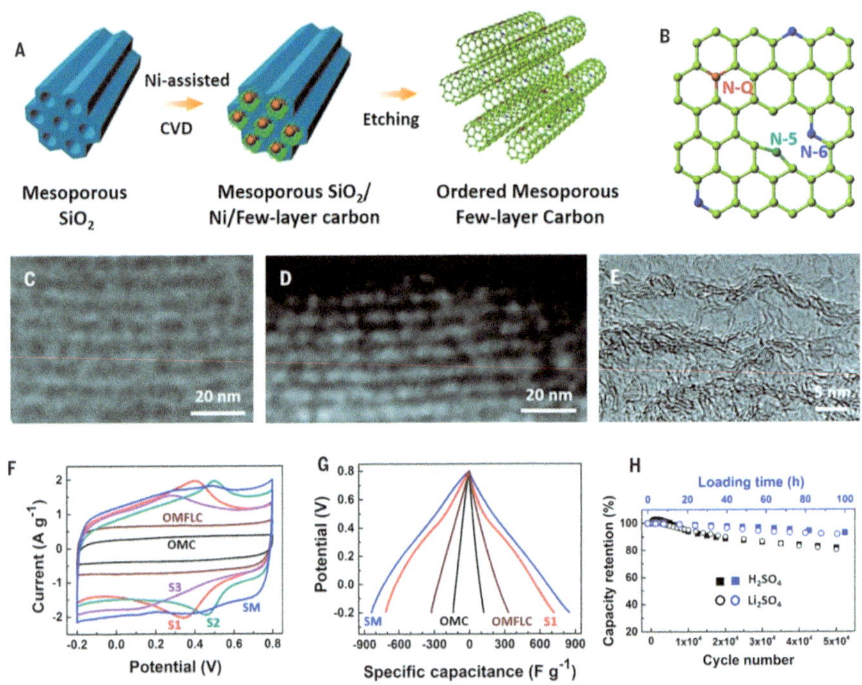

Figure 10.2 (a) Fabrication schematic of ordered mesoporous few-layer carbon (OMFLC). (b) Possible locations for N incorporation into the few-layer carbon network. (c) and (d) High-angle annular dark-field transmission electron microscopy (HAADF) images of ordered mesoporous carbon (OMC) (C) and OMFLC (D); dark regions indicate connected pore channels. (e) High-resolution TEM image of OMFLC; nanoporous walls consist of few-layered carbon sheets. (f) Cyclic voltammetry (CV test, at 2 mV s^{-1}) from the first cycle for OMC, OMFLC, and OMFLC-N (S1 to S3) and for mixed OMFLC-N (SM). (g) Galvanostatic charge/discharge (CC test at 1.0 A g^{-1}) from the first cycle for OMC, OMFLC, and OMFLC-N (S1 and SM). (h) Symmetric electrochemical cell devices retain >92% after 100 h of sustained loading (blue symbols, upper scale) at 1.2 V (in 0.5 M H_2SO_4) and 1.6 V (in 2 M Li_2SO_4), and retain >80% of their initial response after 50 000 cycles (black symbols, lower scale) from 0 to same peak voltages in two electrolytes. From ref. 22. Reprinted with permission from AAAS.

layers paves the way for enhancing capacitance of carbon/graphene-based supercapacitors.

In summary, regulating the distance between each stacked graphene sheet, optimizing the pathway of ion transport and doping other inorganic elements are effective way to utilize the high intrinsic surface areas of graphene along with more electrochemical active sites, further enhancing the performance of graphene-based flexible supercapacitors. Although the limited structural disadvantages of pure 2D graphene could be regulated through these approaches leading to enhanced performance, it is still difficult to precisely regulate the electrical properties of 2D graphene. Accordingly, we anticipate that the inorganic lattice framework in the 2D confinement conditions of inorganic 2D nanomaterials may give more advantages for enhancing the electrochemical performance in flexible supercapacitors.

10.3 Inorganic 2D Nanomaterials-based Flexible Supercapacitors

10.3.1 EDLC Mechanism (Conducting Materials)

10.3.1.1 MXenes-based Supercapacitors

Owing to its intrinsic advantages of conductivity and mechanical properties, graphene represented a remarkable development for flexible devices with a great deal of graphene-based electronic devices being reported in the literature, and even inorganic 2D nanomaterials based electronic devices. Diverse inorganic 2D nanomaterials with a variety of choices for alternative inorganic elements could lead to more versatile properties and functionalities beyond 2D graphene. As electrode materials of flexible supercapacitors, it is of great advantage for inorganic 2D nanomaterials owing to their regulated electrical conductivity, more probability of the extra contribution of abundant active pseudocapacitive centers from 2D inorganic frameworks, and the superior capacity for the electrosorption of electrolyte ions. As a result, displaying the excellent electrochemical performance of flexible supercapacitors based on inorganic 2D nanomaterials. Lukatskaya *et al.* reported 2D Ti_3C_2 MXene layers that combine 2D conductive carbide layers with a hydrophilic, primarily hydroxyl-terminated surface.[12] Specifically, the Ti_3C_2 MXene layer has shown the ability to be intercalated by a variety of cations including Li^+, Na^+, NH_4^+ *etc.* This unique character takes full advantage of its specific surface areas, offering ultrahigh capacitance in aqueous electrolytes (Figure 10.3). As expected, the volumetric capacitance of few-layer $Ti_3C_2T_x$ flexible electrode based supercapacitor reached 340 F cm^{-3} in KOH aqueous electrolyte, which is much higher than activated graphene-based flexible supercapacitors. The flexibility of $Ti_3C_2T_x$ paper with high electric conductivity and unique intercalated character opens the door for developing intercalation electrodes with high performance flexible energy storage devices. Similarly, Ghidiu *et al.* reported a titanium carbide clay produced by a solution of lithium fluoride and hydrochloric acid

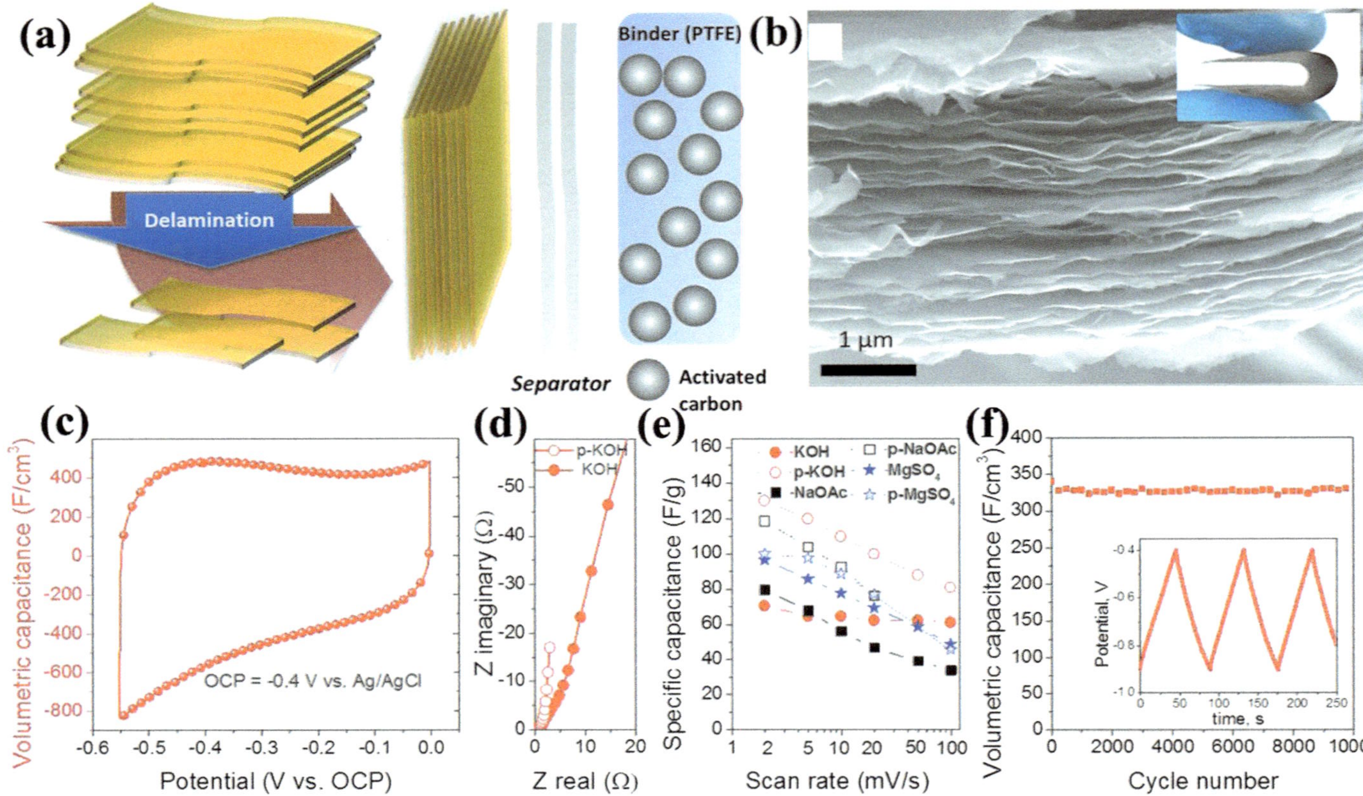

Figure 10.3 Electrochemical performance of binder-free $Ti_3C_2T_x$ paper electrodes. (a) Schematic illustration of electrode fabrication. First, the multilayer $Ti_3C_2T_x$ powders are delaminated to produce few-layer MXene flakes; the resulting colloidal solution is filtered through a porous membrane, producing binder and additive-free $Ti_3C_2T_x$ paper electrodes. (b) SEM images of paper electrode. Inset shows digital photograph of the paper electrode showing its flexibility. (c) CV curve of $Ti_3C_2T_x$ paper in KOH electrolyte. (d) EIS data in KOH for $Ti_3C_2T_x$ electrode (KOH, solid symbols) and $Ti_3C_2T_x$ paper (p-KOH). (e) Rate performance of the $Ti_3C_2T_x$ paper (open symbols) compared with multilayer exfoliated $Ti_3C_2T_x$ electrode (solid symbols) in KOH, $MgSO_4$, and NaOAc containing electrolytes. (f) Capacitance retention test of $Ti_3C_2T_x$ paper in KOH. Inset: galvanostatic charging/discharging test collected at 1 A g^{-1}. From ref. 12. Reprinted with permission from AAAS.

rather than concentrated hydrofluoric acid.[13] The as-obtained electrode material with 5 μm thickness has outstanding volumetric capacitance of up to 900 F cm^{-3} and gram capacitance of up to 245 F g^{-1}. No measurable degradation was found after 10 000 charging/discharging cycles (Figure 10.4). These splendid electrochemical performances of MXenes-based flexible supercapacitors would push energy storage devices to new heights.

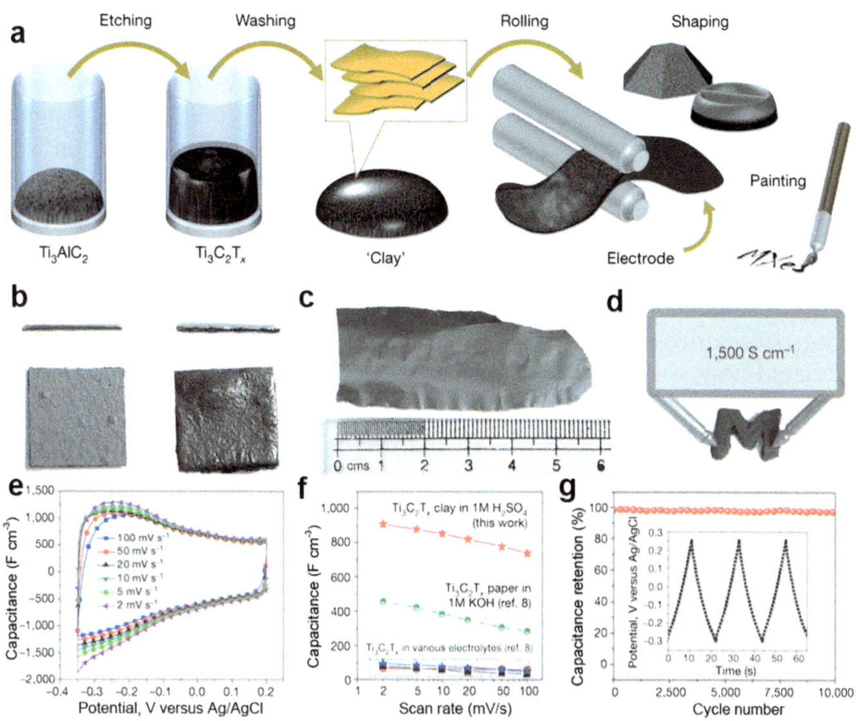

Figure 10.4 (a) MAX phase is etched in a solution of acid and fluoride salt (step 1), then washed with water to remove reaction products and raise the pH towards neutral (step 2). The resulting sediment behaves like a clay; it can be rolled to produce flexible, freestanding films (step 3), molded and dried to yield conducting objects of desired shape (step 4), or diluted and painted onto a substrate to yield a conductive coating (step 5). (b) When dried samples (left, showing cross-section and top view) are hydrated (right) they swell; upon drying, they shrink. (c) Image of a rolled film. (d) 'Clay' shaped into the letter M (~1 cm) and dried, yielding a conductive solid (labelled with the experimental conductivity of 'clay' rolled to 5 mm thickness). The etched material is referred to as $Ti_3C_2T_x$, where the T denotes surface terminations, such as OH, O and F. (e) Cyclic voltammetry profiles at different scan rates for a 5 mm-thick electrode in 1 M H_2SO_4. (f) Comparison of rate performances reported in this work and previously for HF-produced MXene. (g) Capacitance retention test of a 5 mm-thick rolled electrode in 1 M H_2SO_4. Inset shows galvanostatic cycling data collected at 10 A g^{-1}. Reprinted with permission from Macmillan Publishers Ltd; *Nature* (ref. 13), Copyright 2014.

10.3.1.2 TMDs-based Flexible Supercapacitors

Layered TMDs, such as MoS_2, TiS_2, VS_2 *etc.* have been widely investigated as electrocatalysts (*e.g.* HER), spin-electronics devices, sensors, and lithium ion batteries due to their versatile physicochemical properties.[23–25] However, MoS_2 nanosheet was rarely applied to the electrode material of flexible supercapacitors owing to the intrinsically low electrical conductivity of 2H–MoS_2 and relatively low capacitance. It is worth noting that the transition metal atomic layer is sandwiched between two sulfur layers *via* covalent bonding in the TMDs crystallographic structure, accompanied by van der Waals forces between their multi-layers, which allows easy intercalation of foreign cations (H^+, Li^+). Unfortunately, the monolayer structure of TMD nanosheets might be destroyed due to the valence change of transition metal ion with pseudocapacitve reactions during the charging/discharging process, leading to poor cyclability of the supercapacitors. Hence, there remains an open question of how to apply layered TMDs nanosheets as electrode materials for flexible supercapacitors in aqueous electrolyte with high electrochemical performance. To solve the above bottlenecks on TMDs for fabricating high-performance flexible supercapacitors, Acerce *et al.* reported a chemically exfoliated nanosheet of MoS_2 containing a 70% 1T phase concentration,[26] which has both high conductivity and extraordinary efficiency with ultra-high capacitance ranging from 400 to 700 F cm^{-3}. The high concentration of the 1T phase brings high electrical conductivity and hydrophilicity, and the latter character endows this MoS_2 nanosheet with similar characteristic to MXenes (Figure 10.5). Hence, ions such as H^+, Li^+, Na^+ and K^+ could intercalate during the charging process, realizing high capacitance. Besides, this as-obtained MoS_2 nanosheet electrode is also suitable for high-voltage (3.5 V) in organic electrolyte, with a coulombic efficiency of 95% along with stability over 5000 cycles. Following these results, the TMDs, especially MoS_2 nanosheet, in suitable metallic phase through controlled synthesis, may realize favorable electrochemical properties with high conductivity and hydrophilicity.

10.3.2 Pseudo-materials-based Flexible Supercapacitors

As classical 2D layered nanomaterials, 2D transition metal oxides (TMO) and 2D transition metal hydroxides (TMH) ultrathin nanosheets, have received tremendous attention for various properties. To be specific, various kinds of inorganic atoms or ions in their 2D inorganic lattices would lead to an extra contribution of abundant active pseudocapacitive centers. Therefore, the chemical activity of the lattice framework in inorganic 2D nanomaterials also affects the improved electrochemical performance, not just their high specific surfaces. But there are two intrinsic bottlenecks of the TMO and TMH that hinder the practical applications as electrode materials applying for flexible supercapacitors: (i) the electrical conductivity of TMO and TMH is extremely low, hindering electron transport on the electrode surface;

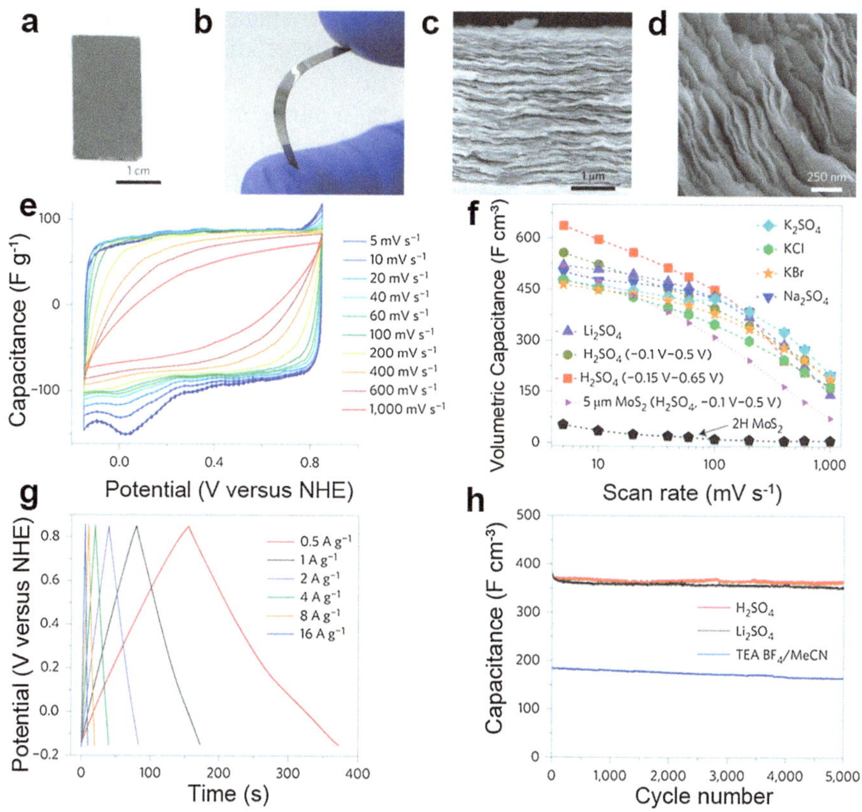

Figure 10.5 (a) and (b) Photographs of electrodes consisting of a thick film of chemically exfoliated 1T MoS_2 prepared by vacuum filtration and transferred onto rigid glass (a) and a flexible polyimide substrate (b). (c) Side view of the electrode observed by scanning electron microscopy (SEM) showing the layered nature of the film made by restacking exfoliated MoS_2 nanosheets. (d) High-magnification image of restacked MoS_2 nanosheets. (e) CVs of 1T phase MoS_2 electrodes in 0.5 M Na_2SO_4 from 5–1000 mV•s^{-1}. The capacitance remains virtually constant up to scan rate of 100 mV•s^{-1}. (f) Evolution of the volumetric capacitance of the 1T phase MoS_2 electrodes with scan rate for different electrolytes and 1 μm and 5 μm-thick films. The concentration of the cations in the electrolyte solutions was fixed at 1 M. (g) Galvanostatic cycles from 0.5–16 A•g^{-1} in K_2SO_4. (h) Capacitance retention after 5000 cycles in 0.5 M Li_2SO_4, H_2SO_4 and 1 M $TEABF_4$ in acetonitrile. Reprinted with permission from Macmillan Publishers Ltd: *Nature Nanotechnology* (ref. 26), Copyright 2015.

(ii) charge/discharge processes would cause the cracking of the electrode in the TMO and TMH nanosheet, leading to poor cyclic performance. Considering these practical problems, hybridizing the structures with conducting materials of compositing electrodes were intensively adopted.[27,28] Among them, a layer-by-layer route was employed to assemble a graphene/pseudomaterials hybrid film, rendering pseudomaterials with reasonable electric conductivity. Xie *et al.* prepared a self-assembled graphene/β-$Ni(OH)_2$ nanocomposite thin-film electrode with layer-by-layer structure *via* a facile one-step solvothermal approach.[29] The conducting graphene layers in nanocomposites act as electron channels for fast electron transfer on pseudomaterials and the latter offer enough electrochemical active sites, leading to a synergic effect, enhancing electrochemical performance. Meanwhile, a thin-film flexible pseudocapacitor in all-solid-state was successfully fabricated by using a Au-sputtered/PET flexible substrate as the counter electrode and PVA/KOH gel electrolyte as both electrolyte and separator (Figure 10.6). Consequently, the as-obtained nanocomposite based flexible supercapacitors device reached a high volume capacitance of 660.8 F cm^{-3} with no measurable capacitance loss even after 2000 cycles, with negligible degradation after 200 bending cycles, showing durability both in electrochemical performance and mechanical strength. This designed structure of the novel LBL graphene/β-$Ni(OH)_2$ nanohybrids provides a prototype for enhancing the electrochemical performance of flexible supercapacitors based on pseudocapacitive electrodes. Analogously, Dong *et al.* reported the fabrication of a thin film electrode of the LBL assembled Co–Al layered double hydroxide nanosheets (Co–Al LDH-NS) and graphene oxide (GO) composite electrodes for flexible supercapacitors.[30] In their work, both GO and Co–Al LDH-NS are exfoliated to an ultrathin nanosheet with electrostatic interaction that brings high surface areas (Figure 10.7). Also, all Co atoms are exposed on the surface of Al LDH-NS, offering enough faradaic redox reaction sites. Electron transport is generated through graphene sheets in the LBL structure, exhibiting high potential of realizing high capacitance. When the flexible Co–Al LDH-NS/GO multilayer films were applied on 1 M KOH aqueous electrolyte, they displayed an outstanding specific capacitance up to 1204 F g^{-1} at the scan rate of 5 mV s^{-1} at the thickness of 40 bilayers. Furthermore, this flexible thin film electrode remained conformal and transparent after being bended and deformed several times. Hence, 2D TMH based electrode with LBL composite structures would show a great promise in flexible supercapacitors.

The reversible redox faradaic reactions occurred on the surface of layered $Ni(OH)_2$ and Co–Al layered double hydroxide nanosheets exhibit potential windows of only about 0.5 V, both of which are much lower than the electrochemical potential window of aqueous solution (1.23 V). The energy density of 2D TMH electrode based flexible supercapacitors is still much lower. In this regard, it is essential to find a new graphene-like layered pseudocapacitive material with higher redox potential, power densities and energy densities to satisfy the practical applications of power supply. In order to solve the challenges around high potential, high power densities and energy densities, Gao *et al.* presented an β-$Co(OH)_2$ material consisting of five-atom layers, which not only possesses a shorter ion diffusion path but also provides 100% exposed hydrogen atoms as the electroactive sites, and thus offers past Faradaic redox reactions, rate capability, and finally achieving large

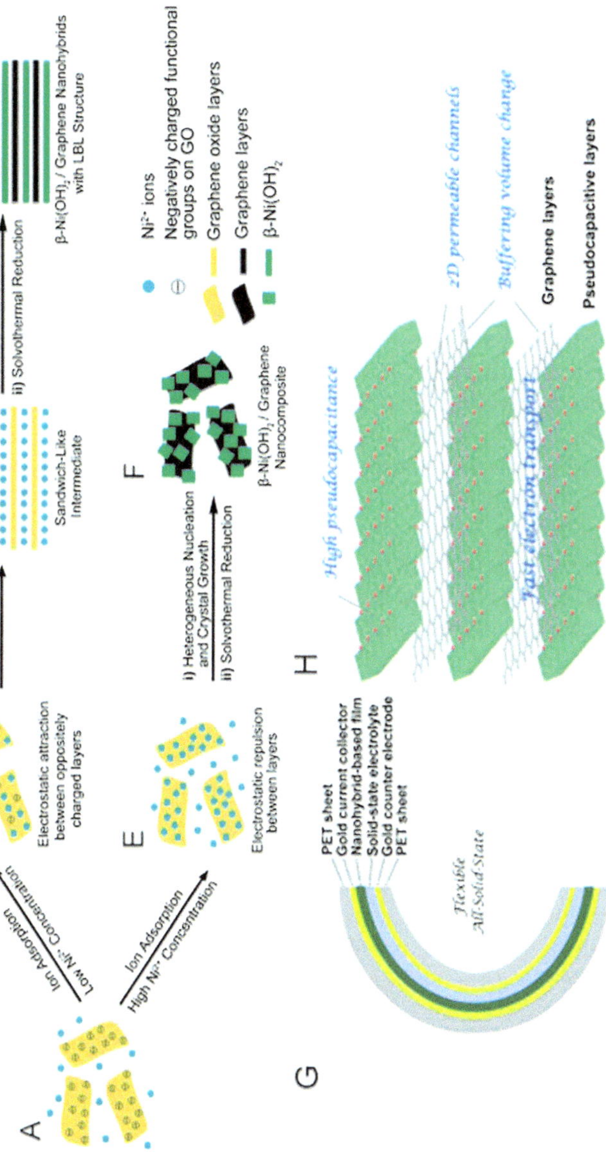

Figure 10.6 Schematic illustration of the formation mechanism of layer-by-layer β-Ni(OH)$_2$/graphene nanohybrids. (a) Negatively charged GO nanosheets and Ni^{2+} ions in the mixed solvent. (b) GO nanosheets with Ni^{2+} ions adsorbed on both sides of the surface, exhibiting positive or negative electrostatic properties due to uneven amount of Ni^{2+} adsorption. (c) Intermediate state with sandwich-like structure assembled by oppositely charged Ni^{2+}-adsorbed GO sheets. (d) β-Ni(OH)$_2$/graphene nanohybrids with lamellar LBL structure formed by solvothermal treatment. (e) Positively charged ion-adsorbed GO sheets with excess Ni^{2+} ions dispersed in the solvent. (f) β-Ni(OH)$_2$/graphene nanocomposites with traditional surface loading type formed by heterogeneous nucleation and subsequent isotropic crystal growth. (g) and (h) Schematic illustration of the flexible all-solid-state thin-film supercapacitor with advantages of layer-by-layer (LBL) structure built by pseudocapacitive β-Ni(OH)$_2$ as active materials and graphene layers as conducting materials. Reprinted from *Nano Energy*, **2**(1), J. Xie *et al.*, Layer-by-layer β-Ni(OH)$_2$/graphene nanohybrids for ultraflexible all-solid-state thin-film supercapacitors with high electrochemical performance, 65–74. Copyright 2013 with permission from Elsevier.

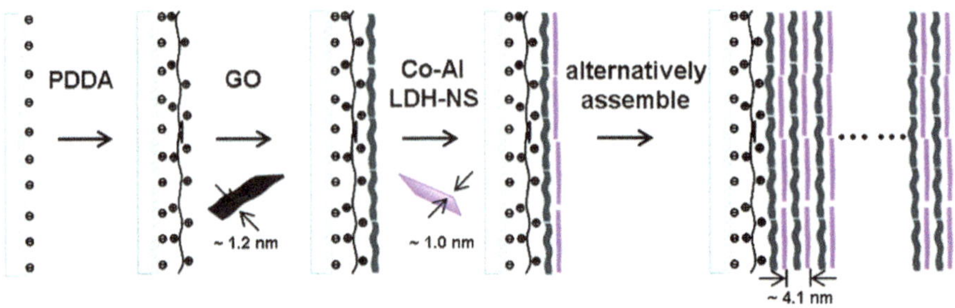

Figure 10.7 Schematic illustration of LBL assembly process for fabricating LBL multilayer films of positively charged Co–Al LDH-NS and negatively charged GO nanosheets. Adapted with permission from ref. 30, Copyright 2011 American Chemical Society.

capacitance along with high working potential.[31] (Figure 10.8) Interesting, the atomically thin β-$Co(OH)_2$ nanosheets have much increased DOS near the Fermi level compared with bulk β-$Co(OH)_2$, revealing that atomically thin β-$Co(OH)_2$ nanosheets have much superior electrical conductivity. More importantly, the increased DOS indicates that the electrons could transport along with the 2D conducting channel of β-$Co(OH)_2$ nanosheets, further reacting with OH^- ions. Above these merits, the single-layer β-$Co(OH)_2$ electrode shows capacitance of 2028 F g^{-1}, which is obviously superior than 525 F g^{-1} of bulk β-$Co(OH)_2$ electrode. The all-solid-state asymmetric supercapacitor from β-$Co(OH)_2$ and N-doped graphene reached a high density of 98.9 Wh kg^{-1} at an exceptional power density of 17 981 W kg^{-1}. The ultrathin β-$Co(OH)_2$ realize ultrahigh performance benefited from its 2D dimensionality offering large electrochemical sites with high electrical conductivity. This rational design offered a general way to optimize the electrochemical performance of TMO/TMH materials through surface modification of 2D ultrathin nanosheet.

Another representative group of inorganic 2D nanomaterials for flexible supercapacitors are layered phosphates. Wu *et al.* reported a simple 2-propanol assisted ultrasonication route to obtain six atomic layers α_1-vanadyl phosphate ($VOPO_4$) ultrathin nanosheets, with an extremely high redox voltage of about 1.0 V approaching the limited electrochemical window of aqueous solution.[32] Despite the fact that the exfoliated $VOPO_4$ nanosheet shows great promise as an electrode material for flexible supercapacitors due to its 2D layered structure and high potential window, the poor electrical conductivity of the $VOPO_4$ nanosheet hinders development. Considering this bottleneck, a LBL graphene/$VOPO_4$ hybrid film with a thickness of about 90 nm was assembled as an electrode with both high conductivity and pseudo capacitance (Figure 10.9). In this case, the existence of hydrogen bonds acts as the interaction force to format the order structure, thus the order stacking of our hybrid structure, achieving the final graphene/$VOPO_4$ hybrid films. Meanwhile, PVA/LiCl gel electrolyte was applied as both solid-state electrolyte and separator, preventing the potential dissolving of the active electrode material

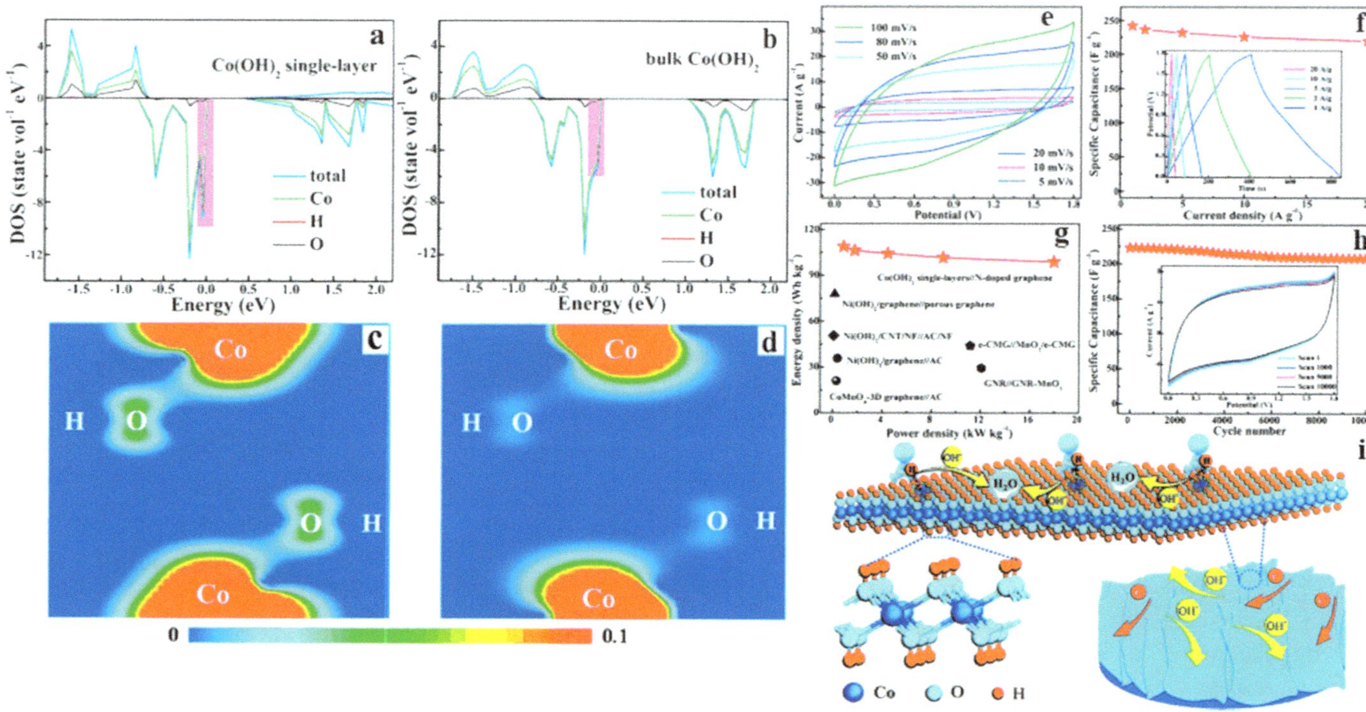

Figure 10.8 (a) and (b) Calculated density-of-states for (a) β-$Co(OH)_2$ single layers and (b) bulk β-$Co(OH)_2$. The shadowed region highlights the increased states density at the valence band edge of the β-$Co(OH)_2$ single-layers. (c) and (d) Charge density waves for the valence band maximum of (c) β-$Co(OH)_2$ single-layers and (d) bulk β-$Co(OH)_2$, plotted from 0 (blue) to 0.1 e Å^{-3} (red). (e) CV curves at various scan rates and (f) galvanostatic charge–discharge curves at different current densities (inset) and the corresponding calculated specific capacitances of the single-layer β-$Co(OH)_2$ all-solid-state asymmetric supercapacitor. (g) Electrochemical performance of the single-layer β-$Co(OH)_2$ all-solid-state asymmetric supercapacitor in comparison with previously reported asymmetric supercapacitors. (h) Cycle performance of the single-layer β-$Co(OH)_2$ all-solid-state asymmetric supercapacitor measured at a scan rate of 20 mV s^{-1}; the inset shows the corresponding CV curves. (i) Schematic illustration of the hydrogen adsorption/desorption process in the single-layer β-$Co(OH)_2$ electrode. Reprinted with permission from S. Gao *et al.*, Ultrahigh Energy Density Realized by a Single-Layer β-Co(OH)2 All-Solid-State Asymmetric Supercapacitator, *Angewandte Chemie International Edition*, John Wiley and Sons, © 2014 WILEY-VCH Verlag GmbH & Co. KGaA, Weinheim.[31]

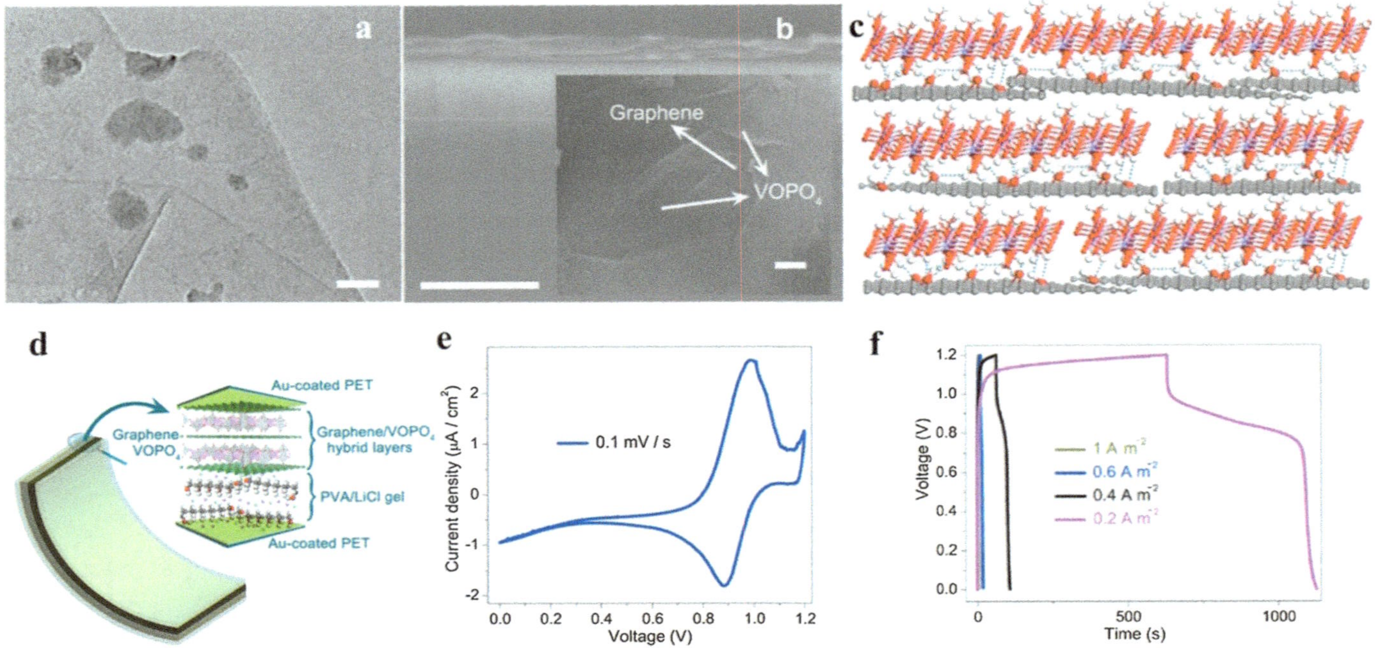

Figure 10.9 (a) TEM image of the 2D $VOPO_4$/graphene hybrid structure, scale bar, 80 nm. (b) The SEM image of the hybrid thin film in cross-sectional view with a thickness of ~90 nm. Scale bar: 500 nm; the inset shows the surface of the hybrid film, of which the graphene was lies under the 2D $VOPO_4$ layers indicated by the white arrows. Inset scale bar: 100 nm. (c) Schematic illustration of the layer-by-layer structure of assembled 2D $VOPO_4$/graphene hybrid thin film. (d) Schematic illustration of the as-fabricated flexible pseudocapacitor in all-solid-state, in which the 2D $VOPO_4$/graphene hybrid layers as the working electrode on the gold-coated PET sheet and PVA/LiCl gel as the electrolytes. (e) CV curves of 2D $VOPO_4$/graphene hybrid film in PVA/LiCl gel electrolyte when working voltage ranging from 0 to 1.2 V at 0.1 mV s^{-1}. (f) Galvanostatic charge/discharge curves at different current densities of 0.2, 0.4, 0.6 and 1 A m^{-2}. Reprinted with permission from Macmillan Publishers Ltd: *Nature Communications* (ref. 32), Copyright 2013.

of $VOPO_4$ nanosheets under PH medium condition.[33] As expected, the LBL hybrid structure brings ultrahigh electrochemical performance of high area capacitance and redox potential (8360.5 μF cm^{-2}/1.0 V), leading to an ultrahigh energy density and power density. (1.7 mWh cm^{-2}/5.2 mW cm^{-2}), which are outstanding among the flexible solid-state supercapacitors along with no measurable degradation of the specific capacitance after 400 bending cycles. The 2D ordered design of the hybrid thin-film structure consisting of highly pseudocapacitive 2D materials composite on highly conductive 2D graphene matrix give rise to this good pseudocapacitive performance.

The construction principles of flexible supercapacitors have been intensively studied recently. Flexible electrode materials include both pure materials such as 2D graphene, MXenes, TMDs and hybrid nanostructures such as graphene/inorganic 2D nanomaterials composites. However, the intrinsic mechanisms and fundamental architecture of electrode materials for flexible supercapacitors have not been fully investigated. Great challenges still exist for further boosting the energy densities of flexible supercapacitors, catering for the great demands of power supply for practical applications. Therefore, next-generation electrode materials are urgently needed with high potential windows and robust cycling stability.

10.4 2D Nanomaterials for Planar Supercapacitors (Microsupercapacitors)

The latest development in electrode materials for flexible supercapacitors gives rise to a trend of searching new alternative materials with both superior electrochemical performance and excellent mechanical properties. In addition to the role of electrode material, the configuration design of a flexible supercapacitor device including electrode materials, electrolyte, and current collectors are also essential. To improve their flexibility and specific capacitance, various methods have been developed for decreasing the thickness of flexible supercapacitors. However, there is no guarantee of preventing problems with short circuiting when there is no separator in the whole device. Of note, when a supercapacitor works at different bending/folded states, it requires the ions to transport smoothly under mechanical strain effects without degradation. Therefore, how to effectively enhance the mechanical strength of flexible supercapacitors in the absence of performance loss remains to be solved. In this regard, new configuration designs of the supercapacitor itself are required to push the performance of flexible supercapacitors to a higher level. Planar supercapacitors have proven to be an alternative option by optimizing the ion transport mechanism in 2D horizontal plane in the charging/discharging processes, usually requiring the three requisite components (electrode material, electrolyte, and current collectors) integrated onto 2D directions. For instance, Jiang *et al.* reported a planar MEMS supercapacitor that employed porous CNT forests as electrode materials.[34] Different from the conventional sandwich-type architecture, the two symmetric electrodes of the MEMS supercapacitor were placed on the same plane. It should be noted that the major advantages of structural design are that the transport distance of ions in the electrolyte could be well controlled in the 2D plane and shortened while eliminating the necessity of a separator. However, due to the dense 3D structures of the CNT forest electrodes, its large surface areas are not fully employed, blocking the access of the ions to the inside surfaces and interfaces of the CNT forests. But there is no denying that it becomes the prototype of planar supercapacitors for future research.

To further explore the inner mechanism on electrode materials for planar supercapacitors, Chmiola *et al.* used carbide-derived carbon (CDC) as the electrode material for fabricating 2D thin-film microsupercapacitors.[35] Of note, when the film thickness decreases from 200 to ~2 μm, a huge increase of volumetric capacitance was observed. For instance, when a CDC film of ~50 μm in $TEABF_4$ electrolyte, the volumetric capacitance (~60 F cm^{-3}) is like what was observed in a ~300 μm traditionally processed electrode from carbon powders (~50 F cm^{-3}). Interestingly, as the coating thickness of electrode material decreases to ~2 μm, the volumetric capacitance increases largely to nearly 180 F cm^{-3} in $TEABF_4$ electrolyte and ~160 F cm^{-3} in 1 M H_2SO_4 aqueous electrolyte. It can be believed that the thin thickness of film electrodes offered respectable gravimetric and volumetric specific capacitance owing to the thin thickness enhanced the utilization of surface areas of electrode materials.

To achieve the superior electrochemical performance of planar supercapacitors, there are three key points as summarized below: (i) constructing ion migration channels for efficient transport of electrolyte ions; (ii) enhancing the utilization of surface areas of electrode material to provide more electrochemical active site; and (iii) reducing the actual thickness of electrode materials to attain high gravimetric/volume capacitance. Therefore, 2D nanomaterials with all the above characteristics are intriguing for planar supercapacitors.

10.4.1 Conductive 2D Nanomaterials for Flexible Planar Supercapacitors

The focus on electrode materials for flexible planar supercapacitors at primary stage was vastly based on 1D nanomaterials (nanowires, nanofibers). Considering that planar supercapacitors demand electrodes with 2D ion migration channels, the application of 2D nanomaterials has been motivated to a large degree by their potential from multiple advantages of high specific surface areas and high mechanical stability. Conducting 2D nanomaterials such as graphene, a part of the TMO/TMD nanosheet with the advent of atomically thin and flat layers have become possible candidates for new designs of thin-film planar supercapacitors with good electrochemical performance. As the prototype of 2D nanomaterial and due to its high electron conductivity and high surface areas, 2D graphene was first studied as an electrode material of planar supercapacitors. For instance, Yoo *et al.* reported an ultrathin flexible supercapacitor based on electrodes composed of multi-layer reduced graphene oxide using an 'in-plane' fabrication approach.[17] It is aforementioned that the stacked 2D graphene nanosheets greatly hindered the electrochemical performance of graphene-based energy storage devices. In this work, the in-plane design of the flexible supercapacitor means that it is not only straightforward to implement and exploit efficiently the surface areas of each graphene layer, but also to render the movements of the electrolyte ions to favor a direction along the planar graphene nanosheet rather than perpendicular to it (Figure 10.10). More importantly, the layer-by-layer (LBL) method was applied to create multilayer graphene oxide films (MGO), which were then chemically reduced by hydrazine, finally forming reduced multilayer graphene oxide films (RMGO). The electrostatic interactions between the cationic polymer (poly ethyleneimine) (PEI) and graphene oxide (GO) allow multi-layer graphene oxide (MGO) films to be obtained instead of stacked geometry. As expected, the 2D in-plane design of flexible supercapacitors takes advantage of the atomic layer thicknesses and flat morphology of graphene when the thickness of the RMGO sample is 10 nm, and is ideal for 2D devices to reach specific capacities up to 394 μF cm^{-2}, about five times higher than that of 80 μF cm^{-2} obtained for graphene. The optimized in-plane design of flexible supercapacitor offers new opportunities to enhance interaction with 2D nanomaterials for the electrolyte ions, and brings full utilization of the high surface area of 2D thin-film, providing a model for 2D nanomaterials-based energy storage nanodevices.

Apart from 2D graphene nanosheets, inorganic 2D nanomaterials are also believed to be promising electrode material platforms for planar supercapacitors with large surface areas and high flexibility. However, for example, owing to their easily oxidized property and poor pseudocapacitive performance,[36] TMDs are inappropriate for fabricating flexible supercapacitors with a sandwiched structure. In that case, planar supercapacitors may appear as a favorable candidate to mitigate this problem. Although conducting inorganic 2D nanomaterials hold promising potential for the construction of flexible supercapacitors, it should be noted that it is relatively difficult to attain conducting inorganic 2D nanomaterials because most of the reported inorganic 2D nanomaterials are either semiconducting or insulating.[37–39] There were

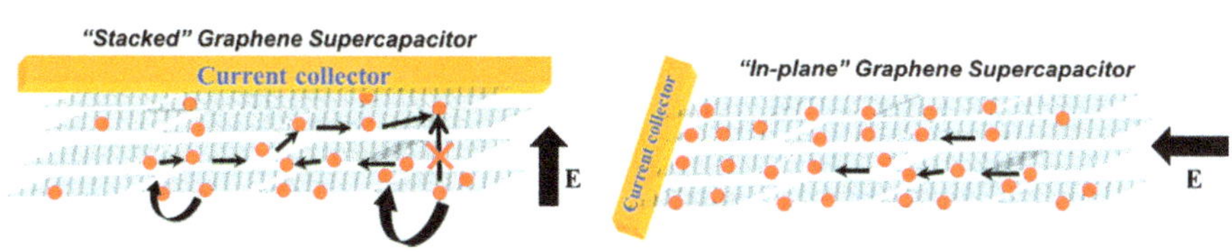

Figure 10.10 Schematic illustration of the stacked geometry used for the fabrication of sandwich-type supercapacitor devices (left) and the in-plane geometry used for planar supercapacitor devices (right). Adapted with permission from ref. 17, Copyright 2011 American Chemical Society.

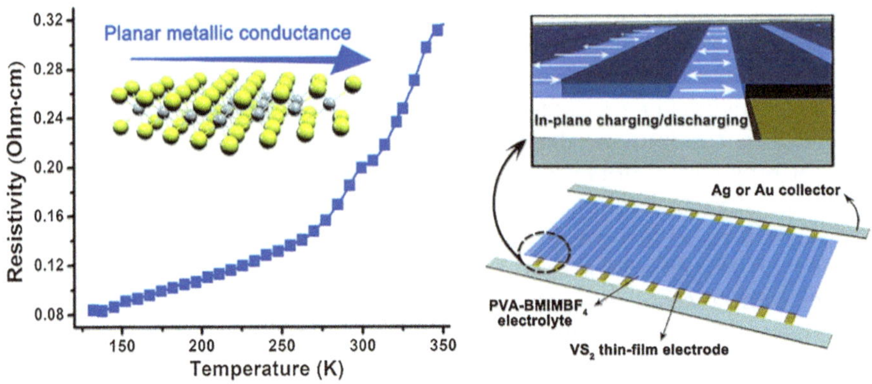

Figure 10.11 Temperature dependence of planar resistivity of VS_2 thin film revealed the planar metallic conductance of VS_2 thin film (left), and schematic illustration of planar ion migration pathways for the in-plane supercapacitor (right). Adapted with permission from ref. 18, Copyright 2011 American Chemical Society.

few reports of metallic inorganic 2D nanomaterials as electrode materials before the investigation of conducting 2D VS_2. As a typical member of conducting TMDs, the 2D VS_2 ultrathin nanosheet is made up of metal V layers sandwiched between two S layers, along with stacked several layers by weak van der Waals interactions (Figure 10.11). Because 2D electron–electron correlations among V atoms of 2D VS_2 may preferably induce more complicated planar electric transport performances, it has the potential to be a reasonable choice for planar supercapacitors. To avoid possible deterioration of VS_2 nanosheet electrodes, Feng *et al.* adopted a solid-state polymer gel electrolyte prepared by configuring the PVA gel with a water-soluble ionic liquid of $BMIMBF_4$ (ref. 18). Of note, the 2D VS_2 ultrathin nanosheet brings a promising planar metallic behavior and high specific surface area, which is an ideal electrode material for realizing electrical double layered capacitors. As expected, the area specific capacitance of as-fabricated 2D VS_2 based planar supercapacitor reached 4760 $\mu F\ cm^{-2}$ in a 150 nm planar configuration, along with no obvious degradation being observed even after 1000 charge/discharge cycles. The 2D VS_2 nanosheet offers a new material platform based on conducting inorganic 2D nanomaterials for a planar supercapacitor with high electrochemical performance.

10.4.2 Pseudo-capacitive 2D Nanomaterials for Flexible Planar Supercapacitors

Even though conducting inorganic 2D nanomaterials have shown great prospects in planar supercapacitors, their electrochemical performance is still far from those of traditional pseudocapacitors. Therefore, a method to introduce extra pseudoreactions into planar supercapacitors has been needed to further improve the electrochemical performance. In this regard, 2D pseudocapacitive nanomaterials are promising materials for fabricating high performance planar supercapacitors. Unfortunately, most 2D pseudocapacitive nanomaterials have poor electron conductivity owing to their insulating state, hindering planar supercapacitors from achieving high capacitance. To achieve the synergic advantages of enhanced electrical conductivity and high pseudocapacitive activity, Peng *et al.* reported a hybrid nanostructure composed of pseudocapacitive 2D ultrathin nanosheets for enhancing the electrochemical performance of the planar supercapacitor. In this case, a 2D δ-MnO_2 nanosheet was integrated on 2D graphene nanosheets as composite electrode materials of the planar structures.[11] (Figure 10.12) The integrated δ-MnO_2 nanosheets serves as active sites for the pseudocapacitive reactions ($MnO_2 + xH^+ + xe^- \leftrightarrow MnOOH_x$), contributing to distinct enhancement of the specific capacitance for planar supercapacitors. Moreover, the 2D δ-MnO_2 nanosheet integrated on 2D graphene nanosheet would potentially tailor the distance between each layer of densely stacked 2D graphene nanosheets, allowing for more electrolyte ions to penetrate efficiently into the interlayer space of the hybridized film. This reasonable design introduces more electrochemically active surfaces for absorption/desorption of the electrolyte ions, bringing extra interface to facilitate charge transport during charging/discharging processes at the hybridized interlayer areas. Therefore, the hybrid 2D nanostructure design improves the capacitance of as-fabricated planar supercapacitors, reaches a high specific capacitance of 267 F g^{-1} with a current density of 0.2 A g^{-1}, and brings good cycling performance with capacitance retention of 92% after 7000 charge/discharge cycles. In addition, with the high microscopic compressibility and macroscopic extensibility of the hybridized 2D planar structure, the as-fabricated planar supercapacitors

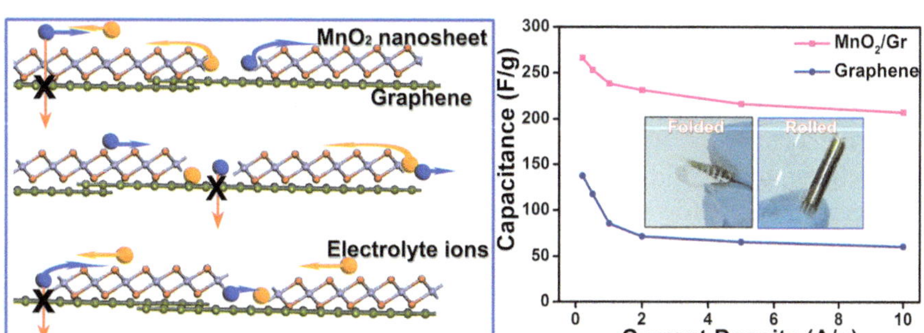

Figure 10.12 Schematic description of the 2D planar ion transport favored within the 2D δ-MnO_2/graphene hybrid structures, and comparison of specific capacitance values for the supercapacitors based on hybrids and based on graphene. Inset: demonstration of two different bending states: folded, and rolled. Adapted with permission from ref. 11, Copyright 2013 American Chemical Society.

display excellent mechanical flexibility undergoing various bending states from being curved, folded to rolled, with >90% capacitance retention after folded thousands of times. In short, hybridization of distinct quasi-2D ultrathin nanosheets provides a prototype to improve the capacitance, electrochemical performance and mechanical strength of planar supercapacitors, representing a promising sign for building next-generation high-performance, flexible energy storage nanodevices.

In summary, the literature reported above has reviewed recent progress in planar supercapacitors (microsupercapacitors). To be specific, the conductive 2D nanomaterials and pseudocapacitive 2D nanomaterials as the ultrathin electrode materials were outlined. It should be noted that conductive 2D nanomaterials lack of pseudocapacitance while pseudocapacitive 2D nanomaterials short of electron conductivity. In this regard, the optimized 2D nanomaterial should simultaneously satisfy high electron conductivity, high voltage window, excellent pseudocapacitive performance and constructing 2D ion migration channel for electrolyte ion transport. Moreover, further works should concentrate on searching for novel 2D nanomaterials as electrode materials for increasing the capacitance of planar supercapacitors.

10.5 Gel Electrolytes for 2D Nanomaterials-based Flexible Supercapacitors

Gel electrolytes in flexible supercapacitors are commonly made of a polymer material as matrix and an electrolytic salt to provide mobile ions. Currently, most gel electrolytes are based on the PVA matrix such as PVA/H_2SO_4,[40,41] PVA/KOH,[29,42] and PVA/LiCl.[32,43] In such kinds of device, the PVA gel electrolyte is introduced as both the solid electrolyte and the separator with a wide range of pH values like the aqueous electrolyte that can be controlled during the process of gel drying. The PVA gel electrolyte can prevent the chemical dissolution of electrode materials by minimizing the water content. Furthermore, the PVA gel electrolyte can also serve as an elastic coating with a certain mechanical strength to avoid structure degradation of electrode materials. The good abilities of avoiding chemical corrosion on electrodes and achieving all-solid-state condition render excellent performance of PVA gel electrolytes.[33] However, the development of polymeric gel electrolyte still faces many challenges. When conventional gel electrolyte is applied, the electrode exceeds 24 h, the gel electrolyte becomes solid-state, thus the water content of the gel electrolyte drops sharply, which leads to a decrease of ion migration as well as losses of capacitance during supercapacitor operation. Therefore, an ideal polymeric gel electrolyte should have a combination of properties including robust water retention ability, high ion migration rate and reasonable mechanical strength, but such a combination of these superior properties is difficult to realize in the current polymeric gel electrolyte systems. Thus, the development of polymeric gel electrolytes is still at an initial stage and the inner electrochemical mechanism remains to be explored.

This gives us plenty of room for optimizing electrochemical performance of flexible supercapacitors *via* design of the gel electrolyte.

10.5.1 Aqueous Gel Electrolyte

Aqueous gel electrolytes are dominantly based on the PVA gel electrolyte. Here, the polyzwitterion, which is a type of unique charged polymer possessing both anionic and cationic groups in one side group, could become a potential alternative for flexible supercapacitors. Generally, the polyzwitterion is highly hydrated with robust water retention ability owing to the strong interactions between the charged groups and water molecules, making the polyzwitterion a kind of good superabsorbent material.[44] Meanwhile, the zwitterionic nature of the polyzwitterion makes the cationic and anionic counterions easily separated during ion migration, leading to high ionic conductivity.[45] Dipole–dipole interactions between the zwitterionic groups of polyzwitterion can easily form a physical gel, providing superior mechanical strength.[46] When compared with conventional polyelectrolyte, the polyzwitterion usually has good solubility in high salt concentration aqueous solutions due to an anti-polyelectrolyte effect.[47] More importantly, the charged and polar groups of zwitterionic gel electrolytes would strengthen the adhesion between the gel electrolyte and the electrodes, enabling the zwitterionic gel electrolyte as a polymer binder to hold the electrodes together. Therefore, the combination of great water retention ability, high ion conductivity, certain mechanical strength, anti-polyelectrolyte effect, and potential polymer binder, make the polyzwitterion a truly promising new type of gel electrolyte for energy storage devices. As per the zwitterionic character of the polyzwitterion, Peng *et al.* have developed a highly efficient zwitterionic aqueous gel electrolyte with unprecedented electrochemical performance.[48] In previously-reported literature, the traditional aqueous gel electrolyte usually lacks water retention ability and ion conductivity in the solid-state. In this case, they focus on how to maximize the electrochemical performance of polymeric gel electrolyte in aqueous system. Therefore, developing a new class of polymeric gel electrolyte PPDP possessing a designed zwitterionic character with robust water retention ability and unique ion migration channel for solid-state supercapacitors is in accordance with the bottlenecks in this field (Figure 10.13). Moreover, the salt $LiCl{\cdot}H_2O$ is chosen and dissolved in our aqueous PPDP gel electrolyte for ion transport, and is commonly used in solid-state supercapacitors with environmental friendly, low cost, no leakage and mitigating evaporation. As expected, the water retention of PPDP gel electrolyte at solid state of 47.1% is higher than that of the PVA gel electrolyte of 40.5%. Meanwhile, the PPDP gel electrolyte has an ionic conductivity of 10.1 mS cm^{-1} in the solid-state, which is evidently higher than the ionic conductivity of 1.6 mS cm^{-1} of the PVA gel electrolyte in the solid-state. Hence, the PPDP gel electrolyte has synergic advantages of robust water retention ability and unique ion migration channel, bringing excellent electrochemical

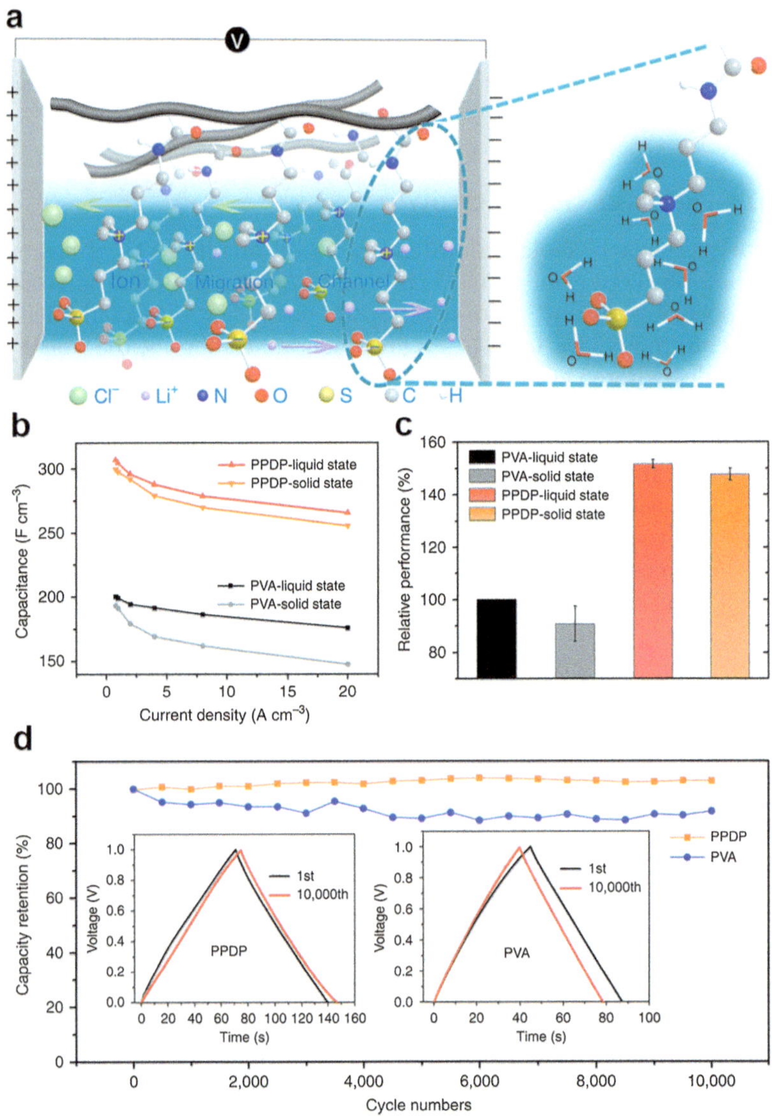

Figure 10.13 (a) Schematic illustration of the PPDP gel electrolyte applied on electrodes. The ion migration channel is formed by applying external electric field. PPDP is strongly hydrated by water molecules with robust water retention ability due to the electrostatic interactions between the zwitterionic groups and water molecules. (b) Comparison of specific capacitance values for graphene-based supercapacitors between PPDP and PVA gel electrolytes at different current densities. (c) Relative performance of PVA and PPDP electrolytes in liquid and solid states, where the error bars are obtained based on the capacitances at different current densities. (d) Cycling performance of graphene electrodes applying PPDP or PVA gel electrolyte at solid state. Inset: comparison of CD curves between the 1st CD cycle and the 10 000th cycle for graphene-based solid-state supercapacitors applying PPDP (left of panel) and PVA (right of panel) gel electrolyte at 4 A cm^{-3}. Reprinted with permission from Macmillan Publishers Ltd: *Nature Communications* (ref. 48), Copyright 2016.

performance of the gel electrolyte in the solid-state, representing the best value among the previously reported graphene-based solid-state supercapacitors. The as-obtained symmetric graphene-based solid-state supercapacitor applying PPDP gel electrolyte reached an ultrahigh volume capacitance of 300.8 F cm^{-3} at 0.8 A cm^{-3} with an excellent rate capacity of only 14.9% capacitance loss as the current density increases from 0.8 A cm^{-3} to 20 A cm^{-3}. The zwitterionic gel electrolyte may become a promising gel electrolyte for solid-state flexible supercapacitors.

10.5.2 Non-aqueous Gel Electrolyte

Nowadays, non-aqueous gel electrolytes, such as block-copolymer-based gel electrolytes and silica-based gel electrolytes, have been developed dissolved in ionic liquid without evaporation issues, achieving the improvement in the ion migration rate along with good mechanical strength, and thus making great advances in electrochemical performance. Kang *et al.* fabricated an electrode using bacterial nanocellulose (BNC) papers coated with carbon nanotubes (CNT), applying ionic triblock-copolymer based gel electrolyte.[49] The designed triblock-copolymer based gel electrolyte is made of a poly (styrene-*block*-ethlene oxide-*block*-styrene) (PS-PEO-PS) triblock copolymer and [EMIM][NTf_2](C-TRI) ionic liquid. Notably, the hydrophobic PS block of (PS-PEO-PS) the triblock copolymer gel electrolyte brought good adhesion to the hydrophobic CNT electrode due to van der Waals interactions. While the PEO block remained a hydrophilic ionic liquid in the polymer matrix. Owing to the adhesive chemical properties of the (PS-PEO-PS) triblock copolymer gel electrolyte, the as-obtained supercapacitor showed excellent mechanical strength, displaying 200 bending cycles to a radius of 3 mm without significant degradation. However, the (PS-PEO-PS) triblock copolymer gel electrolyte shows a good ionic conductivity of ~8 mS cm^{-1} at 7 wt% of polymer. As expected, the BNC/CNT/ion gel supercapacitor demonstrated good electrochemical performance, showing a capacitance of 50.5 F g^{-1} with an energy density of 15.5 mWh g^{-1}. More importantly, the (PS-PEO-PS) triblock copolymer gel electrolyte brought excellent mechanical strength and cycling performance with less than 0.5% change in capacitance over 5000 cycles. The triblock copolymer gel electrolyte may pave a new way for developing next-generation flexible supercapacitors with high mechanical properties.

Silica-based gel electrolytes are also representative non-aqueous gel electrolytes in flexible supercapacitors. Kang *et al.* reported an all-solid-state flexible supercapacitor based on CNT paper and silica-based ionic gel electrolyte.[50] The silica-based gel electrolyte is synthesized by adding fumed silica nanopowder in [EMIM][NTf_2] ionic liquid. Afterwards, the mixture was stirred for 6 h in a nitrogen atmosphere and the silica-based gel electrolyte was successfully prepared. When applying silica-based gel onto the CNT paper electrode, the power and energy density of supercapacitor were much higher than that applying PVA/H_2SO_4 electrolyte. The main reason is that non-aqueous gel electrolyte could offer wider electrochemical window (~3 V) than that of PVA-based aqueous gel electrolyte (~1 V). Meanwhile, the silica channel ensured fast ion transport in the gel electrolyte, leading to higher electrochemical performance. Owing to the facile ion transport channel and high potential of silica-based gel electrolyte, the specific capacitance reached 135 F g^{-1} with a maximum power density of 164 kW kg^{-1}. In addition, silica-based gel electrolytes brought a high cycling performance with 3% capacitance loss after 4000 cycles, and less than 5% loss after 100 bending cycles. Silica-based gels would hold great promise in the application of non-aqueous flexible supercapacitors.

10.6 Conclusions and Outlooks

Here, we have summarized recent progress in 2D nanomaterials as electrode materials for flexible supercapacitors. 2D nanomaterials have shown the synergic advantages of high specific surface area, excellent mechanical strength with high durability and potential electrochemical active sites, providing an ideal platform for construction of flexible supercapacitors with thin film structure. Therefore, finding new 2D nanomaterials with both high redox reactions and high conductivity would be the next breakthrough for flexible supercapacitors with further enhanced electrochemical performances. Besides the selection of electrode materials in 2D nanomaterials, the designed configuration of thin film electrodes has shown the next performance leap for flexible supercapacitors; that is, planar supercapacitors. In planar supercapacitors with all three ingredients of collector, electrode and electrolyte in the same horizon, the assembled thin film of 2D graphene and inorganic 2D nanomaterials enable excellent material platforms for planar configurations, optimizing the 2D ion migration channel. Therefore, 2D nanomaterials are providing future research directions toward the construction of next-generation flexible supercapacitors with enhanced electrochemical performance.

Although considerable progress has been made in the discovery of novel 2D nanomaterials for flexible supercapacitors, current stages are far from being ideal. There is a need to redesign the flexible supercapacitors from electrode, architectures and gel electrolytes. For electrode materials, developing 2D nanomaterials with both high electric conductivity and pseudocapacitive performance remains a challenge. For most electrode materials with pseudocapacitive behavior, they are insulators, which is not satisfied for the fast electron transport during electrochemical reaction processes. Even though hybridizing insulating pseudocapacitive materials with conducting graphene has attracted considerable attention, heterostructural interface problems still existed in the hybrid structure for the weak interaction forces, rather than the crystalline bonding heterostructure. For architectures of flexible supercapacitors, how to effectively deposit the electrode materials in the flexible plastic substrate with lower ohmic contact and how to enhance the actual capacitance of the whole planar supercapacitor devices are still great challenges. For the electrolyte in flexible supercapacitors, although gel electrolytes, such as the zwitterionic gel electrolyte and non-aqueous triblock gel electrolyte have been applied well in flexible supercapacitors, much effort is needed for developing new types of polymer electrolyte with higher ionic conductivity, lower electronic conductivity, higher electrochemical windows and even mitigated self-discharge limits, catering for the next-generation of flexible supercapacitors with superior performance.

References

1. K. S. Novoselov, A. K. Geim, S. V. Morozov, D. Jiang, Y. Zhang, S. V. Dubonos, I. V. Grigorieva and A. A. Firsov, *Science*, 2004, **306**, 666.
2. M. Chhowalla, H. S. Shin, G. Eda, L.-J. Li, K. P. Loh and H. Zhang, *Nat. Chem.*, 2013, **5**, 263.
3. W. Tang, L. Liu, S. Tian, L. Li, Y. Yue, Y. Wu and K. Zhu, *Chem. Commun.*, 2011, **47**, 10058.
4. F. Luan, G. Wang, Y. Ling, X. Lu, H. Wang, Y. Tong, X.-X. Liu and Y. Li, *Nanoscale*, 2013, **5**, 7984.
5. C.-C. Hu, K.-H. Chang, M.-C. Lin and Y.-T. Wu, *Nano Lett.*, 2006, **6**, 2690.
6. X. Lu, G. Wang, T. Zhai, M. Yu, J. Gan, Y. Tong and Y. Li, *Nano Lett.*, 2012, **12**, 1690.
7. X. Lu, T. Zhai, X. Zhang, Y. Shen, L. Yuan, B. Hu, L. Gong, J. Chen, Y. Gao, J. Zhou, Y. Tong and Z. L. Wang, *Adv. Mater.*, 2012, **24**, 938.
8. G. Yu, L. Hu, N. Liu, H. Wang, M. Vosgueritchian, Y. Yang, Y. Cui and Z. Bao, *Nano Lett.*, 2011, **11**, 4438.
9. G. Yu, L. Hu, M. Vosgueritchian, H. Wang, X. Xie, J. R. McDonough, X. Cui, Y. Cui and Z. Bao, *Nano Lett.*, 2011, **11**, 2905.
10. L. Wu, R. Li, J. Guo, C. Zhou, W. Zhang, C. Wang, Y. Huang, Y. Li and J. Liu, *AIP Adv.*, 2013, **3**, 082129.
11. L. Peng, X. Peng, B. Liu, C. Wu, Y. Xie and G. Yu, *Nano Lett.*, 2013, **13**, 2151.
12. M. R. Lukatskaya, O. Mashtalir, C. E. Ren, Y. Dall'Agnese, P. Rozier, P. L. Taberna, M. Naguib, P. Simon, M. W. Barsoum and Y. Gogotsi, *Science*, 2013, **341**, 1502.
13. M. Ghidiu, M. R. Lukatskaya, M. Q. Zhao, Y. Gogotsi and M. W. Barsoum, *Nature*, 2014, **516**, 78.
14. X. Peng, L. L. Peng, C. Z. Wu and Y. Xie, *Chem. Soc. Rev.*, 2014, **43**, 3303.
15. W. Gao, N. Singh, L. Song, Z. Liu, A. L. M. Reddy, L. Ci, R. Vajtai, Q. Zhang, B. Wei and P. M. Ajayan, *Nat. Nanotechnol.*, 2011, **6**, 496.
16. M. Beidaghi and C. Wang, *Adv. Funct. Mater.*, 2012, **22**, 4501.
17. J. J. Yoo, K. Balakrishnan, J. Huang, V. Meunier, B. G. Sumpter, A. Srivastava, M. Conway, A. L. Mohana Reddy, J. Yu, R. Vajtai and P. M. Ajayan, *Nano Lett.*, 2011, **11**, 1423.
18. J. Feng, X. Sun, C. Z. Wu, L. L. Peng, C. W. Lin, S. L. Hu, J. L. Yang and Y. Xie, *J. Am. Chem. Soc.*, 2011, **133**, 17832.

19. X. H. Lu, M. H. Yu, G. M. Wang, Y. X. Tong and Y. Li, *Energy Environ. Sci.*, 2014, **7**, 2160.
20. Y. Zhu, S. Murali, M. D. Stoller, K. J. Ganesh, W. Cai, P. J. Ferreira, A. Pirkle, R. M. Wallace, K. A. Cychosz, M. Thommes, D. Su, E. A. Stach and R. S. Ruoff, *Science*, 2011, **332**, 1537.
21. M. F. El-Kady, V. Strong, S. Dubin and R. B. Kaner, *Science*, 2012, **335**, 1326.
22. T. Q. Lin, I. W. Chen, F. X. Liu, C. Y. Yang, H. Bi, F. F. Xu and F. Q. Huang, *Science*, 2015, **350**, 1508.
23. Z. L. Tao, L. N. Xu, X. L. Gou, J. Chen and H. T. Yuan, *Chem. Commun.*, 2004, 2080.
24. H. S. S. Ramakrishna Matte, A. Gomathi, A. K. Manna, D. J. Late, R. Datta, S. K. Pati and C. N. R. Rao, *Angew. Chem., Int. Ed.*, 2010, **122**, 4153.
25. X. Cao, Y. Shi, W. Shi, X. Rui, Q. Yan, J. Kong and H. Zhang, *Small*, 2013, **9**(20), 3433–3438.
26. M. Acerce, D. Voiry and M. Chhowalla, *Nat. Nanotechnol.*, 2015, **10**, 313.
27. L. Cao, F. Xu, Y. Y. Liang and H. L. Li, *Adv. Mater.*, 2004, **16**, 1853.
28. G. Yu, X. Xie, L. Pan, Z. Bao and Y. Cui, *Nano Energy*, 2013, **2**, 213.
29. J. Xie, X. Sun, N. Zhang, K. Xu, M. Zhou and Y. Xie, *Nano Energy*, 2013, **2**, 65.
30. X. Dong, L. Wang, D. Wang, C. Li and J. Jin, *Langmuir*, 2011, **28**, 293.
31. S. Gao, Y. F. Sun, F. C. Lei, L. Liang, J. W. Liu, W. T. Bi, B. C. Pan and Y. Xie, *Angew. Chem., Int. Ed.*, 2014, **126**, 13003.
32. C. Wu, X. Lu, L. Peng, K. Xu, X. Peng, J. Huang, G. Yu and Y. Xie, *Nat. Commun.*, 2013, **4**, 3431.
33. G. Wang, X. Lu, Y. Ling, T. Zhai, H. Wang, Y. Tong and Y. Li, *ACS Nano*, 2012, **6**, 10296.
34. Y. Q. Jiang, Q. Zhou and L. Lin, Micro Electro Mechanical Systems, 2009, *MEMS 2009, IEEE 22nd International Conference on*, 2009, 587.
35. J. Chmiola, C. Largeot, P.-L. Taberna, P. Simon and Y. Gogotsi, *Science*, 2010, **328**, 480.
36. L. Zhang, H. B. Wu and X. W. Lou, *Chem. Commun.*, 2012, **48**, 6912.
37. C. Lin, X. Zhu, J. Feng, C. Wu, S. Hu, J. Peng, Y. Guo, L. Peng, J. Zhao, J. Huang, J. Yang and Y. Xie, *J. Am. Chem. Soc.*, 2013, **135**, 5144.
38. A. Nag, K. Raidongia, K. P. S. S. Hembram, R. Datta, U. V. Waghmare and C. N. R. Rao, *ACS Nano*, 2010, **4**, 1539.
39. C. N. R. Rao and A. Nag, *Eur. J. Inorg. Chem.*, 2010, **27**, 4244.
40. X. Cao, B. Zheng, W. Shi, J. Yang, Z. Fan, Z. Luo, X. Rui, B. Chen, Q. Yan and H. Zhang, *Adv. Mater.*, 2015, **27**, 4695.
41. G. Wang, H. Wang, X. Lu, Y. Ling, M. Yu, T. Zhai, Y. Tong and Y. Li, *Adv. Mater.*, 2014, **26**, 2676.
42. X. Wang, B. Liu, R. Liu, Q. Wang, X. Hou, D. Chen, R. Wang and G. Shen, *Angew. Chem., Int. Ed.*, 2014, **53**, 1849.
43. C. Zhu, P. Yang, D. Chao, X. Wang, X. Zhang, S. Chen, B. K. Tay, H. Huang, H. Zhang, W. Mai and H. J. Fan, *Adv. Mater.*, 2015, **27**, 4566.
44. R. Lalani and L. Liu, *Biomacromolecules*, 2012, **13**, 1853.

45. C. Tiyapiboonchaiya, J. M. Pringle, J. Sun, N. Byrne, P. C. Howlett, D. R. MacFarlane and M. Forsyth, *Nat. Mater.*, 2004, **3**, 29.
46. V. A. Vasantha, S. Jana, A. Parthiban and J. G. Vancso, *Chem. Commun.*, 2014, **50**, 46.
47. T. Wang, X. Wang, Y. Long, G. Liu and G. Zhang, *Langmuir*, 2013, **29**, 6588.
48. X. Peng, H. Liu, Q. Yin, J. Wu, P. Chen, G. Zhang, G. Liu, C. Wu and Y. Xie, *Nat. Commun.*, 2016, **7**, 11782.
49. Y. J. Kang, S. J. Chun, S. S. Lee, B. Y. Kim, J. H. Kim, H. Chung, S. Y. Lee and W. Kim, *ACS Nano*, 2012, **6**, 6400.
50. Y. J. Kang, H. Chung, C. H. Han and W. Kim, *Nanotechnology*, 2012, **23**, 065401.

CHAPTER 11

Flexible Two-dimensional Nanomaterials for Lithium-ion Batteries Applications

QIUFAN WANG*[a], DAOHONG ZHANG[a] AND GUOZHEN SHEN*[b,c]

[a]Key Laboratory of Catalysis and Materials Science of the State Ethnic Affairs Commission & Ministry of Education, South-Central University for Nationalities, Wuhan, Hubei Province, 430074, China; [b]State Key Laboratory for Superlattices and Microstructures, Institution of Semiconductors, Chinese Academy of Sciences, Beijing 100083, China; [c]College of Materials Science and Opto-Electronics Technology, University of Chinese academy of Sciences, Beijing 100029, China
*E-mail: ygdf@scuec.edu.cn, gzshen@semi.ac.cn

11.1 Introduction

The energy crisis in the 21st century has been, to some extent, the most important current issue faced by humans. In this regard, how to convert and store renewable energy efficiently remains a significant and urgent problem for researchers to resolve. As one of the most important energy-storage devices, lithium-ion batteries (LIBs) have attracted a lot of attention in both academic research and applications, such as cell phones, laptops and digital cameras.[1–5] For LIBs, the charges stored mainly contributed by the intercalation

Smart Materials No. 27
Inorganic Two-dimensional Nanomaterials: Fundamental Understanding, Characterizations and Energy Applications
Edited by Changzheng Wu

Published by the Royal Society of Chemistry, www.rsc.org

of Li^+ on the surface and in the bulk of the active materials, delivering higher energy density (120–180 Wh kg^{-1}) than other energy storage devices.[6] Recent advances in flexible and bendable electronics such as roll-up displays, smart electronics, and wearable devices have promoted the development of flexible lithium-ion batteries due to their high power density.[7–15] Here, all the components (including cathode, anode, electrolyte, separator, and packing material) in the flexible battery need to be flexible and bendable. Hence, there exists a need to develop an up-scalable, low-cost, environmentally friendly process for the sustainable mass production of flexible LIBs with high product recyclability.

The choice of electrode materials for flexible LIBs is of great importance to all. To date, several nanostructured electrode materials such as graphene, transition metal oxide or hydroxides and conductive polymer materials have attracted a wide range of interest. Among them, 2D nanomaterials including nanosheets, nanoflakes and their composites, have become the best alternatives to achieve highly flexible LIBs, owing to their high mechanical integrity that originates from the large overlapping areas when they are stacked layer-by-layer to form a thin film structure, as well as the good electrochemical properties that come from a rich selection of atomic types in the 2D lattice plane, providing further possibilities for highly-active electrochemical sites. Moreover, 2D nanomaterials, whose thickness in the horizontal direction is just a single or several atomic layers, forming confined-dimensional materials, have brought synergic advantages of controllable electrical properties and high specific surface areas as well as superior electrochemical activities.[16–26]

In this chapter, we review the recent progress on typical 2D functional materials for flexible LIBs, including flexible 2D graphene paper, various graphene-based composite cathodes and anodes, 2D inorganic nanosheets. We emphasize the preparation methods for electrodes, their electrochemical performance, and the mechanical flexibility of the devices. Finally, we discuss the future challenges and perspectives of 2D nanomaterials for flexible LIBs. These recent advances in materials preparation and configuration design bring new insights into the future of 2D nanomaterials-based flexible LIBs.

11.2 Paper-like Graphene Nanostructures

The major obstacle to the realization of bendable LIBs is the lack of reliable materials that have superior electrical conductivity, high mechanical flexibility, and high electrochemical stability. Metal foils have been employed as current collectors in conventional LIBs, but their high cost, high weight, and low flexibility have led to the exploration of alternative materials. To replace conventional metal foil current collectors, graphene paper is a functional material, which does not only act as a conducting agent, but also as a current collector.[27] Owing to the large specific surface area (2630 m^2 g^{-1}), which is significantly higher than those of CNT (~1300 m^2 g^{-1}) and graphite (10–20 m^2 g^{-1}), excellent intrinsic mobility of carriers at room temperature

(200 000 $cm^2 V^{-1} s^{-1}$), thermal conductivity (~5000 W $m^{-1} K^{-1}$), high Young's modulus (~1.0 TPa), and excellent optical transmittance (~97.7%),[28–30] graphene itself is a suitable candidate for reversible lithium storage in LIBs, which can provide an ideal platform for the transportation of Li^+ and electrons, and more electro-active sites for energy storage.[31–34] The unique combination of its outstanding properties such as being mechanically strong and electrically conductive makes graphene paper a promising base material for flexible LIBs.

Recently, many effective solution processing methods have been presented to fabricate graphene paper, such as spin-coating,[35] Languir-Blodgett,[36] layer-by-layer deposition,[37] interfacial self-assembly[38] and vacuum filtration.[39,40]

Vacuum filtration is the most widely used approach for the development of graphene papers[41] and the flexible structure of graphene paper makes them potential anodes for free-standing batteries. Wallace and coworkers fabricated graphene paper by a simple vacuum filtration assembly method and investigated the LIB performance of graphene paper.[42] However, during the fabrication process, the individual graphene sheets begin to aggregate and restack owing to the interplanar π–π interactions and van der Waals forces between the graphene layers. This agglomeration reduces the surface area of the graphene films and the diffusion of electrolyte ions, which results in a decrease in the electrochemical performance. Therefore, a lot of strategies were then developed to prevent aggregation of graphene sheets to increase the surface area and promote transport of electrolyte ions, including adding spacers,[43–45] template-assisted growth,[46] and crumpling the graphene sheets.[47]

Recently, graphene oxide (GO) and graphene paper have been successfully fabricated by the flow-directed assembly of individual GO/graphene sheets.[48,49] Ning *et al.*[41] present a novel approach to fabricate flexible graphene papers using chemical vapor deposition (CVD) derived graphene with low surface density and good conductivity, exhibited reversible capacity (1350 mAh g^{-1} at 50 mA g^{-1}) and cycling performance as anodes for lithium rechargeable batteries. Ahn *et al.*[50] demonstrated a simple route to develop mechanically flexible graphene paper electrode. Figure 11.1(a) shows the schematic illustration of a flexible lithium-ion cell using a free-standing, additive free and thick graphene film, and the as-fabricated flexible lithium-ion battery are shown in Figure 11.1(b). The first charge and discharge profiles of the battery ion both the flat and bent state are demonstrated in Figure 11.1(c). The two curves are almost the same, reflecting good stability.

Liu *et al.*[51] successfully fabricated flexible graphene paper made from graphene aerogel pre-synthesized by freeze-drying a GO aqueous dispersion and subsequent thermal reduction. The paper-like product obtained is free-standing and highly flexible. As an anode in a LIB, the discharge/charge profiles at a current density of 100 mA g^{-1} for the graphene paper are presented. From the second cycle, it has a high coulombic efficiency over 98%. The excellent performance can be due to the folded structure of the graphene sheets with fewer layers that provide more lithium insertion active sites, such

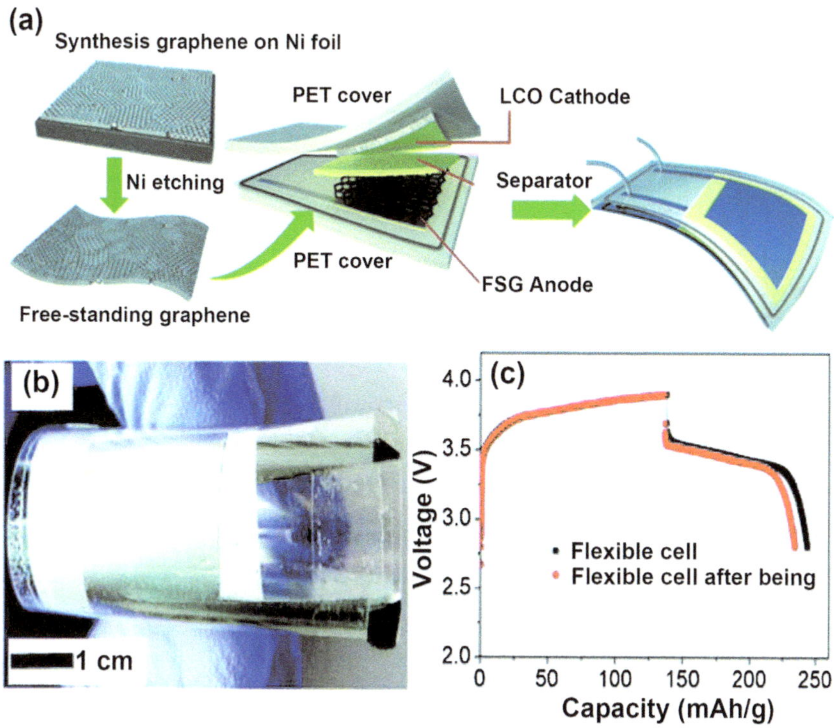

Figure 11.1 (a) Schematic process showing the fabrication of a flexible lithium ion battery using a flexible electrode with PET covering. (b) Camera picture of the flexible lithium ion battery. (c) The first charge and discharge of flexible cell fabricated on plastic substrate in flat and bending state. Reproduced from ref. 50 with permission from The Royal Society of Chemistry.

as edge-type sites and nanopores. The graphene paper can also deliver a good rate capacity. These results demonstrate that the as-fabricated graphene paper is a potential candidate for anode materials with high reversible capacity, good cycle performance, and high rate discharge/charge capability.

Mukherjee *et al.* synthesized free-standing graphene papers (GP) following the photoflash and laser-assisted reduction of GO.[52] Two procedures were adopted for fabricating photothermally reduced free-standing graphene paper, as shown schematically in Figure 11.2(a)–(d). Clearly, a free-standing graphene paper exhibits good bendable features (Figure 11.2(c)), making it a good candidate to be used in flexible batteries. In Figure 11.2(e), these flexible photoflash-reduced graphene electrodes exhibit a good long-term stability at C rates of 20, 40, 120, and 150 C for over 1000 charge/discharge cycles. The GP could deliver a storage capacity of 156 mAh g^{-1} over 1000 cycles at 40 C. Figure 11.2(f) shows the performance of the flash-reduced and laser-reduced graphene anodes in terms of its energy and power densities when compared with traditional carbon-based electrodes for lithium-ion batteries

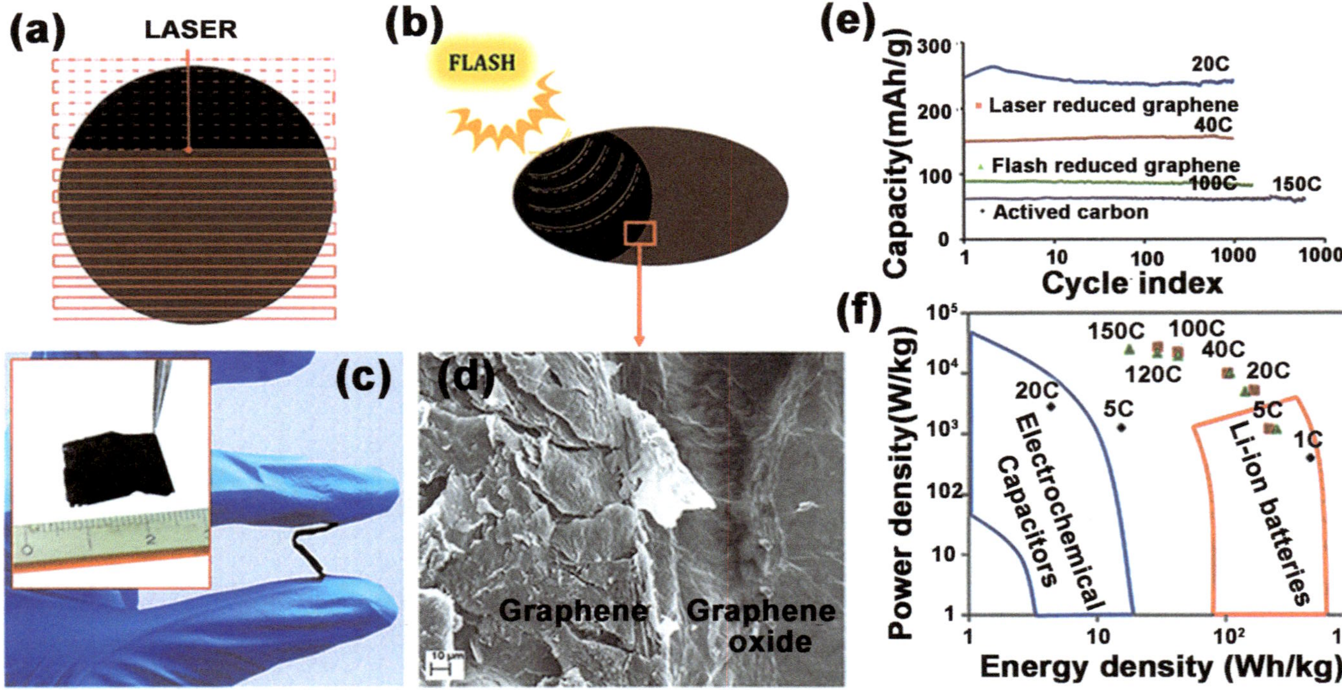

Figure 11.2 Schematic illustrations of the laser reduction system with the raster scan path (a), and (b) the flash reduction of graphene oxide whereby a flash from a digital camera reduces graphene oxide. (c) Photograph of the flash-reduced graphene. (d) Scanning electron microscopy (SEM) image of the two characteristic regions of flash-reduced graphene and graphene oxide. (e) and (f) Long-term stability of photothermally reduced graphene anodes and Ragone plot. Reproduced with permission from ref. 52. Copyright 2012 American Chemical Society.

and electrochemical capacitors. In addition, the photoflash reduced GP retained a Columbic efficiency of ~98.6% of the initial capacity for over 50 cycles. Meanwhile, the thermally reduced graphene showed inferior stability and retained a columbic efficiency of only ~81.8%. Recently, Dao *et al.*[53] fabricated super-tough GP *via* the vacuum filtration of functionalized graphene. In this process, graphene was functionalized with polycaprolactone (PCL) brushes by the polymerization of caprolactone from the functional groups of graphene. It delivered a discharge capacity of ~450 mAh g^{-1} at a current density of 300 mA g^{-1} and a coulombic efficiency of 95% at 60 mA g^{-1} for 50 cycles. In addition, Hu *et al.*[54] synthesized free-standing GP with different thicknesses (~1.5, 3, and 10 μm) using water-suspended reduces GNS, following a vacuum-assisted filtration method. Cyclic voltammetry studies suggest that the thickness of the GP significantly influences the electrochemical performance and the thinner GP exhibited better performance than the thicker GP. At a current density of 100 mA g^{-1}, 1.5 μm thick GP could deliver an initial reversible specific capacity of ~200 mA g^{-1} (first 10 cycles). The inferior performance of the thicker GP was attributed to the lower surface area due to the dense restacking of graphene layers. This stacking of graphene increased the diffusion length through the cross-plane rather than in the plane of the GP.

11.3 Flexible 2D Hybrid Film Paper

Developing flexible free-standing film is crucial for next-generation lightweight and wearable electronic devices, such as soft portable electronic products, roll-up displays, implantable biomedical devices, *etc.* In recent works, self-standing CNF/CNT hybrid paper electrodes or complete cell embedded electrochemical active materials, such as Si,[55] $Li_4Ti_5O_{12}$[56] or $LiFePO_4$,[57] were successfully elaborated using a conventional well-established paper-making route. Cao *et al.*[58] developed a free-standing carbonized cellulose nanofiber (CNF) hybrid film by using a pressured extrusion paper-making process and the subsequent low temperature carbonization. The FE-SEM images of the LCC-A hybrid film were shown in Figure 11.3(a) and (b). It implied that the hybrid film is suitable for the flexible self-standing paper anode with sufficiently electrical conductivity and excellent mechanical robustness. The galvanostatic charge/discharge curves of the LCC-A film at various current rates are shown in Figure 11.3(c). All the curves showed a long flat platform at a stable potential plateau, and the specific capacity decreased from about 160 to 140 mA g^{-1} when current rate increased from 0.5 to 10 C, remaining at 87.5% of the initial capacity of 0.5 C. Cycling performance of the LCC-A film was compared with LCC-U reflected in Figure 11.3(d). After 1000 cycles, the specific capacity of LCC-A remained at 143 mA g^{-1} with negligible decay. The LCC-A paper showed an extraordinary rate and cycling performance. Figure 11.3(e) showed the schematic of the flexible LIB and the prototype LIB encapsulated by a plastic shell. The LIB can light an LED device at different bending angles (Figure 11.3(f) and (h)), revealing significant potential applications in

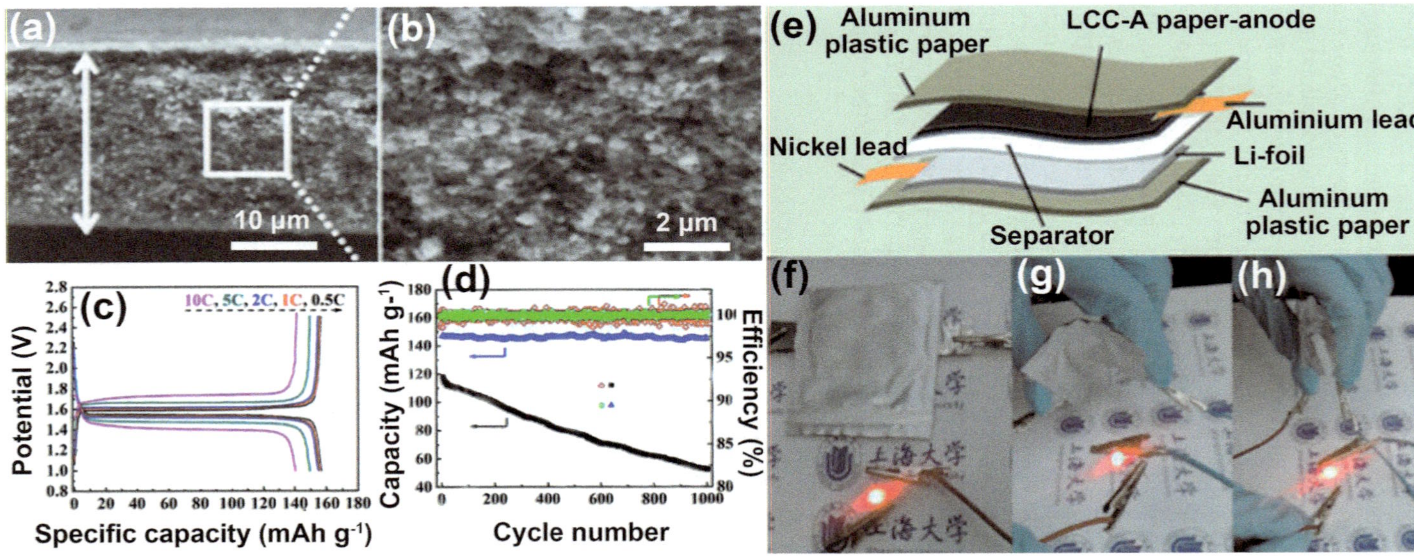

Figure 11.3 (a) and (b) FESEM images of the cross-section of LCC-A at different magnifications. (c) Galvanostatic charge/discharge curves of LCC-A paper anode. (d) Cycling performance of the LCC-U and LCC-A at current rate of 5 C (The left shows the specific capacity and the right shows the coulombic efficiency). (e) Schematic illustration of a flexible LIB assembled using LCC-A paper anode. (f)–(h) Digital pictures showing a lit red LED connected to a flexible LIB being bent. Reproduced with permission from ref. 58. Copyright 2016 American Chemical Society.

flexible electronics, such as bendable displays and other irregularly shaped energy storage devices.

11.4 Flexible 2D Graphene Composite Paper Anodes

Graphene-based hybrid materials are becoming important for electrochemical energy storage devices. Nanostructured composites of graphene with many nanomaterials have been successfully prepared by various methods. Recent advances in various types of graphene and graphene hybrids as electrode materials for LIBs are reviewed in detail. Xie *et al.*[59] synthesized a sandwich-like anode material of ultrafine N-rich porous carbon nanospheres *in-situ* grown on graphene sheets. Ascribed to an ideal combination of hierarchically porous structure, highly conductive graphene platform and high-level nitrogen doping in the sandwich-like electrode, an impressive electrochemical performance is obtained, such as an outstanding rate capability and cycling performance.

11.4.1 Flexible 2D Graphene/Metal Oxides Composite Paper Anodes

As electrode materials, nanostructured metal oxides have some advantages over bulk materials due to their high specific surface area. The high surface area can significantly improve the electrode–electrolyte interaction, and it can provide a shorter diffusion length for Li^+ and electrons. While, the large variation of volume for metal oxides during the cycling process causes pulverization of the electrode, which results in poor reversibility.[60] However, graphene generally suffers from serious agglomeration and restacking, causing poor electrochemical performance. The fabrication of a metal oxides/graphene hybrid can not only increase the space between graphene layers, avoiding the restacking of graphene, but also avoid the volume changes of metal oxides during the cycling process.

Iron oxides, particularly Fe_2O_3 and Fe_3O_4, are one of the most important and promising anode materials due to their high theoretical capacity (Fe_3O_4 of 924 mA g^{-1} and Fe_2O_3 of 1005 mA g^{-1}). Zhou *et al.*[61] synthesized a well-organized flexible interleaved composite of graphene nanosheets decorated with Fe_3O_4 nanoparticles through *in-situ* reduction of iron hydroxide. The composite electrode shows a reversible high specific capacity of 1026 mA g^{-1} as well as excellent rate capability.

Wang *et al.*[62] used a vacuum filtration and thermal reduction process for the preparation of flexible free-standing hollow Fe_3O_4/graphene films, as illustrated in Figure 11.4(a). From the cross-sectional view (Figure 11.4(b) and (c)), the thickness of the film is about 12.3 μm. The inset in Figure 11.4(b) shows that the film has good flexibility. The charge/discharge capacities of the samples at 200 mA g^{-1} over 50 cycles are plotted in Figure 11.4(d). The Fe_3O_4/graphene hybrid film with a graphene content of 39.6% has the

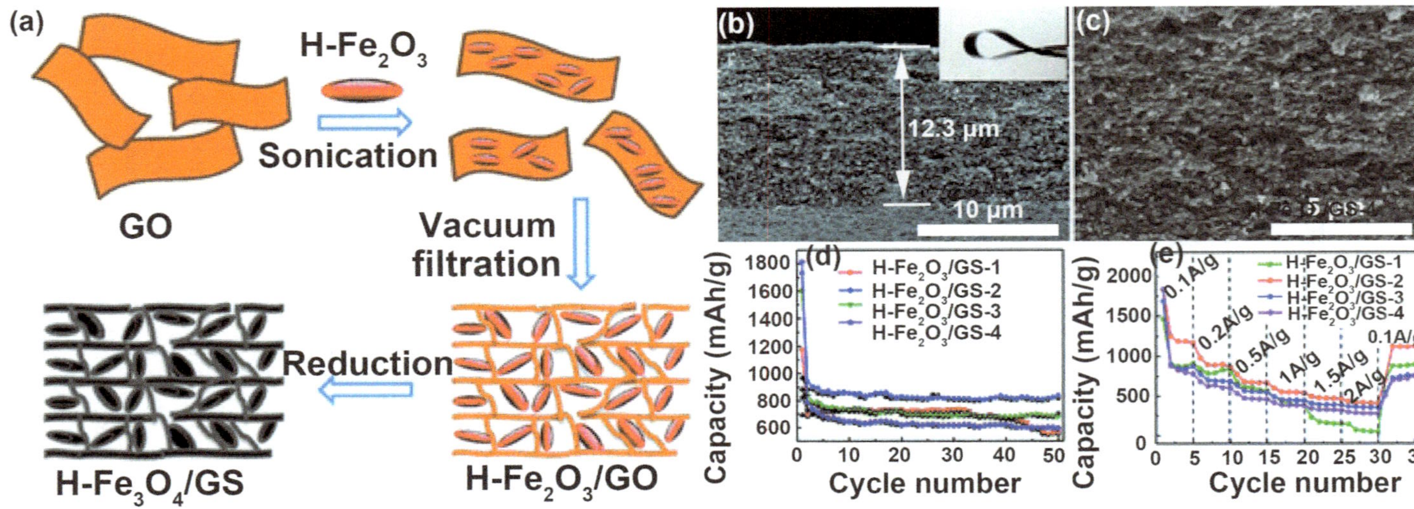

Figure 11.4 (a) Schematic illustration of the synthesis of H–Fe$_3$O$_4$/graphene hybrid film. (b) and (c) Cross-section SEM images of the H–Fe$_3$O$_4$/graphene film; the inset in (b) is a digital photograph of film showing good flexibility. (d) and (e) Cycling performance of H–Fe$_3$O$_4$/graphene samples at a current density of 200 mA g^{-1}. (e) Rate capability of the samples up to 2 A g^{-1}. Reproduced from ref. 62 with permission from The Royal Society of Chemistry.

highest specific capacity, the sample also has a good rate capability (Figure 11.4(e)), due to the graphene network being able to act as a physical barrier to prevent the agglomeration of H–Fe_3O_4 and buffer its volume change during cycling. Cao *et al.*[63] used a two-step annealing method for the preparation of Fe_2O_3 coated on graphene and used it in a lithium-ion battery.

Gong *et al.*[64] have designed porous architectures constructed by a graphene sheet coated with FeO_x for high-performance battery materials. They obtained a long cycle life (1500 cycles for FeO_x with almost 100% capacity retention) at different charge and discharge rates. A bendable graphene@iron oxide hybrid film (GFeF) electrode was also fabricated through a filtration-assisted self-assembly method.[65] The CV curves of the GFeF electrode at a scan rate of 2 mV s^{-1} were taken (Figure 11.5(a)). In the first cycle, a sharp peak at around 0.74 V in the cathodic sweep and a smooth peak at around 1.8 V in the anodic sweep are consistent with previous reports. Additionally, the CV curve of the GFeF electrode in the 50th cycle almost keeps the same profile as that of the 5th cycle, indicating good reversibility of the electrode. The long-term cycling performance was also carried out (Figure 11.5(b)). There was 27% increase during the first 100 cycles, and after 500 cycles, the capacitance retained 93% of its initial value, suggesting its superior cycling stability. The excellent results can be reflected in Figure 11.5(c). The layered graphene sheets could significantly buffer the stress and strain induced by the volume contraction/expansion of metal oxide during long-term cycling. Also, many voids (Figure 11.5(d)) provide a high contact area between the electrode and electrolyte, and each iron oxide nanoparticles are fixed in position by graphene layers. The SEM images of the internal structures of the GFeF electrode before and after charge/discharge cycling were also compared (Figure 11.5(e) and (f)), the results suggest the structure is quite effective in accommodating their volume expansion during the harsh cycling process.

Cobalt oxides are well known as potential alternative anode materials because of their promising high theoretical capacities (Co_3O_4 ~890 mA g^{-1} and CoO ~715 mA g^{-1}). The composition of cobalt oxides with graphene can largely improve the electrochemical performance compared to the pure cobalt oxides that may have large volume variation during the cycling process. Several studies on Co_3O_4/graphene[66–71] and CoO/graphene[72,73] hybrid electrode materials for LIBs have been reported.[74] Huang *et al.* developed a simple layer-by-layer stacking method for the fabrication of larger-area homogeneous CoO/graphene paper. The binder-free electrode showed excellent stability from 0 to 55 °C.

Tin oxide (SnO_2) exhibits a high theoretical Li^+ storage capacity of 728 mA g^{-1}, more than double that of graphite (372 mA g^{-1}). Therefore, it could be a better replacement for the carbon in LIBs. However, during the cycling process, a large volume change of ~300% causes cracking and degeneration of the electrode, which leads to inferior electrochemical performance. Enhanced lithium storage capacities have been achieved in SnO_2/graphene hybrid electrodes compared to bare SnO_2.[75–85] Wang *et al.*[86] developed a ternary self-assembly approach to prepare SnO_2-graphene nanocomposites

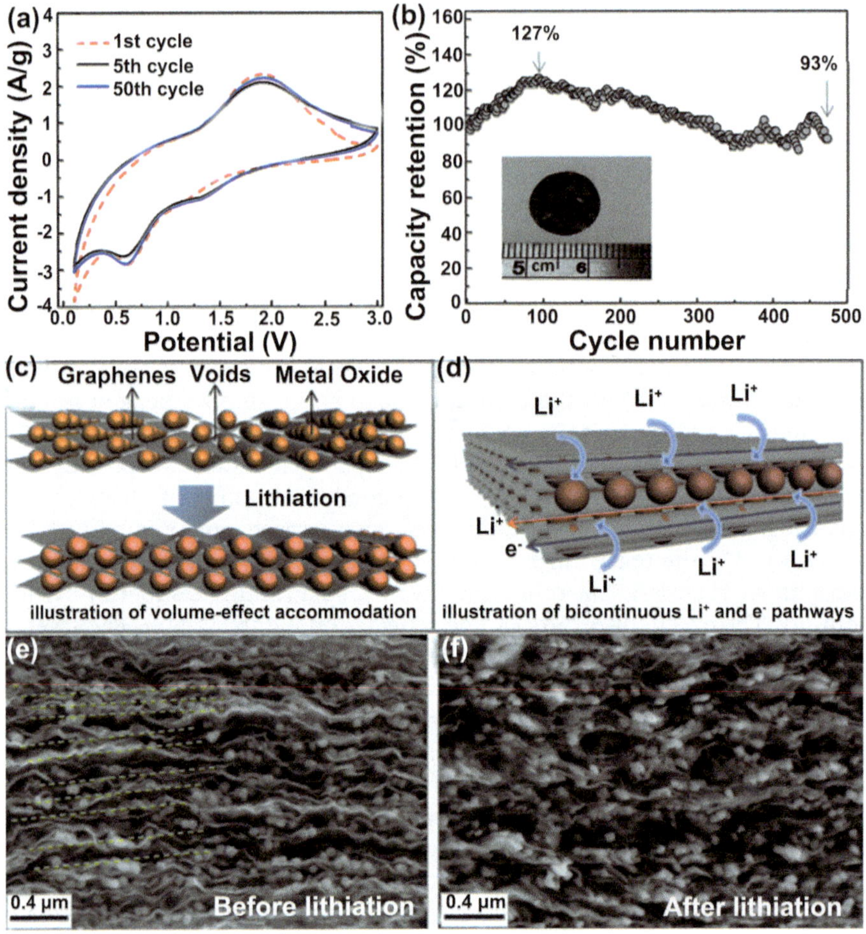

Figure 11.5 (a) CV profiles in the first and reversible steady-state cycles measured at a scan rate of 2 mV s^{-1}. (b) Cyclability of the GFeF-based half-cell. Inset: Photograph of the GFeF freestanding electrode. (c) Illustration of the hybrid structure with volume-effect accommodation during Li+ extraction/insertion. (d) Illustration of the expressway-like structure with bicontinuous ion/electron-transport pathways and large electrode–electrolyte contact area. Morphology comparison of cross-sectional SEM images for GFeF before lithiation (e) and after 500 cycles of charge/discharge (f). Reproduced with the permission from ref. 65. Copyright 2016 American Chemical Society.

with well-controlled nanostructures using both molecular and extended nanoscale building blocks. Figure 11.6(a) shows the SEM image of the robust and flexible SnO_2-graphene nanocomposite film. The SEM image of a polished cross-section (Figure 11.6(b)) reveals typical parallel, wavy-layer architectures. A cross-sectional TEM image of the SnO_2-graphene nanocomposite film (Figure 11.6(c)) shows alternating layers of nanocrystalline SnO_2 and

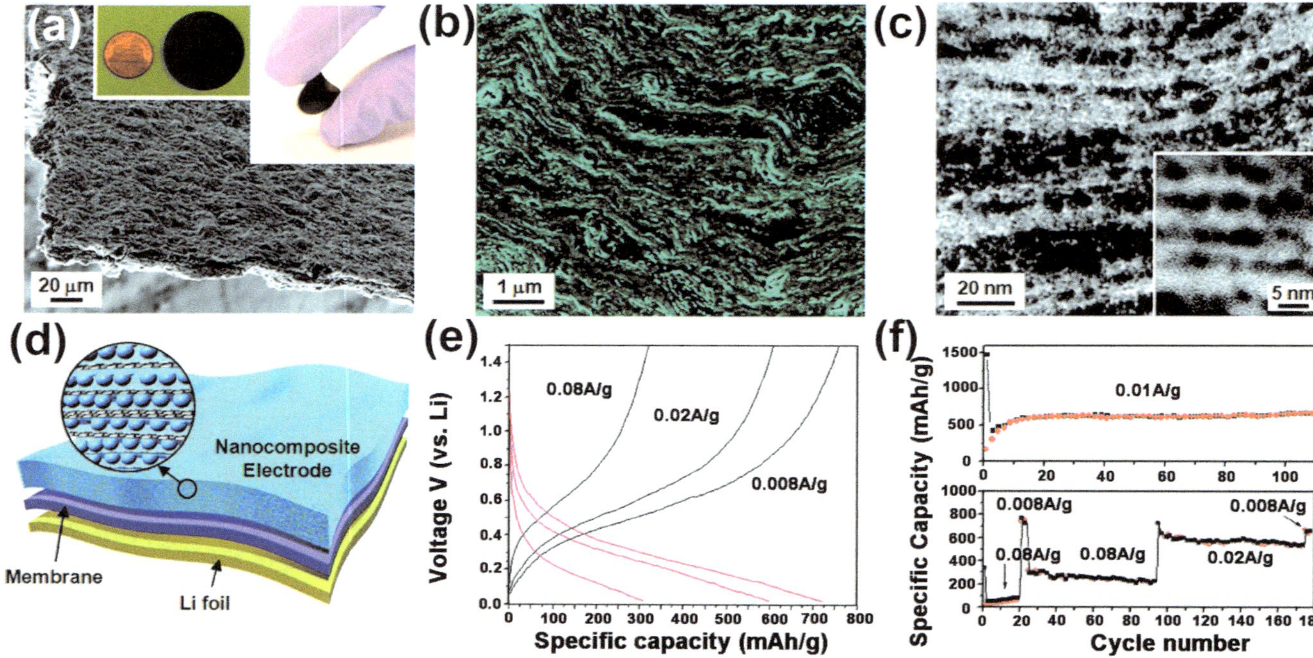

Figure 11.6 (a) Side-view SEM image of a self-assembled free-standing SnO_2-graphene nanocomposite. Photographs in the inset show a disk-like 3 cm diameter SnO_2-graphene nanocomposite electrode on the left and a folded electrode on the right. (b) High-magnification cross-sectional SEM image of the free-standing SnO_2-graphene nanocomposite electrode. The electrode is composed of well-packed wavy layers interspaced by the loosely packed layers through almost the entire cross-section. (c) Cross-sectional TEM images of a SnO_2-graphene nanocomposite film. Inset shows a high-resolution TEM image in the nanocomposite film with alternating layers of nanocrystalline SnO_2 and graphene materials. (d) A Li-ion battery configuration directly using a free-standing metal oxide-graphene nanocomposite film as an electrode. (e) Charge/discharge profiles of a SnO_2-graphene nanocomposite electrode at a current density of 0.008, 0.02, and 0.08 A g^{-1}, respectively. (f) (Top) Specific capacity of SnO_2 as a function of charge/discharge cycles in the SnO_2-graphene nanocomposite electrode at a current density of 0.01 A g^{-1}. (Bottom) Specific capacity of SnO_2 as a function of charge/discharge cycles in the SnO_2-graphene nanocomposite at different charge/discharge current densities of 0.008, 0.08, and 0.02 A g^{-1}, respectively. Note that the nanocomposite electrode is pre-cycled for 20 cycles at 0.08 A g^{-1} to improve the electrolyte wetting and then cycled at 0.008 A g^{-1} to obtain theoretical capacity. The initial lower capacity for the first 20 cycles is attributed to insufficient wetting. Reproduced with the permission from ref. 86. Copyright 2010 American Chemical Society.

graphene materials spanning the entire film. The charge/discharge curves of the SnO_2-graphene nanocomposite film were investigated for use as an anode between 0.02 and 1.5 V (Figure 11.6(d)). The specific capacity of the SnO_2 in the nanocomposite electrode as a function of cycle number is shown in Figure 11.6(e). The discharge capacity drop in the first cycle is attributed to the irreversible conversion of SnO_2 to Sn and Li_2O upon lithiation.[87] Specific capacities of 225 and 550 mA g^{-1} are obtained at current densities of 0.08 and 0.02 A g^{-1} (Figure 11.6(f), bottom), respectively. The results reveal the graphene can improve the electrochemical properties of the nanocomposite by improving the conductivity of the electrode materials and by stabilizing the electrode structure with a good electric contact between SnO_2, and conductive contact between SnO_2 and conductive graphene during the charge/discharge process.

To utilize the unique respective advantages of SnO_2 and graphene, the SnO_2-graphene nanoparticles are embedded into layers of 2D graphene sheets.[88] From Figure 11.7(a)–(e), it can be seen that the fracture edges of the paper exhibits a layered structure through the entire cross-section. The SnO_2 nanoparticles are well embedded into the graphene sheets and the pockets of void spaces are clearly visible, which can work as a buffer to prevent the volume expansion and contraction of SnO_2 nanoparticles during the Li^+ insertion/extraction process. Figure 11.7(f) shows the photograph of the as-prepared hybrid paper, which is free-standing, being very flexible with high strength. Although graphene/SnO_2 composite electrodes exhibits a low capacity during the first few cycles due to the poor electrolyte wetting of the dense graphene/SnO_2 paper, their corresponding discharge capacity still reaches 428 mA g^{-1} after 50 cycles (about 83.4% retention those at 25th cycle), as illustrated in Figure 11.7(g). In contrast, bare SnO_2 electrodes delivered poor cycling stability. Figure 11.7(h) shows that the graphene/SnO_2 electrode in the high-medium frequency region exhibits a much smaller charge transfer resistance than that of the SnO_2 electrode (as revealed by a much smaller diameter of the semicircle in the impedance spectra of the graphene/SnO_2 electrode).

TiO_2 is an environmentally friendly transition metal oxide with a stable structure, which makes it an apt anode material for LIBs, despite a low theoretical capacity of 336 mA g^{-1}.[89] However, the poor rate capability resulting from the inherent poor electrical conductivity (~10^{-12} S m^{-1}) and high electrical resistance between the electrode and current collector as well as low ionic conductivity (~1.2×1020 S^{-1} m^{-2}) limit its practical application. To improve the electrochemical properties, graphene has been used in its hybrid form, which reduces the irreversible heat and internal resistance during the charge/discharge process, thus improving the performance of these hybrid electrode materials.[90–95] A unique hybrid of graphene and TiO_2 as a binder-free paper was fabricated by Hu *et al.*[96] In this structure, TiO_2 forms pillars between the graphene layers and significantly enhances the insertion and extraction rate of Li^+. Thus, when compared to bare TiO_2 and graphene, significantly improved electrochemical performance was

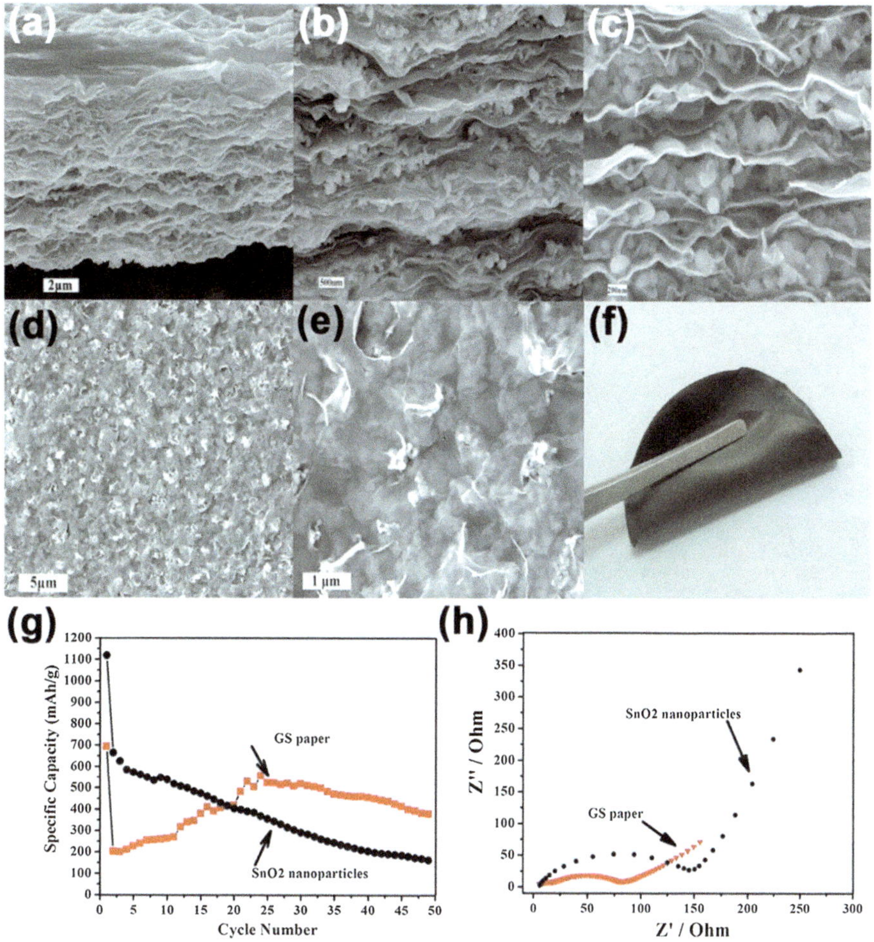

Figure 11.7 FESEM images of cross-section (a)–(c), top view (d) and (e) of the graphene/SnO_2 composite paper. (f) The optical image of as-prepared composite paper. (g) The cycling performance of graphene/SnO_2, and bare SnO_2 electrode. (h) Nyquist plots of graphene/SnO_2, and bare SnO_2 electrode. Reproduced with the permission from ref. 88. Copyright 2012 American Chemical Society.

observed with this hybrid electrode. Li *et al.*[97] fabricated anisotropic and isotropic TiO_2/graphene sandwich paper *via* a vacuum filtration route assisted by hydrothermal reaction. A controlled chemical reaction under a hydrothermal condition led to the formation of TiO_2 nanosheets parallel to the surface of the graphene paper. Figure 11.8 shows the SEM and TEM images of the TiO_2/graphene hybrid material. It was observed that the anisotropic TiO_2/graphene hybrid paper electrode exhibited the capacity of 112 mA g^{-1} at a high rate of 100 C (corresponding to 36 s of charge/discharge), which was found to be three times higher than that of the isotropic TiO_2/graphene

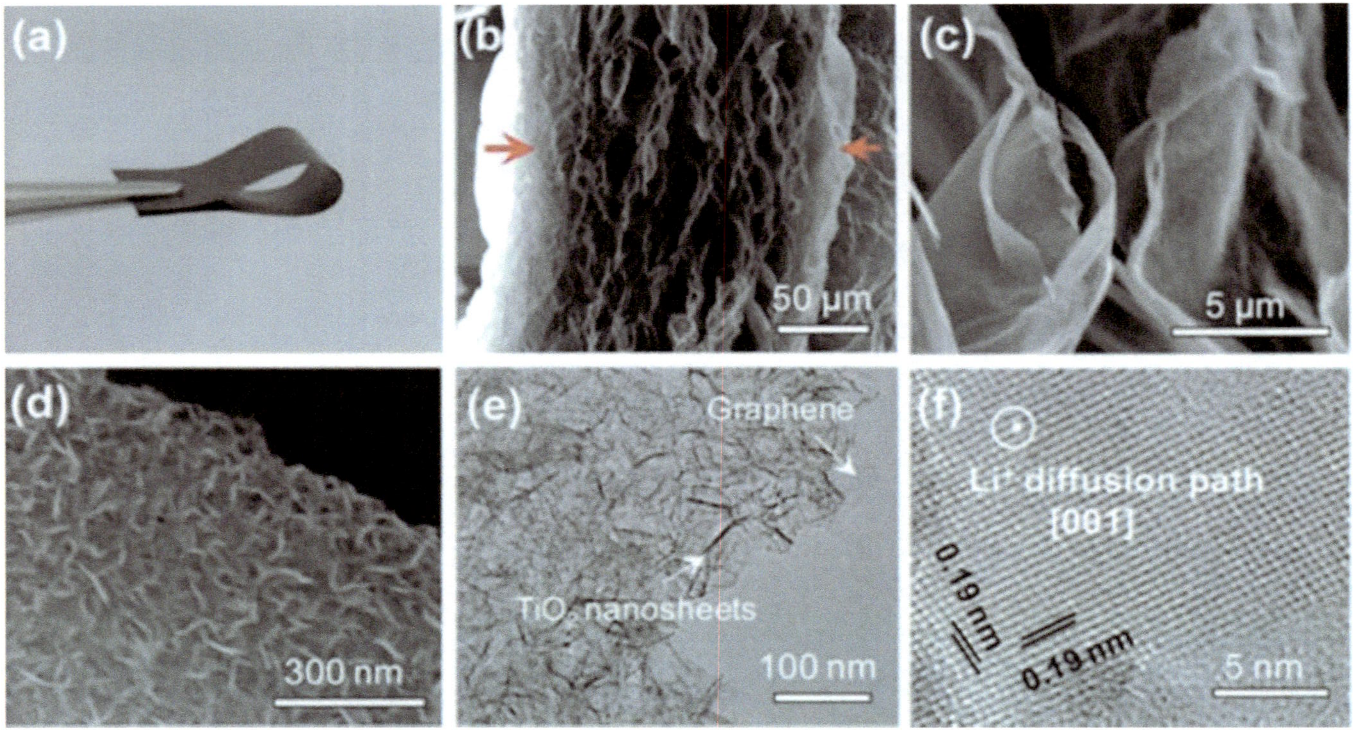

Figure 11.8 Characterization of A-TO/GSP. (a) Photograph of a bent flexible A-TO/GSP. (b) and (c) Cross-sectional SEM images of the A-TO/GSP. (d) High-magnification SEM images of LTO nanosheets in A-TO/GSP. (e) TEM image of A-TO/GSP. (f) HRTEM image of a TiO_2 nanosheet showing lattice fringes with a spacing of 0.19 nm. The lithium diffusion channels are clearly shown to be along [001] (perpendicular to the figure facet). Reproduced from ref. 97 with permission from The Royal Society of Chemistry.

paper electrode. Moreover, the anisotropic TiO_2/graphene paper electrode exhibited stable cyclic performance when tested at 10 C up to 100 cycles, much better than that of the isotropic TiO_2/graphene paper electrode. As shown in Figure 11.9, the anisotropic TiO_2/graphene paper also demonstrated rate capability superior to the isotropic TiO_2/graphene paper electrode. This improved performance was ascribed to the easy intercalation of Li^+ into the reactive surface of TiO_2 of the anisotropic TiO_2/graphene hybrid paper electrode during the electrochemical process. The work opens a door to designing anisotropic electrodes for high-power LIBs. In addition, TiO_2/graphene hybrid paper was fabricated by dipping followed by hydrothermal reaction.[98]

Park *et al.* prepared a highly flexible and free-standing, porous Mn_3O_4/reduced graphene oxide paper by a two-step process: vacuum filtration and thermal treatment.[99] The Mn_3O_4/reduced graphene oxide paper exhibits a first high discharge capacity of 943 mA g^{-1}, which quickly stabilizes and remains at 573 mA g^{-1} even after 100 cycles at 100 mA g^{-1}, which is much higher than the discharge capacity of the corresponding pristine graphene paper.

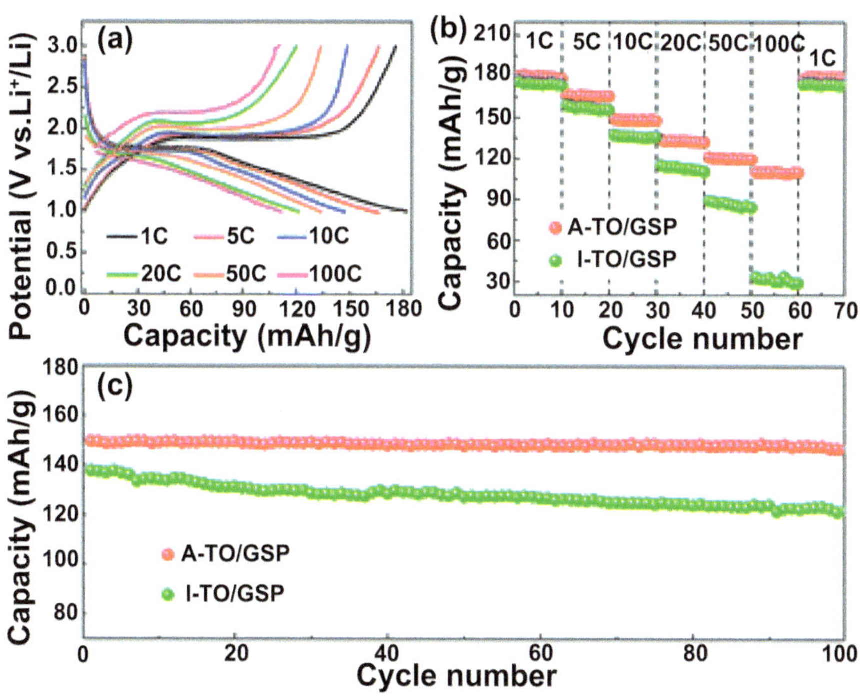

Figure 11.9 (a) Galvanostatic discharge–charge voltage profiles of A-TO/GSP cycled at different rates. Comparison of the (b) specific capacity at different rates and (c) cycling performance at 10 C between the A-TO/GSP and I-TO/GSP electrodes. Reproduced from ref. 97 with permission from The Royal Society of Chemistry.

11.4.2 Flexible 2D Graphene/Metal Sulfide Composite Anodes

The intensive research on graphene has spurred great interest in 2D transition-metal dichalcogenides MX_2 (M = Ti, Nb, Mo, Ta; X = S, Se, Te) because they are promising alternatives for LIB anode materials. Among different types of MX_2, MoS_2 is the most stable and versatile member, exhibiting the highest Li^+ ion capacity. So far, various multilayered MoS_2 nanostructures, such as nanosheet,[100] and nanoflakes[101] have been synthesized and applied as cathodes and anodes for LIBs.[102–104] However, these MoS_2-based anodes suffer from rapid capacity fading due to structure destruction induced by their large volume change in cycling. A possible way to solve this problem is to fabricate hybrid nanostructures.[105–107] Liu *et al.*[108] have developed a facile method for fabricating a freestanding MoS_2-graphene paper, which exhibits high specific capacity of 1240 mA g^{-1}, rate capability and cycle stability, owing to the high capacity of MoS_2, and the good cyclability of graphene.

Rana *et al.*[109] described a facile two-step synthesis process for the preparation of a thick freestanding graphene film (FSG) coated with a model anode material (MoS_2) using the CVD method (Figure 11.10(a)). The fabricated FSG film was obtained by chemical etching of the Ni. Then the FSG was used as a base material on which MoS_2 was grown *via* the hydrothermal method. The as-synthesized hybrid FSG/MoS_2 structure was a freestanding film that could be directly used as a battery anode without any post-processing or the need for additives. The SEM image of FSG film surface after removing the Ni is shown in Figure 11.10(b). An image of the thick FSG film after nickel etching is displayed in the inset of Figure 11.10(b), which indicates that the films are flexible and bendable. The MoS_2 morphology with flower-type nanostructures is shown in the higher magnification images of Figure 11.10(c), where many vertically oriented MoS_2 nanowalls are clearly evident. The wall thickness was found to be in range of 30–40 nm with a gap of a few nanometers. The FSG/MoS_2-based hybrid structure, when used as anode materials, provides a large surface area for good electrode–electrolyte interaction and facilitates smooth charge transfer from the MoS_2 to the FSG. Figure 11.10(d) shows the rate capacity of the FSG/MoS_2 hybrid electrode at two different current rates. A capacity of 498 mA g^{-1} was obtained after the 20th cycle at a current density of 50 mA g^{-1}, whereas a capacity of 385 mA g^{-1} was obtained after the 30th cycle at 150 mA g^{-1}. Good capacity retention was observed as the current density was increased by a factor of three, indicating good rate performance. The columbic efficiency can remain at about 97% for 30 cycles (Figure 11.10(e)). The freestanding and additive-free (FSG/MoS_2) hybrid structure may be useful as an electrode material in flexible energy storage devices.

Gao *et al.*[110] fabricated activated cotton textile embedded with NiS_2 and wrapped with graphene sheets (ACT/NiS_2-graphene) by a simple two-step heat treatment method, as illustrated in Figure 11.11(a). Typically, a piece of cotton textile was first dipped into a $Ni(NO_3)_2$ solution and then dried in the oven to obtain $Ni(NO_3)_2$ loaded cotton textile (Step 1). The $Ni(NO_3)_2$ loaded

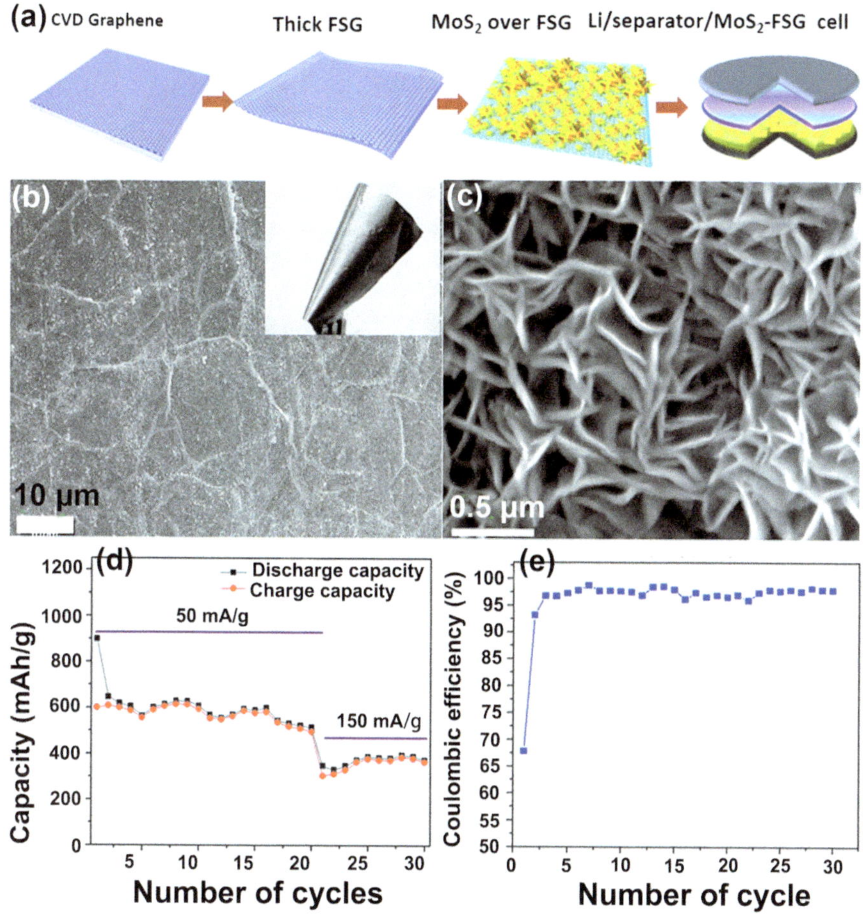

Figure 11.10 (a) Schematic illustration of the synthesis procedure for thick FSG/MoS_2 films. Step 1: thick FSG grown on a Ni substrate by the CVD method. Step 2: FSG film after etching of the Ni. Step 3: MoS_2 growth over FSG by the hydrothermal method at 200 °C. Step 4: fabrication of a coin cell with an FSG/MoS_2 electrode. A cross-sectional view of the coin cell used for electrochemical testing is shown. (b) FE-SEM image of the FSG surface; a photograph of a flexible FSG film is display in the inset. (c) Microstructure of the as-synthesized MoS_2 on FSG; uniform growth and a vertical petal-like morphology are evident. (d) Discharge profile capacity of the cell with the number of cycles for two different current rates. (e) Coulombic efficiency. Reproduced with the permission from ref. 109. Copyright 2014 American Chemical Society.

cotton textile was transferred into a tube furnace and heated at 1000 °C with argon protection for 1 h (Step 2). The as obtained ACT/Ni-graphene composite was mechanically flexible (Figure 11.11(b)), with diameters ranging from 5 to 10 μm (Figure 11.11(c)). Figure 11.11(d) shows the reconstructed 3D images of ACT/Ni-graphene, revealing ACT's tubular fibers. The structure of

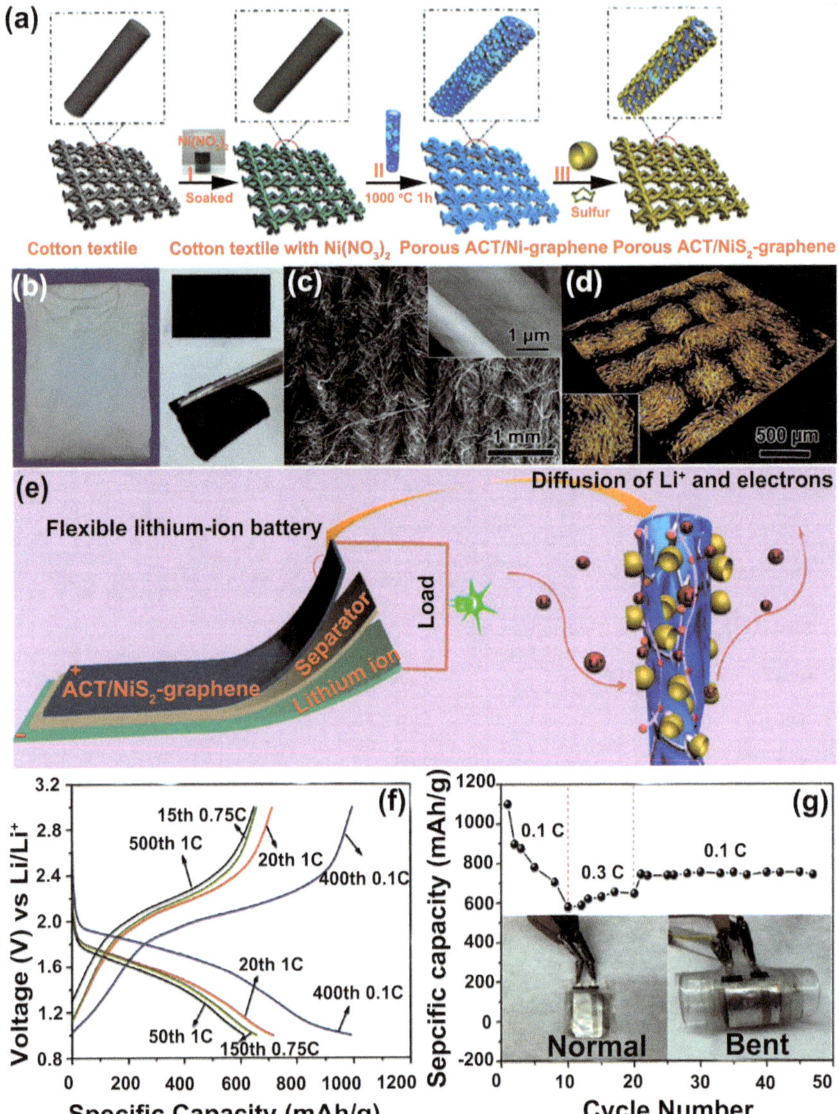

Figure 11.11 Synthesis and characterization of ACT/Ni-graphene composite. (a) Schematic illustration of the fabrication process of porous ACT/NiS_2-graphene composite. (b) Optical photograph of a commercial cotton t-shirt, a piece of ACT/Ni-graphene under normal and folded state, showing its high flexibility. (c) SEM image of cotton textile, inset is the amplified SEM image of a cotton fiber. (d) XCT images of ACT/Ni-graphene composite after removing the Ni nanoparticles by using 2 M HNO_3, inset is the amplified region. (e) Schematical illustration of the assembled flexible lithium-ion battery. Electrochemical performance of the flexible cell assembled with flexible ACT/NiS_2-graphene electrode. (f) 20th, 50th, 150th and 400th charge/discharge voltage-specific capacity curves of ACT/NiS_2-graphene at different charge/discharge rates. (g) Cyclic performance of the flexible cell in normal and bent states. Reproduced with the permission from ref. 110. Copyright 2015 American Chemical Society.

flexible lithium-ion battery was schematically illustrated in Figure 11.11(e). The charge/discharge profiles were shown in Figure 11.11(f). The reversible capacities at the 20th (1 C), 50th (50 C), 150th (0.75 C), and 400th (0.1 C) cycles are about ~710, ~608, ~680, and ~1030 mA g^{-1}, respectively. It can be seen the capacity remained at ~608 mA g^{-1} even when the charge/discharge rate increased 100 times from 0.01 to 1 C. The cyclic performance of the flexible cell at normal and bent states is shown in Figure 11(f). The capacity was maintained at ~740 mA g^{-1} at 0.1 C even after 50 cycles, which is very promising for practical application. The superior performance of the flexible electrode can be ascribed to the following merits: (1) the porous ACT fiber effectively buffering the stress induced by the volume change and maintaining the structure stability during the electrochemical reaction; (2) the tubular structure of ACT shortening the transportation/diffusion path for lithium ions and enabling faster kinetics and high-rate capability; (3) the NiS_2 nanobowls having interior and exterior arc surfaces.

11.4.3 Graphene/Si Composite Anodes

Beside the above binary inorganic materials, silicon has been touted as one of the most promising anode materials for next generation lithium ion batteries due to its abundance in nature, appropriately low working potential, and highest known theoretical specific capacity of ~4200 mA g^{-1}. However, silicon experiences a dramatic volume change during the lithiation and delithiation processes,[111] leading to severe pulverization and subsequent electrical disconnection from the current collector. The combination of Si with graphene is employed in robust and elastic substances to encapsulate silicon, as well as showing potential for advancing bendable free-standing anodes.

Wang *et al.*[112] proposed the fabrication of flexible and paper-like graphene–silicon composite by a one-step *in-situ* filtration method. The flexible graphene-Si composite electrodes showed much higher discharge capacity over 100 cycles (708 mA g^{-1}) than that of the graphene electrodes (304 mA g^{-1}). Here, 2D graphene can serve as a flexible mechanical and efficient electrically conducting support. Jiang *et al.*[113] reported a free-standing Si/graphene paper anode using Si nanoparticles synthesized by acid-etching Al–Si alloy power. The Si nanoparticles were embedded into graphene sheets to form a 'sandwich-like' structure with an inner space of Si nanoparticles and graphene layers, potentially providing enough ion transfer path and buffering space. The digital photo of Si/graphene paper is shown in Figure 11.12(c), clearly showing a free-standing paper with diameter of ~1 cm, which can be readily bend back and forth several times without breaking. As the current densities increase from 200 to 500, 1000, 2000 and 3000 mA g^{-1}, the Si/graphene paper exhibits reversible and stable capacities of 840, 630, 510, 400, 330 mA g^{-1} and goes up to 800 mA g^{-1} when the current density recovers to 200 mA g^{-1}. The Si/graphene paper exhibits high reversibility and electrochemical stability. The excellent electrochemical performance of Si/graphene paper can be

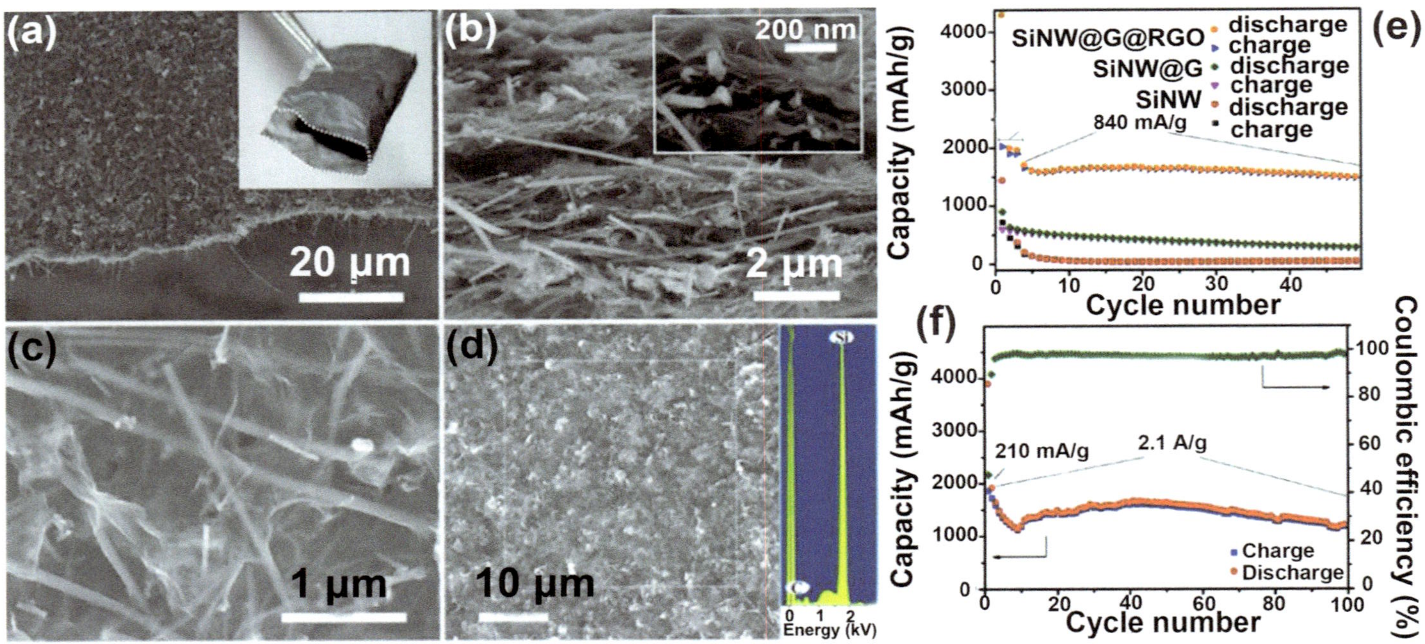

Figure 11.12 Characterization of SiNW@G@RGO: (a) low magnification SEM image. The inset shows an optical image of a bent SiNW@G@RGO film with a dashed line highlighting in edges. (b) Cross-section and (c) top-surface SEM images. The inset of panel b shows an enlarged cross-section view. (d) The inset of d shows the corresponding region for collecting the EDX spectrum. (e) Comparison of capacity retention of different electrodes. (f) Capacity and columbic efficiency of the SiNW@G@RGO cycled at the designated rate for 100 cycles. Reproduced with the permission from ref. 114. Copyright 2013 American Chemical Society.

due to the graphene sheets significantly impede the volume change during the alloying/de-alloying and increase the transportation of electron and Li-ion. Besides, the inner space in the Si nanoparticle clusters plays great role for the volume change accommodation. The as-prepared Si/graphene paper is a promising high stability anode for Li-ion batteries, especially for flexible electronics application.

Wang *et al.*[114] developed a novel kind of adaptable self-supporting binder-free silicon-based anode. Silicon nanowires (SiNW), as a representative mode, are sheathed with overlapped graphene (G) sheets to form core/sheath nanocables (SiNW@G), and then form RGO-sandwiched SiNW@G nanocables (SiNW@G@RGO). The RGO serve as a robust and flexible matrix, accommodating the volume change of embedded SiNW@G nanocables, thus maintaining the structural and electrical integrity of the electrode. The morphologies of as-fabricated SiNW@G@RGO can be found in Figure 11.12(a)–(d). The cyclic performance is shown in Figure 11.12(e). The self-supporting SiNW@G@RGO electrode delivered a significantly high specific capacity of about 1650 mA g^{-1} at 840 mA g^{-1} after 50 cycles, while the SiNW@G and SiNW electrodes remained at 270 mA g^{-1} and 50 mA g^{-1}, respectively. The SiNW@G@RGO electrode also exhibited a high specific capacity and superior capacity retention achieved at a large rate (2.1 A g^{-1}) (Figure 11.12(f)). Thus, the flexible and conductive RGO overcoats accommodate the volume change of embedded SiNW@G nanocables, maintaining the structural and electrical integrity of the electrode.

11.5 Flexible 2D Graphene Composite Cathodes

One of the challenges for improving the performance of LIBs to meet increasingly demanding requirements for energy storage is the development of suitable cathode materials. Graphene is an advantageous carbonaceous material. The exceptional properties of graphene can provide a high specific surface area, higher electrical conductivity, and chemical and mechanical stability to cathode materials. In general, lithium-manganese oxide and vanadium oxides have been widely used with graphene to fabricate hybrid materials for cathodes in LIBs.

Kang *et al.*[115] have constituted a graphene nanosheet based conducting network in the $LiFePO_4$ cathode system of a LIB (~2 μm thickness and ~8000 S m^{-1} conductivity). The ultrathin graphene sheets can 'bridge' active $LiFePO_4$ particles by a 'plane-to-point' mode and moreover, the flexibility of graphene sheets can intimately contact with $LiFePO_4$ particles even in the charge/discharge process. The flexible batteries give superior electrochemical performance (1st discharge capacity = 806.62 mA g^{-1} at a voltage range of 0.02 V and 3.5 V) over the conventional, non-flexible electrode architecture (669.16 mA g^{-1}). Ding *et al.* have successfully prepared graphene/$LiFePO_4$ nanostructures for lithium-ion batteries by a solvent evaporation method.[116] The flexible composite electrode exhibited a high capacity of 163.7 mA g^{-1} at 0.1 C and 114 mA g^{-1} at 5 C without further incorporation of conductive agents.

Gwon *et al.*[117] synthesized graphene-based flexible electrodes based on freestanding graphene paper by employing a pulse laser deposition (PLD) technique. The flexible LIB was assembled by using V_2O_5/graphene paper and graphene paper as cathode and anode, respectively (Figure 11.13(a) and (c)). The flexible V_2O_5/graphene paper electrode demonstrated higher capacity from 10 to 600 μA cm^{-2} (40% retention), while the conventional V_2O_5/Al electrode retained below 10% of its initial capacity with the same experimental conditions (Figure 11.13(d)). Moreover, it can be found that the capacity retention of the V_2O_5/graphene paper is superior to that of a conventional electrode on Al foil (Figure 11.13(e)). These results suggest that the kinetics of lithium insertion/extraction and electron conduction in V_2O_5/graphene paper is significantly improved. In addition, the flexible battery can operate when the battery is rolled up or twisted (Figure 11.13(b)).

Li *et al.*[118] proposed a new design of flexible batteries based on $Li_4Ti_5O_{12}$ (LTO)/graphene foam (GF) and $LiFePO_4$ (LFP)/GF electrodes by hydrothermal deposition of active materials on GF followed by heating in an argon atmosphere (Figure 11.14(a)). The full battery shows excellent flexibility, and no structural failure was observed after repeatedly bending to a radius of 5 mm (Figure 11.14(b)). The flexible battery can power a red light-emitting diode (LED) when bent, as shown in Figure 11.14(c). From Figure 11.14(d) and (e), the discharge/charge curves and cyclic stability of flexible battery are barely affected by external bending stress, thus demonstrating its good flexible electrical stability. The flexible battery shows an excellent high-rate performance and can be operated at a 10 C rate with a capacity of 117 mA g^{-1} (Figure 11.14(f)). Furthermore, the flexible full battery can be cycled over 100 times at a high rate of 10 C with only 4% capacity loss (Figure 11.14(g)).

11.6 Flexible 2D Inorganic Nanosheets-based Electrodes for Li-ion Batteries

11.6.1 Flexible 2D Metal Oxide Nanosheets Composite

Recently, 2D nanoarchitectures have aroused considerable interest for LIBs because of the shortened paths for fast lithium in diffusion and large exposed surface offering more lithium-insertion channels.[119] Chen *et al.*[120] reported a facile synthesis of vertically aligned $Li_4Ti_5O_{12}$ (LTO) nanosheet arrays grown on Ti foil by hydrothermal and subsequent thermal decomposition. The CV curve of the LTO nanosheet arrays shows clear cathodic/anodic peaks that can be attributed to the redox reaction of Ti^{4+}/Ti^{3+}, respectively (Figure 11.15(a)). The self-supported arrays of LTO nanosheets ~14 nm in thickness (LTO-NSA-1) show a higher rate capability with a reversible capacity of 163 mA g^{-1} at 20 C (Figure 11.15(b)), which is close to the theoretical capacity of spinel $Li_4Ti_5O_{12}$ (175 mA g^{-1}). Significantly, the LTO-NSA-1 anode cycled at a rate of 50 C arrived at a stabilized capacity of 134 mA g^{-1} within 100 cycles, and retained a capacity of 124 mA g^{-1} after 3000 cycles with a capacity loss only 7.5%, as shown in Figure 11.15(c). Figure 11.16(a) shows the structure of the assembled flexible

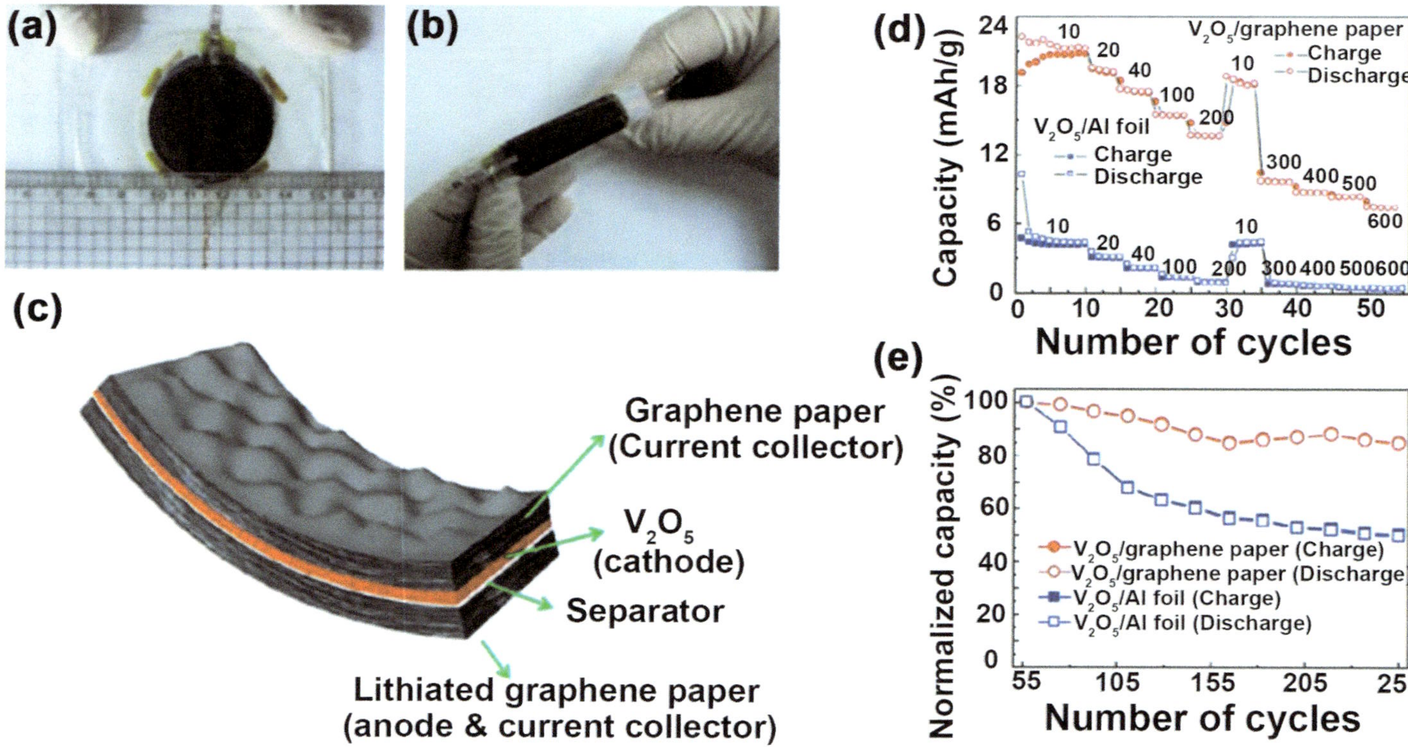

Figure 11.13 (a) The assembled flexible Li battery based on graphene paper. V_2O_5/graphene paper and electrochemically lithiated-graphene paper were use as cathode and anode, respectively. (b) The battery is thin, lightweight, and flexible enough to be rolled up or twisted. (c) Schematic drawing of the flexible Li battery based on graphene paper. (d) The volumetric capacity of V_2O_5/graphene paper and V_2O_5/Al foil at various current densities. (e) Cycling performance of V_2O_5/graphene paper and V_2O_5/Al foil up to 200 cycles at a current density of 20 μA cm^{-2} after testing at various current densities shown in Figure 11.14(d). Reproduced from ref. 117 with permission from The Royal Society of Chemistry.

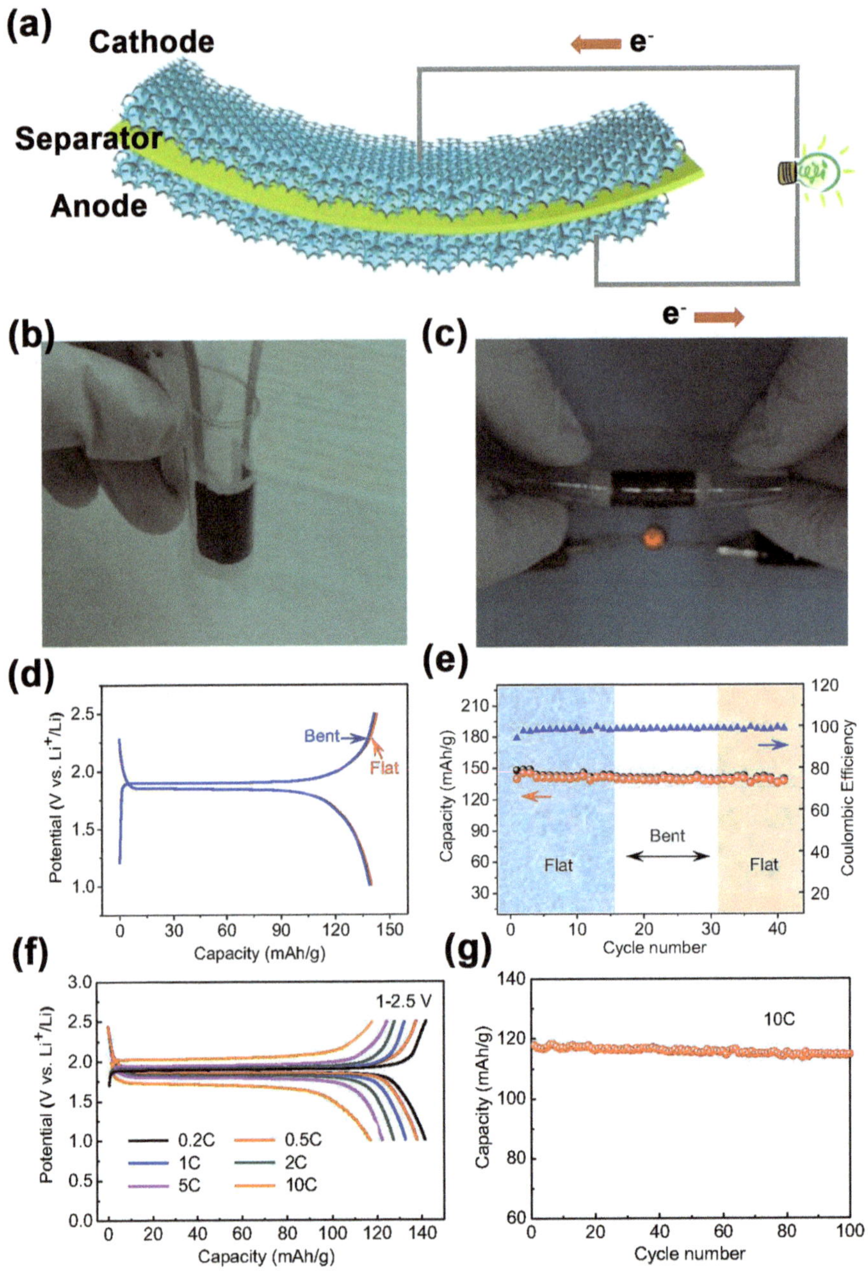

Figure 11.14 (a) Schematic of a flexible battery containing a cathode and an anode made from interconnected GF. (b) Photograph of a bent battery encapsulated by PDMS, showing its good flexibility. (c) Lighting a red LED device under bending. (d) Galvanostatic charging–discharging curves of the battery. Red and blue lines represent the as-fabricated flat battery and the bent battery after repeatedly bending to a radius of 5 mm 20 times, respectively. (e) Cyclic performance of the battery under flat and bent states. (f) Charging–discharging voltage curves of the battery with different current rates. (g) Capacity of the battery charged/discharged at a constant 10 C rate for 100 cycles. Reproduced from ref. 117 with permission from The Royal Society of Chemistry.

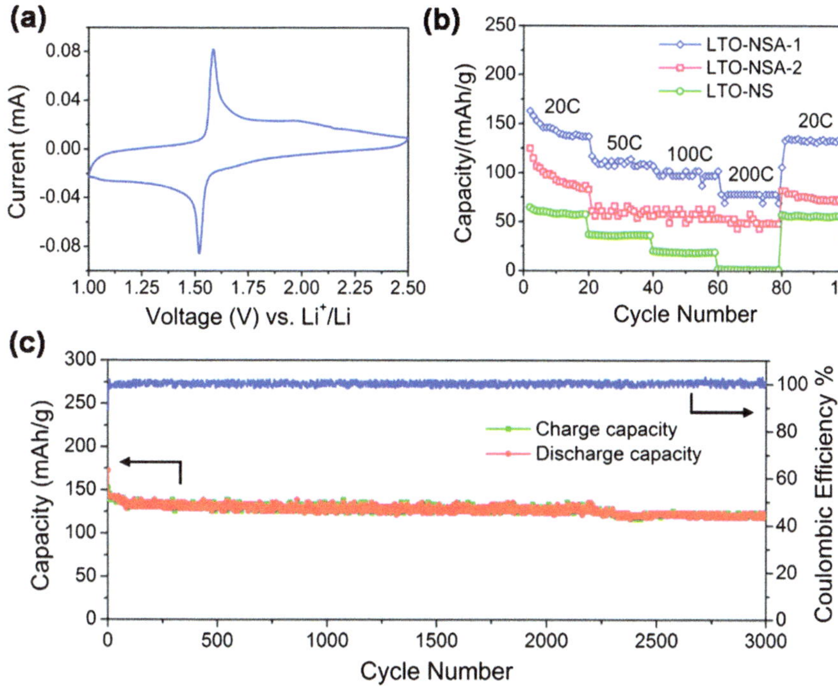

Figure 11.15 (a) Cyclic voltammetry curve at a scan rate of 0.5 mV s^{-1} for LTO-NSA-1. (b) Rate performance of different LTO nanosheet structures. (c) Specific capacity and Coulombic efficiency for 3000 cycles at 50 C for LTO-NSA-1. Reproduced from ref. 120 with permission from The Royal Society of Chemistry.

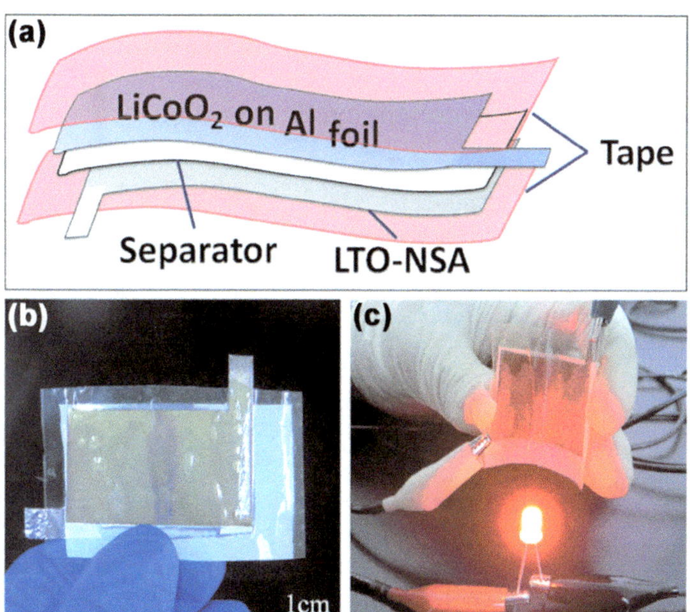

Figure 11.16 (a) Schematic illustration of a flexible lithium ion battery assembled using LTO-NSA on Ti foil as the anode. Optical photographs of a flexible battery (b) and a red LED lit by the flexible battery under bending (c). Reproduced from ref. 120 with permission from The Royal Society of Chemistry.

full battery, which contains LTO nanosheet arrays as the anode, a flexible separator, $LiClO_4$-based electrolyte and commercial $LiCoO_2$ loaded on Al foil as the cathode. The final packaged flexible full battery is displayed in Figure 11.16(b); this flexible battery could be full charged to 3 V within 30 s and was able to light a red LED whether it was flat or bent (Figure 11.16(c)).

Liu *et al.*[121] proposed a free-standing flexible CuO nanosheets (NSs)/r-GO composites fabricated by a vacuum filtration method and hydrothermal reduction process, as illustrated in Figure 11.17(a). CuO NSs can be uniformly incorporated between r-GO layers, which seem like layer-by-layer sandwiched architecture (Figure 11.17(b)). Within, r-GO layers can act as mechanical support and a current collector. Figure 11.17(b) inset demonstrates the photo of as-obtained highly flexible CuO NSs/r-GO composite electrode. Corresponding cyclic performances of CuO NSs/r-GO, bare CuO NSs and bare r-GO electrodes at a current density of 67 mA g^{-1} are shown in Figure 11.17(c). These CuO NSs/r-GO flexible electrode can deliver the highest retaining discharge capacity (736.8 mAh g^{-1}), which is much higher than that of CuO NSs-based electrodes. In addition, the discharge capacity of bare r-GO electrode is close

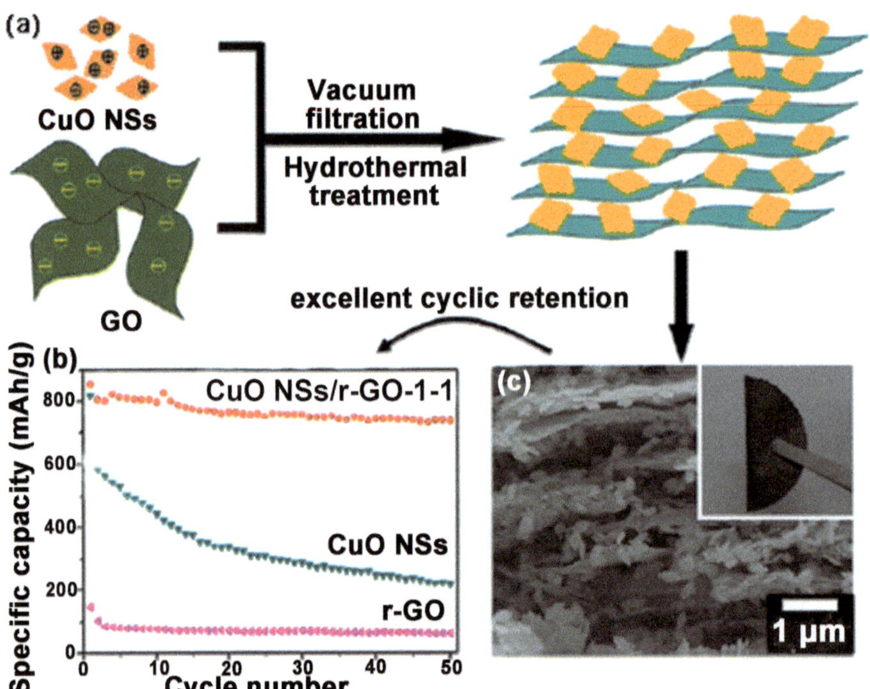

Figure 11.17 (a) Schematic of the synthesis procedure of CuO NSs/r-GO paper. (b) Cross-section SEM images of the CuO NSs/r-GO composites. Inset: an optical image of an as-prepared free-standing CuO NSs/r-GO electrode, showing its good flexibility. (c) Electrochemical characterizations of CuO NSs/r-GO, bare CuO NSs, and bare r-GO electrodes for Li-ion batteries. Reproduced with the permission from ref. 121. Copyright 2013 American Chemical Society.

to 60 mAh g^{-1} after 50 cycles, thus revealing that the total discharge capacity of CuO NSs/r-GO electrode should be mainly attributed to contribution of CuO NSs. The development of high rate lithium ion batteries with high safety and long cycle life is essential for the applications of LIBs in some important fields including fast charging electronics and electric vehicles.

11.6.2 Flexible 2D Metal Sulfide Nanosheets Composites Anode

In addition to the 2D oxide-based nanomaterials, fabrication of many sulfide-based 2D materials applied for rechargeable Li-ion batteries also attracted significant attentions. Herein, MoS_2 has been regarded as a potential candidate for electrode materials in LIBs due to its high theoretical capacity and specific layered structure. Its layered structure enables the efficient intercalation and exfoliation of lithium ions, and it can have much higher capacity than commercially used graphite.[122–125] Thus, many interesting MoS_2 nanosheets-based electrodes were synthesized using hydrothermal method,[126] solvothermal reaction at a low temperature,[127] and template-assisted strategy.[128] Furthermore, SnS/PPy nanosheets were prepared *via* a hydrothermal and *in-situ* polymerization approach.[129]

Li *et al.*[130] prepared the MoS_2 flexible electrode on carbon nanotube paper by a hydrothermal method. The MoS_2 nanosheets were uniformly incorporated with carbon nanotubes, and the hybrid electrode delivered a high capacity of 1053, 1040, 1014, 975, 918, 856, 762, and 684 mA g^{-1} at a current density of 0.05, 0.1, 0.2, 0.5, 1, 2, 5, and 10 A g^{-1}, respectively, which is much higher than the bulk MoS_2 electrode. The MoS_2 nanosheet/carbon tube paper electrode maintained 1091 mA g^{-1} with a capacity retention of 90% for the initial discharge capacity after 100 cycles, while the capacity of bulk MoS_2 electrode decreased and reached 264 mA g^{-1} after 100 cycles. The high performance of MoS_2 nanosheets/carbon tube paper as a novel class of flexible lithium-ion batteries makes it possible to use these electrodes directly in large-scale flexible energy-storage devices.

Yu *et al.*[131] used a hydrothermal method to fabricate MoS_2 nanoflake array/carbon cloths with differing amounts of MoO_3 precursor: 30 mg, 60 mg, and 90 mg, denoted CCMS-1, CCMS-2, and CCMS-3, respectively. Figure 11.18(a) shows a low-magnification SEM image of CCMS-2. The hierarchical MoS_2 nanoflake array/carbon composite fibers have a uniform diameter of about 8.5 μm, as shown in Figure 11.18(b). Figure 11.18(c) shows that each carbon microfiber is covered with numerous highly ordered MoS_2 nanoflake arrays, and void space is present between MoS_2 nanoflakes. The MoS_2 nanoflakes have a uniform width of about 400 nm, as shown in Figure 11.18(d). The discharge/charge capacities and cyclic stability of CCMS-1, CCMS-2, and CCMS-3 at a low current density of 0.15 mA cm^{-2} are shown in Figure 11.18(e). The CCMS-2 shows a higher capacity and good cycling stability.

Liu *et al.*[132] successfully fabricated a 2D hybrid SnS/PPy sandwich-like nanosheet with large interfacial tensions by a facile hydrothermal reduction and *in-situ* polymerization route. The sandwich-like SnS/PPy ultrathin

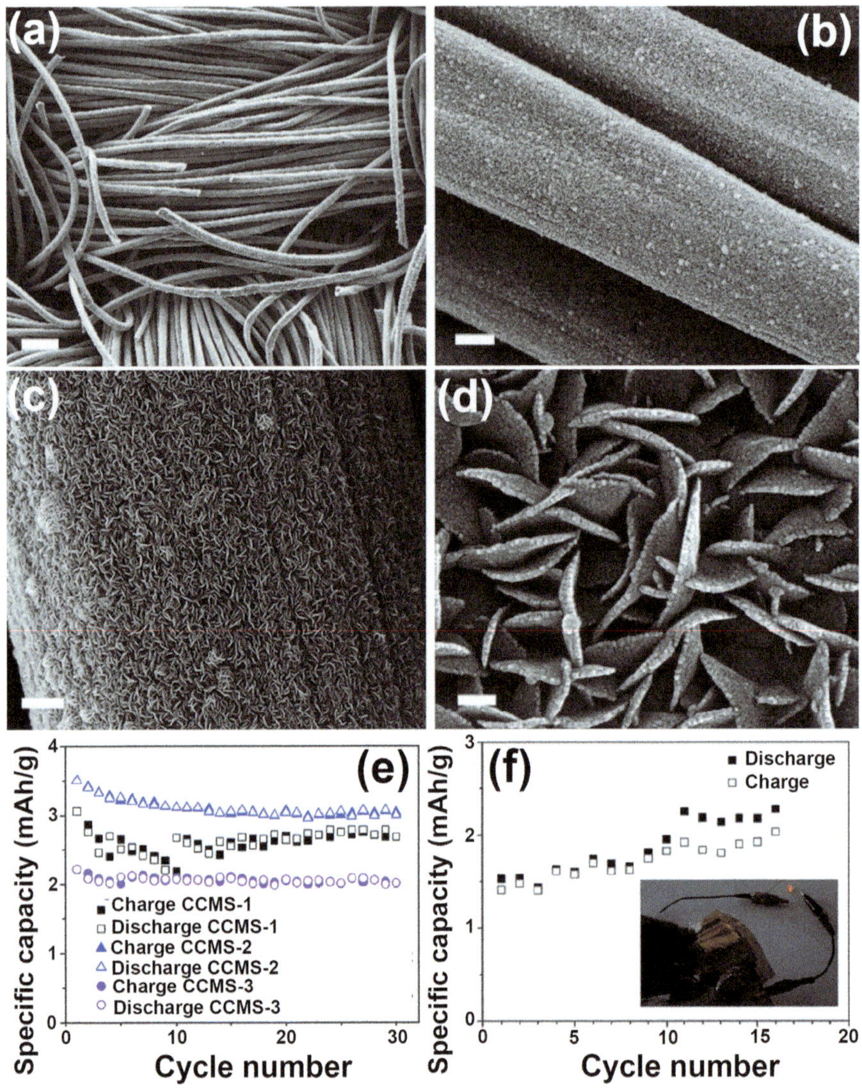

Figure 11.18 (a)–(d) Typical SEM images of MoS_2 nanoflake array/carbon cloth at different magnifications. Scale bar: (a) 50 μm, (b) 3 μm, (c) 1 μm and (d) 100 nm. (e) Cycling performance of CCMS-1, CCMS-2, and CCMS-3, showing the reversible capacity value of 3 mAh cm^{-2} after 30 cycles with a coulombic efficiency of 97.6%. (f) The cycling performance of the full battery, and inset: the flexible full battery can light a red LED even after 50 cycles of bending the full battery. Reproduced from ref. 131 with permission from The Royal Society of Chemistry.

nanosheets not only facilitate fast electron transport and Li^+ diffusion but also offer a stable and flexible framework for electroactive SnS volume expansion, thus can exhibit high performance for Li-ion storage.

11.6.3 Flexible 2D Multiple-compound Nanosheets Composites

A simple and scalable process has been used to fabricate $Li_4Ti_5O_{12}$ nanosheets with N-doped carbon coating (denoted as C-LTO nanosheets) by Li *et al.*[133] Using this material, they developed a flexible film which can be directly used as a flexible electrode for LIBs. The C-LTO nanosheet film electrode has an excellent cycling property. After 100 discharge/charge at 10 C, the C-LTO nanosheet film electrode delivers a specific capacity as high as 152 mA g^{-1}, less than 1% capacity decreases compared with the initial value.

Transition metal molybdates, $MMoO_4$ (M = Co^{2+}, Ni^{2+}, Zn^{2+}, Mn^{2+} *etc.*) have gradually been considered as promising, effective and scalable alternatives to the commercial graphite anode for LIBs in view of their high theoretical capacity, natural abundance, and environmental friendliness.[134,135] As an attractive member in the class of $MMoO_4$ type materials, $CoMoO_4$ has exhibited a higher electrochemical performance than single component oxides due to its multiple oxidation states, higher redox activity and comparative higher electrical conductivity.[136–138] Recently, the rational design and fabrication of electroactive materials with unique nanostructures grown on conductive substrates to be directly used as free-standing and binder-free electrodes is an effective strategy to improve their electrochemical performance. Wang *et al.*[139] demonstrated the high-density assembly of ultrathin mesoporous $CoMoO_4$ nanosheets on flexible carbon fabric with a porous network-like structure by a facile and simple hydrothermal approach combined with a post annealing treatment. The as-fabricated $CoMoO_4$/carbon fabric composites were light-weight and flexible so that they can be folded without any detriment to the structural integrity (Figure 11.19(a)). The TEM image in Figure 11.19(b) showed that the average thickness of an individual nanosheet was about 3–5 nm. The cycling stability and rate capacity of the $CoMoO_4$/carbon fabric were significantly improved compared with the $CoMoO_4$ powder electrode, as shown in Figure 11.19(c) and (d). The excellent electrochemical performance of the $CoMoO_4$/carbon fabric electrode can be attributed to not only the good electrical conductivity and flexible substrate, which can serve as the superb highway for fast electron transportation and buffering matrix to counteract and pulverization problem, but also the ultrathin and mesoporous features of $CoMoO_4$ nanosheets, which can create void spaces between the neighboring nanosheets and provide a shorter pathway for LIB (Figure 11.19(e)).

The VN@SnS_2 nanocomposite was fabricated on a carbon cloth through a two-step hydrothermal process as depicted in Figure 11.20(a).[140] The SEM image (Figure 11.20(c)) revealed that the carbon cloth fibers were uniformly

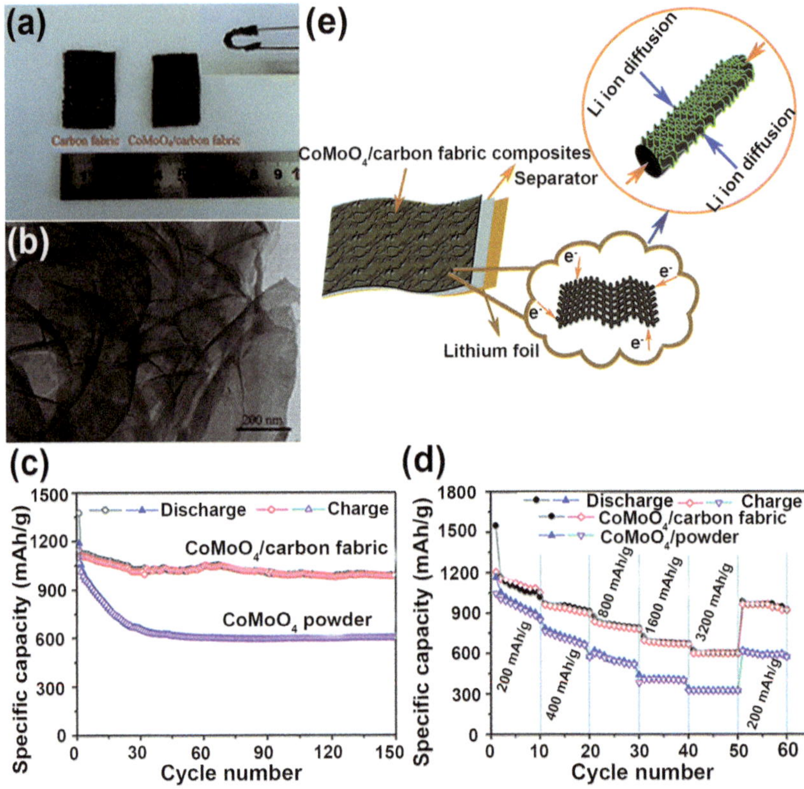

Figure 11.19 (a) Photographs of the primitive carbon fabric and $CoMoO_4$/carbon fabric composites (inset: the as-synthesized $CoMoO_4$/carbon fabric composites under bending). (b) TEM images. (c) Comparison of cycling performance of the $CoMoO_4$/carbon fabric electrode with $CoMoO_4$ powder pasted electrode. (d) Comparison of rate performance of the $CoMoO_4$/carbon fabric electrode with $CoMoO_4$ powder pasted electrode. (e) Schematic illustration of $CoMoO_4$/carbon fabric composites as electrode materials for LIBs. Reproduced with the permission from ref. 139. Copyright 2015 American Chemical Society.

coated by interlaced VN nanowires compared with pure carbon cloth (Figure 11.20(b)). After a two-step hydrothermal process, the VN nanowire was covered with SnS_2 nanosheets (Figure 11.20(d)). The nanosheet structure and the polycrystallline nature of the SnS_2 were confirmed in Figure 11.20(e). The cycling stability and rate capacity of VN@SnS_2 electrode was examined at a current density of 0.65 A g^{-1} for 100 cycles and compared with CC-VN, CC-SnS_2 electrode. The VN@SnS_2 electrode exhibited a high capacity retention (Figure 11.20(f)). Meanwhile, the VN@SnS_2 electrode displayed a discharge capacity of 349 mA g^{-1} at a very high current density of 13 A g^{-1}, revealing an excellent rate capacity (Figure 11.20(g)). The SEM image of VN@SnS_2 electrode after 100 cycles was shown in Figure 11.20(h), which demonstrated the nanocomposite still adhere firmly to the surface of the carbon cloth, indicating the excellent stability of VN@SnS_2 electrode account for its high capacity retention.

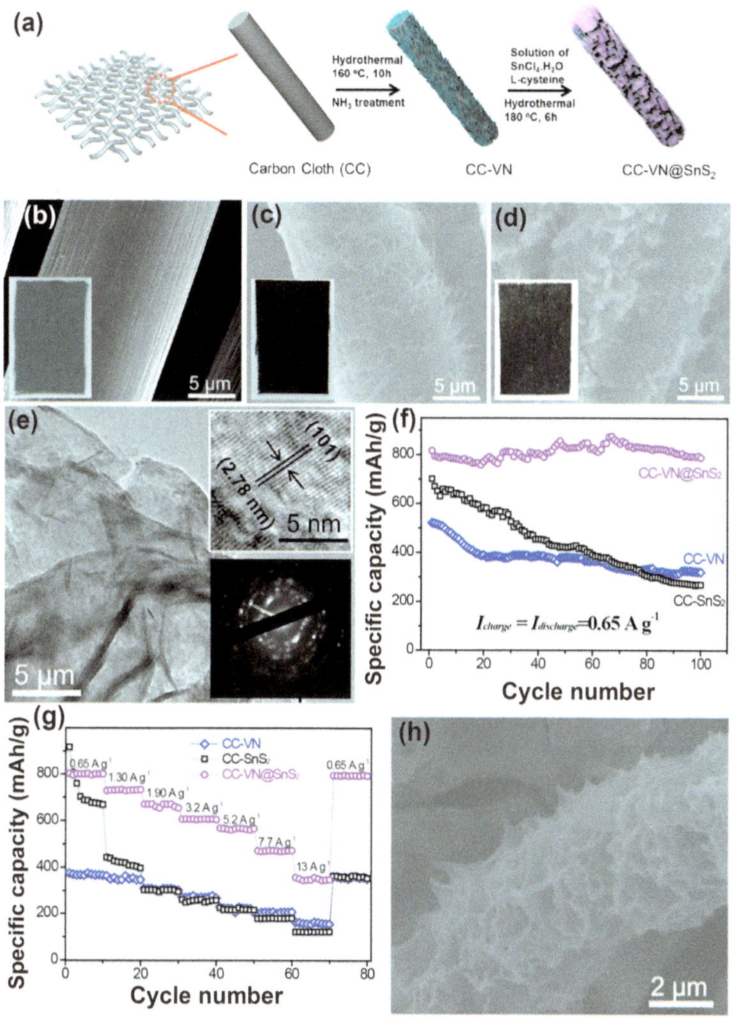

Figure 11.20 (a) Schematic diagram illustrating the two-step growth process for synthesizing the VN@SnS_2 NCs on carbon cloth substrate. SEM images of (b) the bare carbon cloth, (c) VN NWs on the carbon cloth, and (d) VN@SnS_2 on the carbon cloth. (Insets) Pictures of the corresponding carbon cloth that coated the materials. (e) TEM images of the SnS_2 nanosheets; (inset, top) HRTEM and (bottom) corresponding SAED pattern of the nanosheets. (f) Cycling profiles up to 100 cycles at a current density of 0.65 A g^{-1}. (g) Rate capability profiles of the electrodes at different discharge and charge rates. (h) Morphology of the CC-VN@SnS_2 electrode after electrochemical cycling. Reproduced with the permission from ref. 140. Copyright 2015 American Chemical Society.

11.7 Conclusions and Future Study

Flexible LIBs are receiving tremendous attention and research interest due to the increasing need for flexible/wearable energy storage devices. In this chapter, we summarized the recent advances in 2D nanostructured materials and their hybrids, such as graphene, graphene/metal oxides, graphene/metal sulfide, inorganic transition metal oxides and transition metal sulfides nanosheets, that are now receiving increasing attention with respect to energy applications. The integration of 2D hybrid nanomaterials into energy storage devices for rechargeable batteries offers opportunities to address many challenges driven by growing green energy demands.[141,142]

Table 11.1 summaries some state-of-the-art results on the different 2D electrode flexible LIBs. The main advantages of 2D nanostructures can be summarized as: high surface areas and excellent mechanical performance.

Table 11.1 Main parameters of 2D nanomaterials-based electrodes for flexible lithium-ion batteries.

Nanosheet-based electrode	Synthesis method	Cycling life	Capacity retention	Current density	Reference
Graphene paper	Chemical vapor deposition		1350 mAh g^{-1}	50 mA g^{-1}	41
Graphene paper	Freeze-drying	100	560 mAh g^{-1}	100 mA g^{-1}	51
Graphene paper	Photothermal reduction	1000	156 mAh g^{-1}	14 880 mA g^{-1}	52
Graphene paper	Vacuum filtration	50	450 mAh g^{-1}	300 mA g^{-1}	53
$Li_4Ti_5O_{12}$/CNF/CNT hybrid film	Pressured extrusion	1000	143 mAh g^{-1}	10 C	54
Graphene/Fe_3O_4	*In situ* reduction	30	1026 mAh g^{-1}	35 mA g^{-1}	61
Graphene paper/SnO_2 nanoparticles	Filtration	50	438 mAh g^{-1}	100 mA g^{-1}	88
TiO_2 nanosheets/graphene	Hydrothermal	100	147 mAh g^{-1}	1680 mA g^{-1}	97
Mn_3O_4/graphene oxide paper	Filtration	100	573 mAh g^{-1}	100 mA g^{-1}	99
MoS_2-graphene paper	Chemical vapor deposition	20	580 mAh g^{-1}	50 mA g^{-1}	109
Si/graphene paper	Acid-etching	100	1500 mAh g^{-1}	100 mA g^{-1}	113
$Li_4Ti_5O_{12}$ (LTO)/graphene	Hydrothermal	100	117 mAh g^{-1}	10 C	118
TiO_2 nanosheets	Hydrothermal	3000	124 mAh g^{-1}	50 C	120
CuO nanosheets/r-GO	Hydrothermal	50	736.5 mAh g^{-1}	67 mA g^{-1}	121
MoS_2 nanosheets/carbon nanotube	Hydrothermal	100	1091 mAh g^{-1}	100 mA g^{-1}	130
MoS_2 nanosheets/carbon cloth	Hydrothermal	30	3 mAh cm^{-2}	0.15 mA cm^{-2}	131
$Li_4Ti_5O_{12}$ nanosheets/N-doped carbon	Annealing treatment	100	152 mAh g^{-1}	1750 mA g^{-1}	132
$Li_4Ti_5O_{12}$ nanosheets/N-doped carbon	Annealing treatment	100	152 mAh g^{-1}	1750 mA g^{-1}	133
$CoMoO_4$ nanosheets/carbon cloth	Hydrothermal	150	1376 mAh g^{-1}	100 mA g^{-1}	138
$CoMoO_4$ nanosheets/carbon cloth	Hydrothermal	150	1376 mAh g^{-1}	100 mA g^{-1}	139
VN@SnS_2 nanosheets/carbon cloth	Hydrothermal	100	791 mAh g^{-1}	0.65 A g^{-1}	140

Despite the achievements discussed in this chapter, there exists substantial room for the development of high performance flexible LIBs and the increasing interest in flexible electronics provides great opportunities.

(1) More attention should be paid to controlling characteristics such as the morphology, defects, and functionalities to obtain high quality 2D nanomaterials. Furthermore, studying the charge/discharge process of 2D nanomaterial-based flexible LIBs is required.
(2) The synthesis and modification of hybrid nanostructures with high capacity and cycling stability must be designed.
(3) As flexible LIBs are being developed for future practical application, some effort is needed to improve mechanical flexibility.
(4) Lightweight, thin, flexible, stable packaging materials with exceptional barrier properties are also needed to protect battery materials, electrolytes and electrochemical reactions from being affected by the external environment.
(5) As a power sources, the lack of variety in flexible LIB shape and architecture imposes severe limitations on flexibility. Hence, overcoming the design limitations of flexible LIBs would boost the development in portable and even wearable electronics.
(6) In general, the trend for the development of flexible LIBs is to integrate or embed them into flexible electronic devices to create multi-functional or self-powered hybrid systems. Some effort should be made to combine flexible LIBs with different electronics, such as solar cells, electrochromic devices and nanogenerators.

Acknowledgements

We gratefully acknowledge the financial support of Central University Special Project (CZQ16004), the National Natural Science Foundation of China (61625404) and the Beijing Natural Science Foundation (4162062).

References

1. J.-M. Tarascon and M. Armand, *Nature*, 2001, **414**, 359–367.
2. P. Poizot, S. Laruelle, S. Grugeon, L. Dupont and J. Tarascon, *Nature*, 2000, **407**, 496–499.
3. M. S. Whittingham, *Chem. Rev.*, 2004, **104**, 4271–4302.
4. Y.-K. Sun, S.-T. Myung, B.-C. Park, J. Prakash, I. Belharouak and K. Amine, *Nat. Mater.*, 2009, **8**, 320–324.
5. M. Armand and J.-M. Tarascon, *Nature*, 2008, **451**, 652–657.
6. W. W. Li, X. F. Wang, B. Liu, S. J. Luo, Z. Liu, X. J. Hou, Q. Y. Xiang, D. Chen and G. Z. Shen, *Chem.–Eur. J.*, 2013, **19**, 8650–8656.
7. L. Ji, Z. Tan, T. Kuykendall, E. J. An, Y. Fu, V. Battaglia and Y. Zhang, *Energy Environ. Sci.*, 2011, **4**, 3611.

8. J. Chmiola, C. Largeot, P. Taberna, P. Simon and Y. Gogotsi, *Science*, 2010, **328**, 480–483.
9. D. Pech, M. Brunet, H. Durou, P. Huang, V. Mochalin, Y. Gogotsi, P. Taberna and P. Simon, *Nat. Nanotechnol.*, 2010, **5**, 651–654.
10. J. Feng, X. Sun, C. Wu, L. Peng, C. Lin, S. Hu, J. Yang and Y. Xie, *J. Am. Chem. Soc.*, 2011, **133**, 17832–17838.
11. J. Sung, S. Kim and K. Lee, *J. Power Sources*, 2004, **133**, 312–319.
12. J. Yoo, K. Balakrishnan, J. Huang, V. Meunier, B. Sumpter, A. Srivastava, M. Conway, A. L. M. Reddy, J. Yu, R. Vajtai and P. Ajayan, *Nano Lett.*, 2011, **11**, 1423–1427.
13. B. Liu, D. S. Tan, X. F. Wang, D. Chen and G. Z. Shen, *Small*, 2013, **9**, 1998–2004.
14. B. Liu, B. Y. Liu, X. F. Wang, X. Wu, W. Zhao, Z. Xu, D. Chen and G. Z. Shen, *Adv. Mater.*, 2014, **26**, 4999–5004.
15. X. J. Hou, B. Liu, X. F. Wang, Z. R. Wang, Q. F. Wang, D. Chen and G. Z. Shen, *Nanoscale*, 2013, **5**, 7831–7837.
16. K. S. Novoselov, A. K. Geim, S. V. Morozov, D. Jiang, Y. Zhang, S. V. Dubonos, I. V. Grigorieva and A. A. Firsov, *Science*, 2004, **306**, 666–669.
17. O. C. Compton and S. T. Nguyen, *Small*, 2010, **6**, 711–723.
18. K. S. Novoselov, D. Jiang, F. Schedin, T. J. Booth, V. V. Khotkevich, S. V. Morozov and A. K. Geim, *Proc. Natl. Acad. Sci. U. S. A.*, 2005, **102**, 10451–10453.
19. S. Guo and S. Dong, *Chem. Soc. Rev.*, 2011, **40**, 2644–2672.
20. X. Huang, Z. Yin, S. Wu, X. Qi, Q. He, Q. Zhang, Q. Yan, F. Boey and H. Zhang, *Small*, 2011, **7**, 1876–1902.
21. J. Kim, L. J. Cote and J. Huang, *Acc. Chem. Res.*, 2012, **45**, 1356–1364.
22. S. Z. Butler, S. M. Hollen, L. Cao, Y. Cui, J. A. Gupta, H. R. Gutierrez, T. F. Heinz, S. S. Hong, J. Huang, A. F. Ismach, E. J. -Halpertin, M. Kuno, V. V. Plashnitsa, R. D. Robinson, R. S. Ruoff, S. Salahuddin, J. Shan, L. Shi, M. G. Spencer, M. Terrones, W. Windl and J. E. Goldberger, *ACS Nano*, 2013, **7**, 2898–2926.
23. M. Xu, T. Liang, M. Shi and H. Chen, *Chem. Rev.*, 2013, **113**, 3766–3798.
24. M. Chhowalla, H. S. Shin, G. Eda, L.-J. Li, K. P. Loh and H. Zhang, *Nat. Chem.*, 2013, **5**, 263–275.
25. Q. Tang and Z. Zhou, *Prog. Mater. Sci.*, 2013, **58**, 1244–1315.
26. V. Nicolosi, M. Chhowalla, M. G. Kanatzidis, M. S. Strano and J. N. Coleman, *Science*, 2013, **340**, 1226419.
27. H. Gwon, H.-S. Kim, K. U. Lee, D.-H. Seo, Y. C. Park, Y.-S. Lee, B. T. Ahn and K. Kang, *Energy Environ. Sci.*, 2011, **4**, 1277–1283.
28. A. K. Geim and K. S. Novoselov, *Nat. Mater.*, 2007, **6**, 183–191.
29. H. Y. Mao, Y. H. Lu, J. D. Lin, S. Zhong, A. T. S. Wee and W. Chen, *Prog. Surf. Sci.*, 2013, **88**, 132–159.
30. S. Pang, Y. Hernandez, X. Feng and K. Mullen, *Adv. Mater.*, 2011, **23**, 2779–2795.
31. Y. Zhu, S. Murali, W. Cai, X. Li, J. W. Suk, J. R. Potts and R. S. Ruoff, *Adv. Mater.*, 2010, **22**, 3906–3924.

32. D. Pan, S. Wang, B. Zhao, M. Wu, H. Zhang, Y. Wang and Z. Jiao, *Chem. Mater.*, 2009, **21**, 3136–3142.
33. A. Gerouki, M. Goldner, R. Goldner, T. Haas, T. Liu and S. Slaven, *J. Electrochem. Soc.*, 1996, **143**, L262–L263.
34. X. Jia, J. Campos-Delgado, M. Terrones, V. Meunier and M. S. Dresselhaus, *Nanoscale*, 2011, **3**, 86–95.
35. H. A. Becerril, J. Mao, Z. Liu, R. M. Stoltenberg, Z. Bao and Y. Chen, *ACS Nano*, 2008, **2**, 463–470.
36. X. Li, G. Zhang, X. Bai, X. Sun, X. Wang, E. Wang and H. Dai, *Nat. Nanotechnol.*, 2008, **3**, 538–542.
37. F. Gunes, H.-J. Shin, C. Biswas, G. H. Han, E. S. Kim, S. J. Chae, J.-Y. Choi and Y. H. Lee, *ACS Nano*, 2010, **4**, 4595–4600.
38. S. Gan, L. Zhong, T. Wu, D. Han, J. Zhang, J. Ulstrup, Q. Chi and L. Niu, *Adv. Mater.*, 2012, **24**, 3958–3964.
39. D. A. Dikin, S. Stankovich, E. J. Zimney, R. D. Piner, G. H. B. Dommett, G. Evmenenko, S. T. Nguyen and R. S. Ruoff, *Nature*, 2007, **448**, 457.
40. D. Li, M. B. Muller, S. Gilje, R. B. Kaner and G. G. Wallace, *Nat. Nanotechnol.*, 2008, **3**, 101–105.
41. G. Q. Ning, C. G. Xu, Y. M. Cao, X. Zhu, Z. M. Jiang, Z. J. Fan, W. Z. Qian, F. Wei and J. S. Gao, *J. Mater. Chem. A*, 2013, **1**, 408–414.
42. C. Wang, D. Li, C. O. Too and G. G. Wallace, *Chem. Mater.*, 2009, **21**, 2604.
43. G. K. Wang, X. Sun, F. Y. Lu, H. T. Sun, M. P. Yu, W. L. Jiang, C. S. Liu and J. Lian, *Small*, 2012, **8**, 452–459.
44. L. Qiu, X. Yang, X. Gou, W. Yang, Z.-F. Ma, G. G. Wallace and D. Li, *Chem.–Eur. J.*, 2010, **16**, 10653–10658.
45. Z. Lei, N. Christov and X. S. Zhao, *Energy Environ. Sci.*, 2011, **4**, 1866–1873.
46. J. L. Vickery, A. J. Patil and S. Mann, *Adv. Mater.*, 2009, **21**, 2180–2184.
47. C. Liu, Z. Yu, D. Neff, A. Zhamu and B. Z. Jang, *Nano Lett.*, 2010, **10**, 4863–4868.
48. O. C. Compton, D. A. Didin, D. W. Putz, L. C. Brinson and S. T. Nguyen, *Adv. Mater.*, 2010, **22**, 892–896.
49. D. Li, M. B. Muller, S. Gilje, R. B. Kaner and G. G. Wallace, *Nat. Nanotechnol.*, 2008, **3**, 101.
50. K. Rana, S. D. Kim and J.-H. Ahn, *Nanoscale*, 2015, **7**, 7065–7071.
51. F. Liu, S. Y. Song, D. F. Xue and H. J. Zhang, *Adv. Mater.*, 2012, **24**, 1089–1094.
52. R. Mukherjee, A. V. Thomas, A. Krishnamurthy and N. Koratkar, *ACS Nano*, 2012, **6**, 7867–7878.
53. T. D. Dao, J. E. Hong, K. S. Ryu and H. M. Jeong, *Chem. Eng. J.*, 2014, **250**, 257–266.
54. Y. Hu, X. Li, D. Geng, M. Cai, R. Li and X. Sun, *Electrochim. Acta*, 2013, **91**, 227–233.
55. S. M. Cao, X. Feng, Y. Y. Song, X. Xue, H. J. Liu, M. Miao, J. H. Fang and L. Y. Shi, *ACS Appl. Mater. Interfaces*, 2015, **7**, 10695–10701.
56. F. Ronci, P. Reale, B. Scrosati, S. Panero, A. V. Rossi, M. M. Di, P. Perfetti and J. M. Merino, *J. Phys. Chem. B*, 2002, **106**, 3082–3086.

57. S. Leijonmarck, A. Cornell, G. Lindbergh and L. Wagberg, *J. Mater. Chem. A*, 2013, **1**, 4671–4677.
58. S. M. Cao, X. Feng, Y. Y. Song, H. J. Liu, M. Miao, J. H. Fang and L. Y. Shi, *ACS Appl. Mater. Interfaces*, 2016, **8**, 1073–1079.
59. Z. Q. Xie, Z. Y. He, X. H. Feng, W. W. Xu, X. D. Cui, J. H. Zhang, C. Y. M. A. Carreon, Z. Liu and Y. Wang, *ACS Appl. Mater. Interfaces*, 2016, **8**(16), 10324.
60. S. W. Kim, D. H. Seo, X. Ma, G. Ceder and K. Kang, *Adv. Energy Mater.*, 2012, **2**, 710–721.
61. G. M. Zhou, D.-W. Wang, F. Li, L. L. Zhang, N. Li, Z.-S. Wu, L. Wen, G. Q. Lu and H.-M. Cheng, *Chem. Mater.*, 2010, **22**, 5306–5313.
62. R. H. Wang, C. H. Xu, J. Sun, L. Gao and C. C. Lin, *J. Mater. Chem. A*, 2013, **1**, 1794.
63. X. H. Cao, B. Zheng, X. H. Rui, W. H. Shi, Q. Y. Yan and H. Zhang, *Angew. Chem., Int. Ed.*, 2014, **53**, 1404–1409.
64. Y. J. Gong, S. B. Yang, Z. Liu, L. L. Ma, R. Vajtai and P. M. Ajayan, *Adv. Mater.*, 2013, **25**, 3979–3984.
65. M. Li, F. Pan, E. S. G. Choo, Y. B. Lv, Y. Chen and J. M. Xue, *ACS Appl. Mater. Interfaces*, 2016, **8**, 6972–6981.
66. G. Wang, J. Liu, S. Tang, H. Li and D. Cao, *J. Solid State Electrochem.*, 2011, **15**, 2587–2592.
67. B. Li, H. Cao, J. Shao, G. Li, M. Qu and G. Yin, *Inorg. Chem.*, 2011, **50**, 1628–1632.
68. J. Zhu, Y. K. Sharma, Z. Zeng, X. Zhang, M. Srinivasan, S. Mhaisalkar, H. Zhang, H. H. Hng and Q. Yan, *J. Phys. Chem. C*, 2011, **115**, 8400–8406.
69. L. Li, G. Zhou, X. Y. Shan, S. Pei, F. Li and H. M. Cheng, *J. Power Sources*, 2014, **255**, 52–58.
70. H. Kim, D. H. Seo, S. W. Kim, J. Kim and K. Kang, *Carbon*, 2011, **49**, 326–332.
71. K. H. Kim, D. W. Jung, V. H. Pham, J. S. Chung, B. S. Kong, J. K. Lee, K. Kim and E. S. Oh, *Electrochim. Acta*, 2012, **69**, 358–363.
72. M. Zhang, M. Jia, Y. Jin and X. Shi, *Appl. Surf. Sci.*, 2012, **263**, 573–578.
73. C. Peng, B. Chen, Y. Qin, S. Yang, C. Li, Y. Zuo, S. Liu and J. Yang, *ACS Nano*, 2012, **6**, 1074–1081.
74. X.-L. Huang, R.-Z. Wang, D. Xu, Z.-L. Wang, H.-G. Wang, J.-J. Xu, Z. Wu, Q.-C. Liu, Y. Zhang and X.-B. Zhang, *Adv. Funct. Mater.*, 2013, **23**, 4345–4353.
75. J. Yao, X. Shen, B. Wang, H. Liu and G. Wang, *Electrochem. Commun.*, 2009, **11**, 1849–1852.
76. M. Zhang, D. Lei, Z. Du, X. Yin, L. Chen, Q. Li, Y. Wang and T. Wang, *J. Mater. Chem.*, 2011, **21**, 1673–1676.
77. H. T. Wang, D. W. He, Y. S. Wang, H. P. Wu and J. G. Wang, *Adv. Mater. Res.*, 2012, **465**, 108–111.
78. J. Liu, J. Huang, L. Hao, H. Liu and X. Li, *Ceram. Int.*, 2013, **39**, 8623–8627.
79. S. M. Paek, E. J. Yoo and I. Honma, *Nano Lett.*, 2009, **9**, 72–75.

80. A. Dhanabalan, X. Li, R. Agrawal, C. Chen and C. Wang, *Nanomaterials*, 2013, **3**, 606–614.
81. Q. H. Wu, C. Wang and J. G. Ren, *Ionics*, 2013, **19**, 1875–1882.
82. H. Zhang, P. Xu, Y. Ni, H. Geng, G. Zheng, B. Dong and Z. Jiao, *J. Mater. Res.*, 2014, **29**, 617–624.
83. T. Gao, K. Huang, X. Qi, H. Li, L. Yang and J. Zhong, *Ceram. Int.*, 2014, **40**, 6891–6897.
84. Q. Guo, S. Chen and X. Qin, *Mater. Lett.*, 2014, **119**, 4–7.
85. H. Yang, Z. Hou, N. Zhou, B. He, J. Cao and Y. Kuang, *Ceram. Int.*, 2014, **40**, 13903–13910.
86. D. H. Wang, R. Kou, D. Choi, Z. G. Yang, Z. M. Nie, J. Li, L. V. Saraf, D. H. Hu, J. G. Zhang, G. L. Graff, J. Liu, M. A. Pope and I. A. Aksay, *ACS Nano*, 2010, **4**, 1587–1595.
87. I. A. Courtney and J. R. Dahn, *J. Electrochem. Soc.*, 1997, **144**, 2045–2052.
88. J. F. Liang, Y. Zhao, L. Guo and L. Li, *ACS Appl. Mater. Interfaces*, 2012, **4**, 5742–5748.
89. X. Su, Q. L. Wu, X. Zhan, J. Wu, S. Wei and Z. Guo, *J. Mater. Sci.*, 2012, **47**, 2519–2534.
90. D. Wng, D. Choi, J. Li, Z. Yang, Z. Nie, R. Kou, D. Hu, C. Wang, L. V. Saraf, J. Zhang, I. A. Aksay and J. Liu, *ACS Nano*, 2009, **3**, 907–914.
91. J. Wang, Y. Zhou, B. Xiong, Y. Zhao, X. Huang and Z. Shao, *Electrochim. Acta*, 2013, **88**, 847–857.
92. H. C. Tao, L. Z. Fan, X. Yan and X. Qu, *Electrochim. Acta*, 2012, **69**, 328–333.
93. H. Cao, B. Li, J. Zhang, F. Lian, X. Kong and M. Qu, *J. Mater. Chem.*, 2012, **22**, 9759–9766.
94. X. Zhang, P. S. Kumar, V. Aravindan, H. H. Liu, J. Sundaramurthy, S. G. Mhaisalkar, H. M. Duong, S. Ramakrishna and S. Madhavi, *J. Phys. Chem. C*, 2012, **116**, 14780–14788.
95. L. Dong, M. Li, L. Dong, M. Zhao, J. Feng, Y. Han, J. Deng, X. Li, D. Li and X. Sun, *Int. J. Hydrogen Energy*, 2014, **39**, 16116–16122.
96. T. Hu, X. Sun, H. Sun, M. Yu, F. Lu, C. Liu and J. Lian, *Carbon*, 2013, **51**, 322–326.
97. N. Li, G. M. Zhou, R. P. Fang, F. Li and H.-M. Cheng, *Nanoscale*, 2013, **5**, 7780.
98. T. Hu, X. Sun, H. T. Sun, M. P. Yu, F. Y. Lu, C. S. Liu and J. Lian, *Carbon*, 2013, **51**, 322–326.
99. S.-K. Park, C.-Y. Seong, S. Yoo and Y. Z. Piao, *Energy*, 2016, **99**, 266–273.
100. S. Ding, J. S. Chen and X. W. Lou, *Chem.–Eur. J.*, 2011, **17**, 13142–13145.
101. C. Feng, J. Ma, H. Li, R. Zeng, Z. Guo and H. Liu, *Mater. Res. Bull.*, 2009, **44**, 1811–1815.
102. Y. Miki, D. Nakazato, H. Ikuta, T. Uchida and M. Wakihara, *J. Power Sources*, 1995, **54**, 508–510.
103. U. K. Sen and S. Mitra, *ACS Appl. Mater. Interfaces*, 2013, **5**, 1240–1247.

104. M. Wang, G. Li, H. Xu, Y. Qian and J. Yang, *ACS Appl. Mater. Interfaces*, 2013, **5**, 1003–1008.
105. K. Bindumadhavan, S. K. Srivastava and S. Mahanty, *Chem. Commun.*, 2013, **49**, 1823–1825.
106. S.-K. Park, S.-H. Yu, S. Woo, B. Quan, D.-C. Lee, M. K. Kim, Y.-E. Sung and Y. Piao, *Dalton Trans.*, 2013, **42**, 2399–2405.
107. K. Chang and W. Chen, *ACS Nano*, 2011, **5**, 4720–4728.
108. Y.-T. Liu, X.-D. Zhu, Z.-Q. Duan and X.-M. Xie, *Chem. Commun.*, 2013, **49**, 10305–10307.
109. K. Rana, J. Singh, J.-T. Lee, J. H. Park and J.-H. Ahn, *ACS Appl. Mater. Interfaces*, 2014, **6**, 11158–11166.
110. Z. Gao, N. N. Song, Y. Y. Zhang and X. D. Li, *Nano Lett.*, 2015, **15**, 8194–8203.
111. U. Kasavajjula, C. Wang and A. J. Appleby, *J. Power Sources*, 2007, **163**, 1003–1039.
112. J.-Z. Wang, C. Zhong, S. L. Chou and H.-K. Liu, *Electrochem. Commun.*, 2010, **12**, 1467–1470.
113. H. Jiang, X. Zhou, G. G. Liu, Y. H. Zhou, H. Q. Ye, Y. Liu and K. Han, *Electrochim. Acta*, 2015, **188**, 777–784.
114. B. Wang, X. L. Li, X. F. Zhang, B. Luo, M. H. Jin, M. H. Liang, S. A. Dayeh, S. T. Picraux and L. J. Zhi, *ACS Nano*, 2013, **7**, 1437–1445.
115. F.-Y. Su, C. H. You, Y.-B. He, W. Lv, W. Cui, F. M. Jin, B. H. Li, Q.-H. Yang and F. Y. Kang, *J. Mater. Chem.*, 2010, **20**, 9644–9650.
116. Y. H. Ding, H. M. Ren, Y. Y. Huang, F. H. Chang and P. Zhang, *Mater. Res. Bull.*, 2013, **48**, 3713.
117. H. Gwon, H.-S. Kim, K. U. Lee, D.-H. Seo, Y. C. Park, Y.-S. Lee, B. T. Ahn and K. Kang, *Energy Environ. Sci.*, 2011, **4**, 1277–1283.
118. N. Li, Z. P. Chen, W. C. Ren, F. Li and H.-M. Cheng, *Proc. Natl. Acad. Sci.*, 2012, **109**, 17360–17365.
119. J. Liu and X.-W. Liu, *Adv. Mater.*, 2012, **24**, 4097.
120. S. Chen, Y. L. Xin, Y. Y. Zhou, Y. R. Ma, H. H. Zhou and L. M. Qi, *Energy Environ. Sci.*, 2014, **7**, 1924–1930.
121. Y. Liu, W. Wang, L. Gu, Y. W. Wang, Y. L. Ying, Y. Y. Mao, L. W. Sun and X. S. Peng, *ACS Appl. Mater. Interfaces*, 2013, **5**, 9850–9855.
122. C. Q. Feng, J. Ma, H. Li, R. Zeng, Z. P. Guo and H. K. Liu, *Mater. Res. Bull.*, 2009, **44**, 1811.
123. H. Hwang, H. Kim and J. Cho, *Nano Lett.*, 2011, **11**, 4826.
124. S. J. Ding, D. Y. Zhang, J. S. Chen and X. W. Lou, *Nanoscale*, 2012, **4**, 95.
125. G. D. Du, Z. P. Guo, S. Q. Wang, R. Zeng, Z. X. Chen and H. K. Liu, *Chem. Commun.*, 2010, **46**, 1106.
126. S. Ding, J. Chen and X. W. Lou, *Chem.–Eur. J.*, 2011, **17**, 13142–13145.
127. M. Wang, G. Li, H. Xu, Y. Qian and J. Yang, *ACS Appl. Mater. Interfaces*, 2013, **5**, 1003–1008.
128. L. Zhang, H. B. Wu, Y. Yan, X. Wang and X. W. Lou, *Energy Environ. Sci.*, 2014, **7**, 3302–3306.

129. J. Liu, M. Z. Gu, L. Z. Ouyang, H. Wang, L. C. Yang and M. Zhu, *ACS Appl. Mater. Interfaces*, 2016, **8**(13), 8502–8510.
130. H. S. Li, X. Wang, B. Ding, G. Pang, P. Nie, L. F. Shen and X. G. Zhang, *ChemElectroChem*, 2014, **1**, 1118–1125.
131. H. L. Yu, C. L. Zhu, K. Zhang, Y. J. Chen, C. Y. Li, P. Gao, P. P. Yang and Q. Y. Ouyang, *J. Mater. Chem. A*, 2014, **2**, 4551.
132. J. Liu, M. Z. Gu, L. Z. Qoyang, H. Wang, L. C. Yang and M. Zhu, *ACS Appl. Mater. Interfaces*, 2016, **8**, 8502–8510.
133. T. Yang, H. Zhang, Y. luo, L. Mei, D. Guo, Q. Li and T. Wang, *Electrochim. Acta*, 2015, **158**, 327.
134. J. Haetge, I. Djerdj and T. Brezesinski, *Chem. Commun.*, 2012, **48**, 6726.
135. X. Yu, B. Lu and Z. Xu, *Adv. Mater.*, 2014, **26**, 1044.
136. L.-Q. Mai, F. Yang, Y.-L. Zhao, X. Xu, L. Xu and Y. Z. Luo, *Nat. Commun.*, 2011, **2**, 381.
137. H. Yu, C. Guan, X. Rui, B. Ouyang, B. Yadian, Y. Huang, H. Zhang, H. E. Hoster, H. J. Fan and Q. Yan, *Nanoscale*, 2014, **6**, 10556.
138. B. Wang, S. M. Li, X. Y. Wu, J. H. Liu, W. M. Tian and J. Chen, *New J. Chem.*, 2016, **40**, 2259–2267.
139. B. Wang, S. M. Li, X. Y. Wu, J. H. Liu, W. M, Tian and J. Chen, *New J. Chem.*, 2016, **40**, 2259.
140. M. S. Balogun, W. T. Qiu, J. H. Jian, Y. C. Huang, Y. Luo, H. Yang, C. L. Liang, X. H. Lu and Y. X. Tong, *ACS Appl. Mater. Interfaces*, 2015, **7**, 23205–23215.
141. R. Raccichini, A. Varzi, S. Passerini and B. Scrosati, *Nat. Mater.*, 2015, **14**, 271–279.
142. F. Bonaccorso, L. Colombo, G. Yu, M. Stoller, V. Tozzini, A. C. Ferrari, R. S. Ruoff and V. Pellegrini, *Science*, 2015, **347**, 1246501.

CHAPTER 12

Two-dimensional Nanomaterials—An Ideal Platform to Understand Photocatalysis

WENTUAN BI AND CHANGZHENG WU*

Department of Chemistry & Hefei National Laboratory for Physical Sciences at the Microscale & Collaborative Innovation Center of Chemistry for Energy Materials, University of Science and Technology of China, No.96, JinZhai Road, Hefei, Anhui, 230026, P. R. China
*E-mail: czwu@ustc.edu.cn

12.1 Introduction

Environmental crises, along with the sustained growth in global energy consumption has stimulated a worldwide quest for renewable energy resources such as hydroelectric power, biomass and geothermal energy. Among them, solar energy, as a fundamental yet most abundant renewable energy, is predicted to be able to satisfy the current world energy demand.[1–5] Nevertheless, the intermittent nature of solar energy has long been noted as the stumbling block for making it practical and dependable. Therefore, since the discovery of water splitting on a TiO_2 photoanode in 1972, researchers and engineers

Smart Materials No. 27
Inorganic Two-dimensional Nanomaterials: Fundamental Understanding, Characterizations and Energy Applications
Edited by Changzheng Wu

Published by the Royal Society of Chemistry, www.rsc.org

have been infatuated with the idea of converting sunlight directly into readily storable fuels, which provides a simple and economically competitive approach to overcome the fluctuations of solar energy over time. However, despite the remarkable advancement achieved in photocatalysts in recent decades, relatively low energy conversion efficiency is still the most critical factor hindering practical application of this technology. To overcome this bottleneck, an in-depth understanding of the physical and chemical processes during artificial photosynthesis is urgently needed.

Lessons from nature have inspired the development of artificial solar fuels. It has been revealed that natural photosynthesis in plants occurs *via* a series of step-wise proton coupled electron-transfer processes creating sufficient energy for water splitting and carbohydrate production. Natural photosynthesis involves a sophisticated Z-scheme diagram with two photosystems connected in series with an electron transfer chain. The oxidation of water and reduction of $NADP^+$ occur simultaneously in time but separately in space, thus resulting in considerably high quantum efficiency under optimal conditions.[6] Similarly, artificial solar fuels production takes advantage of photo-redox catalysis, which primarily contains three successive processes: light-harvesting, charge separation and transport, and the ensuing surface redox reaction (see Figure 12.1). The overall conversion efficiency is simultaneously determined by these three processes. Therefore, numerous efforts have been devoted to exploring novel catalyst systems with optimized photon capture, charge separation, and surface redox reaction kinetics, especially in the fields of water splitting, CO_2 reduction, and organic synthesis.

These photocatalysts are either heterogeneous or homogeneous, depending on whether the catalyst exists in a different phase from the reactants. When compared with homogeneous catalysts, heterogeneous catalysts have been widely studied due to their superior stability and relative ease of catalyst separation that benefits the continuous chemical processes. As is known, heterogeneous catalytic reactions take place on the surface of a catalyst.

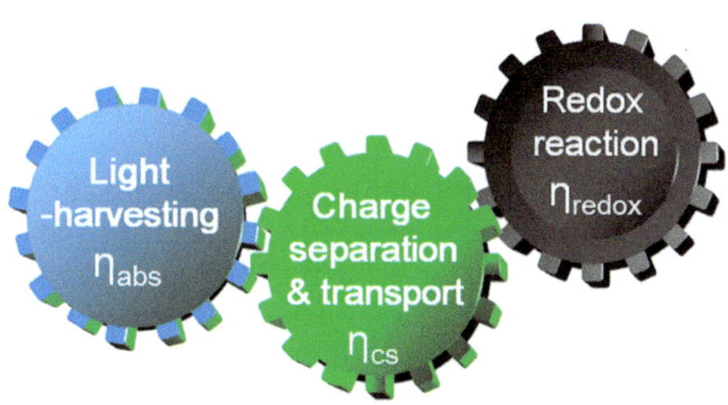

Figure 12.1 The primary process of artificial fuels production: light-harvesting, charge separation and transport, and the ensuing surface redox reaction. The overall conversion efficiency is simultaneously determined by these three processes.

Increasing the surface area is considered a straight-forward way of gaining high efficiency. Therefore, extensive efforts have been devoted to maximize the surface area of the catalysts *via* designing various nanostructures.

Two-dimensional (2D) nanomaterials draw attention due to their remarkably high surface-to-volume ratio, and relative ease of assembly for practical application because of their lateral dimension of several hundred nanometers to several micrometers. More importantly, 2D nanomaterials benefit the investigation of artificial photosynthesis in the following three ways: (1) with atomically well-defined local structure, 2D nanomaterials enable the identification of active sites down to the atomic-scale. (2) Due to confinement effects and the spatial decoupling of upper and lower atoms, 2D nanomaterials generally exhibit novel atomic structures and electronic states, providing new opportunities to manipulate the three primary processes in photocatalysis. (3) In view of the atomic-scale thickness, surface atoms of 2D nanomaterials become dominant as compared with those in the bulk counterparts; thus, the influence of surface effects on the whole system becomes more remarkable accordingly, making it feasible to tailor the intrinsic properties of 2D nanomaterials *via* surface chemical modifications. Overall, with these advantages, 2D nanomaterials stand out as an ideal platform to investigate artificial photocatalysis.

Recently, development based on elegantly-designed 2D nanostructures has greatly advanced our understanding of artificial photocatalysis and consequently enabled higher solar-to-fuel conversion efficiency. In this section, we summarize recent progress in optimizing the three primary processes of artificial photosynthesis: light-harvesting, charge separation and transport, and surface redox reaction kinetics. Finally, we discuss current challenges and future research trends in this field.

12.2 Light Harvesting

Light harvesting is the first step of solar energy conversion. When incident light with specific wavelengths passes through a semiconductor, its intensity decays in an exponential way, as expressed by the following equation:[7]

$$I = I_0 \exp(-\alpha l) \tag{12.1}$$

where I_0 is the intensity of the incident light, and α and l are the absorption coefficient, the penetration depth of incident light, respectively. As per this rule, efficient light harvesting requires semiconductors with superior absorption coefficient or sufficient thickness. Intriguingly, despite the atomic thickness, 2D nanostructures are typically promising light-absorbers because of their unique morphology and electronic structures. For instance, graphene displays an extraordinary absorbance of 2.3% in the visible region,[8] comparable to that of approximately 20 nm thick Si, the most commonly used absorber in solar cells. Atomically thin transition metal dichalcogenides (TMDs) such as MoS_2, $MoSe_2$, and WS_2 also exhibit strong light-matter interaction, capturing up to 5–10% incident sunlight.[9–11] According to theoretical calculations, this extraordinary absorption is attributed to the resonance effect derived from band nesting. Moreover, with decreasing thickness, MoS_2 undergoes an indirect-to-direct band gap transition.[12] Benefiting from this transition, MoS_2 monolayers can harvest sunlight without changing the momentum of excited electrons, which brings about a higher absorption cross-section and hence higher absorption coefficient. Moreover, given that highly anisotropic nanosheets are randomly dispersed in solution during the catalytic process, multiple reflection and scattering at the nanosheets/electrolyte interface can substantially increase the path of photons and thus facilitates light harvesting.[13]

Light harvesting by semiconductors converts sunlight into potential energy, then driving the redox reactions in fuel production. From a thermodynamics viewpoint, the band edges of semiconducting nanostructures must straddle the redox potentials involved in the photocatalytic reaction (see detail in Figure 12.2).[14] Taking water splitting as an example, it is an uphill reaction that involves both hydrogen evolution and oxygen evolution reactions. In this sense, the conduction band minimum of the semiconductors for overall water splitting must be more negative than $E_{H2/H2O}$ while the valence band maximum should be more positive than the water oxidation potential $E_{H2O/O2}$, thus requiring significant energy input (the change standard Gibbs free energy ΔG = 237 kJ mol^{-1} or 1.23 eV). Moreover, besides the thermodynamic potential, an electrochemical overpotential is inevitably needed to overcome the kinetic barriers in the actual catalytic systems. Therefore, semiconductors with appropriate band gaps and band edge positions become an indispensable prerequisite for developing highly efficient photocatalysts.

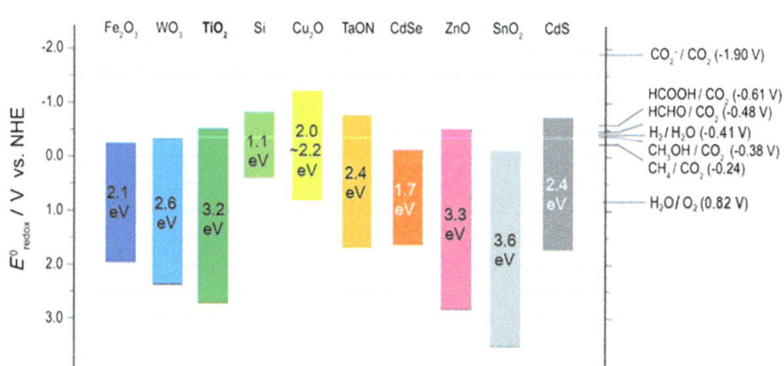

Figure 12.2 Band-edge positions of various semiconductors and the relative redox couples (*vs.* NHE, at pH 7), reproduced with permission from S. N. Habisreutinger *et al.*, Photocatalytic reduction of CO_2 on TiO_2 and other semiconductors, *Angewandte Chemie International Edition*, John Wiley and Sons, Copyright © 2013 WILEY-VCH Verlag GmbH & Co. KGaA, Weinheim.[14]

Since the electronic band engineering of semiconductors is critical to photocatalysis, intensive efforts have been devoted to optimizing the factors closely related to electronic band structure, including composition, structure, and more recently, the dimensionality. As the thickness is reduced, especially when the thickness is comparable to or below the Bohr exciton radius, a quantum confinement effect emerges and provides new opportunities to tailor the band gaps and band edge positions. As an example, for CdSe nanoplatelets with the thickness smaller than the Bohr exciton radius (approximately 5.6 nm),[15] noticeable blue-shifts in optical absorption and photoluminescence (PL) spectra are observed relative to their bulk counterparts.[16] As shown in Figure 12.3, the absorption edge is further blue-shifted as the thickness is reduced from the 7 monolayer to the 4 monolayer. The blue-shift of the absorption edge leads to a notable upshift of the conduction band minimum, which even spans the proton reduction potential, consequently bringing about remarkable photocatalytic activity; whereas bulk CdSe is inactive for photocatalytic H_2 evolution.[17,18] Similar phenomena is widespread in other 2D semiconductor systems.[19–21] In the case of all-inorganic cesium lead-halide perovskite $CsPbX_3$ (X = Cl, Br, I), the PL emission peaks of 488, 477, 462, 435, and 405 nm are assigned to $CsPbBr_3$ nanoplates with thicknesses from 5 to 1 perovskite unit cells, respectively. Notably, the resulting five-monolayer thick nanoplates exhibit excellent photoluminescent quantum yields (PLQYs), as high as 84.4 ± 1.8%. A further anion

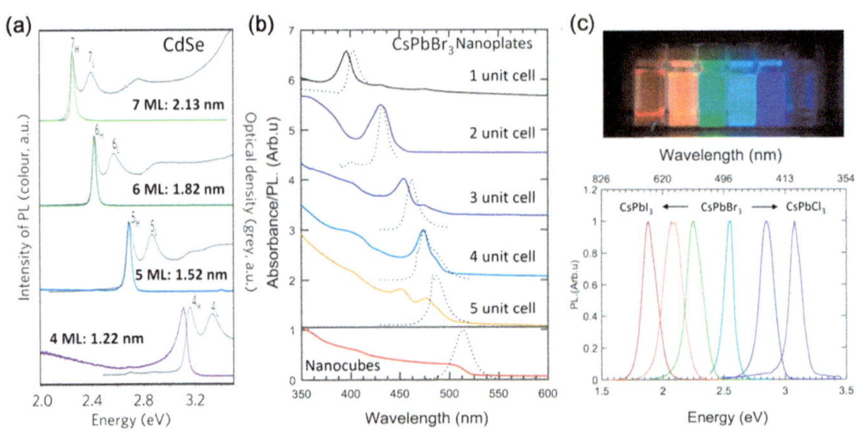

Figure 12.3 (a) Emission and absorption spectra of CdSe nanoplatelets with different thicknesses. Reprinted with permission from Macmillan Publishers Ltd: *Nature Materials* (ref. 16), Copyright 2011. (b) Absorption (solid lines) and emission (dashed lines) spectra of $CsPbBr_3$ NPLs with different thicknesses (1–5 unit cells) and nanocubes for comparison. (c) Colloidal solution of anion exchanged NPLs in hexane under UV illumination (λ = 365 nm) and the corresponding PL spectra of the halide-anion exchanged samples. Part (b) and (c) are adapted with permission from ref. 19, Copyright 2015 American Chemical Society.

exchange process enables fine regulation of their composition and leads to emission spectra that cover the entire visible spectrum.

Besides the increase of band gaps and the concomitant changes of band edge positions, more novel physics arises from 2D quantum confinement effect. As mentioned above, TMDs such as MoS_2 undergo an indirect-to-direct band gap transition when the thickness is down to the monolayer. With the aid of micrometer-scale angle-resolved photoemission spectroscopy, it is experimentally verified that the valence band maximum shifts from Γ to K in the first Brillouin zone, leading to a crossover to a direct-gap material in monolayer.[22]

Apart from sufficient energy to drive the redox reactions involved in solar fuel production, semiconductors that are capable of harvesting light over a broadband spectrum are highly desirable. However, the widely studied oxide-based photocatalysts, which show superior chemical and thermal stability, are generally wide band gap semiconductors with photoactivity only in the UV region. To develop visible-light-responsive catalysts, doping is the most conventional strategy. In general, the conduction bands of oxide-based semiconductors are mainly composed of transition metal orbitals, while the valence bands mainly consist of O 2p orbitals.[23] Upon exotic cation doping, intermediate energy levels could be introduced in the forbidden band, thus extending the light absorption to the visible region. In contrast, anion doping usually contributes to the conduction band and upshifts it to narrow the band gap.[24] Moreover, in addition to the type of dopants, special attention should be paid to controlling the dopant concentration in that heteroatoms of high concentration probably become recombination centers of electron–hole pairs, which is detrimental to the photocatalytic activity.[25,26] Overall, band gap engineering *via* doping provides an effective way to improve light absorption, regardless of the dimensionality of semiconductors. Nevertheless, as the thickness is reduced to the atomic-scale, dopants in nanosheets can be expected to directly participate in the catalytic reaction, thus greatly reducing the probability of charge carrier recombination. This will be addressed in detail in Section 12.3. On the other hand, the high reactivity of surface atoms in 2D nanosheets facilitates exotic element doping. In contrast to the conventional N-doping process under harsh conditions using NH_3, Matsumoto *et al.* reported N-doped niobium oxide nanosheets under UV illumination at room temperature.[27]

In view of the atomic-scale thickness, surface atoms of 2D nanomaterials become dominant as compared with those in the bulk counterparts; thus, the influence of surface effects on the whole system becomes more remarkable accordingly. In this regard, various strategies directed to the surface atoms, including defect engineering, exotic atom incorporation and ligand/molecule adsorption, have been employed to effectively tailor the intrinsic properties of 2D nanomaterials. As is known, defects inevitably exist in nanomaterials, yet they provide a versatile approach to control the fundamental properties. Recently, Xiong *et al.* reported that oxygen vacancies in ultrathin

WO_3 nanosheets form an intermediate energy level below the conduction band minimum, successfully extending the absorption to the near-infrared region.[28] As shown in Figure 12.4, numerous small pits are observed on the surface of WO_3 nanosheets, which is consistent with the electron spin resonance (ESR) signal at $g = 2.002$, manifesting the electron trapped by oxygen vacancies. With the introduction of oxygen vacancies, the band gaps of WO_3 nanosheets decrease from 2.61 to 2.09 eV, and remarkable NIR light harvesting is observed. Furthermore, the work functions and valence band maxima measured by synchrotron radiation photoemission spectroscopy (SRPES) clearly depict the difference of electronic band structures tailored by defect engineering. Likewise, MoO_{3-x} nanosheets also display intense absorption in a broad range from the visible to the near-infrared region, which probably originates from the localized surface plasmon resonances.[29]

The incorporation of exotic atoms or units into the lattice provides another approach to regulate the electronic structure of 2D nanomaterials, especially for these systems with open pores in their lattices. For example, carbon nitride (C_3N_4), a layered polymeric framework that consists of conjugated tri-s-triazine units, is an ideal support for adsorption or incorporation of exotic atoms (Figure 12.5). Recently, Wu *et al.* reported a simple 'bottom-up' way to incorporate titanium-oxide (TiO) unit into the C_3N_4 network by polycondensation of precursors containing dicyandiamide and $TiCl_4$.[30] HAADF-STEM

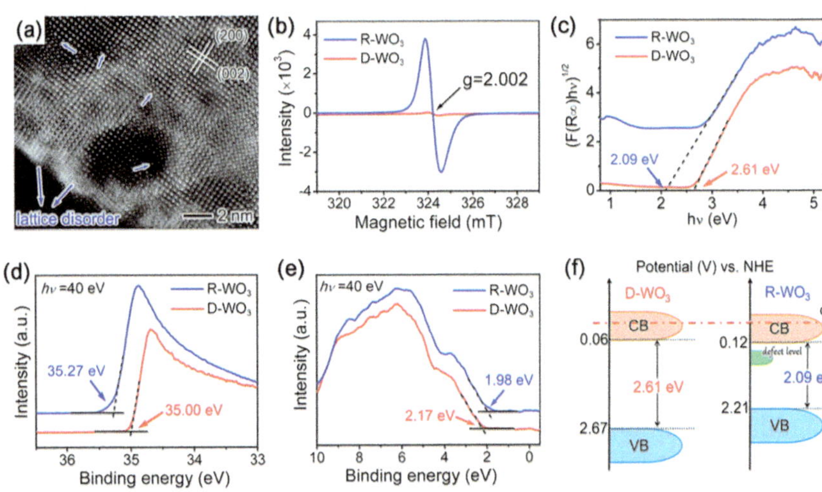

Figure 12.4 (a) Atomic-resolution HAADF-STEM image of deficient WO_3 nanosheets. (b) Room-temperature ESR spectra of the defect-controlled WO_3 samples. (c) Optical band gaps determined by UV-vis diffuse reflectance spectra. (d) Secondary electron cutoff (E_{cutoff}) of the SRPES spectra. (e) Valence-band spectra measured by SRPES. (f) Schematics illustrating the electronic band structures. The reduction level for free $O^{2\bullet-}$ radicals is marked by the red dash line. D-WO_3 and R-WO_3 present defect-deficient, defect-rich WO_3, respectively, adapted with permission from ref. 28, Copyright 2016 American Chemical Society.

image shows that TiO units with atomic size are homogeneously dispersed in the C_3N_4 network. The coordination number determined by EXAFS further confirms that the isolated TiO units are indeed incorporated into the subnanopores. Upon the incorporation of TiO units, more electrons are injected into the C_3N_4 framework, delocalizing π-electron in the conjugated system. Therefore, a remarkable red shift of the absorption band edge, which corresponds to the bandgap from 2.83 eV to 2.23 eV, is observed. Moreover, metal-to-ligand charge transfer (MLCT) is found to contribute visible-near-infrared absorption, as in the case of incorporation of Pt^{2+} or Cu^{2+} ions into the C_3N_4 network.[31] As shown in Figure 12.5(d), the absorption of C_3N_4-Pt^{2+} gradually increases with the concentration of Pt^{2+} from 0.04 wt% to 0.18 wt%, even extends to near-infrared region. Theoretical calculation reveals that the incorporation of Pt^{2+} ions into the C_3N_4 network restructures the electronic structure, giving rise to new electronic states with an energy gap of 1.8 eV, corresponding to the MLCT process. Such a MLCT-induced photoexcitation process would also contribute to the absorption of photons with energy above 1.8 eV.

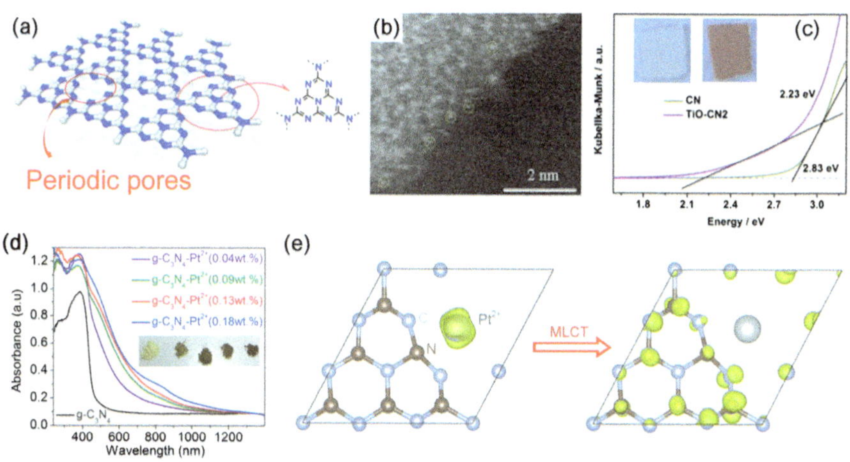

Figure 12.5 (a) The structure of C_3N_4 with periodic pores formed by conjugated tri-s-triazine units. (b) HAADF-STEM image of the TiO-C_3N_4 nanosheet. (c) The plot of transformed Kubelka-Munk function *vs.* photon energy for pure C_3N_4 and TiO-C_3N_4, adapted from ref. 30 with permission from The Royal Society of Chemistry. (d) UV-vis absorption spectra of C_3N_4 and C_3N_4-Pt^{2+} at various Pt concentrations. (e) Photoexcited charge density transition from the Pt^{2+}-induced hybrid HOMO states (in the range of 0–1 eV below the Fermi level) to the LUMO of C_3N_4-Pt^{2+} (in the range of 2–3 eV above the Fermi level), indicating the MLCT process. The yellow bubble represents electron population. Part (d) and (e) are adapted with permission from Y. Li *et al.*, Implenting Metal-to-Ligand Charge Transfer in Organic Semiconductor for Improved Visible-Near-Infrared Photocatalysis, *Advanced Materials*, John Wiley and Sons, © 2016 WILEY-VCH Verlag GmbH & Co. KGaA, Weinheim.[31]

Owing to a thickness down to the monolayer, 2D nanomaterials possess a large fraction of coordinatively unsaturated surface atoms as compared with those in the bulk; as a result, surface effects become more remarkable on the whole system accordingly, making it feasible to tailor the intrinsic properties *via* surface chemical modifications, such as ligand/molecule chemisorption, and metal ion grafting.[32–34] Moreover, note that surface atoms, especially these in intrinsic non-layered 2D nanomaterials, are typically coordinatively unsaturated and thus exhibit high chemical reactivity, facilitating the formation of new interfacial chemical bonds. For intrinsic layered systems, the increased reactivity of surface atoms arises from unpaired electrons along the edges or defects in the planar surface. For instance, the reversible surface ligand exchanges between L-type amine and Z-type metal carboxylates M(oleate)$_2$ (M = Cd, Zn) has been found to effectively tune the absorption of 2D II–VI semiconductors.[35] As shown in Figure 12.6, the absorption of CdSe quantum belts is reversibly shifted to lower energy by Cd(oleate)$_2$ passivation with respect to amine passivation. These shifts are attributed to the changes of both strain states and confinement dimensions upon surface ligand exchange. More importantly, the large fraction of surface and the confined dimension of CdSe quantum belts magnify this surface ligand exchange effect as compared with CdSe of other morphologies, leading to remarkable energy shift up to 140 meV.

For intrinsic layered systems, such as TMDs, sulfur vacancies, as common defect sites, are widely located along the edges or in the planar surface, providing potential targets for surface modification and functionalization. Nguyen *et al.* reported the surface chemical functionalization of 2D MoS$_2$ using thiols with different chain lengths and electron withdrawing capabilities.[36] The electron donating effect of thiols is found to induce a mild *n*-type doping of MoS$_2$, which significantly increases the Fermi level and meanwhile shifts both the valence and conduction bands of MoS$_2$ flakes. As shown in

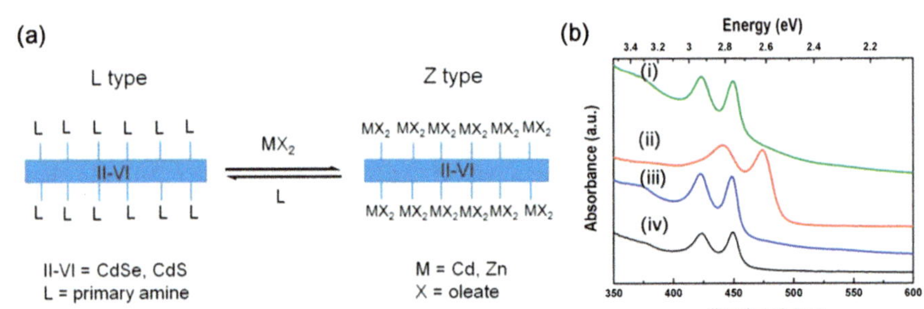

Figure 12.6 (a) Exchange of Z- and L-Type Surface Passivation in 2D II–VI Nanomaterials. (b) UV-vis extinction spectra (in toluene dispersion at RT) of (i) starting {CdSe[*n*-octylamine]$_x$}, (ii) {CdSe[Cd(oleate)$_2$]$_y$}, (iii) back-exchanged {CdSe[*n*-octylamine]$_x$}, and (iv) back-exchanged {CdSe [oleylamine]$_z$}, adapted with permission from ref. 35, Copyright 2015 American Chemical Society.

Figure 12.7, the magnitude of the band edge shifts is closely dependent on the Lewis basicity of the thiols. In this regard, surface functionalization indeed presents a facile approach to tune the intrinsic electronic structure of 2D nanomaterials and thus bring new opportunities to regulate light absorption.

Electronic band engineering plays an important role in determining the applicability of semiconductors. Obviously, this restriction limits the choice of materials to wide band gap semiconductors if only one semiconductor is used for light harvesting. To circumvent this problem, a Z-scheme system with two-photon absorption that mimics the natural photosynthetic process has been developed. Such a system, which is connected by redox shuttles, allows semiconductors to act as photoanodes for oxidation and photocathodes for reduction independently. Other light trapping strategies to improve the light-harvesting efficiency including the construction of periodic nanostructures and the incorporation of plasmonic metal nanostructures have

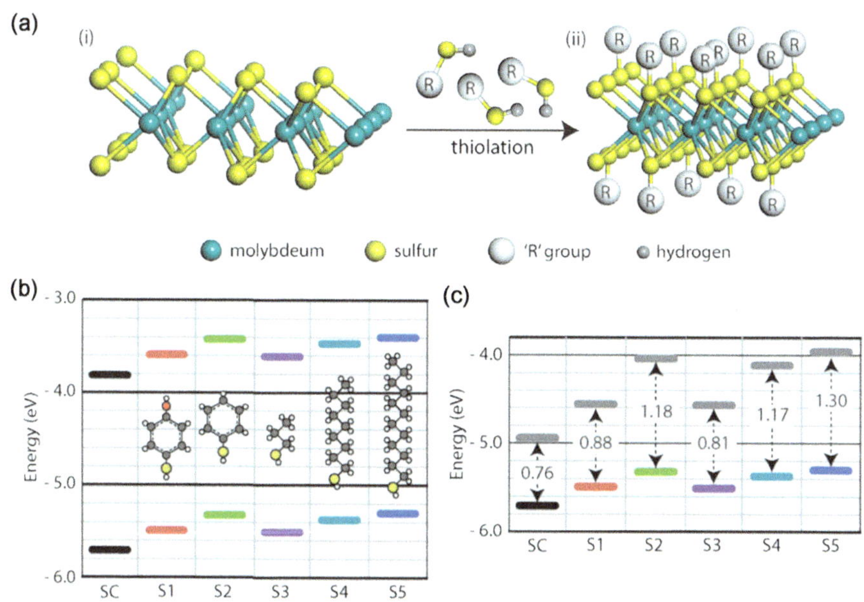

Figure 12.7 (a) Defective MoS_2 monolayer showing sulfur vacancies, and 'R' group functionalized MoS_2 monolayer. (b) Energy diagram of the as exfoliated and thiol functionalized MoS_2. (c) The valence band positions measured using PESA and the Fermi levels estimated using XPS. The thiols used are *p*-mercaptophenol (S1), thiophenol (S2), 1-propanethiol (S3), 1-nonanethiol (S4), and 1-dodecanethiol (S5), respectively. Thiol structures are drawn within the respective band gap of the functionalized samples. White atoms refer to hydrogen, grey to carbon, yellow to sulfur, and red to oxygen, adapted with permission from E. P. Nguyen *et al.*, Electronic Tuning of 2D MoS_2 through Surface Functionalization, *Advanced Materials*, John Wiley and Sons, © 2015 WILEY-VCH Verlag GmbH & Co. KGaA, Weinheim.[36]

been discussed in detail elsewhere.[37–39] Overall, benefiting from a large fraction of surface atoms and thin confinement dimension, 2D nanostructures enable incredible flexibility for tailoring the intrinsic electronic structure, thus bringing new opportunities to achieve high-efficient light harvesting.

12.3 Interfacial Charge Carrier Dynamics

Among all the steps of artificial photocatalysis, efficient interfacial charge carrier separation and transport is undoubtedly the most challenging one.[40–42] In principle, upon excitation by photons with the energy above the band gap, electron–hole pairs, which are bound by the Coulombic attraction force, form in a semiconductor. Initially, the electron and hole with excessive kinetic energy, the so-called hot charge carriers, undergo intraband relaxation *via* electron–phonon coupling and these processes usually occur on a time scale of several tens to hundreds of fs. Once the electron and hole relax to the bottom of the CB and the top of the VB, respectively, they may recombine in tens of ps either radiatively, thereby emitting a photon, or non-radiatively, with heat production. However, considering that intermediate states are practically inevitable in semiconductors due to the intrinsic surface states or trap states, charge trapping by these states becomes considerably important and it takes place on a time scale of hundreds of fs to tens of ps, which is usually faster than that of band edge recombination. Following trapping, electron and hole can either undergo additional trapping, *e.g.* from shallow trapping centers to deeper ones in the bandgap or further recombine radiatively or non-radiatively. These key dynamic processes and related time scales are summarized in Figure 12.8.[43] However, it needs to be emphasized that, in contrast to the short-lived excited state, a lot of events of interest in photocatalysis, such as diffusions and radical reactions, typically occur at the time scale ranging from femtoseconds to microseconds or even longer. Therefore, extracting the photogenerated carriers competitively with their recombination has long been recognized as the main challenge in heterogeneous photocatalysis.

As is known, photocatalytic reactions occur on the surface of a semiconductor. Only the photogenerated charges that reach the surface can participate in photocatalytic reactions. 2D nanomaterials with a thickness down to a few nanometers or less typically possess high surface-to-volume ratio and thus they can effectively reduce the diffusion distance of photogenerated electron–hole pairs, suppressing their recombination. More importantly, the unique structural features in the atomic scale give rise to new intermediate states that can significantly affect the charge carrier dynamics. Here we summarize the major advantages and recent advances of 2D nanosheets in tailoring the charge carrier dynamics. Special emphasis will be paid to the charge separation behaviors of 2D nanosheets as co-catalysts. Experimental techniques for tracking the charge carrier dynamics, particularly the time-resolved pump-probe spectroscopy are also discussed.

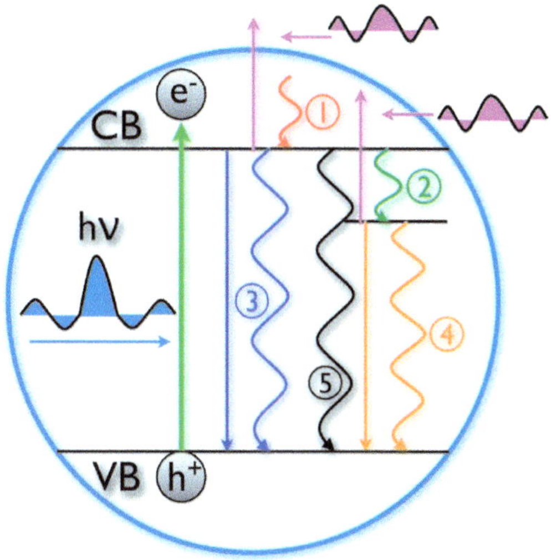

Figure 12.8 Illustration of key dynamic processes in a QD and related pump-probe transient absorption measurement. The excitation or pump is represented as a green arrow and the probe is represented as fuchsia-colored arrows. Different dynamic processes shown include: (1) electron relaxation through electron-phonon coupling within the conduction band (a similar process occurs for the hole in the valence band) following excitation across the bandgap, tens to hundreds of fs; (2) trapping of electrons–holes into trap states due to defects or surface states, hundreds of fs to tens of ps; (3) radiative and non-radiative band edge electron–hole (or exciton) recombination, tens of ps; (4) radiative and non-radiative trapped electron–hole (or relaxed exciton) recombination, hundreds of ps to a few ns; and (5) non-linear and non-radiative exciton–exciton annihilation, hundreds of fs to tens of ps, adapted with permission from D. A. Wheeler and J. Z. Zhang, Exciton Dynamics in Semiconductor Nanocrystals, *Advanced Materials*, John Wiley and Sons, Copyright © 2013 WILEY-VCH Verlag GmbH & Co. KGaA, Weinheim.[43]

When semiconductors are in contact with an electrolyte solution, charge transfer takes place due to the discrepancy between the Fermi level of the semiconductors and the redox potential of electroactive species in solution, eventually leading to band bending under equilibrium. However, in the case of nanomaterials, the band bending is negligibly small and thus charge transport occurs dominantly *via* diffusion, in a spatially random fashion along the optical path. In this condition, a random walk model is proposed to describe the motion of the charge carriers:[7]

$$t_d = r^2/\pi^2 D \tag{12.2}$$

where t_d is the average diffusion time from the inside of the nanoparticle to the surface, r is the particle size, π is a mathematical constant, and D is the diffusion coefficient of the charge carrier. As an example, the average transit time of TiO_2 with a radius of 6 nm is only 3 ps, which is comparable to the carrier recombination time (1–10 ps). In the case of Fe_2O_3, which suffers from short hole diffusion length (only 2–3 nm), only holes generated within the diffusion length can migrate to the surface and participate in the oxidation reaction.[44] In this sense, reducing the size of the semiconductor becomes a direct strategy to effectively shorten the time required for carrier migration, thereby inhibiting their recombination.

Atomically thin nanosheets possess high specific surface areas and the consequent large fraction of uncoordinated surface atoms. According to the above diffusion model, the atomically thickness of 2D nanosheets facilitates rapid carrier transport. Meanwhile, the large fraction of uncoordinated surface atoms allows intimate contact with electrolyte, thus accelerating the charge transfer to the adsorbate or electrodes in photocatalytic reactions. Benefiting from these advantages, Wu *et al.* developed a molecular co-catalyst strategy to accelerate the hole transfer. The trifluoroacetic acid (TFA) molecule, by virtue of its reversible redox couple $TFA•/TFA^-$ and highly active intermolecular radical reactions, provides an additional charge separation channel across the catalyst/solution interface.[45] Combined *in-situ* ESR and energy level alignment, it is confirmed that photogenerated holes in the VB of $K_4Nb_6O_{17}$ nanosheets can readily react with the adsorbed TFA molecule to yield TFA radical and immediately the highly reactive radicals transfer holes to methanol, forming a highway for hole transport. Moreover, transient absorption (TA) spectroscopy is applied to track the charge carrier dynamics of nanosystems. As shown in Figure 12.9, in the presence of TFA, the average time describing the overall electron transfer from CB to defect states (e-DS) is increased by a factor of ~2.48 (312 ± 37 ps for $K_4Nb_6O_{17}$-TFA *vs.* 126 ± 11 ps for $K_4Nb_6O_{17}$) and meanwhile the lifetime of the overall radiative electron–hole recombination is also increased by a factor of *ca.* 2.73 (8.2 ± 0.4 ns for $K_4Nb_6O_{17}$-TFA *vs.* 3.0 ± 0.1 ns for $K_4Nb_6O_{17}$). The increased life-times of both TA and PL spectra are understandable because effective vacation of hole in either VB or defect states *via* TFA anion-accelerated hole transfer will eventually decelerate the process of electron–hole recombination and consequently hinder electron transfer from CB to e-DS.

In addition to the high specific surface areas, the unique structural features of 2D nanomaterials in the atomic scale bring about new electronic states that can significantly alter the charge carrier dynamics. Recent studies have revealed that the atomically thin sheets typically possess much increased electronic states near the band edges as compared with their bulk counterparts, which provide new opportunities to tune the electrical properties, such as the carrier concentration, carrier mobility, *etc.* As an example, Xie *et al.* reported SnS_2 single-layers with three atom thickness using formamide assisted exfoliation method.[46] Synchrotron radiation XAFS reveal that bond length and degree of structural distortion were increased in these single-layer SnS_2. This surface atomic distortion influences its electronic properties, as confirmed by corresponding first-principles calculations. The calculated densities of states (DOS) showed an obviously increased density

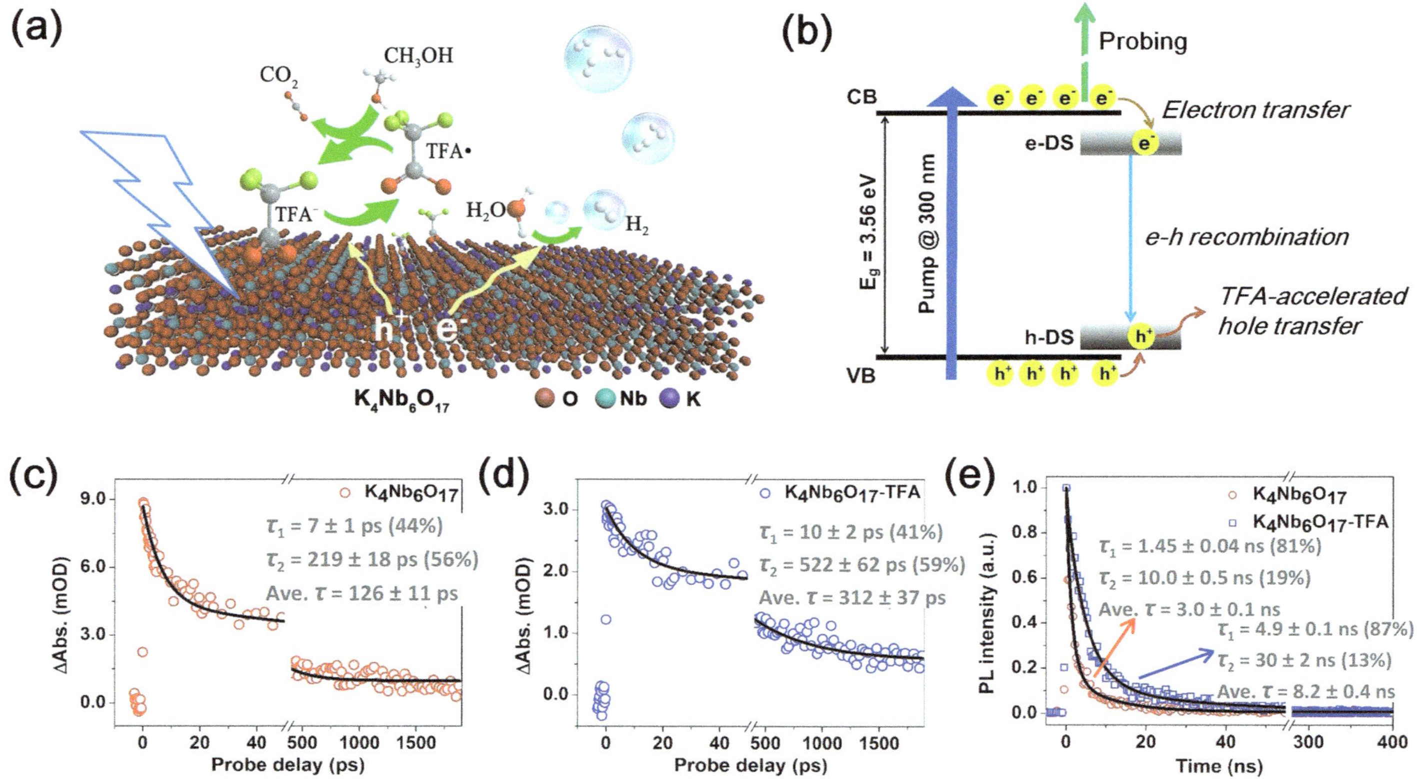

Figure 12.9 (a) Schematic illustration for the molecular co-catalysts strategy for accelerating hole transfer. (b) Schematic illustration for the mechanisms involved. Representative ultrafast TA kinetics probed at 500 nm (pump at 300 nm) for $K_4Nb_6O_{17}$ nanosheets in the (c) absence and (d) presence of TFA. (e) Time-resolved PL spectra for both $K_4Nb_6O_{17}$ nanosheets and $K_4Nb_6O_{17}$-TFA. Reprinted with permission from Macmillan Publishers Ltd: *Nature Communications* (ref. 45), Copyright 2015.

of states at valence band maximum compared with the bulk counterpart, which indicates higher carrier mobility, thus remarkably enhancing visible light water splitting properties. As an outcome, the photocurrent of single-layer SnS_2 under visible light is over 70 times larger than that of bulk material. Atomically thin sheets with intrinsic non-layered structure typically accompany more dramatic structural changes, thus leading to unique physical and chemical properties. Xie *et al.* synthesized four-atom-thick ZnSe nanosheets *via* an organic–inorganic lamellar hybrid intermediate strategy.[47] Extended XAFS reveal that the bond length and bond angle were distorted relative to bulk ZnSe, which in turn significantly influenced its electronic properties. The density functional theory calculations demonstrate that single-layer ZnSe displays an obviously enhanced DOS at the conduction band minimum, which ensures a higher carrier migration rate, hence resulting in improved electron–hole separation efficiency. Thus, the photocurrent density of ZnSe single layers is 195 times higher than that of the bulk ZnSe. Meanwhile, the incident photon-to-current conversion efficiency (IPCE) of single-layer ZnSe (42.5%) is also strikingly higher than that of the bulk counterparts (0.25%).

Xie and co-workers also developed a lamellar Bi-oleate hybrid intermediate method to prepare single-unit-cell Bi_2WO_6 layers with the thickness of 1.65 nm.[48] As shown in Figure 12.10, DFT calculations show that atomically-thin Bi_2WO_6 layers have an increased density of states at the conduction band edge compared with bulk Bi_2WO_6. Furthermore, the distribution of charge density near the Fermi level indicates that most electron density derives from the surface atomic layer. Moreover, temperature-dependent resistivity curve demonstrates that electronic conductivity of the single-unit-cell Bi_2WO_6 layers is dramatically higher than that of bulk counterpart. Both theoretical calculation and experiments demonstrate that reducing the dimensionality from 3D bulk to 2D atomic thickness can effectively improve its electrical conductivity. The highly increased electronic conductivity and atomic layer may promote the rapid transfer of photo-generated carriers from the interior to surface, which reduce the electron–hole recombination and increase the lifetime of carriers. As expected, time-resolved PL decay spectra show that Bi_2WO_6 with single-unit-cell thickness increases the carriers' lifetime from 14.7 to 83.2 ns, indicating an effectively reduced electron–hole recombination rate. Thus, photocatalytic methanol formation rate of the single-unit-cell Bi_2WO_6 is 125-times higher than that of the bulk one. Hu *et al.* also found that the exfoliation of Ni–Fe layered double hydroxides into single-layer nanosheets can remarkably improve the electronic conductivity, which lead to smaller Tafel slope for oxygen evolution reaction.[49]

Defects inevitably exist in nanomaterials and cause the formation of trap states in the band gap, thus significantly affecting fundamental properties, such as light absorption, charge transport and separation, as well as the reactivity. Oxygen vacancy, as the most common defect in oxide semiconductors, has long been used to engineer the electronic properties of semiconductors. However, the role of oxygen vacancies in the photocatalytic process is

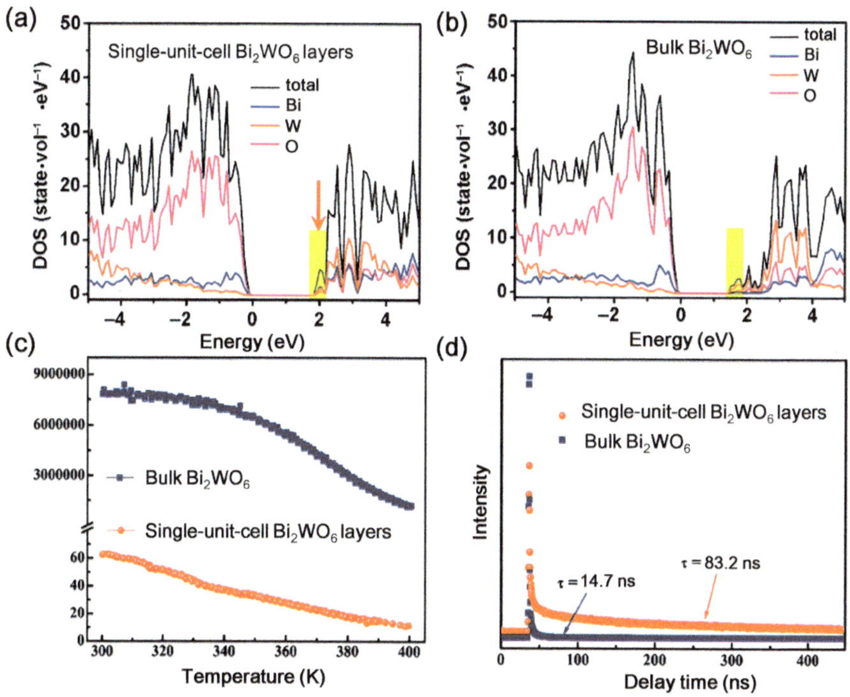

Figure 12.10 Calculated density of states of (a) single-unit-cell Bi_2WO_6 layer slab and (b) bulk Bi_2WO_6 slab. (c) temperature-dependent resistivities and (d) fluorescence emission decay spectra for the single-unit-cell Bi_2WO_6 layers and bulk Bi_2WO_6. Adapted with permission from L. Liang *et al.*, Single Unit Cell Bismuth Tungstate Layers Realizing Robust Solar CO_2 Reduction to Methanol, *Angewandte Chemie International Edition*, John Wiley and Sons, © 2015 WILEY-VCH Verlag GmbH & Co. KGaA, Weinheim.[48]

still an open issue, probably due to the neglect of the influence of oxygen vacancies in different spatial location.[50] As shown in Figure 12.11, oxygen vacancies distribute both in the internal and on the surface of bulk materials. Nevertheless, as the thickness is reduced to the atomic-scale, the surface atoms of 2D nanomaterials become dominant as compared with those in the bulk counterparts; accordingly, defects in ultrathin nanosheets can be considered to locate on the surface only. In this regard, ultrathin nanosheets provide an ideal model for studying surface oxygen vacancies. Inspired by this unique structural feature, Xie *et al.* systemically studied the influence of spatial location of oxygen vacancies on the interfacial charge carrier dynamics. Taking $K_4Nb_6O_{17}$ nanosheets with the thickness of half-single-unit-cell as an example, it was found that both surface and bulk oxygen vacancies can serve as charge carrier traps, whereas charge carriers trapped by bulk oxygen vacancies are prone to recombine due to the long charge migration path. Conversely, benefiting from the negligible charge migration paths, surface oxygen vacancies also serve as reaction sites where charge carriers can

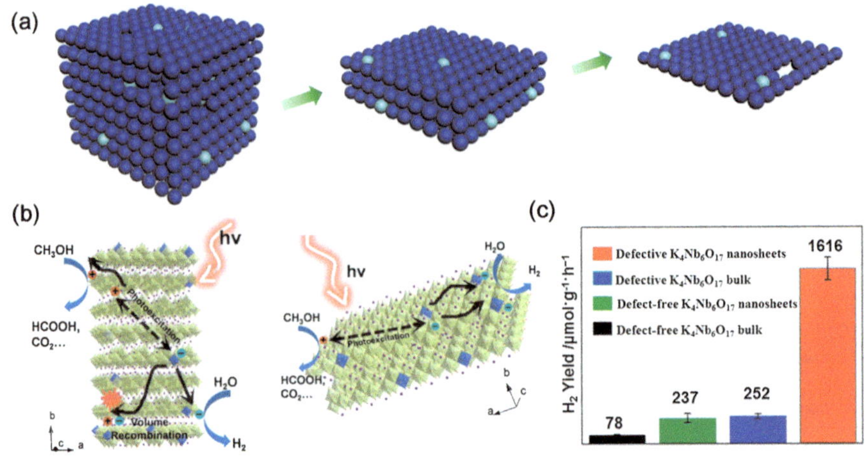

Figure 12.11 (a) Schematic illustration of the spatial distribution of defects in bulk and in ultrathin nanosheets. (b) Photo-generated carrier migration paths in the defective $K_4Nb_6O_{17}$ bulk and nanosheets. (c) A comparison of H_2 evolution activity between the defective $K_4Nb_6O_{17}$ nanosheets, the defect-free $K_4Nb_6O_{17}$ nanosheets, the defect free $K_4Nb_6O_{17}$ bulk and the defective $K_4Nb_6O_{17}$ bulk. Parts (b) and (c) are adapted with permission from W. Bi *et al.*, Spatial Location Engineering of Oxygen Vacancies for Optimized Photocatalytic H_2 Evolution Activity, *Small*, John Wiley and Sons © 2014 WILEY-VCH Verlag GmbH & Co. KGaA, Weinheim.[50]

directly participate in the redox reaction, thus effectively reducing the probability of hole–electron recombination.

Furthermore, Xie *et al.* reported five-atom-thick In_2O_3 porous sheets with different oxygen vacancies concentration *via* fast heating the ultrathin $In(OH)_3$ nanosheets intermediate under air and oxygen atmosphere, respectively.[51] First-principle calculations show that the five-atom-thick In_2O_3 pore sheets, which are rich in oxygen vacancies, exhibit a new donor level and increased density of states at the conduction band minimum and at valence band maximum, which indicate a smaller band gap and higher carrier concentration, hence enhancing the solar light absorption and the electron–hole separation efficiency. Therefore, the photocurrent of the oxygen vacancies-rich ultrathin In_2O_3 porous sheets-based photoelectrode is 2.5 and 15 times larger than that of the oxygen vacancies-poor sheets and bulk In_2O_3-based photoelectrodes, respectively.

As discussed in Section 12.2, exotic element doping is the most conventional strategy for designing visible-light-responsive photocatalysts. Unfortunately, random fluctuation induced by bulk dopant inevitably increases the probability of electron–hole recombination, leading to reduced catalytic activity. Intriguingly, as the thickness is reduced to the atomic-scale, dopants in nanosheets can be expected to directly participate in the catalytic reaction, which can promote the utilization of charge carriers and thus effectively reduce the risk of charge recombination. In addition, given that 2D nanomaterials possess a large fraction of coordinatively unsaturated surface atoms as compared with those in the bulk counterparts, surface modification becomes a feasible way to precisely tailor the surface electronic states and thus manipulate the charge carrier dynamics.

Taking advantage of the highly active surface atoms of atomically thin sheets, Wu and co-workers proposed a surface sulfur incorporation strategy to achieve metallic $Ni(OH)_2$ nanosheets.[52] $Ni(OH)_2$ ultrathin nanosheets are treated under H_2S flow to enable sulfur incorporation. Ni K-edge XAFS and ^1H solid-state NMR spectra demonstrate that sulfur substitute oxygen atoms. Considering that sulfur atoms have a bigger atomic radius and a weaker electronegativity than that of oxygen atoms, sulfur substitution would result in weaker covalent bonding and delocalize immobilized electrons, offering possibility to tailor the electronic structure. The DFT calculations show that sulfur incorporated $Ni(OH)_2$ nanosheets possess zero band gap, indicating metallic conductivity. The temperature-dependent resistivity curve and Hall coefficient measurements imply that sulfur incorporated $Ni(OH)_2$ nanosheets indeed exhibit a metallic characteristic with 3.13×10^{-4} Ωm, and the main carriers are electrons, whose concentration is up to 10^{20} cm^{-3}. It is anticipated that the metallic $Ni(OH)_2$ ultrathin nanosheets can facilitate the electron transfer and thus open versatile paths to design Ni based heterogeneous photocatalysts.

Additionally, Ajayan *et al.* achieved precise control of the surface electronic states of 2D metal chalcogenides by Lewis acid–base chemistry.[53] As is known, the surface atoms of 2D metal chalcogenides usually terminate with lone pair electrons and thus they can readily react with Lewis acids, forming stable coordination complexes for further functionalization. Take 2D InSe as an example: this is a layered structure in which the planes of In atoms are sandwiched between two atomic layers of Se by strong in-plane covalent bonding, while layers are vertically stacked together *via* weak van der Waals interactions. Under the molecular orbital approximation, three of the four sp^3 orbitals form In–Se covalent bonding and the remaining one is occupied by a lone pair electrons, featuring *n*-type semiconductor (Figure 12.12). But a Lewis acid with empty electron orbitals, such as Ti^{4+} can accept the lone pair electrons of 2D InSe, forming a planar *p*-type InSe–Ti coordination complex, thus regulating the Fermi level effectively. The InSe–Ti/InSe p–n junction formed *via* this method allows effective charge carrier separation, as confirmed by the obvious rectification and photovoltaic effects. Moreover, the Lewis acid-base can serve as a bridge to connect functional molecules to the atomic layers, realizing various applications. For instance, attaching ruthenium-based dye molecule (N719) to the InSe–Ti coordination complex increase the Fermi level, which enhance the photoresponsivity ~20 times.

Graphite and the family of TMDs, which are bound by weak van der Waals interactions, possess high mobility and/or excellent catalytic activity. Consequently, combining these two classes of systems with light-absorbers and

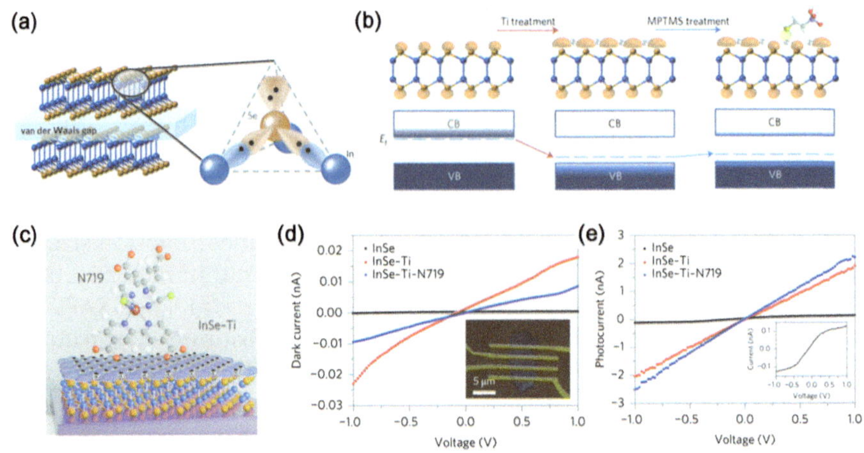

Figure 12.12 (a) Crystal structure model of double-layered InSe. Under the molecular orbital approximation, Se atoms have sp^3 hybridization configuration and the lone pair electrons on Se terminate the lattice plane. (b) The lone pair electrons from the Se atoms enter the empty orbitals of Ti^{4+}, and the original Fermi level (E_f) goes down, resulting in *p*-type doping. After introducing the 3-mercaptopropyltri-methoxysilane (MPTMS), the lone pair electrons enter the orbital of Ti^{4+} and repel the Se lone pair electrons. Consequently, the Fermi level moves upwards, rendering a weaker *p*-type doping. (c) The schematic of the N719-functionalized InSe–Ti complex. The –COO group anchors the N719 dye to the surface of the InSe flake *via* a Ti bridge. (d) The dark currents of the InSe flake (inset) before treatment (black), after Ti treatment (red) and after N719 treatment (blue), respectively. (e) The photocurrent of InSe before (black), after Ti treatment (red) and N719 treatment (blue). Inset: magnified InSe photocurrent. Reprinted with permission from Macmillan Publishers Ltd: *Nature Nanotechnology* (ref. 53), Copyright 2016.

understanding the photoexcited carrier dynamics in these hybrid nanostructures becomes a vigorous area of research. Heinz *et al.* studied the nonradiative energy transfer (NRET) between CdSe/CdZnS core–shell quantum dots (QDs) and adjacent graphene and MoS_2 with different thickness (Figure 12.13).[54] The PL of QDs is quenched 90% on the 2D monolayers and the quenching on graphene is greater than on MoS_2. This phenomenon indicates the existence of efficient photoexcited carriers transfer channels. Also, time-resolved photoluminescence (TRPL) measurements indicate that the PL lifetime is decreased by an order of magnitude in the presence of the 2D materials. However, for graphene, the PL lifetime of the QD decreases as the thickness increases, while the MoS_2 shows the opposite trends. The opposite trends of NRET in graphene and MoS_2 can be understandable in terms of the differences in the local electric field. The layer-dependent changes of electric field within graphene are weaker. However, in the case of MoS_2, the screening of the electric field with each added layer is so strong that it leads to a pronounced decrease in the overall NRET with the increase of the thickness.

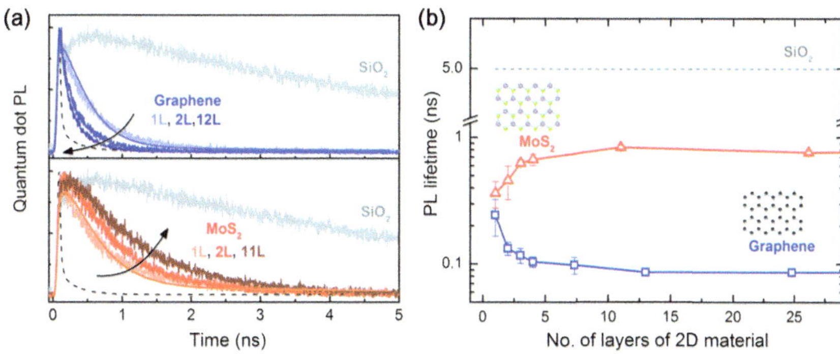

Figure 12.13 (a) TRPL traces on graphene layers are shown in the top panel and on MoS_2 in the bottom panel. Exemplary fits are overlaid on the data in solid blue and red lines. The grey trace corresponds to the SiO_2 substrate and the instrument response function is the dashed black line. (b) The QD lifetimes as a function of thickness of 2D layers are shown. Each data point corresponds to an average decay lifetime over multiple samples; the error bars represent the standard deviation of the measurements. The grey dashed line at 5 ns denotes the average lifetime of the dots on the SiO_2 substrate. The solid red and blue lines are guides to the eye. Adapted with permission from ref. 54, Copyright 2016 American Chemical Society.

The NRET behavior of 2D materials helps to understand the near-field coupling of light-absorbers to a range of semiconducting and metallic materials, and thus opens versatile paths to manipulate the charge carrier dynamics.

2D TMDs which contain transition metals with unfilled d-orbitals are rich in composition and structure, thus exhibiting excellent optical, electrical, and catalytic properties. Different from 2H-MoS_2, 1T-MoS_2 exhibits metallic characteristic due to the Mo–S octahedral coordination. Benefiting from its metallic conductivity, the ultrathin 1T-MoS_2 nanosheet has been confirmed as a promising co-catalyst to facilitate charge separation and simultaneously provide additional active sites for photocatalytic H_2 evolution. Song *et al.* used a bottom-up method to fabricate an intimate nanojunction consisting of few layer 1T-MoS_2 on the surface of CdS nanorods (see Figure 12.14).[55] Steady-state PL spectroscopy shows dramatic quenching of band edge and trap state emission of CdS nanorods after the introduction of 1T-MoS_2. Time-resolved PL spectroscopy indicates that the loading of 1T-MoS_2 reduces the average lifetime of the CdS NRs by a factor of approximately 1.82 (ave 2.67 ns for 1T-MoS_2@CdS *versus* 4.87 ns for bare CdS NRs). Further insight into the charge carrier dynamics was obtained by ultrafast TA spectroscopy. As shown in Figure 12.14(e), obvious bleach corresponding to the excitonic band of CdS is observed. The exciton relaxation kinetics is characterized by two time constants (t_1 and t_2), which is related to the consecutive relaxation to two trap states with different trap depths. For 1T-MoS_2@CdS, the shortening in both t_1 and t_2 should be attributed to the opening of additional electron transfer channel, *i.e.* electron transfer from both trap states to the loaded 1T-MoS_2, which remarkably promotes charge separation efficiency and consequently

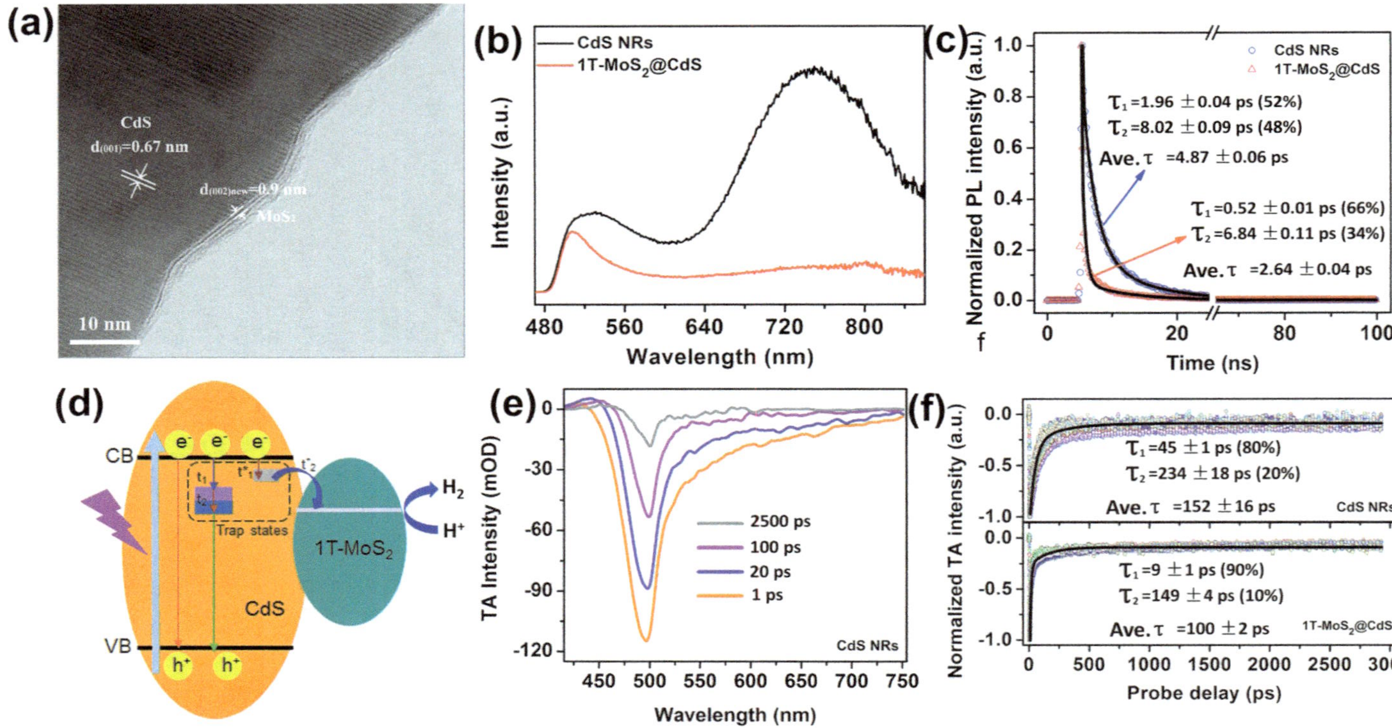

Figure 12.14 (a) TEM image of 1T-MoS_2@CdS, showing the folded edges of 1T-MoS_2 layers on the CdS NRs. (b) Steady-state PL spectra (excitation at $\lambda = 400$ nm) of the CdS NRs and 1T-MoS_2@CdS. (c) Time-resolved PL spectra (excitation at $\lambda = 400$ nm, emission at $\lambda = 515$ nm) of the CdS NRs and 1T-MoS_2@CdS. (d) Schematic illustration of the 1T-MoS_2 cocatalyst strategy for accelerating electron transfer. (e) TA spectra of 1T-MoS_2 @CdS (excitation at $\lambda = 400$ nm) showing the evolution of the dominant 1S excitonic bleach of 1T-MoS_2@CdS. The signal (*i.e.*, the absorbance changes) is given in mOD, for which OD is the optical density. (f) The relaxation kinetics of the CdS NRs and 1T-MoS_2@CdS 1S excitonic bleach. Adapted with permission from Q. Liu *et al.*, *In-situ* integration of a metallic 1T-MoS_2/CsS Heterostructure as a Means to Promote Visible-Light-Driven Photocatalytic Hydrogen Evolution, *ChemCatChem*, John Wiley and Sons © 2016 WILEY-VCH Verlag GmbH & Co. KGaA, Weinheim.[55]

reduces undesirable charge recombination. Thus, the visible-light driven photocatalytic H_2 rate of the 1T-MoS_2@CdS heterojunction is 39-fold relative to that of bare CdS nanorods with only 0.2 wt% 1T-MoS_2 loading.

Owing to the quantum coupling between the 2D layers, the heterostructure generally possess fascinating new phenomena. Different kinds of atomically thin 2D layers, including TMDs, graphene and hexagonal boron nitride, provide rich material supports to fabricate various heterostructures, such as in-plane lateral heterojunctions, vertical heterojunctions and the corresponding superlattice.[56,57] Theory predicts that WS_2 and MoS_2 monolayers have different band gaps of 2.31 eV and 2.39 eV, respectively. The conduction band minimum and valence band maximum of WS_2 is slightly higher than that of MoS_2, thus forming a type II semiconductor heterojunction, as shown in Figure 12.15. Accordingly, the photoexcited electrons prefer to

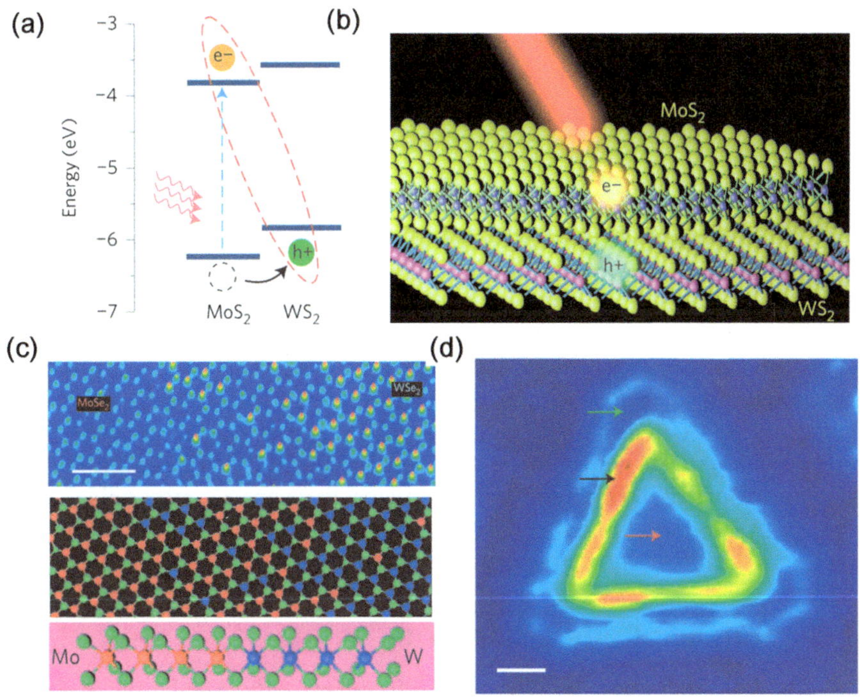

Figure 12.15 (a) The theoretically predicted band alignment of a MoS_2/WS_2 vertical heterostructure, which forms a type II heterojunction. (b) Illustration of a MoS_2/WS_2 vertical heterostructure with a MoS_2 monolayer lying on top of a WS_2 monolayer. Electrons and holes created by light are shown to separate into different layers. Reprinted with permission from Macmillan Publishers Ltd: *Nature Nanotechnology* (ref. 58), Copyright 2014. (c) 3D color map and scattered electron intensities of the metal sub-lattice sites of the $MoSe_2$–WSe_2 in-plane heterojunction. (d) 2D PL intensity map of a triangular lateral heterostructure. Reprinted with permission from Macmillan Publishers Ltd: *Nature Materials* (ref. 59), Copyright 2014.

reside in the MoS_2 layer, while holes stay in the WS_2 layer. The spatial charge separation can effectively suppress charge carrier recombination. Wang *et al.* reported the first experimental observation of ultrafast charge transfer in atomically thin MoS_2/WS_2 vertical heterostructure.[58] Photoluminescence of the MoS_2/WS_2 heterostructure is strongly quenched relative to the isolated monolayer MoS_2 and WS_2, indicating the existence of efficient charge separation in this heterostructure. TA spectra of MoS_2/WS_2 heterostructures reveal that the photoexcition in MoS_2 modify the exciton transition in WS_2. It also shows that holes in the MoS_2 layer can transfer into the WS_2 layer within 50 fs upon photoexcitation. Such ultrafast charge transfer in van der Waals heterostructures will benefit solar energy conversion applications, such as optoelectronics and photocatalysis. Xu *et al.* reported high-quality $MoSe_2$ and WSe_2 lateral heterojunctions *via* in-plane heteroepitaxy using physical vapor transport.[59] Aberration-corrected ADF STEM images verify that all the atoms lie on a single undistorted MX_2 honeycomb lattice. A sharp and asymmetric interface is clearly observed due to the intensity contrast of Mo and W atoms, as shown in Figure 12.15. The substantially diverse electronic properties between $MoSe_2$ and WSe_2 monolayers bring about fascinating new charge carrier dynamics across the interface. As shown in the micro-PL maps at room temperature, the emission intensity of the heterojunctions is remarkably increased as compared with that of either $MoSe_2$ or WSe_2 monolayers, indicating the enhanced radiative recombination at the interface. However, considering that WS_2 is an *n*-type semiconductor and WSe_2 is a *p*-type semiconductor, the WSe_2–WS_2 lateral heterostructure can form a *p*–*n* junction, which is confirmed by their rectification characteristic.[60] Moreover, the response time of the photodiode fabricated with WSe_2–WS_2 *p*–*n* junction is less than 100 μs, indicating the efficient photocarrier transport and separation. PL mapping reveals that the photocurrent response mainly arises from the lightly doped WS_2 and the interface region. Beyond the typical 2D layered semiconductors, heterojunctions constructed by other 2D materials enable more abundant electronic properties across the interface, which will open a new door to control the charge carrier dynamics.

12.4 Surface Redox Reaction

Photocatalytic reactions usually occur on surface atoms with unsaturated coordination, such as steps, edges, and corner atoms that are often called 'active sites'.[61] However, to date, a comprehensive understanding of the interplay between active sites and catalytic activity at the atomic level is still challenging, owing to the entangled structural information from the bulk. Fortunately, atomically-thin 2D sheets could serve as a promising model system to reveal the structure–catalytic activity relationship, considering their atomically well-defined local structure arising from the ultrahigh fraction of surface atoms that are comparable to the overall atoms. In this sense, numerous theoretical and experimental investigations concentrating on the catalytic application of 2D nanosheets have been carried out in recent years, and there is substantial agreement on the fact that the impressive catalytic activity of 2D nanosystems derives from their unusual surface atomic structure and the concomitant changes of electronic structure. Here we summarize the major advantages and opportunities of 2D nanosheets in improving the surface redox reactions of photocatalysis. Special emphasis will be paid to recent advances in fine characterization and controllable tailoring of the local atomic and electronic structure of 2D nanosheets, aiming to provide an in-depth understanding of the structure–catalytic activity relationship at the atomic level.

Since photocatalytic reactions occur on the surface, increasing the surface area is considered a straight-forward way of improving the efficiency. To this end, extensive efforts have been devoted to improve the surface-to-volume ratio of the catalysts *via* designing various nanostructures. 2D nanomaterials with the thickness down to a few nanometers or less typically possess high specific surface areas. As an example, graphene has a theoretical specific surface area of 2630 $m^2\ g^{-1}$. After activation by KOH, the microwave exfoliated graphite oxide even achieves a specific surface area of up to 3100 $m^2\ g^{-1}$.[62] The specific surface area of oxide nanosheets is generally estimated to be several hundred square meters per gram.[63] As a widely studied photocatalyst, TiO_2 nanosheets with the average thickness about 1 nm exhibit an ultrahigh specific surface area of 336 $m^2\ g^{-1}$, which is approximately six times higher than that of the commercial P25 TiO_2 (50 $m^2\ g^{-1}$).[64] As expected, the large surface area endows more adsorptive and reactive sites to improve the surface redox reactions. For photocatalysis, as discussed in Section 12.3, high specific surface area also reduces the diffusion length of photogenerated electron–hole pairs, suppressing their recombination to allow more charge carriers to participate in the photocatalytic reactions.

In addition to the ultrahigh specific surface area, the novel electronic structures arising from the unique structural features in the atomic scale are more prevalent yet critical to manipulating the surface redox reactions. Structural defects such as vacancies and the associated distortions seem inevitable in 2D nanomaterials when the thickness is down to a few nanometers or less, which can strongly affect their electronic structure and hence tune the reactivity. Owing to the unsaturated coordination, defects on the surface generally serve as chemisorption sites and the concomitant electron transfer leads to activation of the adsorbed species. For 2D nanomaterials, surface atoms become dominant as compared with those in the bulk counterparts. Accordingly, defects in ultrathin nanosheets can be considered to directly participate in chemisorption and the ensuing redox reactions. As shown in Figure 12.16, an oxygen molecule chemisorbed at the coordinatively unsaturated W site of defective WO_3 nanosheets exhibits an end-on configuration, accompanying with electron transfer from WO_3 to oxygen molecule and consequently activating the oxygen species. However, the perfect WO_3 surface can hardly chemisorb oxygen species.[28]

Xie and co-workers experimentally achieved high solar photocatalytic activity in ultrathin BiOCl nanomaterials.[65] In their study, the dominate defects

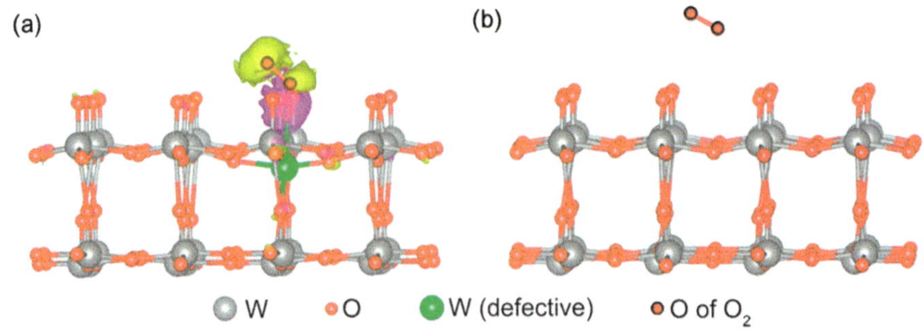

Figure 12.16 (a) Oxygen molecule chemisorbed at a coordinatively unsaturated W site of defective WO_3 nanosheets. (b) Oxygen molecule adsorbed at the surface of perfect WO_3 nanosheets. The purple and olive colors represent increase and decrease in electron density, respectively. Adapted with permission from ref. 28, Copyright 2016 American Chemical Society.

change from isolated vacancy V_{Bi}''' to triple vacancy associates $V_{Bi}'''V_O^{\bullet\bullet}V_{Bi}'''$ when reducing the thickness of the BiOCl catalysts to atomic scale (see Figure 12.17). It is well-known that the type and quantity of defects in the semiconductors strongly affect their properties. Because of the emergence of negatively charged vacancy associates, it was found that the adsorption capability of cationic dye molecules and separation efficiency of electron–hole pairs are enhanced. Moreover, the changes of defects type bring about evident impacts on the energy band structures of BiOCl nanomaterials, which lead to up-shift of band edges and the narrowing of band gap, thus improving the visible-light-driven photocatalytic performance. In short, this study reveals that the thickness dependent changes of defects in 2D materials also bring new possibilities for exploring highly efficient photocatalysts.

As is well known, surface properties are crucial to catalysis. Consequently, the catalytic activity is sensitively dependent on the exposed facets, which possess diverse geometric and electronic structures. To this end, numerous efforts are devoted to the controlled fabrication of single-crystalline semiconductors with well-defined morphologies and their facet-dependent catalytic activity. 2D nanomaterial with the thickness down to the few nanometers or less is considered to expose only a specific facet, thus providing an ideal platform to study the face-dependent catalytic activity. For example, BiOCl is a semiconductor with a unique layered structure, which is characterized by [Bi_2O_2] slabs interleaved with double slabs of halogen atoms. Recent advances made by Jiang and co-workers on the controllable synthesis of BiOCl single-crystalline nanosheets have achieved the selective growth of (001) and (010) facets *via* a facile hydrothermal route.[66] Consider that the intrinsic internal electric field is perpendicular to the nanosheets with exposed (001) facets and parallel to those with exposed

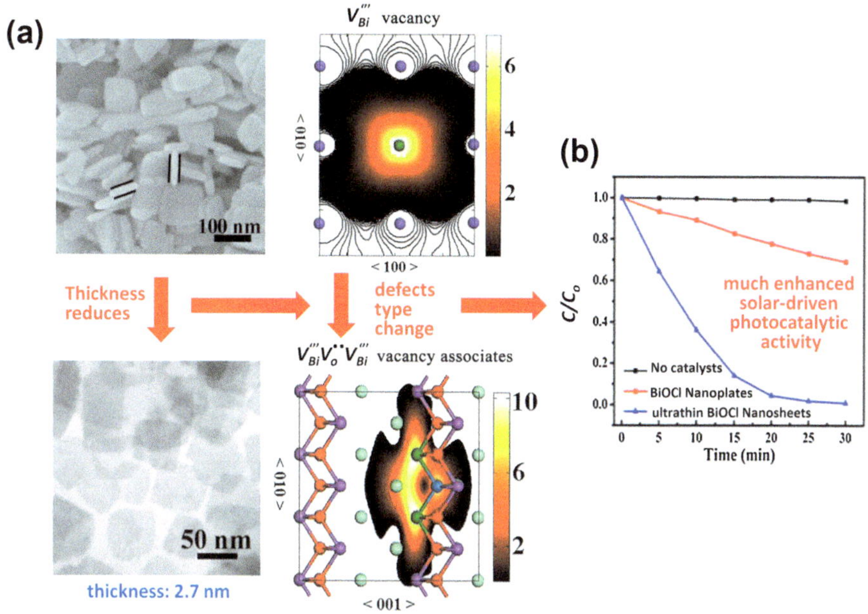

Figure 12.17 (a) Morphology characterizations of BiOCl nanoplates and ultrathin nanosheets and the corresponding schematic representations of trapped positrons of isolated vacancy V_{Bi}''' and triple vacancy associates $V_{Bi}'''V_O^{\bullet\bullet}V_{Bi}'''$, respectively. (b) Comparison of the photodecomposition of Rhodamine B. Adapted with permission from ref. 65, Copyright 2013 American Chemical Society.

(010) facets (see Figure 12.18). The BiOCl nanosheets with exposed (001) facets enable more effective separation of the photoinduced electron–hole pairs. Accordingly, the catalytic results indicate that BiOCl single-crystalline nanosheets with exposed (001) facets possess higher photocurrent and activity for pollutant degradation *via* direct semiconductor photoexcitation under UV light. Meanwhile, due to the different local structure of oxygen vacancies *in-situ* generated on different surfaces, BiOCl single-crystalline nanosheets can activate oxygen molecules through different paths.[67] As revealed by ESR spectra, the changes of intensity before and after SOD addition indicate that the amount of $\cdot O^{2-}$ generated on (001) facets is much greater than that generated on (010) facets, indicating that nanosheets with exposed (010) facets favors O_2 reduction through one-electron transfer, while those exposed (010) facets prefers to reduce O_2 to O_2^{2-} *via* two-electron transfer. These findings will deepen the understanding of facet-dependent photoreactivity of semiconductor nanomaterials and expand their potential application in photocatalytic reactions.

Despite the fact that coordinatively unsaturated surface atoms generally serve as the active sites in the catalytic reaction, their quantity and reactivity

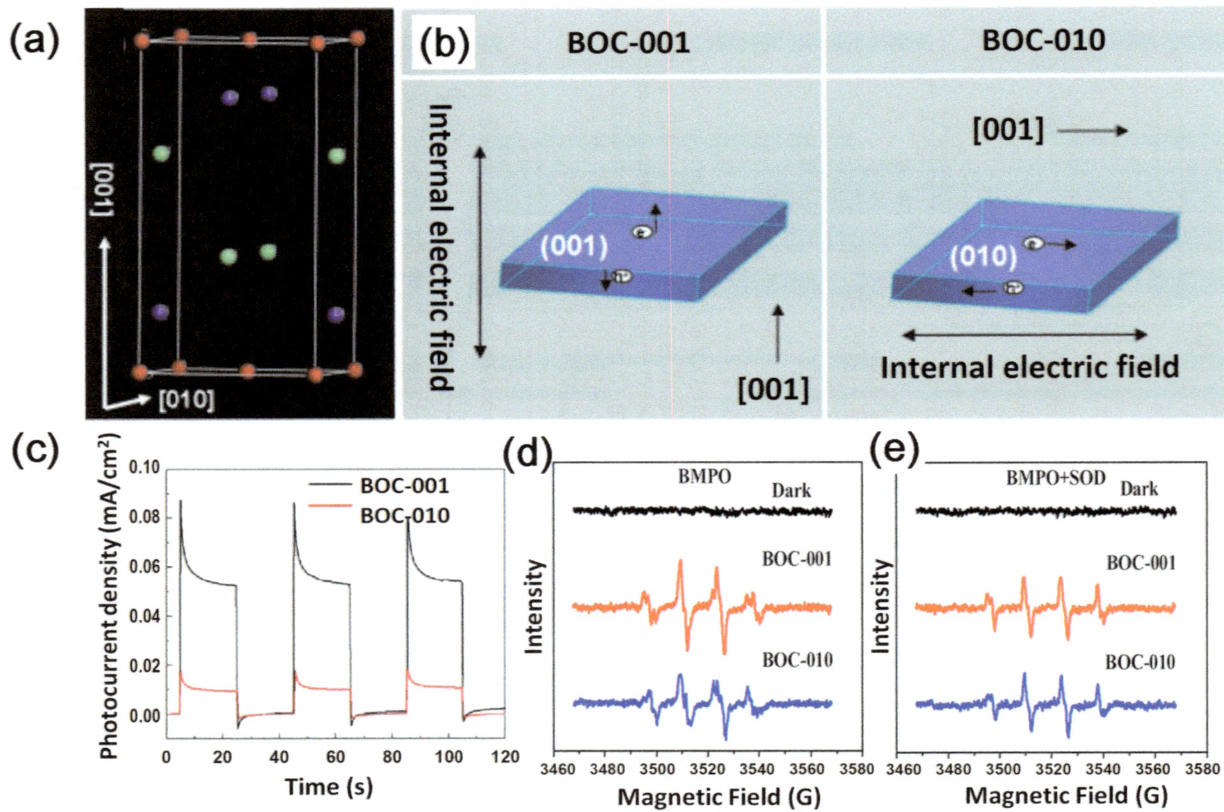

Figure 12.18 (a) Crystal structure of BiOCl. (b) Model showing the direction of the internal electric field in each of the BiOCl single-crystalline nanosheets with exposed (001) facets and (010) facets, respectively. (c) Comparison of the photocurrent responses in 0.5 M Na_2SO_4 aqueous solutions under UV-vis irradiation, adapted with permission from ref. 66, Copyright 2012 American Chemical Society. ESR spectra of active oxygen radicals generated during photocatalysis (d) before and (e) after SOD addition, adapted with permission from ref. 67, Copyright 2013 American Chemical Society.

are still limited. Constructing hybrid systems to exert the advantages of each component becomes an inevitable trend to develop high-efficient photocatalysis. Noble metals can easily activate the adsorbed species, forming highly active intermediates, and therefore show excellent catalytic activity. From this perspective, new opportunities will arise in solar energy storage and conversion when ultrathin 2D nanosheets meet the active noble metals. As an example, Ida and co-workers reported isolated single atom Rh dopants in 2D titanium oxide nanosheets can effectively act as co-catalysts for water splitting (see Figure 12.19).[68] First-principles calculations show that the single atom Rh significantly reduces the dissociation energy of water and therefore accelerates the catalytic reaction. Single atom dopant, which maximizes the utilization of co-catalysts and meanwhile reduces the cost, is expected to extend to other catalytic systems.

Different from doping, Wu and co-workers developed a pure surface modification strategy, which grafts isolated Pt atoms onto C_3N_4 *via* adsorption and annealing.[69] As shown in Figure 12.20, the HAADF-STEM image reveals that Pt species are anchored on C_3N_4 network with high dispersion and stability, even after the photocatalytic reaction. Extended X-ray absorption fine structure spectroscopy indicates that isolated Pt atoms are located on the top of the five-membered rings of the N/C-coordinating framework. This unique structure enables single-atom Pt to change the surface trap state of C_3N_4, leading to an approximately two-fold longer lifetime of the photo-generated electrons, as demonstrated by TA spectroscopy. The increased lifetime upon the introduction of single-atom Pt co-catalyst offers more opportunities for photo-generated electrons to participate in the catalytic reaction, thus, significantly improving the catalytic activity of C_3N_4, which is up to 50 times of that for bare C_3N_4, and 8.6 times higher than that of C_3N_4 with Pt nanoparticles of equivalent weights.

As mentioned above, co-catalysts not only act as additional active sites, but also effectively suppress the recombination of charge carriers, and therefore they have become an indispensable part of high-efficient photocatalysts, especially in photocatalytic water splitting. Nevertheless, the most efficient co-catalysts until now remain limited to noble metals which suffer from scarcity and high cost. Intriguingly, as the thickness is reduced to the monolayer, TMDs exhibit notable catalytic activity for hydrogen evolution; whereas the bulk counterparts show negligible activity. Moreover, considering their natural abundance, chemical stability, and potential reactivity, TMDs become promising alternatives to noble metal in photocatalysis. MoS_2, which is the most representative member of 2D TMDs materials, has been extensively investigated recently. Zhang and co-workers put forward a facile one-pot wet-chemical method to prepare CdS nanocrystals hybridized with single-layer TMDs nanosheets.[70,71] As shown in Figure 12.21, the heterogeneous, single-layer TMDs nanosheets selectively grow on the Cd-rich (0001) surface, and the thickness of the TMD nanosheets is only about 0.4 nm. Due to the large number of active sites introduced by

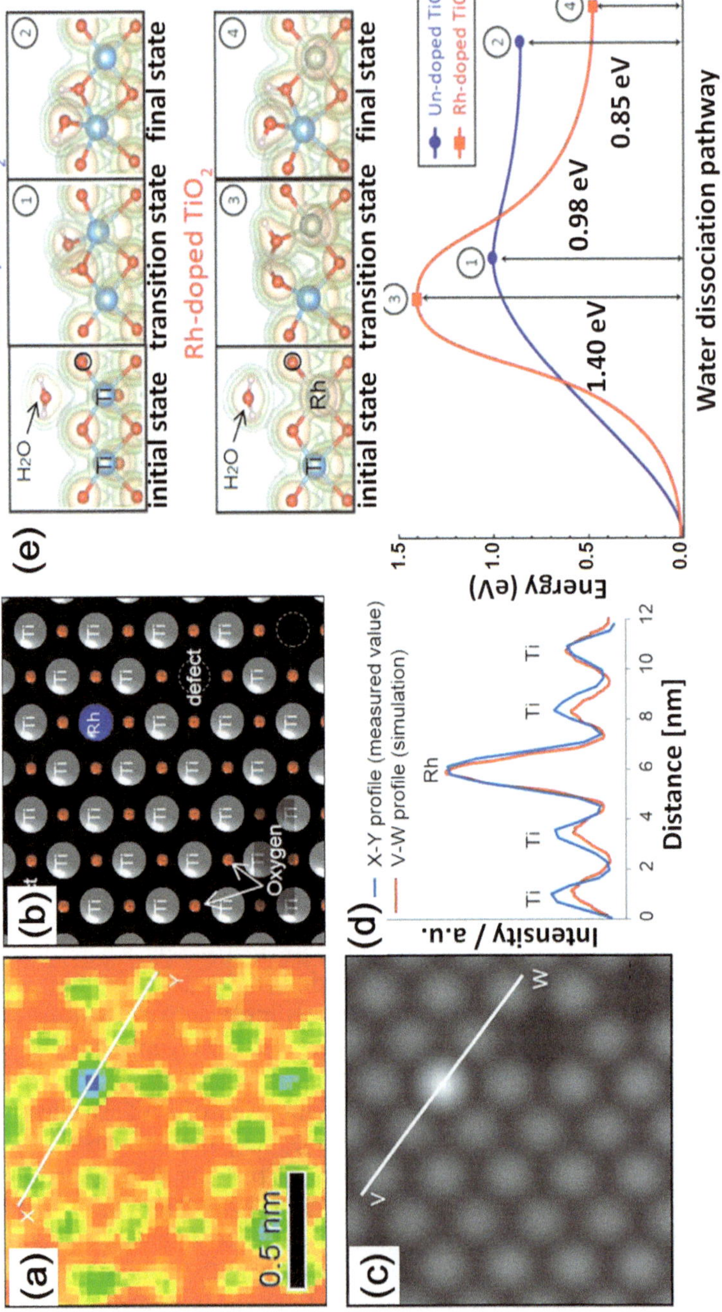

Figure 12.19 (a) Color-coded intensity map derived from HAADF-STEM image of Rh ($x = 0.026$)-doped $Ti_{1.82-x}Rh_xO_4$ nanosheet and (b) the corresponding structural model. (c) HAADF-STEM image simulated from the structural model using frozen phonon model, and (d) intensity profiles along the line X–Y in the observed image and the line V–W in the simulated image. (e) Charge density and dissociation energetics for undoped and Rh-doped titania nanosheet. Adapted with permission from ref. 68, Copyright 2015 American Chemical Society.

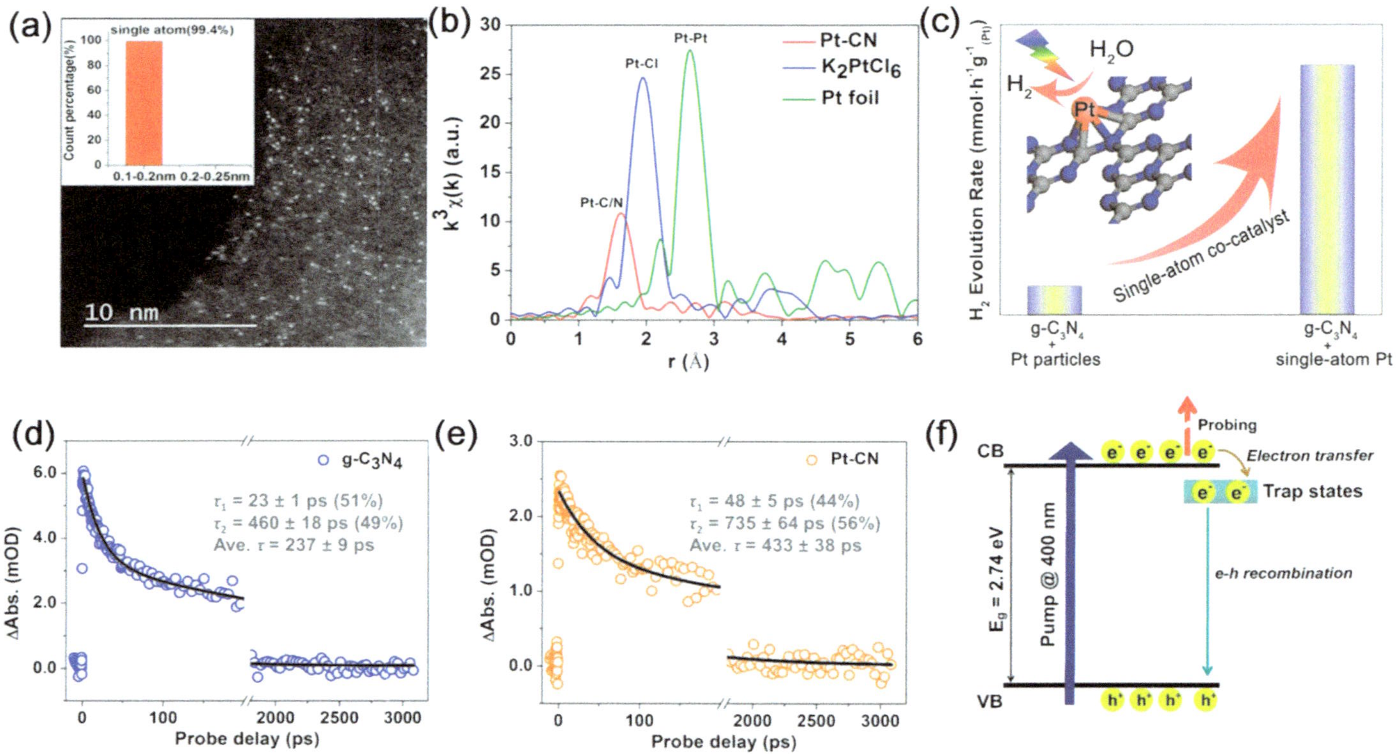

Figure 12.20 (a) HAADF-STEM image of Pt-CN. Inset is the size distribution of the bright spots. (b) Fourier transforms of the Pt L_3-edge EXAFS oscillations of Pt-CN, K_2PtCl_6, and Pt foil. (c) Photocatalytic activity comparison of g-C_3N_4 with single-atom Pt and Pt nanoparticles inset is the schematic models of Pt-CN. Representative ultrafast TA kinetics probed at 750 nm (pump at 400 nm) for (d) g-C_3N_4 and (e) Pt-CN. The TA signal (*i.e.*, the absorbance changes, or ΔAbs. in short) is given in mOD where OD stands for optical density. (f) Schematic illustration of the mechanisms involved. Adapted with permission from X. Li, Single-Atom Pt as Co-Catalyst for Enhanced Photocatalytic H_2 Evolution, *Advanced Materials*, John Wiley and Sons, © 2016 WILEY-VCH Verlag GmbH & Co. KGaA, Weinheim.[69]

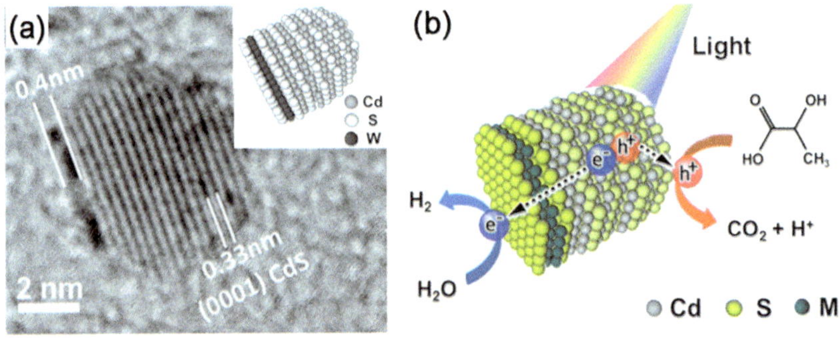

Figure 12.21 (a) HRTEM images of WS_2–CdS nanohybrids viewed perpendicularly to the *c* axis of CdS nanocrystals. (b) Schematic illustration of the photocatalytic process of MS_2–CdS nanohybrids in the lactic acid solution. Adapted with permission from J. Chen, One-pot Synthesis of CdS Nanocrystals Hybridized with Single-Layer Transition-Metal Dichalcogenide Nanosheets for Efficient Photocatalytic Hydrogen Evolution, *Angewandte Chemie International Edition*, John Wiley and Sons © 2015 WILEY-VCH Verlag GmbH & Co. KGaA, Weinheim.[70]

2D single-layer TMD nanosheets as well as the inherent *p*–*n* heterojunction formed between TMD and CdS, the nanohybrids catalysts exhibit excellent catalytic water splitting activity under visible light irradiation. Additionally, the considerably small sizes of the heterogeneous photocatalysts (from 4 to 11 nm) shorten the diffusion distance of the charge carriers and effectively suppress the recombination of charge carries, both of which further enhance the photocatalytic performance.

Moreover, it is worth emphasizing that TMDs materials generally present two different polytypes depending on the coordination of the metal atoms: the 1T and 2H phase with octahedral or trigonal prismatic coordination, respectively. For 2H MoS_2, only the edge sites are highly active, while the basal plane remains catalytically inert. However, 1T MoS_2 has catalytic activity in both edge and basal plane. In this regard, phase engineering plays an important role in designing high-efficient photocatalytic systems. Recently, Chang and co-workers describe a novel method to accurately synthesize different phases of MoS_2 by controlling the calcined temperature.[72] As is shown in Figure 12.22, upon the introduction of 1T-MoS_2, TiO_2 (P25) exhibit drastically enhanced photocatalytic hydrogen evolution rate of 228.2 μmol h^{-1}, while there is no obvious improvement in the case of 2H-MoS_2. However, when 2H-MoS_2 monolayer is loaded on semiconductors with more negative CB position, such as CdS, the photocatalytic activity is also significantly improved. In this regard, besides the intrinsic catalytic properties of 2D TMDs, band alignment between TMDs and the photo-harvesting semiconductors should also be taken into consideration for the design and construction of highly efficient photocatalysts.

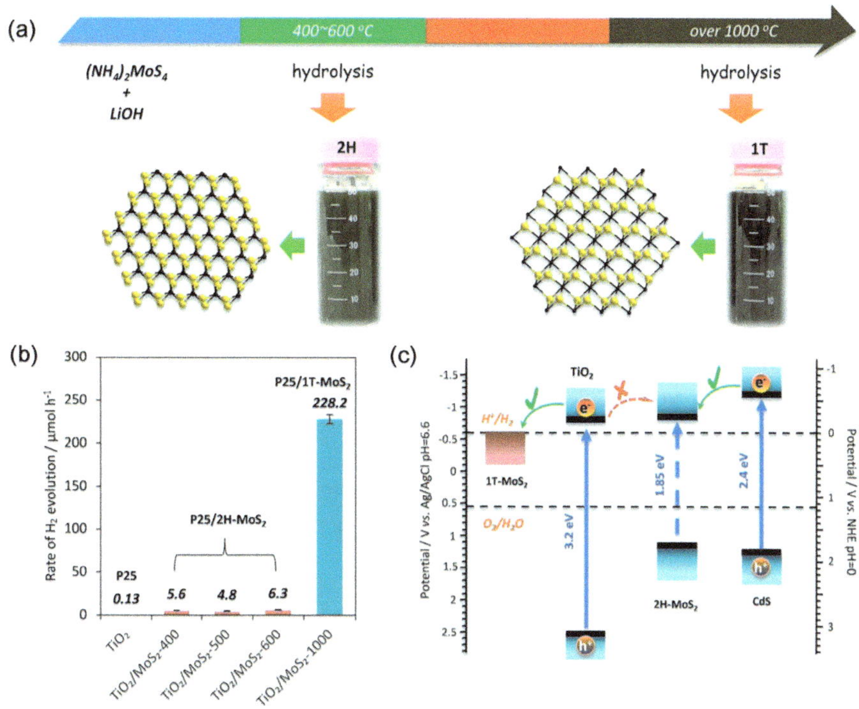

Figure 12.22 (a) Schematic illustrations of hydrolysis exfoliation of 2H and 1T MoS_2 at different calcination temperatures. (b) Photocatalytic H_2 evolution activity of 1T- and 2H-MoS_2 over TiO_2(P25). (c) Schematic energy-level diagrams of 1T-, 2H-MoS_2, TiO_2, and CdS in comparison with the potentials for water reduction and oxidation. Adapted with permission from K. Chang *et al.*, Targeted Synthesis of 2H- and 1T-Phase MoS_2 Monolayers for Catalytic Hydrogen Evolution, *Advanced Materials*, John Wiley and Sons © 2016 WILEY-VCH Verlag GmbH & Co. KGaA, Weinheim.[72]

References

1. M. Gratzel, *Nature*, 2001, **414**, 338.
2. N. S. Lewis, *Science*, 2016, **351**, 353.
3. M. G. Walter, E. L. Warren, J. R. McKone, S. W. Boettcher, Q. Mi, E. A. Santori and N. S. Lewis, *Chem. Rev.*, 2010, **110**, 6446.
4. K. Maeda and K. Domen, *J. Phys. Chem. Lett.*, 2010, **1**, 2655.
5. T. Hisatomi, J. Kubota and K. Domen, *Chem. Soc. Rev.*, 2014, **43**, 7520.
6. Y. Tachibana, L. Vayssieres and J. R. Durrant, *Nat. Photonics*, 2012, **6**, 511.
7. A. Hagfeldt and M. Graetzel, *Chem. Rev.*, 1995, **95**, 49.
8. R. R. Nair, P. Blake, A. N. Grigorenko, K. S. Novoselov, T. J. Booth, T. Stauber, N. M. R. Peres and A. K. Geim, *Science*, 2008, **320**, 1308.
9. M. Bernardi, M. Palummo and J. C. Grossman, *Nano Lett.*, 2013, **13**, 3664.

10. D. Kozawa, R. Kumar, A. Carvalho, K. Kumar Amara, W. Zhao, S. Wang, M. Toh, R. M. Ribeiro, A. H. Castro Neto, K. Matsuda and G. Eda, *Nat. Commun.*, 2014, **5**, 4543.
11. A. Carvalho, R. M. Ribeiro and A. H. Castro Neto, *Phys. Rev. B*, 2013, **88**, 115205.
12. K. F. Mak, C. Lee, J. Hone, J. Shan and T. F. Heinz, *Phys. Rev. Lett.*, 2010, **105**, 136805.
13. M. Zhou, X. W. Lou and Y. Xie, *Nano Today*, 2013, **8**, 598.
14. S. N. Habisreutinger, L. Schmidt-Mende and J. K. Stolarczyk, *Angew. Chem., Int. Ed.*, 2013, **52**, 7372.
15. R. W. Meulenberg, J. R. I. Lee, A. Wolcott, J. Z. Zhang, L. J. Terminello and T. van Buuren, *ACS Nano*, 2009, **3**, 325.
16. S. Ithurria, M. D. Tessier, B. Mahler, R. P. S. M. Lobo, B. Dubertret and A. L. Efros, *Nat. Mater.*, 2011, **10**, 936.
17. F. A. Frame and F. E. Osterloh, *J. Phys. Chem. C*, 2010, **114**, 10628.
18. J. Zhao, M. A. Holmes and F. E. Osterloh, *ACS Nano*, 2013, **7**, 4316.
19. Y. Bekenstein, B. A. Koscher, S. W. Eaton, P. Yang and A. P. Alivisatos, *J. Am. Chem. Soc.*, 2015, **137**, 16008.
20. J. Liu, Y. Xue, Z. Wang, Z.-Q. Xu, C. Zheng, B. Weber, J. Song, Y. Wang, Y. Lu, Y. Zhang and Q. Bao, *ACS Nano*, 2016, **10**, 3536.
21. Q. A. Akkerman, S. G. Motti, A. R. Srimath Kandada, E. Mosconi, V. D'Innocenzo, G. Bertoni, S. Marras, B. A. Kamino, L. Miranda, F. De Angelis, A. Petrozza, M. Prato and L. Manna, *J. Am. Chem. Soc.*, 2016, **138**, 1010.
22. W. Jin, P. C. Yeh, N. Zaki, D. Zhang, J. T. Sadowski, A. Al Mahboob, A. M. van der Zande, D. A. Chenet, J. I. Dadap, I. P. Herman, P. Sutter, J. Hone and R. M. Osgood, *Phys. Rev. Lett.*, 2013, **111**, 106801.
23. R. Asahi, T. Morikawa, T. Ohwaki, K. Aoki and Y. Taga, *Science*, 2001, **293**, 269.
24. Y. Ma, X. Wang, Y. Jia, X. Chen, H. Han and C. Li, *Chem. Rev.*, 2014, **114**, 9987.
25. W. Choi, A. Termin and M. R. Hoffmann, *J. Phys. Chem.*, 1994, **98**, 13669.
26. H. Irie, Y. Watanabe and K. Hashimoto, *J. Phys. Chem. B*, 2003, **107**, 5483.
27. Y. Matsumoto, M. Koinuma, Y. Iwanaga, T. Sato and S. Ida, *J. Am. Chem. Soc.*, 2009, **131**, 6644.
28. N. Zhang, X. Li, H. Ye, S. Chen, H. Ju, D. Liu, Y. Lin, W. Ye, C. Wang, Q. Xu, J. Zhu, L. Song, J. Jiang and Y. Xiong, *J. Am. Chem. Soc.*, 2016, **138**, 8928.
29. H. Cheng, T. Kamegawa, K. Mori and H. Yamashita, *Angew. Chem., Int. Ed.*, 2014, **53**, 2910.
30. X. Lu, K. Xu, S. Tao, Z. Shao, X. Peng, W. Bi, P. Chen, H. Ding, W. Chu, C. Wu and Y. Xie, *Chem. Sci.*, 2016, **7**, 1462.
31. Y. Li, Z. Wang, T. Xia, H. Ju, K. Zhang, R. Long, Q. Xu, C. Wang, L. Song, J. Zhu, J. Jiang and Y. Xiong, *Adv. Mater.*, 2016, **28**, 6959.
32. M. Nolan, A. Iwaszuk, A. K. Lucid, J. J. Carey and M. Fronzi, *Adv. Mater.*, 2016, **28**, 5425.
33. H. Tetsuka, A. Nagoya, T. Fukusumi and T. Matsui, *Adv. Mater.*, 2016, **28**, 4632.
34. Y. Guo, K. Xu, C. Wu, J. Zhao and Y. Xie, *Chem. Soc. Rev.*, 2015, **44**, 637.

35. Y. Zhou, F. Wang and W. E. Buhro, *J. Am. Chem. Soc.*, 2015, **137**, 15198.
36. E. P. Nguyen, B. J. Carey, J. Z. Ou, J. van Embden, E. D. Gaspera, A. F. Chrimes, M. J. S. Spencer, S. Zhuiykov, K. Kalantar-zadeh and T. Daeneke, *Adv. Mater.*, 2015, **27**, 6225.
37. H. A. Atwater and A. Polman, *Nat. Mater.*, 2010, **9**, 205.
38. Y. Luo, D. Y. Lei, S. A. Maier and J. B. Pendry, *Phys. Rev. Lett.*, 2012, **108**, 023901.
39. X. Zheng and L. Zhang, *Energy Environ. Sci.*, 2016, **9**, 2511.
40. A. H. Zewail, *J. Phys. Chem. A*, 2000, **104**, 5660.
41. A. H. Zewail, *Angew. Chem., Int. Ed.*, 2000, **39**, 2586.
42. J. Tang, J. R. Durrant and D. R. Klug, *J. Am. Chem. Soc.*, 2008, **130**, 13885.
43. D. A. Wheeler and J. Z. Zhang, *Adv. Mater.*, 2013, **25**, 2878.
44. J. H. Kennedy and K. W. Frese, *J. Electrochem. Soc.*, 1978, **125**, 709.
45. W. Bi, X. Li, L. Zhang, T. Jin, L. Zhang, Q. Zhang, Y. Luo, C. Wu and Y. Xie, *Nat. Commun.*, 2015, **6**, 8647.
46. Y. Sun, H. Cheng, S. Gao, Z. Sun, Q. Liu, Q. Liu, F. Lei, T. Yao, J. He, S. Wei and Y. Xie, *Angew. Chem., Int. Ed.*, 2012, **51**, 8727.
47. Y. Sun, Z. Sun, S. Gao, H. Cheng, Q. Liu, J. Piao, T. Yao, C. Wu, S. Hu, S. Wei and Y. Xie, *Nat. Commun.*, 2012, **3**, 1057.
48. L. Liang, F. Lei, S. Gao, Y. Sun, X. Jiao, J. Wu, S. Qamar and Y. Xie, *Angew. Chem., Int. Ed.*, 2015, **54**, 13971.
49. F. Song and X. Hu, *Nat. Commun.*, 2014, **5**, 4477.
50. W. Bi, C. Ye, C. Xiao, W. Tong, X. Zhang, W. Shao and Y. Xie, *Small*, 2014, **10**, 2820.
51. F. Lei, Y. Sun, K. Liu, S. Gao, L. Liang, B. Pan and Y. Xie, *J. Am. Chem. Soc.*, 2014, **136**, 6826.
52. X. Zhu, X. Dou, J. Dai, X. An, Y. Guo, L. Zhang, S. Tao, J. Zhao, W. Chu, X. C. Zeng, C. Wu and Y. Xie, *Angew. Chem., Int. Ed.*, 2016, **55**, 12465.
53. S. Lei, X. Wang, B. Li, J. Kang, Y. He, A. George, L. Ge, Y. Gong, P. Dong, Z. Jin, G. Brunetto, W. Chen, Z.-T. Lin, R. Baines, D. S. Galvão, J. Lou, E. Barrera, K. Banerjee, R. Vajtai and P. Ajayan, *Nat. Nanotechnol.*, 2016, **11**, 465.
54. A. Raja, A. Montoya–Castillo, J. Zultak, X.-X. Zhang, Z. Ye, C. Roquelet, D. A. Chenet, A. M. van der Zande, P. Huang, S. Jockusch, J. Hone, D. R. Reichman, L. E. Brus and T. F. Heinz, *Nano Lett.*, 2016, **16**, 2328.
55. Q. Liu, Q. Shang, A. Khalil, Q. Fang, S. Chen, Q. He, T. Xiang, D. Liu, Q. Zhang, Y. Luo and L. Song, *ChemCatChem*, 2016, **8**, 2614.
56. D. Deng, K. S. Novoselov, Q. Fu, N. Zheng, Z. Tian and X. Bao, *Nat. Nanotechnol.*, 2016, **11**, 218.
57. C. Tan and H. Zhang, *J. Am. Chem. Soc.*, 2015, **137**, 12162.
58. X. Hong, J. Kim, S. F. Shi, Y. Zhang, C. Jin, Y. Sun, S. Tongay, J. Wu, Y. Zhang and F. Wang, *Nat. Nanotechnol.*, 2014, **9**, 682.
59. C. Huang, S. Wu, A. M. Sanchez, J. J. P. Peters, R. Beanland, J. S. Ross, P. Rivera, W. Yao, D. H. Cobden and X. Xu, *Nat. Mater.*, 2014, **13**, 1096.
60. X. Duan, C. Wang, J. C. Shaw, R. Cheng, Y. Chen, H. Li, X. Wu, Y. Tang, Q. Zhang, A. Pan, J. Jiang, R. Yu, Y. Huang and X. Duan, *Nat. Nanotechnol.*, 2014, **9**, 1024.

61. R. Narayanan and M. A. El-Sayed, *Nano Lett.*, 2004, **4**, 1343.
62. Y. Zhu, S. Murali, M. D. Stoller, K. J. Ganesh, W. Cai, P. J. Ferreira, A. Pirkle, R. M. Wallace, K. A. Cychosz, M. Thommes, D. Su, E. A. Stach and R. S. Ruoff, *Science*, 2011, **332**, 1537.
63. Z. Sun, T. Liao, Y. Dou, S. M. Hwang, M.-S. Park, L. Jiang, J. H. Kim and S. X. Dou, *Nat. Commun.*, 2014, **5**, 3813.
64. G. Xiang, T. Li, J. Zhuang and X. Wang, *Chem. Commun.*, 2010, **46**, 6801.
65. M. Guan, C. Xiao, J. Zhang, S. Fan, R. An, Q. Cheng, J. Xie, M. Zhou, B. Ye and Y. Xie, *J. Am. Chem. Soc.*, 2013, **135**, 10411.
66. K. Zhao, L. Zhang, J. Wang, Q. Li, W. He and J. J. Yin, *J. Am. Chem. Soc.*, 2013, **135**, 15750.
67. J. Jiang, K. Zhao, X. Xiao and L. Zhang, *J. Am. Chem. Soc.*, 2012, **134**, 4473.
68. S. Ida, N. Kim, E. Ertekin, S. Takenaka and T. Ishihara, *J. Am. Chem. Soc.*, 2015, **137**, 239.
69. X. Li, W. Bi, L. Zhang, S. Tao, W. Chu, Q. Zhang, Y. Luo, C. Wu and Y. Xie, *Adv. Mater.*, 2016, **28**, 2427.
70. J. Chen, X. J. Wu, L. Yin, B. Li, X. Hong, Z. Fan, B. Chen, C. Xue and H. Zhang, *Angew. Chem., Int. Ed.*, 2015, **54**, 1210.
71. Q. Lu, Y. Yu, Q. Ma, B. Chen and H. Zhang, *Adv. Mater.*, 2016, **28**, 1917.
72. K. Chang, X. Hai, H. Pang, H. Zhang, L. Shi, G. Liu, H. Liu, G. Zhao, M. Li and J. Ye, *Adv. Mater.*, 2016, **28**, 10033.

CHAPTER 13

Two-dimensional Nanomaterials as Promising Candidates for Thermoelectric Applications

JING PENG, JIAJING WU AND CHANGZHENG WU*

Department of Chemistry & Hefei National Laboratory for Physical Sciences at the Microscale & Collaborative Innovation Center of Chemistry for Energy Materials, University of Science and Technology of China, No.96, JinZhai Road, Hefei, Anhui, 230026, P. R. China
*E-mail: czwu@ustc.edu.cn

13.1 Introduction

With the increasing depletion of fossil fuels, energy issues have become a focus of attention around the world. In 2011, the global energy consumption was about 12.27 billion tons of standard oil, of which fossil energy (oil, coal and gas) accounted for about 87%. At the current rate of consumption, known oil reserves will only last, for human use, for about 50 years. Therefore, the search for clean and safe new energy to improve the efficiency of existing energy use is needed. According to the US Department of Energy estimates of primary energy consumption, more than 55% of the energy in the form of waste heat was eventually released into the environment.[1] Therefore, it is of

Smart Materials No. 27
Inorganic Two-dimensional Nanomaterials: Fundamental Understanding, Characterizations and Energy Applications
Edited by Changzheng Wu

Published by the Royal Society of Chemistry, www.rsc.org

great significance for researchers to study waste heat reuse for improving the efficiency of energy use, reduce the dependence on fossil fuels and mitigate carbon dioxide emissions.

Thermoelectric materials, as a new type of clean energy material, can directly realize the conversion between thermal energy and electrical energy with the advantage of small volume, no noise, long life, and no pollution to the environment. Its application to industrial waste heat and the use of waste heat and solar thermal compound power generation is expected to improve energy utilization efficiency and hence relieve the problem of environmental pollution, providing a comprehensive and coordinated choice. Therefore, since the energy crisis has become even more serious in the 21st century, thermoelectric materials research has aroused wide interest among researchers. However, the energy conversion efficiency of thermoelectric materials has still only been around 10% for several decades, which is still a considerable distance away from theoretical predictions. Theoretical calculations show that the energy conversion efficiency of thermoelectric materials can approach infinite Carnot cycle efficiency. Therefore, it is important to find a mechanism and select a new thermoelectric material system to further improve the performance of thermoelectric materials and to break the restriction of low energy conversion efficiency on a large-scale practical application. The thermoelectric effect is the general term for the electrical effect caused by the temperature difference and the thermal effect caused by a current. It includes three interrelated effects: Seebeck effect,[2] Peltier effect[3] and Thomson effect.[4]

In 1823, Seebeck first discovered that when two different conductors constitute a closed loop, the two contacts will have electromotive force and form a current if the temperature of the two contacts is different, this is called the Seebeck effect. When two different metallic conductors (or semiconductors) are in contact with each other, the difference in the internal electron density is eliminated at the contact surface by diffusion. Since the diffusion rate of electrons is proportional to the temperature of the contact zone, a constant voltage can form between the two species if there is a temperature difference between the two metals (or semiconductors). Meanwhile, the Seebeck effect can also be produced in the same material. With two ends of a material at different temperatures, the temperature difference is caused by the carrier material at both ends of the non-equilibrium distribution. Higher energy hot-side carriers transfer to the cold end of the diffusion, bringing about a potential difference formed at both ends of the material.

The Peltier effect is a phenomenon where heat and heat absorption can emerge when the current through two different metals (or semiconductors) form a junction. This was found by Peltier in 1834. When the carrier from one material transfers into another material, it will cause the energy and non-uniform charge in both sides of the material to contact. To maintain the conservation of energy and charge, it is necessary to absorb energy *via* interaction with the lattice (thermal vibration) near the junction and release of energy at the other junction, which appears in the Peltier effect. The Peltier effect is the inverse of the Seebeck effect. However, unlike the Seebeck effect, the Peltier effect is a typical nodal phenomenon that can occur only through the connection between different materials.

The thermoelectric effect is essentially a reversible cross-coupling effect of electric and heat transport (Figure 13.1), and hence we can use the Seebeck and Peltier effects to realize thermoelectric power generation and thermoelectric refrigeration, respectively. In a thermoelectric generator, the *n*-type and *p*-type semiconductors form a loop where both holes and electrons flow from the high-temperature side to the low-temperature side when maintaining a temperature difference ΔT between the two junctions. When equilibrium is reached, excess holes accumulate at the cold end of the *p*-type semiconductor and electrons accumulate at the cold junction of the *n*-type semiconductor, creating a potential difference across the cold junctions of the *p*- and *n*-type semiconductors. In the case of a thermoelectric cooler, the carriers in both arms (electrons in the *n*-type and holes in the *p*-type) can

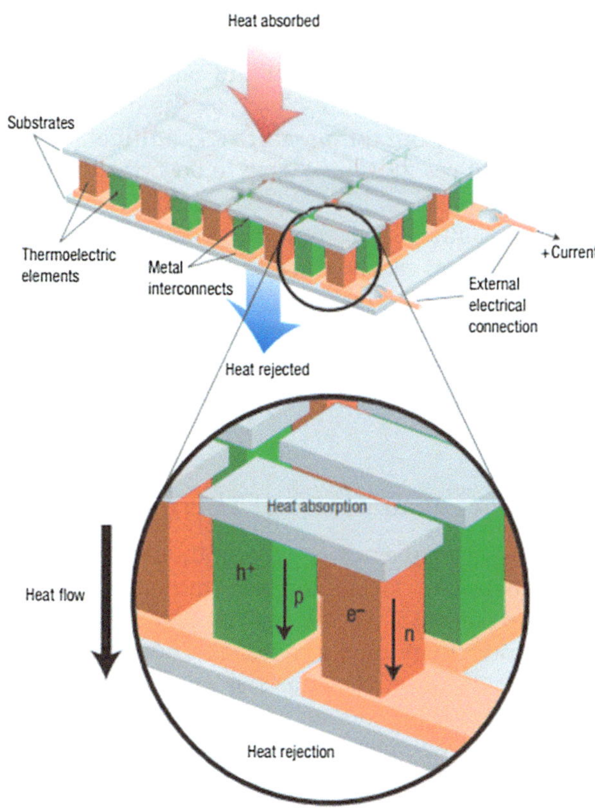

Figure 13.1 Thermoelectric scheme revealing the working principle of cooling and power generation. Reprinted with permission from Macmillan Publishers Ltd: *Nature Materials* (ref. 5), Copyright 2008.

absorb heat from the junction and carry heat when the current flows through the circuits of the two semiconductors, forming a cooling effect.

By 1856, after a thermodynamic analysis of the Seebeck and Peltier effects, Thomson predicted that there should be a third thermoelectric phenomenon and later found this by experiment. He found that if there is a temperature gradient in a uniform conductor through the current, the conductor not only produces Joule heat, but also absorbs or releases a certain amount of heat, a phenomenon known as the Thomson effect.

The above three effects laid the foundation of thermoelectric theory, and showed broad prospects for the practical application of thermoelectric conversion materials. Since then and until the beginning of the 20th century, the German scientist Altenkirch proposed an important and relatively complete thermoelectric cooling and power generation theory, which for today's thermoelectric research gives clear direction. He pointed out that a good thermoelectric material must have a large Seebeck coefficient S, to ensure a more significant effect of temperature difference. At the same time, it also needs to have a higher conductivity σ for smaller Joule heat generation. In addition, there should be a lower thermal conductivity κ, so that heat can be maintained near the joint. These values reflect the overall performance of the thermoelectric power and can be specifically expressed by a unified thermoelectric power Z, $Z = S^2\sigma/\kappa$, its dimension is K^{-1}, so it is the product of the absolute temperature ZT, a dimensionless value. In practice, the dimensionless ZT is also used to describe the performance of thermoelectric materials.

The theoretical calculations show that the energy conversion efficiency of the thermoelectric material can be close to the Carnot efficiency. However, after decades of development, the energy conversion efficiency of thermoelectric materials has been hovering around 10%, and there is still a considerable distance from its theoretical prediction. This is mainly related to the lack of breakthrough to improve the relevant performance of the thermoelectric material.

13.1.1 Performance Parameters of Thermoelectric Materials

From the above discussion, the power generation efficiency and the cooling efficiency of thermoelectric devices are determined by the thermoelectric figure-of-merit, which depends on the Seebeck coefficient, the conductivity and the thermal conductivity of the material. Therefore, to obtain a high thermoelectric figure-of-merit, the material should have a large Seebeck coefficient and high electrical conductivity as well as low thermal conductivity. The large Seebeck coefficient ensures a large thermoelectric potential at the same temperature difference and high conductivity reduces the heat loss caused by Joule heat, while low thermal conductivity maintains a high temperature difference across the material. Therefore, by fully understanding the physical nature of these three physical parameters, we can effectively optimize the thermoelectric properties of materials.

The Seebeek coefficient and conductivity of a material are known as the electrical properties of the material. The Seebeek coefficient and conductivity describe the transport properties and interactions of the carrier during the electrical transport of the material, which is closely relevant to the carrier concentration of the material, the carrier mobility and band structure. The Seebeck effect is essentially the excitation of the electrons at the hot end of the material to a higher energy level, with more electrons above the Fermi level and fewer electrons below the Fermi level as per the Fermi–Dirac distribution. To balance this energy gap, the high-energy electrons at the hot end are reduced by diffusion at the cold junction to reduce their energy so that a potential difference occurs between the two ends. As per the Boltzmann transport theory, the Seebeck coefficient and conductivity can be expressed separately:[5,6]

$$S = \frac{8\pi^2 k_B^2}{3eh^2} m^* T \left(\frac{\pi}{3n} \right)^{2/3} \tag{13.1}$$

$$\sigma = ne\mu \tag{13.2}$$

where kB is the Boltzmann constant, e is the electron charge, h is the Planck constant, m^* is the carrier effective mass, n is the carrier concentration and μ is the carrier mobility.

Heat conduction is the transport of heat within a material. From a microscopic view, the heat conduction is mainly through lattice vibration (phonon) and the carrier motion synergetic completing. Therefore, for semiconductor materials in the extrinsic excitation region, thermal conductivity is mainly composed of lattice thermal conductivity κ_l and carrier thermal conductivity κ_c, that is, $\kappa = \kappa_l + \kappa_c$. The carriers in the crystal are the supporter of charge and energy and it will not only contribute to the current, but also heat conduction when under directional migration. The contribution of the carrier to thermal conductivity follows the Wiedemann–Franz law:[7]

$$\kappa_c = L\sigma K \tag{13.3}$$

where L is the Lorentz constant. For strongly coupled degenerate states, the Lorentz constant L is a solid-independent constant, $L = 2.45 \times 10^{-8}$ W Ω K^{-2}.

From Figure 13.2, we can clearly see that the three parameters for determining the performance of thermoelectric materials—Seebeck coefficient, conductivity and thermal conductivity—are strongly coupled together by carriers, and restraint with each other, which makes it very difficult to optimize the ZT value of thermoelectric materials. In fact, the co-modulation of electrothermal transport has been a historic problem. Although in research efforts, the optimization of thermoelectric materials has made great progress, the ZT value is still at a low level. Therefore, it is of great theoretical and practical significance to realize the coordination and even decouple the three parameters of thermoelectric materials and to break the restriction of this kind of low energy conversion efficiency on large-scale practical application.

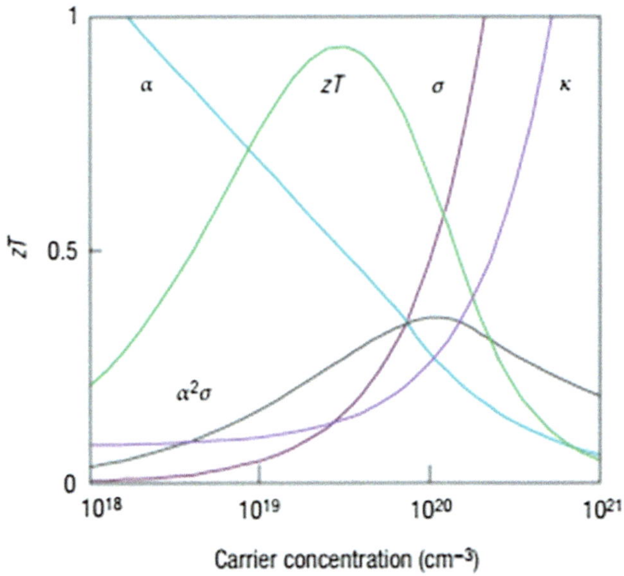

Figure 13.2 Optimizing the efficiency (ZT) of a thermoelectric material involves its carrier concentration. (Thermal conductivity κ plotted on the y axis from 0 to 10 Wm^{-1} K^{-1} and the Seebeck coefficient α from 0 to 500 μv K^1 with electrical conductivity σ from 0 to 5000 Ω^{-1} cm^{-1}). Reprinted with permission from Macmillan Publishers Ltd: *Nature Materials* (ref. 5), Copyright 2008.

13.1.2 The Modulation of Thermoelectric Parameters

13.1.2.1 Enhancement of the Seebeck Coefficient

From eqn (13.1) we can see that the Seebeck coefficient is not only related to the carrier concentration, but also to the carrier effective mass. This indicates that to obtain a large Seebeck coefficient, the material should have a large carrier effective mass. However, as the effective mass of carrier increases, the mobility of the carrier is inevitably reduced, leading to a decrease in the conductivity. The exact relationship between carrier effective mass and mobility is very complex and is dependent on the electronic structure of the material. For the Seebeck coefficient, the relationship of the electronic structure can be more intuitive, shown in the Mott formula expressed as:[8]

$$S \approx \frac{\pi^2}{3}\frac{k^2T}{e}\left\{\frac{\mathrm{d}\ln\sigma(E)}{\mathrm{d}E}\bigg|E = E_f\right\} \tag{13.4}$$

As is known, $\sigma(E)$ is only proportional to the density of states (DOS) at point E if the electron scattering is independent of the energy, where $\sigma(E)$ is the electrical conductivity (a function of the Fermi surface energy). It can be seen from the Mott formula that the Seebeck coefficient of the material is proportional to the slope of the energy band near the Fermi surface: the larger the slope, the larger the Seebeck coefficient. Therefore, theoretical calculations point out that the Seebeck coefficient can be significantly increased through the distortion of the electron density by increasing the level of impurity. In fact, in some special cases, a resonance level can be produced when the impurity level is very close to the Fermi level and hence the Seebeck coefficient can be enhanced.[9–15]

The energy filtering effect is also an effective method for improving the Seebeck coefficient.[16,17] We know that the Seebeck effect is essentially the diffusion of hot-side carriers to the cold junction. In the actual process, the contribution of low energy carriers to the total Seebeck coefficient is negative, thus reducing Seebeck. When a high-density interface or nanostructure is introduced into the system, an additional energy barrier will appear for carrier transport. When carriers pass through these energy barriers, a part of the carriers will be filtered by the energy barrier, and the average energy of the carriers leaping over the energy barriers is increased, thereby increasing the Seebeek coefficient of the material. Although the energy filtering effect will inevitably reduce the conductivity due to the loss of carriers, since the Seebeck coefficient is a square term in the power factor expression, the power factor can be effectively increased if the interface barrier height is high, thereby increasing ZT.

In the previous discussion, it was assumed that the motions of carriers and phonons are independent of each other, regardless of their interactions after the change. In fact, this change cannot be ignored. The phonons will diffuse with the carrier during the directional diffusion from the hot junction to the cold junction, changing the equilibrium distribution of the carriers acting as scattering centers; the momentum of some carriers in the direction of the phonon will increase, and the result is that these carriers are dragged to the phonon in the direction of movement, which is called the phonon drag effect.[18,19] Besides the internal electric field caused by carrier diffusion in the material inside the material, an internal electric field is also generated by the carrier diffusion caused by the phonon traction when the system reaches equilibrium. Therefore, the internal electric field can also contribute to the Seebeck coefficient. In fact, the Seebeck coefficient of most materials is the sum of the contribution of carrier diffusion and phonon, while the phonon drag effect is usually observed only at low temperatures and is often ignored. Recent studies have shown that phonon drag effects, which are usually observed only at low temperatures, can also occur at room temperature or even at high temperatures when the scale of the material is reduced to the nanoscale. This has led to the development of practical applications for high-efficiency thermoelectric materials under ambient- and high-temperatures.

13.1.2.2 Lowering the Thermal Conductivity

As per the Boltzmann theory and the Wiedemann–Franz law, we can clearly see that only a material's lattice thermal conductivity is independent of the conductivity. Therefore, lowering the lattice thermal conductivity becomes the preferred way for improving *ZT*. Lattice thermal conductivity is controlled by the phonon, which is the vibration between the chemical bond in solid atoms, and heat conduction is observed as the phonon carrying heat from the hot side to the cold side of the diffusion. The solid theory has found that lattice vibrations are characterized by quantization. Therefore, the phonons will be affected by various scattering mechanisms in the process of motion. With regards to the collision theory of gas molecular motion, if V is the average velocity of the phonons and C_V is the specific heat of the solid material, the average free path of the phonons in the two random collisions is l, then the lattice thermal conductivity can be expressed as:[20]

$$\kappa_l = \frac{1}{3} C_V l V_S \tag{13.5}$$

To improve the performance of thermoelectric materials by reducing the thermal conductivity, it is natural to first think of materials with intrinsically low lattice thermal conductivity. In fact, over the past several decades, several substances with intrinsically low thermal conductivities have been discovered, which usually have large unit cells or heavy atoms, so that lower specific heat and phonon velocity can be obtained. The lead chalcogenides, particularly PbTe, usually have intrinsically low lattice thermal conductivity. This can be understood by the low speed of sound caused by the soft bonds from large and heavy atoms like lead and tellurium. In addition, the relatively high anharmonicity of phonon vibrations, probably resulting from the high coordination number and the unusual structural features relating to the near ferroelectric instability, may also lead to the phenomenon. In addition, materials with complex structures often have abnormally low lattice thermal conductivity and thus have a high *ZT* value, such as Zn_4Sb_3 (1.3 at 670 K)[21] and $Yb_{14}MnSb_{11}$ (~1.0 at 900 K).[22] Taking Zn_4Sb_3 as an example, the abnormally low lattice thermal conductivity (room temperature, ~0.3W/mK) is associated with the random distribution of partial Zn ions at multiple sites. The disordered distribution of some Zn ions in the lattice leads to serious lattice distortions in the whole material, destroying the coordinated vibrations between the atoms and blocking the phonon transfer. Na_xCoO_2 is another oxide thermoelectric material with an ordered CoO_2 layer (3 atomic layers) and a disordered Na metal ion layer (1 atomic layer) (*ZT* is about 1).[23] When the number of atomic layers in the ordered layers is further reduced in the superlattice consisting of ordered and disordered layers, the lattice thermal conductivity of the corresponding materials may be further reduced, possibly with a higher *ZT*.

Although some materials exhibit intrinsically low lattice thermal conductivities due to low specific heat and phonon rates of their complex crystal structures, in most cases, for a given compound, the specific heat and phonon velocity are not very sensitive to structural changes, making it difficult to modulate. Thus, for these materials, the route for obtaining high thermoelectric properties by reducing the lattice thermal conductivity is focused on reducing the mean free path of the phonons. The average free path of the phonon is determined by two aspects: the collision between phonons, and the scattering in the phonon transport process. Point defects, such as interstitial ions, vacancies, and alloying, can increase the probability of scattering in the phonon transport process and reduce lattice thermal conductivity, as evidenced by numerous experiments. However, for a material, the internal phonons can be distributed by wavelengths. As per Rayleigh scattering laws, different crystal defects scatter phonons that are within the same characteristic length. Therefore, the point defects can effectively scatter the short phonon or the edge phonon of the Brillouin zone but little effect on the long phonon, thus limiting the lattice thermal conductivity. The effective scattering of the long-phonon phonon can be realized by nanocrystallization to increase the grain boundary of the material.

In traditional bulk thermoelectric materials composed of large grains, the grain boundary content is relatively small, and for anisotropy-less compounds, the electrical properties generally show the intrinsic behavior of single-crystal materials. However, when the grain size is close to nanometer scale, the content of grain boundary is greatly increased. The state of the atomic arrangement at these interfaces is different from the regular periodic field inside the grains, and the distribution of electrons at the interfaces different from that inside the grains will produce an additional barrier, which will play an important role in the transportation of the carriers and phonons. The grain boundaries have a certain scattering effect on the long-wavelength phonons, which is beneficial to the decrease of the thermal conductivity. Poudel *et al.* reported the experimental results of increasing the pyroelectric performance of *p*-type $Bi_{0.5}Sb_{1.5}Te_3$ bulk materials by increasing grain boundaries. They use high-energy ball mill first $Bi_{0.5}Sb_{1.5}Te_3$ ground into an average particle size of about 20 nm of nanopowder. From the X-ray diffraction pattern analysis, the powder is $Bi_{0.5}Sb_{1.5}Te_3$ single phase. Over the whole temperature range, the electrical conductivity of the bulk material obtained by hot pressing sintering after ball milling is improved to a certain extent when compared with the unmilled sample. The Seebeck coefficient is lower than that of unalloyed alloy below 175 °C, and when above 175 °C, it is higher than that of the unalloyed alloy. This is also due to the energy filtering effect of grain boundaries. Thermal conductivity in the whole temperature range has a greater reduction, enhancing the maximum material *ZT* value from 1.0 to 1.4. High-resolution transmission electron microscopy (TEM) shows that some of the anomalously grown grains contain some Sb-rich quantum dots. These nanocrystals and the quantum dots in the crystal grains have an essential impact in phonon scattering and hence the thermal conductivity of the material decreases. However, quantum dots also play the role of carrier compensation so that the conductivity of the material has increased simultaneously.[24]

13.1.3 Dimensionality Effect in Thermoelectric Materials

Dimensionality has a very important effect on the physical and chemical properties of materials. When the material is thinned down from the bulk phase to 2D, their electronic and lattice structure changes. For example, the layered transition metal chalcogenide MoS_2, if it is exfoliated into few-layer nanosheets, its band gap increases gradually as its thickness decreases;[25] and more importantly, an indirect to direct band gap transition will be realize when it is thinned down to monolayer thickness with large carrier mobility and on-off ratio, which provides a promising opportunity to substitute silicon as the next-generation foundation of electronic devices.[26,27] As another example, single-layer $NbSe_2$ has Ising superconducting characteristics, whereas the bulk material does not.[28] Similarly, dimensionality also plays a critical role in the thermoelectric properties of materials and the performance in many thermoelectric system can be enhanced by reducing the dimensionality. It even largely improves its *ZT* value. These structures may significantly enhance *Z* since the electrons are confined in 2D.[29] In addition, the nanosheets may reduce the phonon thermal conductivity and therefore increase *Z* as phonons can now be scattered by the interfaces and grain boundary between layers. As shown in Figure 13.3, by reducing the dimensionality, the effect on the thermoelectric material can be divided into the following:

1. Increasing the density of states near the Fermi level, thus increasing the Seebeck coefficient.
2. Improving the carrier mobility of the material due to quantum confinement and modulation doping effect.

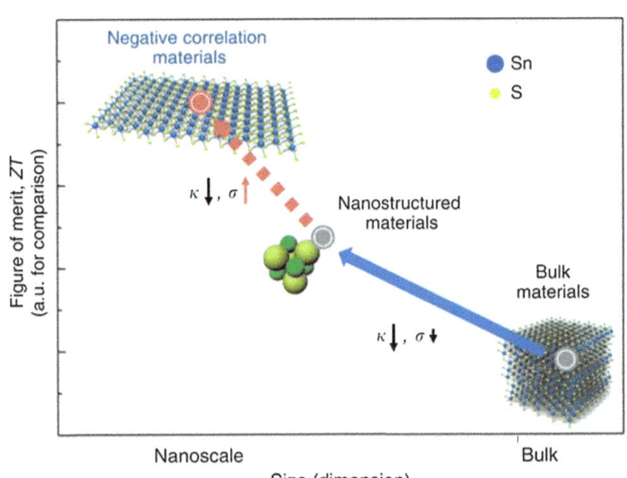

Figure 13.3 Progress for high *ZT* in thermoelectric materials with dimensional modulation. Reprinted with permission from Macmillan Publishers Ltd: *Nature Communications* (ref. 29), Copyright 2016.

3. Making better use of multi-energy valley semiconductor Fermi surface anisotropy.
4. Increasing the boundary scattering of phonons on the surface of potential well wall, which reduces the lattice thermal conductivity.

Actually, S^2 for quantum-well structures enhances linearly with decreasing thickness due to the quantum confinement effect according to the Hicks–Dresselhans model. Therefore, true atomic thickness would generate a 2D electron gas in the monolayer, which would improve the Seebeck coefficient for the nanosheet composite. Furthermore, the potential barriers at the numerous interfaces and boundaries of the nanosheet composite would be expected to filter out low-energy carriers (cold carriers) and transmit high-energy carriers (hot carriers), thus resulting in enhanced S. Additionally, high σ is usually associated with large κ, as the theoretical calculation points out, which unfortunately offsets the former's impact on the thermoelectric performance. As is known, phonon coherent lengths, including wavelengths and mean free paths, are usually much longer than those of electrons, enabling selective scattering of phonons rather than electrons. This has been demonstrated in some recent reports, which clarified that the above contradiction can be resolved by suppressing phonon transport at the interfaces, grain boundaries, and defects, which reduces κ whilst maintaining σ and is called phonon glass electron crystal. As such, the enormous grain boundaries and interfaces in the nanosheet composite could be benefit for the scattering of mid/long-wavelength phonons, thus contributing to a reduction in κ with a minimal detrimental effect on the electrical conductivity. In addition, ultrathin inorganic single layers usually exhibit surface structural disorder, allowing for efficient scattering of short-wavelength phonons, which synergetic decreases κ. Therefore, further breakthroughs in the design and synthesis of novel two-dimensional thermoelectric materials having higher σ and S along with lower κ hold the key to the development of high-efficiency thermoelectric materials.

13.2 Thermoelectric Materials with 2D Characteristics

Thermoelectric conversion technology is a technology that directly uses semiconductor materials to achieve thermal energy and electrical energy conversion. With the increasingly serious environmental protection situation, the research and development of clean energy has become a focus of global scientific research. Among them, thermoelectric conversion technology, by virtue of its system small size, high reliability, lack of pollutant emission, applicability for a wide temperature range *etc.*, is the focus of attention. Abundant papers and reviews on thermoelectric material systems are available due to their promising application and they are a hot research interest.[3,37–40]

Peierls has argued that the long-range order of a 2D solid will be destroyed by the thermal motion of long-wavelength phonons, in a sense that the mean square deviation of an atom from its equilibrium position enhances logarithmically with the size of the system, and the Bragg peaks of the diffraction pattern formed by the system are broad instead of sharp. Additionally, Hicks and Dresselhaus reported a theoretical calculation, in which S^2 for quantum-well (QW) structures increases linearly with thinning down QW thickness when the thickness is less than a few nanometers. They assume that the enhancement of S^2 arises mainly from an enhancement in the density of states near the conduction band edge when the carrier electrons are confined in such a space. Therefore, searching for the thermoelectric properties of materials with 2D properties triggers much interest, since such systems provide promising opportunities to discover thermoelectric materials with high *ZT* values.

Herein, we will present an overview including recent research progress in layered material systems for their thermoelectric properties, which are classified into four parts here. IVA-VIA and VA-VIA layered thermoelectric materials, typically SnSe and Bi_2Te_3, will be discussed in the first two parts, with attention to the chemistry, crystal structure, physical properties, and optimization of thermoelectric performance. Recent experimental investigations of the thermoelectric performance of transition metal dichalcogenide materials will be discussed in the third part as well as our current understanding of this class of materials. An intriguing newcomer to the field of thermoelectricity is the class of metal oxides, which, will be discussed in the fourth part.

13.2.1 IV–VI Thermoelectric Materials

When it comes to conventional thermoelectric materials, PbTe-based materials are the premier ones. They have been extensively studied for applications in the military and space in the last century, and they have already been used in industrial production. SnSe, another IV–VI compound like PbTe, demonstrates a 2D crystal structure. Monocrystalline and polycrystalline SnSe is not a newly discovered thermoelectric material. As early as the 1960s, researchers studied its thermoelectric properties at low temperatures, but did not make any great progress until it was found that the high temperature thermoelectric figure-of-merit rises rapidly with increasing temperature above 700 K and reaches a peak value of 2.6 at 923 K. After 1999, the reported *ZT* value of SnSe exceeded 2.0 mainly for the superlattice structure, which is difficult to achieve in mass production.[30–35]

SnSe crystallizes in a layered orthorhombic structure at room temperature, including two-atom-thick SnSe slabs in the *b*–*c* plane with strong Sn–Se bonds in the plane, which are stacked through weaker Sn–Se interactions in the *a* direction (as shown in Figure 13.4).[30] Taking into account the layered crystal structure of SnSe, Zhao Lidong confirmed that the carrier mobility in SnSe single crystal is five times that of polycrystalline, due to a better conductive property within the layers than interlayers. It was reported that SnSe single crystals showed enhanced *ZT* of 2.6 ± 0.3 and 2.3 ± 0.3 along the *b* axis and *c* axis, respectively, but a reduced *ZT* of 0.8 ± 0.2 along the *a* axis, and

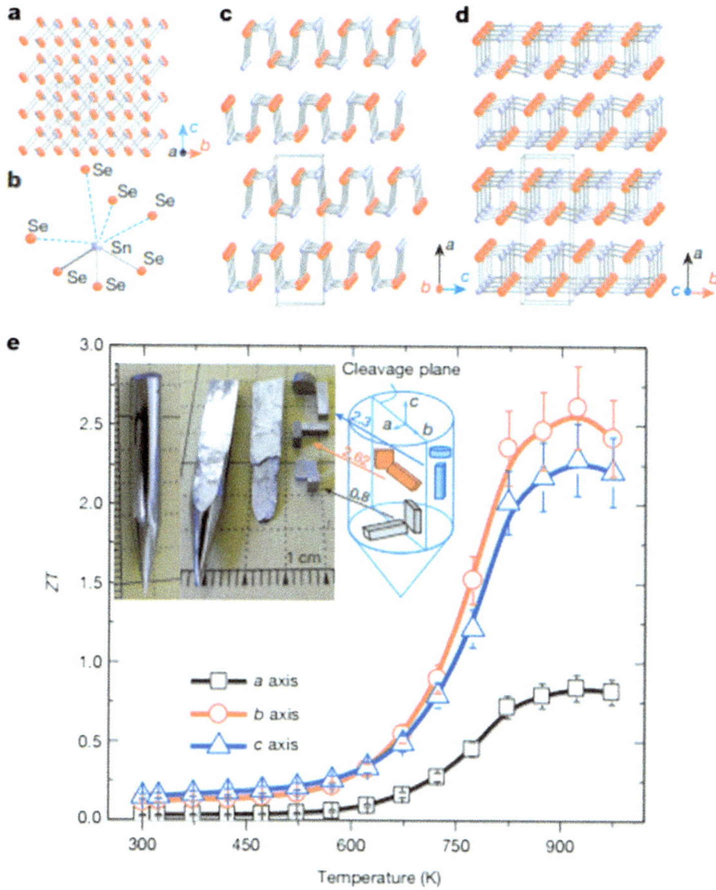

Figure 13.4 (a), (b), (c) and (d) SnSe crystal structure. (e) *ZT* values of SnSe along *a*, *b*, and *c* axes. Reprinted with permission from Macmillan Publishers Ltd: *Nature* (ref. 30), Copyright 2014.

as per the author, the remarkably high *ZT* along the *b* axis was attributed to the relatively low lattice thermal conductivity of the crystal (as shown in Figure 13.5).[30,31] It was speculated that the main reason was that there was a high-temperature phase transition near 750 K for SnSe crystals. After the phase transition, it had a smaller band gap, which may increase the carrier mobility, thus optimizing the electrical transport properties of the material.

However, the *ZT* value of the layered SnSe in the temperature range of 300–773 K is low, resulting in a low average *ZT* value, which severely limits the use of SnSe in the important temperature range of 300–773 K because the thermoelectric conversion efficiency is mainly determined by the average *ZT* value. The basic idea of improving the *ZT* of SnSe in the low temperature region is to improve the conductivity and Seebeck coefficient of SnSe to obtain a high electric transporting performance in the temperature range

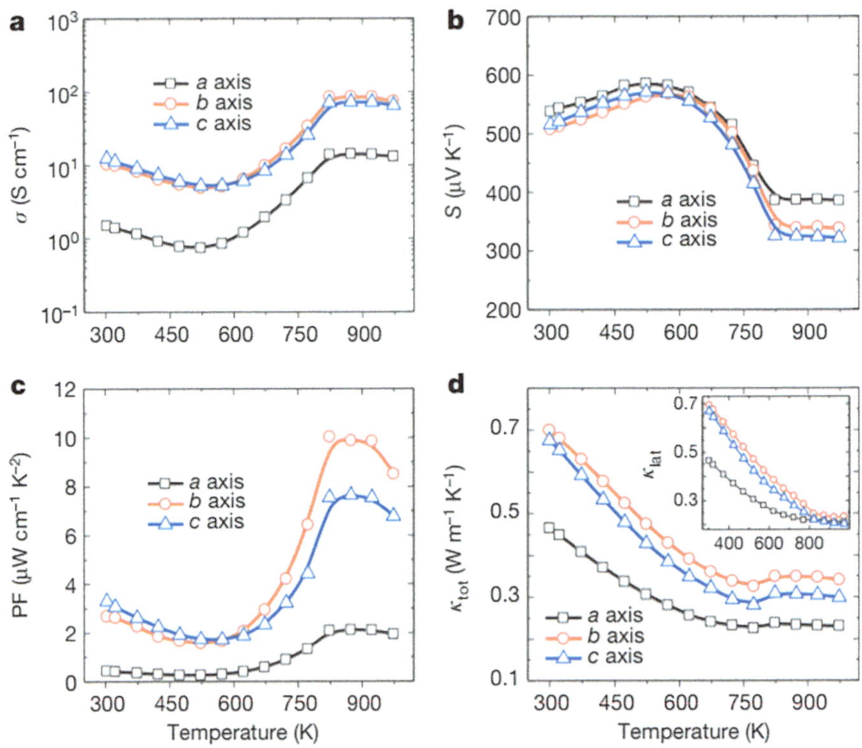

Figure 13.5 (a) Electrical conductivity. (b) Seebeck coefficient. (c) Power factor. (d) Thermal conductivity of SnSe as a function of temperature. Reprinted with permission from Macmillan Publishers Ltd: *Nature* (ref. 30), Copyright 2014.

of 300–773 K. The energy band structure is an effective method to control the conductivity and Seebeck coefficient of thermoelectric materials,[36,37] as in the PbTe system there are two energy similar valence bands, thus energy alignment can increase the effective mass, thereby improving the Seebeck coefficient. However, the electronic band structure in SnSe is more complex, because there are six valence bands close in energy, but fortunately, if the Fermi level successfully enters the valence band, it can be shown that a number of valence bands would at the same time make contributions to electron transportation (as shown in Figure 13.6).[31] Moving the Fermi level not only maintains relatively high carrier mobility, but also increases the Seebeck coefficient by a factor of five, allowing the layered SnSe to have a substantial increase in *ZT* over the entire temperature range. The *ZT* value in the temperature range of 300–773 K increases from 0.1–0.9 to 0.7–2.0 in in hole-doped single-crystal of SnSe (as shown in Figure 13.7).[31] Selecting 300 K and 773 K as the low temperature end and the high temperature end, respectively, layered SnSe as the *p*-type material of thermoelectric device with the same performance *n*-type material can produce 16.7% theoretical power

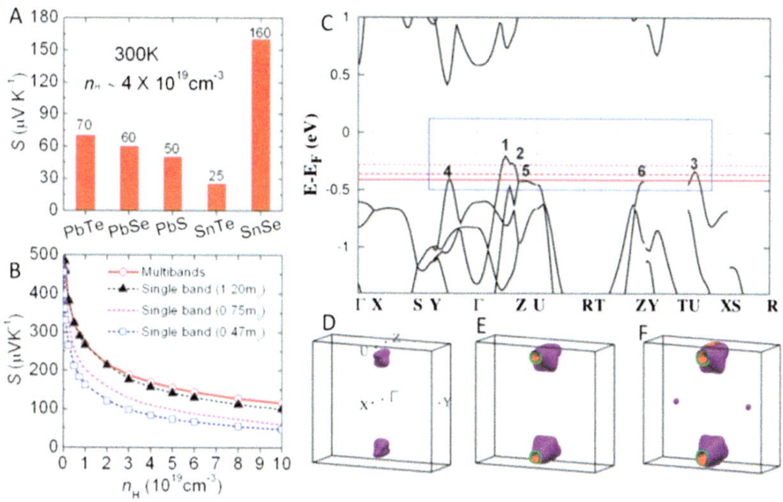

Figure 13.6 (a) Seebeck coefficient comparisons of PbTe, PbSe, PbS, SnTe, SnSe at room temperature. (b) Calculated Seebeck coefficients with different carrier densities at room temperature. (c) DFT electronic band structure of hole doped SnSe showing non-parabola. (d), (e) and (f) The Fermi surface of SnSe with different carrier density. From ref. 31, reprinted with permission from AAAS.

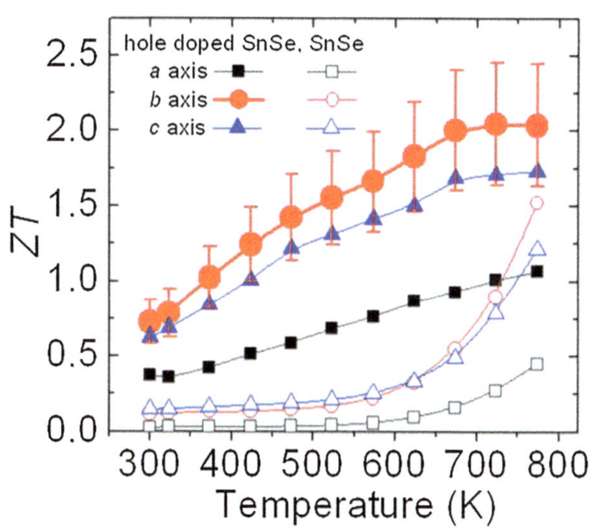

Figure 13.7 *ZT* values of hole-doped SnSe under *a*, *b*, *c* axes; the *ZT* measurement uncertainty is about 20%. From ref. 31, reprinted with permission from AAAS.

generation efficiency. This result means that it is possible to develop a thermoelectric energy source material with excellent performance, abundant reserves and environmental friendliness.

13.2.2 V–VI Thermoelectric Materials

13.2.2.1 Bi_2Te_3

Bi_2Te_3 compounds and their solid solution alloy are the earliest and most mature of the thermoelectric materials, demonstrating the best *ZT* value at room temperature. Most current refrigeration components use such materials. Bi_2Te_3 is a typical 2D layered material, of which the atomic layers can also be divided into three kinds, Te(I) layers, Te(II) layers and Bi layers. The Te(I)–Bi–Te(II)–Bi–Te(I) five layers make up the periodic repetition of the crystal in the *c*-axis direction. The interlayer distance between Te(I)–Bi is 0.174 nm, which is the combination of covalent bond and ionic bond. The interlayer distance between Te(II)–Bi is 0.204 nm, and is a covalent bond combination; Te(I)–Te(I) distance was 0.260 nm, which was combined with van der Waals forces.[36–41]

Though Bi_2Te_3 performs poorly in the pristine state, there is great enhancement through dimensionality reducing, superlattice or nanostructuring, and chemical doping *etc.* In the last century, Bi_2Te_3 and Bi_2Te_3 alloys presented the highest thermoelectric value. However, for both Bi_2Te_3 and Bi_2Te_3 alloys, very slow progress was made in enhancing the *ZT* value since 1960, and all Bi_2Te_3 based materials were produced in bulk. From 1960 to 1990, little attention was paid to the field of thermoelectrics globally, while the $(Bi_{1-x}Sb_x)_2(Se_{1-y}Te_y)_3$ alloy family maintained the most commercialized material with *ZT* of about 1.[42] Until 1993, Hicks L. D. using theoretical calculations successfully predicted that it was not only possible but also effective to improve the *ZT* of the layered materials by preparing them into superlattice structure with quantum wells.[43] They also found that anisotropy has a high effect on the improvement of the quality factor. The calculated results showed that the layering contributed a significant increase in the high anisotropy, such as the quality factor of Bi_2Te_3, with the potential of a superlattice multilayer film made in a specific direction. When compared to its bulk counterparts, the layered Bi_2Te_3 quantum-well structures demonstrated a much higher *ZT* value. In the Bi_2Te_3–Sb_2Te_3 quantum-well superlattices, with a periodic repetition of 6 nm, a high *ZT* value of 2.4 was reported, obviously superior to its bulk couterparts.[44]

In 2000, it was reported that if doped appropriately, $CsBi_4Te_6$ could exhibit a high *ZT* value below room temperature reaching a max *ZT* value of 0.8 at 225 K.[45] Its outstanding performance could even match and exceed the reported $Bi_{2-x}Sb_xTe_{3-y}Se_y$ alloys. Soon, the *ZT* value record in Bi_2Te_3 based materials would be broken by the *p*-type Bi_2Te_3/Sb_2Te_3 superlattice devices with a maximum observed factor of 2.4. This enhancement is achieved by controlling the transportation of phonons and electrons in this superlattice.[46] Chemical doping is another practical strategy to enhance the thermoelectric efficiency in layered Bi_2Te_3 systems. In Bi_2Te_3 based materials, the addition of a small amount of electron acceptors (Cd, Cu, Ag) could reduce the lattice thermal conductivity (as it is shown in Figure 13.8) and suppress the intrinsic excited state, and successfully improved the thermoelectric quality factor to 1.0–1.4.[44–46] Besides, the thermoelectric performance peaks were significantly

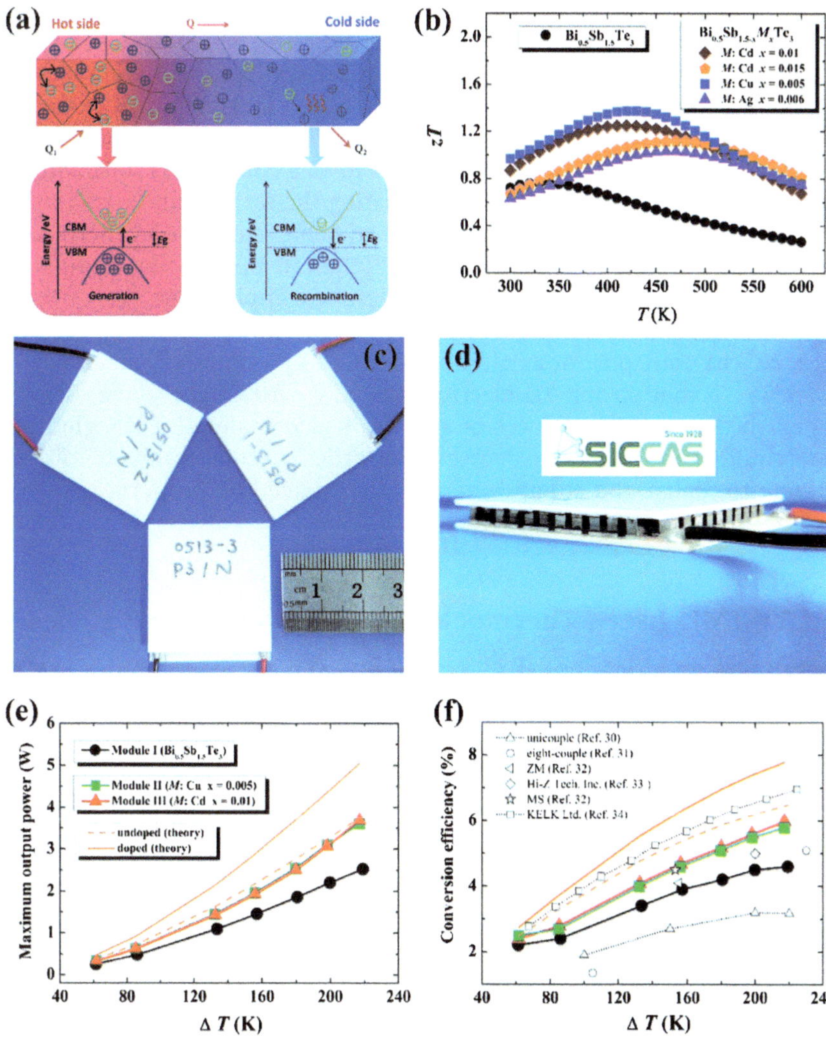

Figure 13.8 (a) Schematic diagram of the excitation in a narrow band gap semiconductor. (b) Thermoelectric figure-of-merit as a function of temperature for Cd, Cu, and Ag doped $Bi_{0.5}Sb_{1.5}Te_3$. (c) and (d) Modules of thermoelectric devices. (e) and (f) Power outputs and energy conversion efficiency for three thermoelectric modules. Adapted from ref. 44 with permission from The Royal Society of Chemistry.

broadened, with the average figure of merit increasing to 1.0–1.2 at 180–300 °C. Thermoelectric generation models based on these materials showed a 6% increase in the thermoelectric conversion rate, which is 30% higher than that of the non-optimized Bi_2Te_3-based thermoelectric materials.[44–47]

13.2.2.2 Bi_2Se_3

Layered Bi_2Se_3 has the advantage in low thermal conductivities, low cost and double polarization effect when compared to Bi_2Te_3, which has been widely used in industry, so it has important scientific significance and application potential for carrying on the study of Bi_2Se_3-based thermoelectric materials. Bi_2Se_3 and Bi_2Te_3 have the same hexagonal crystal structure. These binary alloy materials can be mixed in any ratio to form a continuous solid solution alloy, which can be obtained by finely adjusting the carrier concentration and reducing the lattice thermal conductivity to achieve enhanced thermoelectric properties.[48] Therefore, Bi_2Se_3 is one of the most promising thermoelectric materials suitable for near room temperature application.

The layered structural of Bi_2Se_3 results in anisotropy of the transport properties, *i.e.* the transport of carriers is confined in a plane perpendicular to the *c* axis, so that strong 2D electron gas may exist in the plane. Besides, in the van der Waals interlayer of a Bi_2Se_3 based material, entangled host atoms could be introduced to decrease the lattice thermal conductivity of the material, to realize the enhancement of thermoelectric properties, such as $Bi_{1.98}Ca_{0.02}Se_3$.[49]

13.2.3 TMD-based Thermoelectric Materials

Recently, 2D materials based on transition metal dichacogenides (TMDs) posed potential importance in energy conversion and device functionalities owing to a variety of novel physical and chemical properties. TMD research is currently a hot topic due to their special chemical and physical properties, and especially for their promising potential in energy storage and conversion and information transfer devices. Progress in the thermoelectric field for certain TMD materials has been made, because the electronic state and the scattering mechanism of TMDs can make great contributions to significantly large power factors due to 2D electronic state properties.

TiS_2 is a typically layered structured 2D TMD material, which can intercalate a wide range of atoms or molecules into its van der Waals gap between the layer slabs. In 2001, H. Imai investigated TiS_2 for its thermoelectric and transport properties and discovered that TiS_2 in a nearly stoichiometric composition state demonstrated a large power factor of 37.1–52.3 mW/K^2 cm over a wide temperature range,[50] allowing for the 'phonon-glass and electron-crystal' concept. These results were quite comparable to that of the well commercialized Bi_2Te_3 at that time. A series of research on intercalating exotic elements or organic cations followed and 2D TiS_2 could further enhance its thermoelectric performance. Very recently, Chunlei Wan

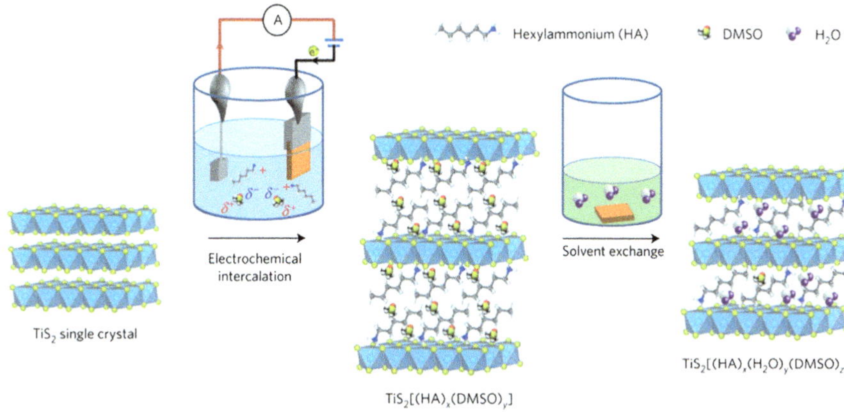

Figure 13.9 Electrochemical synthesis scheme of TiS_2-based inorganic/organic superlattices. Reprinted with permission from Macmillan Publishers Ltd: *Nature Materials* (ref. 51), Copyright 2015.

proposed a method for the synthesis of a hybrid superlattice of TiS_2/[(hexylammonium)$_x$(H_2O)$_y$(DMSO)$_z$],[51] a *n*-type flexible thermoelectric system, by electrochemical intercalating organic cations into TiS_2 monolayers followed by a solvent exchange, shown in Figure 13.9. Through the organic cations in the van der Waals interlayers, the externally injected electrons could be stabilized between the inorganic layers and then provide the *n*-type carriers for current and energy transport. Even though the mismatched superlattice yielded a remarkable enhanced electrical mobility, its thermal conductivity decreased after intercalation. The mixed superlattice was obtained with a significantly decreased thermal conductivity that is two orders of magnitude smaller than that of TiS_2 in monolayer and bulk. In addition to the enhanced performance, a facile composites synthesis method would make great progress in obtaining *n*-type thermoelectric materials, as is shown in Figure 13.10.

Besides, a series of misfit layer compounds $(MS)_{1+x}(TiS_2)_2$ (M = Pb, Bi, Sn), with MS layers and paired trigonal anti-prismatic TiS_2 layers alternatively stacking with van der Waals gap intercalation, were suggested as bulk thermoelectric materials showing a *ZT* value of 0.28–0.37 at 700 K.[52]

13.2.4 Layered Oxide-based Thermoelectric Materials

As a metallic 2D transition metal oxide, $NaCo_2O_4$ was proposed to be a potential thermoelectric material by Terasaki *et al.* in 1997,[53] after which several studies improved its thermoelectric performance to higher *ZT* values.[54,55] This material presented highly anisotropic properties of its in- and out-of-plane directions. Moreover, the layered structure made great contributions to its thermopower potential because of its striking in-plane thermoelectric power of about 100 uV/K at 300 K (Figure 13.11) which was almost ten times higher than that of typical metals. $NaCo_2O_4$ presents a bronze-type compound structure expressed as A_xBO_2 $(0.5 < x < 1)$, which was first ascertained by Jansen

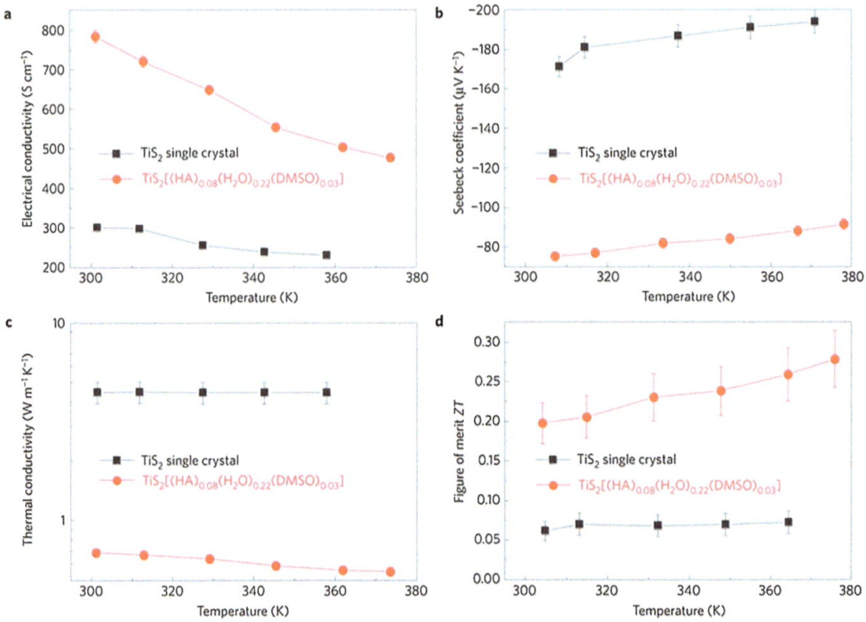

Figure 13.10 The in-plane thermoelectric properties of TiS_2-based inorganic/organic superlattices. (a) Electrical conductivity. (b) Seebeck coefficient. (c) Thermal conductivity. (d) Thermoelectric figure-of-merit. Reprinted with permission from Macmillan Publishers Ltd: *Nature Materials* (ref. 51), Copyright 2015.

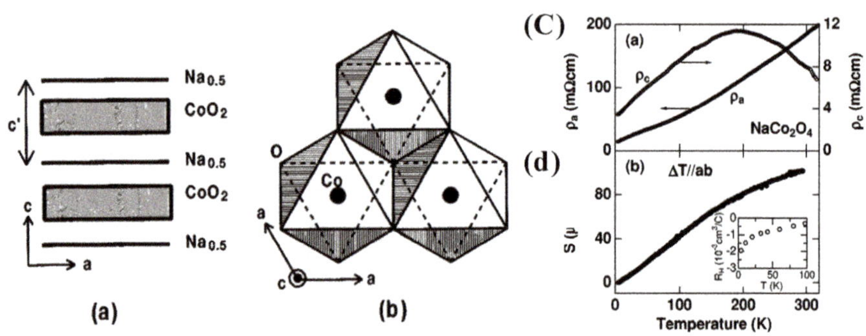

Figure 13.11 Schematic picture of lattice structure of $NaCo_2O_4$. (a) The layered structure. (b) The CoO_2 layer. (c) The electric conductivity. (d) The Seebeck coefficient of $NaCo_2O_4$ single crystal. Reprinted with permission from I. Terasaki *et al.*, *Physical Review B*, **56**, R12685–R12687, 1997. Copyright 1997 by the American Physical Society.

and Hoppe as is schematically shown in Figure 13.11.[56] $NaCo_2O_4$ crystalizes in 50% occupied Na and 2D triangle latticed CoO_2 layers alternately stacked along the *c* axis showing highly 2D properties. In this system, the CoO_2 layer plays a conductive role, and the $Na_{0.5}$ layer with a half-atom vacancy is disordered and makes great contributions to the scattering of the phonons. In fact, it ideally corresponds to the new phonon-glass electronic crystal concept. Trasaki further proposed that $NaCo_2O_4$ is a strong electron correlation system.[55] In this system, the Coulomb repulsion among the electrons can cause the common electron band structure to split, so that the material parameters may go beyond the traditional band theory calculations, which was confirmed by its heat capacity at low temperature. To further improve the thermoelectric properties of $NaCo_2O_4$, research on $NaCo_2O_4$ doping and different Na content studies were carried out. The current doping method includes Na site and Co site element doping. As in $Na_{1-x}Ba_xCo_2O_4$, Ba^{+2} doping could increase the effective mass of the material, thus enhancing the phonon scattering and reducing the phonon thermal conductivity, and in the meantime this method further improved the Seebeck coefficient of the material.[56] $Na_{1-x}Ag_xCo_2O_4$ and $Na_{1-x}Ca_xCo_2O_4$ also performs well in the thermoelectric energy conversion field.[57–59]

13.3 Thermoelectric Performance of 2D Nanosheets

In recent years, with the increasingly rapid development of low-dimensional nanotechnology, nanostructures appear to be widely applied in the field of energy conversion, which also has made great progress in the thermoelectric field. Despite the findings of Hicks and Dresselhaus,[60] the figure-of-merit of thermoelectric materials has increased by a factor of three or four in the last 20 years, and the increase was not exclusively due to the modification of the Seebeck coefficient, but to other factors related to the lattice thermal conductivity like interface roughness or, in general, surface effects.

In the introduction section, we introduced some methods for optimizing the thermoelectric parameters and for improving the properties of thermoelectric materials. Although these methods have some improvement on the thermoelectricity, due to the strong correlation between the three parameters, one or two parameters are optimized while the other parameters are affected in the opposite way, which restricts the large enhancement of the thermoelectric figure-of-merit. How to realize the decoupling of the three parameters has been a historic problem in the field of thermoelectrics. The emergence of nanomaterials provides us with new opportunities. As mentioned earlier, nanomaterials have completely different band structures from bulk materials and exhibit many novel properties. Using this unusual band structure, a large Seebeck coefficient can be obtained. Additionally, the nanocrystalline material has the quantum boundary, which can effectively diffuse the phonons, so that the lower thermal conductivity can be obtained. However, the scattering of phonons at these grain boundaries inevitably results in the scattering of the carriers, leading to a decrease in the conductivity. Therefore, it is key to search for a route to utilize the large Seebeck coefficient and low thermal conductivity of nanostructures as well as maintaining the conductivity without it decreasing, or even with it improving. As described previously, some materials are endowed with special structures, properties and behavior which causes them to exhibit large intrinsic Seebeck coefficients or high electrical conductivity or low thermal conductivity. If we can combine these intrinsic structures, properties and behaviors with nanostructures, it will be possible to realize the thermoelectric decoupling of the three parameters of independent regulation, and greatly enhance the thermoelectric properties.

By reducing the dimension of traditional materials, the *ZT* value of thermoelectric materials has been greatly improved, with the nano-superlattice structure of thin film materials being the most representative.[61–63] The superlattice of 2D thin films present anisotropy in properties, in which electrons are limited by periodic boundaries showing greatly improved electronic state density, and leading to the improvement of Seebeck coefficient. Meanwhile, the scattering effect of the interface on the phonons further reduces the lattice thermal conductivity.[63] However, in the preparation of nanostructured bulk materials, the retention of high thermoelectric properties remains challenging due to the nanocrystallization effect in bulk materials. In 2001, Bi_2Te_3/Sb_2Te_3 superlattice thin films were prepared by chemical vapor deposition of metal–organic materials for the first time by Venkatasu-bramanian, with *ZT* increasing to 2.4, which was much higher than that of the bulk alloy materials with the same composition.[61]

The increase in the Seebeck coefficient and the decrease of thermal conductivity at the same time can greatly improve the thermoelectric properties of the materials, which has aroused extensive attention.[64–73] In 1993, Hicks and Dresselhaus first considered the effect of superlattice quantum-well structures on thermoelectric properties. They used 2D Bi_2Te_3 layered model calculations, and the results show that the superlattice material can be much higher than the *ZT* value of the bulk material, and with the quantum-well width decreases, *ZT* value monotonically increased. The quantum-well is a potential trap of electrons or holes with distinct quantum confinement effect, arranged by two kinds of different semiconductor materials. The effect of quantum-well structures is beneficial for enhancing the thermoelectric figure-of-merit. Recently, L. D. Hicks experimentally identified the enhanced effect of quantum-well structures on the thermoelectric *ZT* value by using $PbTe/Pb_{1-x}Eu_xTe$ multiple-quantum-well structures, in which the PbTe was used as the quantum well and the $Pb_{1-x}Eu_xTe$ as the barrier material, grown with molecular-beam epitaxy method.[74] They found that there was a symmetric offset between conduction and valence bands. Its energy gap increased with the content of Eu, and only a content of 6% Eu could present large confinement energies and result in electrons transporting in the quantum wells of PbTe, while a mobility decrease accompanied the increasing of the Eu content. They deduced that the increase in *ZT* was due to the 2D effects that arose mainly from the thermoelectric power factor, and considering the well width was thicker than 10 Å, their lattice thermal conductivity was assumed to be similar to the bulk value, as per the experimental data shown in Figure 13.12.

A great number of experimental studies have confirmed the correctness of this theoretical prediction on quantum-wells, but because of the lack of simple and inexpensive means of synthesis, it is difficult to obtain sufficient thermal conductivity bulks, and is therefore difficult to commercialize. Therefore, looking for easily synthesizable bulk materials with a natural superlattice structure is expected to break through this bottleneck and bring

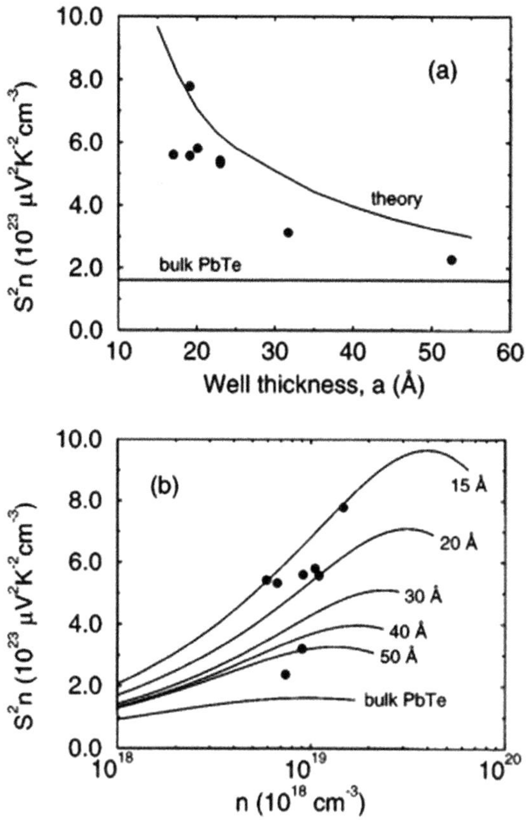

Figure 13.12 S^2n results for PbTe/$Pb_{0.927}Eu_{0.073}$Te MQW's as a function of (a) well thickness, a, and (b) carrier density, n, at 300 K, respectively. Reprinted with permission from L. D. Hicks *et al.*, *Physical Review B*, **53**, R10493–R10496, 1996. Copyright 1996 by the American Physical Society.

new opportunities to the development of thermoelectric materials. BiCuSeO is a highly efficient thermoelectric material with a natural superlattice structure that has been discovered in recent years.[75] BiCuSeO has a crystal structure like that of Co-based oxide or Fe-based superconductors (Figure 13.13)[76] and is alternately stacked by $(Bi_2O_2)^{2+}$ and $(Cu_2Se_2)^{2-}$ layers, where $(Bi_2O_2)^{2+}$ is the insulating layer that acts as a 'charge bank' and the $(Cu_2Se_2)^{2-}$ layer serves as a conductive layer for carrier transport. This natural superlattice structure is very advantageous for phonon scattering, however, the intrinsic conductivity of BiCuSeO is low: only 400 S m^{-1} at 373 K. Recent research has shown that the Bi^{3+} -doping or Cu^{2+} -site defects can greatly improve its electrical transport properties and optimize its thermoelectric properties.

A reduction in the dimension of traditional materials has proven to be practical in thermoelectric enhancement. A large and tunable Seebeck coefficient was achieved in a single layered semiconducting molybdenum disulfide (MoS_2) by an external electric field, which indicated a promising application potential of TMD nanosheets in the thermoelectric field.[77,78]

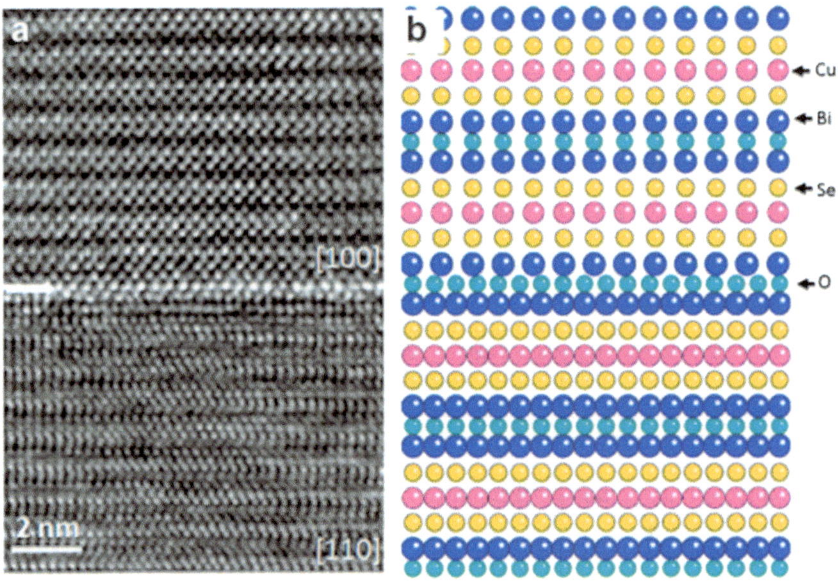

Figure 13.13 (a) The lattice structure of the interface between two grains of the BiCuSeO sample. (b) Atomic model of the two grains along with [100] and [110] directions. Reprinted with permission from Macmillan Publishers Ltd: *NPG Asia Materials* (ref. 76), Copyright 2013.

Very recently, ultrathin WSe_2 single crystals were also reported to achieve optimization of power factor not only in the hole but also in the electron side with gating tuning.[79] Xie *et al.* obtained monolayer nanosheets of Bi_2Se_3 by a hydrothermal method.[80] It was found that the thickness of the monolayer was only one unit cell to maintain the original strong 2D electron gas, showing a higher Seebeck coefficient than the bulk, as shown in Figure 13.14.[80] At the same time, due to the reduction of the thickness to the atomic size, the crystal structure contains a certain amount of distortion, which enhances phonon scattering, and is coupled with a large number of grain boundaries to significantly increase phonons scattering and reduce the thermal conductivity. In addition, the single-layer nanosheets with the atomic layer thickness have a high electronic density and a high concentration of 2D electron gas, which effectively compensates the conductivity of the grain boundary caused by carrier scattering.[80] Therefore, dimensionality reducing in layered materials into atomic layer thickness presents an effectiveness for achieving the decoupling control of the strong correlated three thermoelectric parameters, demonstrating much higher thermoelectricity *ZT* value than bulk materials.

Additionally, the three parameters can be decoupled in the nanosheet by combination with a polymer. Polymers have ultralow thermal conductivity, but their conductivity is poor; while the general thermal conductivity of inorganic materials compared to the polymer is relatively high, but if combined with the polymer, its high conductivity and Seebeck coefficient, combined with

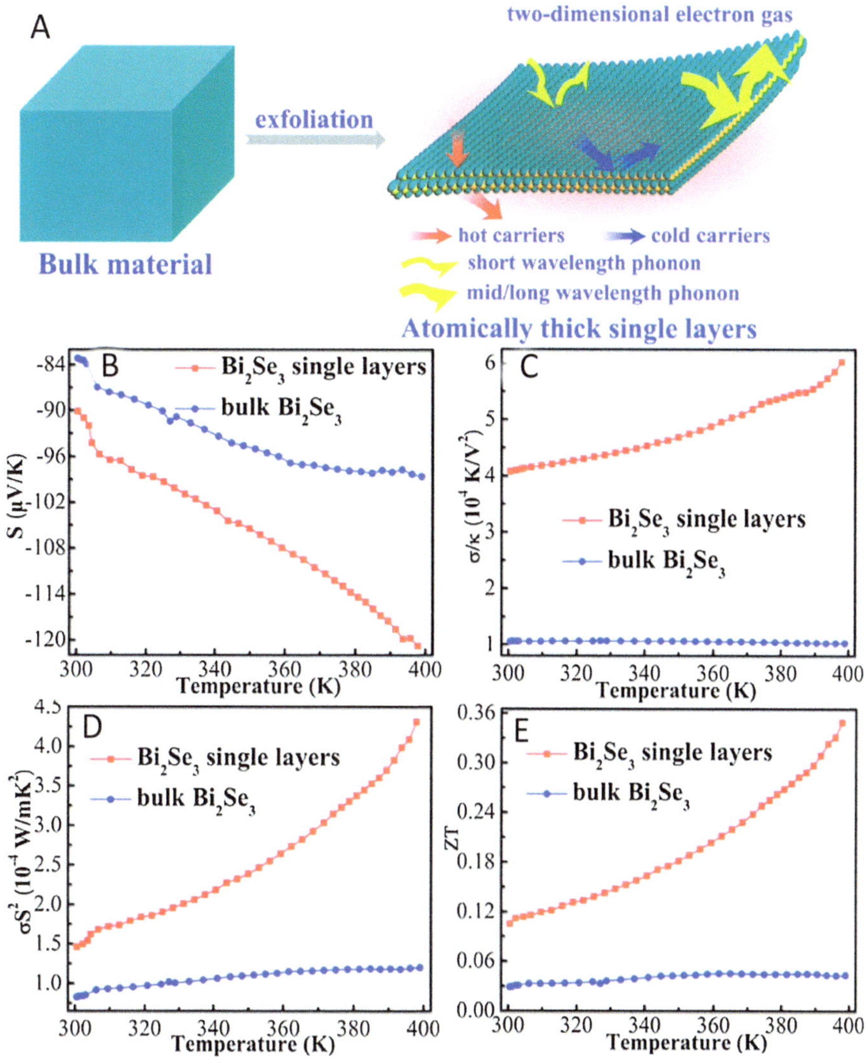

Figure 13.14 The morphology and thermoelectric properties of the freestanding Bi_2Se_3 single layer. (A) Schematic lattice structure. (B) to (E) Thermoelectric properties. Adapted with permission from ref. 80, Copyright American Chemistry Society 2012.

the low thermal conductivity of the polymer, will significantly improve the thermoelectric figure-of-merit. Tin selenide (SnSe) nanosheets (NSs) are synthesized by hydrothermal lithium-intercalation and a subsequent exfoliation process from a SnSe bulk. The conducting polymer poly(3,4-ethylenedioxythiohene):poly-(styrenesulfonate) (PEDOT:PSS)-based thermoelectric composites are self-assembled with varying SnSe NSs content (shown in Figure 13.15). The exfoliated SnSe nanosheets show thin 2D structures that are evenly distributed in the PEDOT:PSS matrix. The significantly enhanced power factor

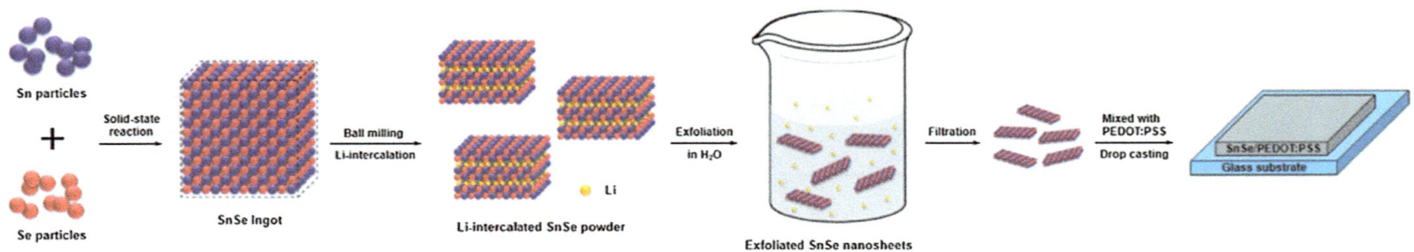

Figure 13.15 The schematic fabrication of SnSe NS/PEDOT:PSS composite films. Adapted with permission from ref. 81, Copyright American Chemistry Society 2016.

($S^2\sigma$) of the SnSe NS/PEDOT:PSS composites with increasing SnSe NSs content can be explained by the potential difference at the interface between the SnSe and PEDOT:PSS. The fabricated SnSe NS/PEDOT:PSS composites reveal a maximum figure-of-merit (*ZT*) of 0.32 at a SnSe NSs loading of 20 wt%, in Figure 13.16. Combining inorganic nanoparticles with a conducting polymer matrix forms products with extremely low thermal conductivities, which is a promising strategy for the realization of thermoelectric parameter decoupling and thus enhancement of thermoelectric performance.

13.4 Conclusion

Thermoelectric materials, as a new kind of energy conversion material, can directly convert heat energy and electric energy. They are gradually becoming one of the hot research fields in clean energy. However, due to their inter-related inverse-proportional coupling of the three performance parameters it is difficult to achieve alone or with collaborative optimization, limiting their substantial increase in *ZT* value and practical applications.

In this chapter, we have summarized recent progress in 2D thermoelectric materials and the strategy to modulate the relation of thermoelectric parameters. Theoretical and experimental studies have shown that 2D nanocrystallization of thermoelectric materials is expected to significantly improve the thermoelectric superiority of these materials. However, this route improves a single parameter merit, such as enhancement of the Seebeck coefficient caused by the increase in the density of states near the Fermi level and electric conductivity resulting from surface 2D gas, and a decrease of thermal conductivity arising from the surface electron scattering. The decoupling of the three thermoelectric parameters is realized in the 2D material systems optimizing a parameter performance will not bring about significant negative effects to the others. This is of great benefit to the improvement of the thermoelectric value of the thermoelectric material and is an important step for approaching to the goal of phonon glass electron crystal.

Although the 2D material-based devices have not yet been commercialized, they have stimulated interest in the application of 2D materials for thermoelectric applications. In the future, research on 2D materials in the application of thermoelectrics will focus on the following aspects: (1) more in-depth research and exploration of the 2D nano-thermoelectric material's thermoelectric transmission mechanism and improving the thermoelectric theory system, providing a reasonable theoretical guidance to seek higher *ZT* values; (2) strengthening the preparation of a variety of research efforts and combining a variety of optimization techniques and integrated treatment to synthesize thermoelectric materials with better performance; (3) deepening the research for device fabrication and design technology to achieve a 2D thermoelectric material industry; and (4) the challenge that, since 2D nanostructures, especially nanosheets of thermoelectric materials, have only a few nanometers or even single layer thickness, the conventional performance testing technology is not reasonable, which severely constrains the development of 2D thermoelectric materials.

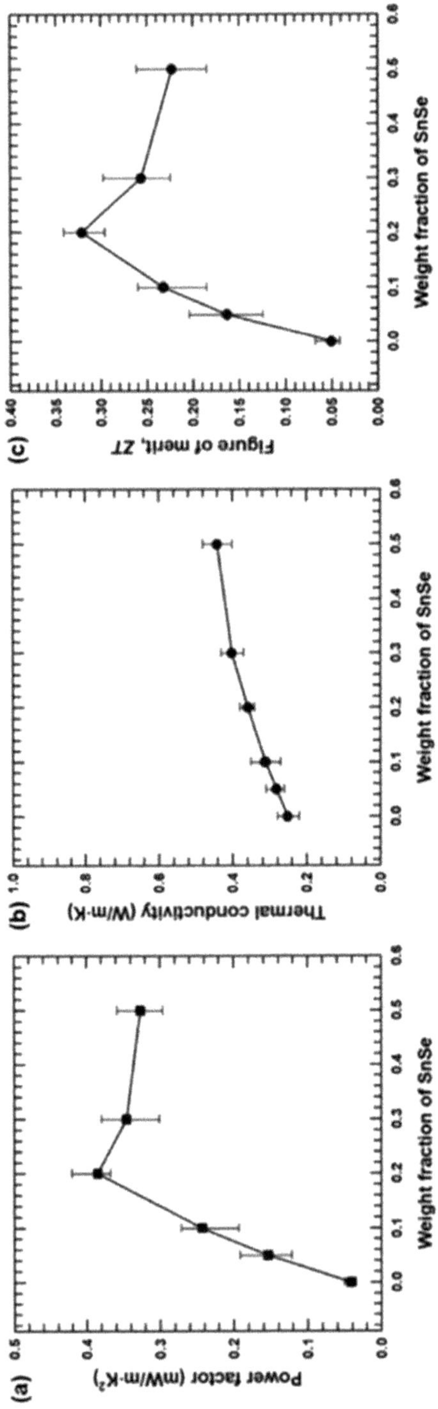

Figure 13.16 The thermoelectric properties of SnSe NS/PEDOT:PSS composite films. Adapted with permission from ref. 81, Copyright American Chemistry Society 2016.

References

1. M. Wise, K. Calvin and A. Thomson, *et al.*, *Science*, 2009, **324**(5931), 1183–1186.
2. T. J. Seebeck, *Abhandlungen der Deutschen Akademie der Wissenschaften zu Berlin*, 1822, p. 265.
3. J. C. Peltier, *Ann. Chim. Phys.*, 1834, **LV1**, 371.
4. E. Altenkirsch, *Phys. Z.*, 1909, **10**, 560.
5. G. J. Snyder and E. S. Toberer, *Nat. Mater.*, 2008, **7**, 105.
6. A. Shakouri, *Annu. Rev. Mater. Res.*, 2011, **41**, 399.
7. G. D. Mahan and M. Bartkowiak, *Appl. Phys. Lett.*, 1999, **74**, 953.
8. N. F. Mott and H. Jones, *The Theory of the Properties of Metals and Alloys*, Dover Publications, New York, 1958.
9. D. Bilc, S. D. Mahanti, E. Quarez, K.-F. Hsu, R. Pcionek and M. G. Kanatzidis, *Phys. Rev. Lett.*, 2004, **93**, 146403.
10. S. Ahmad, K. Hoang and S. D. Mahanti, *Phys. Rev. Lett.*, 2006, **96**, 056403.
11. J. P. Heremans, V. Jovovoc, E. S. Toberer, A. Saramat, K. Kurosaki, A. Charoenphakdee, S. Yamanaka and G. J. Snyder, *Science*, 2008, **321**, 554.
12. C. M. Jaworski, B. Wiendlocha, V. Jovovic and J. P. Heremans, *Energy Environ. Sci. Technol.*, 2011, **4**, 4155.
13. C. M. Jaworski, J. Tobola, E. M. Levin, K. Schmidt-Rohr and J. P. Heremans, *Phys. Rev. B*, 2009, **80**, 125208.
14. K. Esfarjani, G. Chen and Z. F. Ren, *Environ. Sci. Technol.*, 2012, **5**, 5246.
15. J. P. Heremans, B. Wiendlocha and A. M. Chamoire, *Environ. Sci. Technol.*, 2012, **5**, 5510.
16. D. Vashaee and A. Shakouri, *Phys. Rev. Lett.*, 2004, **92**, 106103.
17. J. P. Heremans, C. M. Thrush and D. T. Morelli, *Phys. Rev. B*, 2004, **70**, 115334.
18. C. Herring, T. H. Geballe and J. E. Kunzler, *Phys. Rev.*, 1958, **111**, 36.
19. A. I. Hochbaum, R. Chen, R. D. Delgado, W. Liang, E. C. Garnett, M. Najarian, A. Marjumbar and P. Yang, *Nature*, 2008, **451**, 163.
20. L. D. Hicks, T. C. Harman and M. S. Dreeselhaus, *Appl. Phys. Lett.*, 1993, **63**, 3230.
21. T. Caillat, J. P. Fleurial and A. Borshchevsky, *J. Phys. Chem. Solids*, 1997, **58**, 1119–1125.
22. S. R. Brown, S. M. Kauzlarich, F. Gascoin and G. Snyder, *J. Mater. Chem.*, 2006, **18**, 1873–1877.
23. K. Fujita, T. Mochida and K. Nakamura, *Jpn. J. Appl. Phys., Part 1*, 2011, **40**, 4644–4647.
24. B. Poudel, Q. Hao and Y. Ma, *et al.*, *Science*, 2008, **320**, 634–638.
25. A. Splendiani, L. Sun, Y. Zhang, T. Li, J. Kim, C. Y. Chim, G. Galli and F. Wang, *Nano Lett.*, 2010, **10**, 1271–1275.
26. B. Radisavljevic, A. Radenovic, J. Brivio, V. Giacometti and A. Kis, *Nat. Nanotechnol.*, 2011, **6**, 147–150.
27. O. Lopez-Sanchez, D. Lembke, M. Kayci, A. Radenovic and A. Kis, *Nat. Nanotechnol.*, 2013, **8**, 497–501.

28. X. Xi, Z. Wang, W. Zhao, J.-H. Park, K. T. Law, H. Berger, L. Forró, J. Shan and K. F. Mak, *Nat. Phys.*, 2015, **12**, 139–143.
29. M. J. Lee, J. H. Ahn, J. H. Sung, H. Heo, S. G. Jeon, W. Lee, J. Y. Song, K. H. Hong, B. Choi, S. H. Lee and M. H. Jo, *Nat. Commun.*, 2016, **7**, 12011.
30. L. D. Zhao, S. H. Lo, Y. Zhang, H. Sun, G. Tan, C. Uher, C. Wolverton, V. P. Dravid and M. G. Kanatzidis, *Nature*, 2014, **508**, 373–377.
31. L.-D. Zhao, G. Tan, S. Hao, J. He, Y. Pei, H. Chi, H. Wang, S. Gong, H. Xu, V. P. Dravid, C. Uher, G. J. Snyder, C. Wolverton and M. G. Kanatzidis, *Science*, 2016, **351**, 141–144.
32. S. Sassi, C. Candolfi, J.-B. Vaney, V. Ohorodniichuk, P. Masschelein, A. Dauscher and B. Lenoir, *Appl. Phys. Lett.*, 2014, **104**, 212105.
33. R. Guo, X. Wang, Y. Kuang and B. Huang, *Phys. Rev. B*, 2015, **92**, 115202.
34. B. Poudel, Q. Hao, Y. Ma, Y. Lan, A. Minnich, B. Yu, X. Yan, D. Wang, A. Muto, D. Vashaee, X. Chen, J. Liu, M. S. Dresselhaus, G. Chen and Z. Ren, *Science*, 2008, **320**, 634–638.
35. J. Carrete, N. Mingo and S. Curtarolo, *Appl. Phys. Lett.*, 2014, **105**, 101907.
36. X. Zhang and L.-D. Zhao, *J. Materiomics*, 2015, **1**, 92–105.
37. L.-D. Zhao, V. P. Dravid and M. G. Kanatzidis, *Energy Environ. Sci.*, 2014, **7**, 251–268.
38. D. L. Greenaway and G. Harbeke, *J. Phys. Chem. Solids*, 1965, **26**, 1585.
39. J. R. Drabble and C. H. L. Goodman, *J. Phys. Chem. Solids*, 1958, **5**, 142–144.
40. S. Nakajima, *J. Phys. Chem. Solids*, 1963, **24**, 479.
41. W. R. Bekebrede and O. J. Guentert, *J. Phys. Chem. Solids*, 1962, **23**, 1023.
42. G. J. Snyder and E. S. Toberer, *Nat. Mater.*, 2008, **7**, 105–114.
43. L. D. Hicks and M. S. Dresselhaus, *Phys. Rev. B*, 1993, **47**, 12727–12731.
44. F. Hao, P. Qiu, Y. Tang, S. Bai, T. Xing, H.-S. Chu, Q. Zhang, P. Lu, T. Zhang, D. Ren, J. Chen, X. Shi and L. Chen, *Energy Environ. Sci.*, 2016, **9**, 3120–3127.
45. D.-Y. Chung, T. Hogan, P. Brazis, M. Rocci-Lane, C. Kannewurf, M. Bastea, C. Uher and M. G. Kanatzidis, *Science*, 2000, **287**, 1024–1027.
46. R. Venkatasubramanian, E. Siivola, T. Colpitts and B. O'Quinn, *Nature*, 2001, **413**, 597–602.
47. X. Tang, W. Xie, H. Li, W. Zhao, Q. Zhang and M. Niino, *Appl. Phys. Lett.*, 2007, **90**, 012102.
48. X. Yan, B. Poudel, Y. Ma, W. S. Liu, G. Joshi, H. Wang, Y. Lan, D. Wang, G. Chen and Z. F. Ren, *Nano Lett.*, 2010, **10**, 3373–3378.
49. Y. S. Hor, A. Richardella, P. Roushan, Y. Xia, J. G. Checkelsky, A. Yazdani, M. Z. Hasan, N. P. Ong and R. J. Cava, *Phys. Rev. B*, 2009, **79**, 195208.
50. H. Imai, Y. Shimakawa and Y. Kubo, *Phys. Rev. B*, 2001, **64**, 235305.
51. C. Wan, X. Gu, F. Dang, T. Itoh, Y. Wang, H. Sasaki, M. Kondo, K. Koga, K. Yabuki, G. J. Snyder, R. Yang and K. Koumoto, *Nat. Mater.*, 2015, **14**, 622–627.
52. C. Wan, Y. Kodama, M. Kondo, R. Sasai, X. Qian, X. Gu, K. Koga, K. Yabuki, R. Yang and K. Koumoto, *Nano Lett.*, 2015, **15**, 6302–6308.
53. I. Terasaki, Y. Sasago and K. Uchinokura, *Phys. Rev. B*, 1997, **56**, R12685–R12687.

54. H. Yakabe and K. Kikuehi, Thermoelectric porperties of transition metal oxide NaCo2O4 system (C), *16th International Coference on Thermoelectrics*, 1997, p. 523.
55. Y. Ando, N. Miyamoto, K. Segawa, T. Kawata and I. Terasaki, *Phys. Rev. B*, 1999, **60**, 10580–10583.
56. Y. Ando, N. Miyamoto, K. Segawa, T. Kawata and I. Terasaki, *Phys. Rev. B*, 1999, **60**, 10580–10583.
57. H. Yakabe and K. Kikuehi, Thermoelectric porperties of transition metal oxide NaCo2O4 system (C), *16th International Coference on Thermoelectrics*, 1997, p. 513.
58. T. Itoh and T. Kawata, Thermoelectric properties in (Na, Ca)Co2O4 cermics(C), *17th International Conference on Thermoelectrics*, 1998, p. 595.
59. H. Yakabe, K. Fujita and K. Nakamura, *17th International Conference on Thermoelectrics*, 1998, p. 551.
60. F. Li, J. F. Li and L. D. Zhao, *Environ. Sci. Technol.*, 2012, **5**(5), 7188–7195.
61. L. D. Hicks and M. S. Dresselhaus, *Phys. Rev. B*, 1993, **47**, 12727–12731.
62. R. Venkatasubramanian, E. Siivola, T. Colpitts and B. O'Quinn, *Nature*, 2001, **413**, 597–602.
63. J. R. Sootsman, D. Y. Chung and M. G. Kanatzidis, *Angew. Chem.*, 2009, **48**, 8616–8639.
64. T. C. Harman, P. J. Taylor and M. P. Walsh, *et al.*, *Science*, 2002, **297**(5590), 2229–2232.
65. T. Ikeda, L. A. Collins and V. A. Ravi, *et al.*, *Chem. Mater.*, 2007, **19**(4), 763–767.
66. Y. Zhao, J. S. Dyck and B. M. Hernandez, *et al.*, *J. Am. Chem. Soc.*, 2010, **132**(14), 4982–4983.
67. J. Chen, G. Zhang and B. W. Li, *Nano Lett.*, 2010, **10**(10), 3978–3983.
68. M. Scheele, N. Oeschler and I. Veremchuk, *et al.*, *ACS Nano*, 2010, **4**(7), 4283–4291.
69. P. F. P. Poudeu, A. Güeguen and C. I. Wu, *et al.*, *Chem. Mater.*, 2010, **22**(3), 1046–1053.
70. Y. C. Zhang, H. Wang and S. Kraemer, *et al.*, *ACS Nano*, 2011, **5**(4), 3158–3165.
71. A. Soni, Y. Y. Zhao and L. G. Yu, *et al.*, *Nano Lett.*, 2012, **12**(3), 1203–1209.
72. A. Soni, Y. Q. Shen and M. Yin, *et al.*, *Nano Lett.*, 2012, **12**(8), 4305–4310.
73. R. J. Mehta, Y. L. Zhang and C. Karthik, *et al.*, *Nat. Mater.*, 2012, **11**(3), 233–240.
74. L. D. Hicks, T. C. Harman, X. Sun and M. S. Dresselhaus, *Phys. Rev. B*, 1996, **53**, R10493–R10496.
75. R. J. Mehta, Y. L. Zhang and C. Karthik, *et al.*, *Nat. Mater.*, 2012, **11**(3), 233–240.
76. Y. L. Pei, J. Q. He and J. F. Li, *NPG Asia Mater.*, 2013, **5**, e47.
77. M. Buscema, M. Barkelid, V. Zwiller, H. S. J. van der Zant, G. A. Steele and A. Castellanos-Gomez, Large and tunable photothermoelectric effect in single-layer MoS2, *Nano Lett.*, 2013, **13**, 358–363.
78. J. Wu, H. Schmidt, K. K. Amara, X. Xu, G. Eda and B. Ozyilmaz, *Nano Lett.*, 2014, **14**, 2730–2734.

79. M. Yoshida, T. Iizuka, Y. Saito, M. Onga, R. Suzuki, Y. Zhang, Y. Iwasa and S. Shimizu, *Nano Lett.*, 2016, **16**, 2061–2065.
80. Y. Sun, H. Cheng, S. Gao, Q. Liu, Z. Sun, C. Xiao, C. Wu, S. Wei and Y. Xie, *J. Am. Chem. Soc.*, 2012, **134**, 20294–20297.
81. H. Ju and J. Kim, *ACS Nano*, 2016, **10**, 5730–5739.

Subject Index